GAS TURBINES

SECOND EDITION

FUNDAMENTALS OF GAS TURBINES

SECOND EDITION

WILLIAM W. BATHIE

Iowa State University of Science and Technology

John Wiley & Sons, Inc.

New York • Chichester • Brisbane • Toronto • Singapore

Acquisitions Editor	Cliff Robichaud
Assistant Editor	Catherine Beckham
Marketing Manager	Debra Riegert
Production Editor	Ken Santor
Designer	Laura Ierardi/Kevin Murphy
Manufacturing Coordinator	Dorothy Sinclair
Illustration Coordinator	Gene Aiello

This book was set in 10.5/12 Times Roman by ATLIS Graphics and Design, and printed and bound by Hamilton Printing.

Recognizing the importance of preserving what has been written, it is a policy of John Wiley & Sons, Inc. to have books of enduring value published in the United States printed on acid-free paper, and we exert our best efforts to that end.

The paper on this book was manufactured by a mill whose forest management programs include sustained yield harvesting of its timberlands. Sustained yield harvesting principles ensure that the number of trees cut each year does not exceed the amount of new growth.

Library of Congress Cataloging in Publication Data:

Bathie, William W.

Fundamentals of gas turbines / William W. Bathie. — 2nd ed.

p. cm.

Includes bibliographical references.

ISBN 0-471-31122-7 (cloth : alk. paper)

1. Gas-turbines. I. Title.

TJ778.B34 1996

621.43'3—dc20

95-32998

CIP

PREFACE TO THE SECOND EDITION

The Second Edition retains the basic objectives of the First Edition, which were directed towards the undergraduate student who has a knowledge of elementary thermodynamics, fluid mechanics and combustion. Both Système International (SI) and English units are used in the example problems. The numerical values in the example problems are presented in a way that allows the reader to select only SI, only English, or both SI and English. Problems using both SI and English units are presented at the end of the chapters.

The main changes are to aid in the clarification of the material, to update the material and include additional material in several of the chapters. The chapter on the history of the gas turbine engine has been updated to include major events that have occurred in the last fifteen years. The material on combined cycles has been expanded to include typical steam cycle values and dual-pressure cycles. The example problems continue to be presented in a way that the reader can see the effect of changing one of the parameters. Many problems use values from a previous problem. These are summarized at the beginning of each example problem so that the reader does not have to refer back to a previous example problem.

The chapters on compressors and turbines have been extensively revised and expanded. The example problems are written so that the reader is aware of the effect of changing each variable and to make the reader aware of what limits the pressure change across the component.

Software for shaft-power cycles and compressors is available. These allow the reader to quickly see the effect of changing one of the variables. Several end-of-chapter problems are included that encourage the reader to write a computer program to solve the general problems associated with gas turbine engines.

In closing, I want to express my appreciation to my students and colleagues who encouraged me to write a second edition. Special thanks go to those who reviewed my manuscript and to Martha Clifford, who assisted in the preparation of the manuscript. Finally, I thank my wife, Shirley, and my two children, Mark and Belinda, for their encouragement and patience during the writing of this second edition.

William W. Bathie
May, 1995

PREFACE TO THE FIRST EDITION

The theory of the gas turbine and the ways that it should function were established long before the necessary materials and the detailed knowledge of flow mechanisms were available.

The basic gas turbine was clearly delineated, in concept, in the 1791 patent of John Barber. The development of a useful gas turbine did not occur until the 1930s when Brown Boveri demonstrated the first sizeable gas turbine power plant at the Swiss National Exhibition at Zurich in 1939. The first turbojet-powered flight occurred on August 27, 1939.

Today, the gas turbine is used exclusively to power all new commercial airplanes and most business aircraft. It is used for electric power generation and for gas pipeline compressor drives and has been extensively tested in boats, trains, automobiles, and trucks.

The aim of this book is to present the fundamentals of the gas turbine including cycles, components, component matching, and environmental considerations. It is directed toward the undergraduate engineering student who has a knowledge of elementary thermodynamics, fluid mechanics, and combustion and should be useful to any graduate student desiring an overview of the gas turbine. I am hopeful it will serve as a text in both undergraduate and graduate courses as well as a reference book for engineers in industry wanting an understanding of the various gas turbine cycles, components, component matching, and environmental considerations.

This book is the outgrowth of many years of teaching a course on the fundamentals of the gas turbine and several years experience in the gas turbine industry.

The text is essentially self-contained. Many tables are included in the book. The reader who wants more extensive tables will find that the equations used in developing the tables are included. Also, a Suggested Reading section at the end of most chapters is provided for those who want another viewpoint or a more detailed analysis.

There are numerous example problems with state-of-the-art values. Both *Système International* (SI) and English units are used in the example problems. The numerical values in the example problems are presented in a way that allows the user to select only SI, only English, or both SI and English units. Example problems are similar to show the effect of changing one or more of the cycle parameters. Problems using both SI and English units are included at the end of the chapters.

The book may be considered as consisting of six parts or areas.

The first part (Chapter 1) gives a brief review of the history of the gas turbine. Users are urged to read as many of the references cited as possible to obtain an understanding and appreciation of the development of the gas turbine.

The second part (Chapters 2 to 4) consists of a review of thermodynamics, fluid mechanics, and combustion. It presents the terms, concepts, and equations used in later chapters.

The third part (Chapters 5 and 6) discusses the various gas turbine cycles and variables that influence the cycle performance. It is assumed in these chapters that the pressure ratio, turbine inlet temperature, component efficiencies, and air flow have single values that do not change.

The fourth part (Chapters 7 to 9) discusses components and component performance, showing the variation in the parameters such as air flow and component efficiency. It gives readers the fundamentals of component performance including the different ways of constructing the various components.

The fifth part (Chapter 10) investigates component matching and what happens to the steady-state operation of a gas turbine at "off-design" conditions.

The sixth part (Chapter 11) examines gas turbine air and noise emissions, current air and noise regulations, and engine modifications that can be made to reduce the quantity of air pollutants and noise emitted by a gas turbine engine.

I express my appreciation to the many students and colleagues who have provided me with the encouragement to undertake and complete this endeavor and their constructive criticism during its preparation. I thank the following reviewers for their contribution: Joe D. Hoffman, Purdue University, Gordon C. Oates, University of Washington, and Walter F. O'Brien, Jr., Virginia Polytechnic Institute and State University. I also thank the many people who helped in the typing of this manuscript. Finally, I thank my wife, Shirley, and my two children, Mark and Belinda, for their encouragement and patience during the writing of this book.

William W. Bathie

CONTENTS

3 FLUID MECHANICS 47

4 COMBUSTION 70

5 SHAFT-POWER GAS TURBINES 90

1

HISTORY OF THE GAS TURBINE

1.1 INTRODUCTION

The theory of the gas turbine and the way that it should function were established long before the necessary materials and the detailed knowledge of flow mechanisms were available to the personnel involved. Engineers had to wait for the development of materials that could withstand high temperatures and improved component efficiencies to make the gas turbine a useful powerplant.

A gas turbine is an engine designed to convert the energy of a fuel into some form of useful power, such as mechanical (shaft) power or the high-speed thrust of a jet. A gas turbine consists basically of a gas generator section and a power conversion section. The gas generator section consists of a compressor, combustion chamber, and turbine, the turbine extracting only sufficient power to drive the compressor. This results in a high-temperature, high-pressure gas at the turbine exit. The different types of gas turbines result by adding various inlet and exit components to the gas generator.

Three different types of gas turbine engines are illustrated in Figures 1.1 through 1.3. Figure 1.1 illustrates a gas turbine that operates on the simple (basic) cycle, Figure 1.2 a turbojet engine with an afterburner, and Figure 1.3 a nonmixed turbofan engine.

These pictures are included at this point to familiarize the reader with three of the several types of gas turbines currently being manufactured. These, along with the other types of gas turbines, are discussed in Chapters 5 and 6. The reader may find it beneficial to come back to this chapter after reading the material on cycle analysis and components.

Today, the gas turbine is widely used. During the period from the granting of the first patent in 1791 to the present, the gas turbine has been developed into a very reliable, versatile engine with a high power-to-weight ratio. It is used exclusively to power all new commercial airplanes. Most business aircraft are powered by one form or another of a gas turbine. The gas turbine has been used to power boats and trains, has been widely accepted for electric power generators and gas pipeline compressor

Figure 1.1. Cutaway of AlliedSignal's TF40 gas turbine engine. (Courtesy of AlliedSignal)

drives, and is being tested for use in buses and trucks. It has been tested extensively as a power plant for an automobile.

This chapter traces the history of the gas turbine from 1791 to the present. The purpose here is to give an outline of the development of the gas turbine, listing historic events with dates. The reader who wants a more detailed history is referred to the references at the end of the chapter.

1.2 1791 TO 1930

The development of the gas turbine was characterized by a number of patents. Many of the early designs were never built, and those that were built usually produced extremely low power outputs if any at all.

The initial step in the development of the gas turbine occurred when John Barber obtained the first patent in 1791. Although it is doubtful that Barber's turbine was ever built, it did serve as the basis for further development (1).

The next important development came in 1872, when Stolze obtained a patent. His design consisted of a multistage axial-flow compressor (probably the first of its kind), a multistage reaction turbine on the same shaft, a heat exchanger, and a combustion chamber (2). His design showed many more stages in the turbine than in the compressor.

The first U.S. patent covering a complete gas turbine was issued on June 24, 1895, to Charles G. Curtis (3).

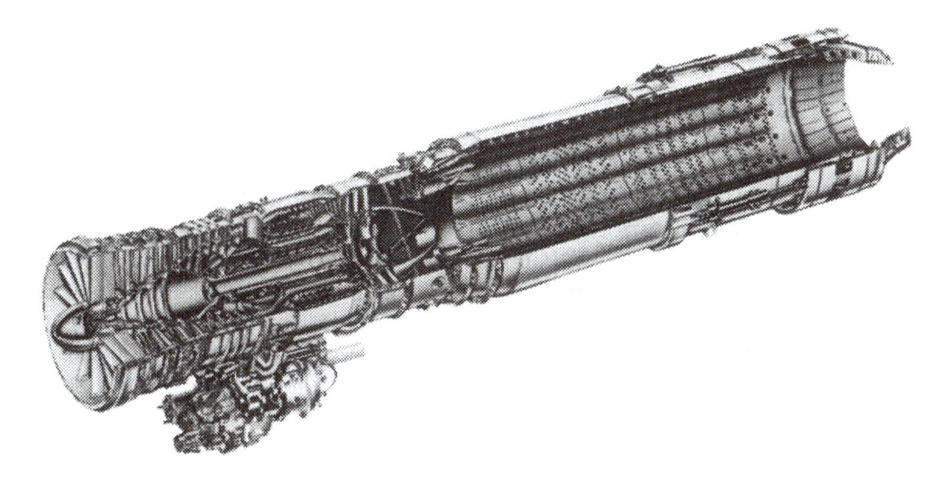

Figure 1.2. Cutaway of General Electric's J85-21 turbojet engine with afterburner. (Courtesy of General Electric Co.)

The first gas turbine that actually operated was designed and built in France by Stolze (3). Testing of this unit began in 1900, but the results were very discouraging because of low component efficiencies.

About the same time, a serious attempt was made to build a gas turbine on a large scale by the two Armangand brothers in Paris. Their first tests were made with a 25-hp de Laval turbine using compressor air from the Paris compressed air mains.

A later unit consisted of a Rateau compressor operating at a pressure ratio of approximately 4, a Curtiss turbine wheel 37.4 in. in diameter operating at 4250 rpm, and a turbine inlet temperature of 1040°F. This setup supplied compressed air instead of mechanical power and had an estimated thermal efficiency of 3% according to Stodola (4).

The next major work apparently was done by H. Holzwarth. To him belongs the credit for having built the first economically practical gas turbine. The Holzwarth

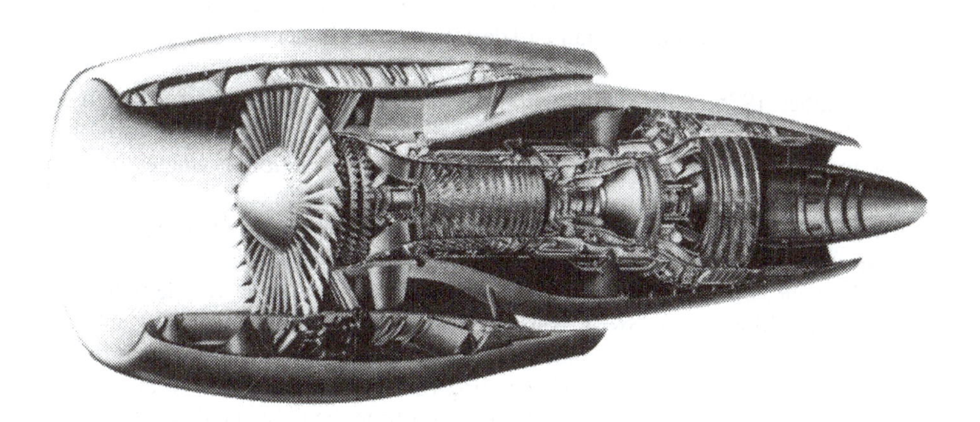

Figure 1.3. Cutaway of General Electric's CF6-50 turbofan engine. (Courtesy of General Electric Co.)

gas turbine operates on an explosion cycle without precompression. The principal here is a purely rotating machine with intermittent combustion. The first of his turbines was built and tested in Hanover. A thorough discussion of this type of gas turbine is contained in a book by Stodola (4).

At the same time, Stanford Moss operated the first gas-driven turbine wheel in the United States. He, like others, had little success because of limitations on materials and component efficiencies.

Dr. S. Moss did considerable work on the development of supercharged reciprocating engines. In 1918, a Liberty engine equipped with a GE supercharger was tested at sea level and at Pikes Peak at an altitude of 14,109 ft. The engine developed 350 hp at sea level, 230 hp at Pikes Peak without a supercharger, and 356 hp with the GE supercharger (5).

1.3 1930 TO 1940

Development in the 1930s resulted in the Velox boilers, the Houdry process cracking stills, and the construction of the first gas turbine successfully operated expressly for power generation, all of these being accomplished by Brown Boveri. Extensive work was done on the development of the gas turbine to propel an airplane by the British, Italian, and German governments, which led to the flight of the first gas turbine-powered airplane.

Brown Boveri Effects

Brown Boveri must be given credit for pioneering the construction of gas turbines to generate electrical energy in power stations and for industrial applications.

The first Brown Boveri "gas turbines" were those used to drive the combustion air compressors of the Velox boilers. According to Seippel (2), the Velox boilers were based on conclusions drawn by Noack of Brown Boveri from a test of a 500-kW oil-fired Holzwarth turbine built in 1927 by Stodola and Schule. The first Velox boiler turbine was built in 1932; it needed some auxiliary power because the requirements of the compressor exceeded the output of the turbine. Later, the charging sets were able to deliver excess power. The standard design of Velox boilers for ships is discussed in (6).

In November 1936, a new application for the charging sets appeared when the Sun Oil Co. of Philadelphia called for superchargers to burn carbon residues in their Houdry process cracking stills under pressure with the production of whatever power was possible (7).

Brown Boveri in 1939 built the first sizable gas turbine power plant expressly for power generation, which had a terminal output of 4000 kW. This unit was based on the simple cycle, as it was designed for standby power only. It was demonstrated in running condition at the Swiss National Exhibition at Zurich in 1939 and was commissioned in the Neuchâtel underground power station in 1940. According to Pfenninger (7), it had only 1200 h of operating time by 1953, this low total resulting from the fact that it was run only a few hours at a time, being a standby set, and not because of mechanical trouble.

TABLE 1.1. **Results of Tests Run by Stodola on the Brown Boveri Gas Turbine Installed at Neuchâtel (8)**

Item	Test I	Test II	Test III
Load (kW)	Light	4,021	3,057
Fuel	Fuel oil	Fuel oil	Fuel oil
Compressor pressure ratio	3.82	4.38	4.28
Compressor efficiency (%)	86.4	86.6	86.9
Compressor speed (rpm)	3,020	3,020	3,030
Compressor air flow, lb/h	499,620	498,176	498,049
Turbine inlet temperature (°F)	705.2	1,067	987.8
Turbine efficiency (%)	85.4	88.4	88.4
Fuel consumption (lb/kWh)	—	1.078	1.193
Thermal efficiency (%)	—	18.04	16.37

Brown Boveri invited Stodola, aged 80, to conduct the official tests on this first machine. The important results of these tests are given in Table 1.1.

The Brown Boveri gas turbine at Neuchâtel, Switzerland was dedicated as An International Historic Mechanical Engineering Landmark on September 2, 1988 (9).

The first closed-cycle gas turbine power station was also commissioned in 1939 (10). This gas turbine was fueled on oil, had a 2-MW rating, and had approximately 6000 h of running time by March 1977.

British Efforts

Two independent groups in England began work on the construction and testing of gas turbines for power plants in an airplane during this time: One group under the direction of Whittle worked on a turbojet using a centrifugal-flow compressor, and a second group under the direction of Griffith and Constant worked on constructing and testing axial-flow compressors.

Whittle (11) applied for and was granted his first patent in 1930. He applied for assistance to the Air Ministry and several private firms but was tuned down by all of them, some because it was a long-term research project, others because they did not have the money to commit to a project of this type.

Whittle, lacking the necessary financial support required to perform any actual test work, continued to do paper work from 1930 to 1936 when Power Jets, Ltd., was formed. The design considered was a simple jet engine having a single-stage centrifugal compressor with bilateral intakes, a single-stage turbine coupled directly to the compressor, and a single combustion chamber.

Testing of the first engine began on April 12, 1937, and continued for 11 days, at which time it became apparent that the combustion chamber presented a major problem and the compressor performance was far below expectations.

A redesigned engine was started on April 16, 1938, and continued until May 6 when the engine was severely damaged by a turbine blade failure. A third model of the engine was placed on the test stand in October 1938.

In the summer of 1939, the Air Ministry placed a contract with Power Jets, Ltd., for a flight engine known as the W1. Details of this engine are included in the next section, because testing of the W1 testing did not begin until 1941.

TABLE 1.2. **Leading Particulars of the He S-3b Turbojet Engine (12)**

Compressor	Centrifugal
Engine weight	795 lb
Engine thrust (static)	1100 lb
Thrust specific fuel consumption	1.6 lb/(lb h)

In 1936 a second group in England, under the direction of Griffith and Constant at the Royal Aircraft Establishment began work on constructing and testing axial-flow compressors.

German Efforts

Shortly before 1935, Hans von Ohain [see (12)] became interested in gas turbines for propulsion and obtained one or more patents for a turbojet engine with a centrifugal compressor. He was hired by Ernst Heinkel, president of Ernst Heinkel A.G., in 1936 and put in charge of a turbojet engine. Work began in April 1936 on a demonstration engine that was run in March 1937. The first flight engine was tested in 1938 and after a redesign, particularly to the combustion chamber, the He S-3b engine was ready for testing in 1939. Leading particulars for this engine are shown in Table 1.2.

The He S-3b engine was installed in the He-178 airplane and the first turbojet-powered flight was made on August 27, 1939.

During this period two other important developments occurred.

In Germany, a second line was begun independent of the work of von Ohain by the Junkers Airplane Company. This was under the direction of H. Wagner. After looking at various forms of gas turbines (including turboprops, turbojets, and turbine- and piston-driven ducted fans), the emphasis (prior to 1938) was placed on the turbojet engine using an axial-flow compressor. The test engine built and tested during 1938 was extremely small in diameter and length and low in weight.

By the end of 1939, Junkers was working on an axial turbojet, the 004; Bramo was working on a more advanced axial turbojet, the 003; and Heinkel was developing two turbojets, a new centrifugal engine, the 001, designed by von Ohain from what had been learned from the He S-3, and an advanced axial turbojet designated the 006.

Italian Efforts

In 1933, S. Campini, an Italian engineer, proposed using a reciprocating engine driving a compressor to produce a high-velocity jet stream. In this design, the airframe was a large cylinder for the power plant. The aircraft, designed by G. Caproni, flew for the first time on August 27, 1940 (5).

U.S. Efforts

Development of the gas turbine in the United States was devoted mainly to studying the gas turbine with the result that the turbine and compressor efficiencies were too low to justify its development. Considerable time was spent on development of the turbosupercharger, which is essentially a gas turbine with the combustion chamber

TABLE 1.3. **Leading Particulars of the W1 Engine (12)**

Compressor impeller			
Tip diameter (in.)			19
Tip width (in.)			2
Number of blades			29
Type			Centrifugal
Turbine			
Mean diameter of blades (in.)			14
Blade length (in.)			2.4
Number of blades			72
Typical performance at 16,500 rpm	Design	Test	
Air flow (lb/s)	22.6	20.6	
Delivery pressure (psig)	33	31.5	
Thrust (lb)	900	850	
Exhaust temperature (°C)	530	560	
Thrust specific fuel consumption (lb/lb h)	1.19	1.39	

replaced by a reciprocating engine. Typical efficiencies for the compressors and turbines used in turbosuperchargers were of the order of 60 to 65% for the compressor and 65% to 70% for the turbine.

The first production gas turbines in the United States were those produced by Allis-Chalmers Manufacturing Company for use in Houdry process cracking stills under a Brown Boveri license.

1.4 1940 TO 1945

From 1940 to 1945 extensive work was done on making the turbojet engine a production engine. At this time also, the first gas turbine was used in a locomotive in operation. Only two turbojet engines were in production before the end of World War II, the British Welland engine, which used a centrifugal-flow compressor, and the German Jumo 004 engine, which used an axial-flow compressor. The discussion below centers around these historic events.

British Efforts

Power Jets, Ltd., under the guidance of Whittle and the contract placed by the Air Ministry in 1939, proceeded with development of the W1 engine. The leading particulars of this engine, which was first run on April 12, 1941, are listed in Table 1.3.

It is of interest to note that the turbine on this engine was water-cooled.

The British made their first flight with an airplane powered by a gas turbine on May 15, 1941, when the Gloucester E28/39 aircraft powered by the W1 turbojet became airborn. The success of this flight led to considerable time and money being spent on the development of the gas turbine, especially in the form of a turbojet engine.

Power Jets, Ltd., under the guidance of Whittle, built the W1A engine, which was designed for a thrust of 1450 lb and used air cooling for the turbine instead of water cooling; the W2 engine, which was unsuccessful; and the W2B engine, which was the parent design of the Rolls-Royce Welland engines. All of these engines used centrifugal compressors.

Rolls-Royce, which had been doing gas turbine research and development work on its own, was assigned the responsibility for production of the W2B engine, initially known as the W2B/23 and later as the Welland. The Welland was first flown in the Meteor airplane on June 12, 1943, and was placed into production late in 1943 with deliveries in May 1944. Production engines had a thrust of approximately 1600 lb and a thrust-specific fuel consumption of 1.12 lb/lb-h.

Rolls-Royce also produced several other engines. One, the Derwent I, eliminated the reverse-flow combustion chamber of the W2B/23. Known originally as the B/26, the Derwent I was first flown in March 1944. Another, the Nene, was the first Rolls-Royce engine that did not follow closely the W2B design. A scaled-down version of the Nene, known as the Derwent V, was tested in June 1945, with production commencing in September 1945. This engine had a thrust rating of 3500 lb and a thrust-specific fuel consumption of approximately 1.00 lb/lb-h.

Other British engines built during this period were the deHavilland Goblin and the RAE-Metrovick F-2 engines.

The deHavilland H-1, later known as the Goblin, was a single-sided centrifugal compressor engine. It was the engine used in the first flight of the Meteor in March 1943.

The RAE and Metropolitan Vickers Company worked on the F-2 turbojet engine, which employed an axial-flow compressor. The engine was first run in December 1941 and, after modifications to the combustion system and bearings, a modified Meteor was flown in November 1943. It was once again redesigned and designated as the F-2/4 engine and was cleared for flight once again in November 1945.

German Efforts

The Germans, unlike the British, concentrated mainly on the turbojet engine with an axial compressor even though the engine used in their first flight had a centrifugal compressor.

From 1939 until 1942, German development of jet engines continued along the lines established in 1939 with the less promising ones eliminated. Intensive development of the von Ohain 001 centrifugal turbojet was continued until late in 1941. At this point it was dropped by Heinkel, as it was clearly behind the 003 and 004 axial-flow engines. Therefore, at the start of 1942 only the 003, 004, and 006 engines, all axial-flow, were under development.

The Junkers 004 (Jumo 004) engine was placed on the test stand in November 1940 (004A engine). The first flight with the 004A was made in an Me 110 in March 1942, and two 004As powered an Me 262 in July 1942.

Various versions of the 004 engine were produced before the end of World War II, this along with the Welland engine being the only production engines before the end of the war.

The BMW 003 engine was first run in 1940. Development continued on this engine but, because of the superior performance of the Junkers 004 near the end of 1943, the 004 design was frozen and placed in production with development work continuing on the 003.

TABLE 1.4. **Comparison of the Junkers 004 and the Rolls-Royce Welland (12)**

	Junkers 004B		Welland
Compressor			
Type	Axial		Centrifugal
Pressure ratio	3.1		4
Compressor efficiency (%)	80		75
Turbine inlet temperature		About the same	
Turbine			
Efficiency (%)	80		87
lb weight/lb thrust	0.83		0.53
Frontal area/lb thrust (in^2/lb)	0.46		0.94
Fuel consumption (lb/h lb)	1.40–1.48		1.12

A comparison between the Junkers 004 and the Welland, the two production gas turbines of this period, is given in Table 1.4.

The reliability of the Welland turbojet was considerably higher than that of the Junkers 004. The Welland had undergone repeated 100-h tests prior to going into production, whereas the 004 was scheduled for overhaul for every 25 h of operation, this being raised near the end of the war to 35 h. The weakest part of the 004 was the combustion chamber. It must be pointed out that the time (labor-hours) required to produce the 004 was considerably lower than that of the Welland.

U.S. Efforts

Early in the 1940s, several proposals were made to build a gas turbine in the United States. A discussion of each type is given below.

One was a 2500-hp turboprop engine known as the Turbodyne. This engine was first proposed to the Army and Navy in 1940, with a joint Army–Navy development contract being awarded to Northrup Aircraft, Inc., in June 1941. Another was the L-1000, a turbojet engine proposed by Lockheed Aircraft. NACA, based on work done in the 1930s by S. Campini, began construction of ducted axial-flow fan powered by a reciprocating engine. Westinghouse Electric Corporation, through a Navy contract, worked on an axial-flow compressor turbojet known as the 19.

Wright Aeronautical Corporation, upon learning of the Power Jets, Ltd., development, tried unsuccessfully to obtain an American license for the manufacture of the Whittle engine.

The Power Jets, Ltd., W1X engine and drawings of the W2B were sent to the General Electric Company in 1941. After making several mechanical changes, General Electric tested their first engine, known as the I engine. A modified engine, the I-A, was flown in the Bell P-59A in October 1942.

The General Electric Company constructed other models known as the I-14, I-16, and I-40, which produced static thrusts of 1400, 1600, and 4000 lb, respectively. The I-40, first tested in January 1944 and flown in a Lockheed XP-80A in June 1944, was the only jet fighter-engine combination considered for production by the United States prior to the end of World War II. The I-40 eventually became known as the J33 and was mass-produced by the Allison Division of General Motors.

The other General Electric engine was started before the end of 1945. This was the TG-180 (later known as the J35), and was an axial-flow turbojet engine.

Because of the deep involvement in the production of air-cooled reciprocating engines and because it was a private venture on their part, Pratt & Whitney Aircraft was not deeply involved in gas turbine work. As partial payment for Lend-Lease during World War II, Pratt & Whitney Aircraft was given the plans and details for the Rolls-Royce Nene engine (which, when produced by Pratt & Whitney became the J42) and a layout of the Rolls-Royce TAY engine. This later engine was essentially developed by Pratt & Whitney and became known as the J48.

Other Efforts

Although the main emphasis during this period was on the application of the gas turbine to airplane propulsion, one historic nonaircraft installation occurred. This was the installation of a 2200-hp gas turbine in a locomotive ordered by the Swiss Federal Railways. This locomotive was first run in January 1941 and, at its most economical output of 1700 hp, had an efficiency of 18.4%. It should be noted that this was a gas turbine with a regenerator (13).

Table 1.5 lists some of the operating parameters of gas turbine engines that existed in 1946.

1.5 1945 TO 1950

From 1945 to 1950, numerous companies that had been interested in the applications of the gas turbine and had been performing studies before or during World War II or that had been investigating components for use in a gas turbine reentered the field. A new type of power plant was ready for development. Government financial support was plentiful and could be readily obtained. Many companies, with visions on the profits that could be reaped from a successful venture into this field, started or continued research and development of the gas turbine for many applications. As with any new field, only a few survived the test, as can be seen by a comparison of a list of companies that were working on it in 1945 and that are now active in the field.

At this point it becomes impossible to discuss all the companies involved and the many applications for gas turbines. In this and successive periods, the major accomplishments for the various areas are discussed. Each application is discussed separately. Some of the applications overlap two areas, and, in the case of government work, it is hard to pin down when they occurred: that is, the first running of an engine, the first flight of an engine, and so on.

Automotive

The first public demonstration of a gas turbine-powered automobile took place during this period at Silverstone, Northamptonshire, on March 9, 1950 (15). The gas turbine was installed in a Rover company car and consisted of a centrifugal compressor, single-stage turbine driving the compressor, and a separate power turbine. Although this engine had been arranged to include a heat exchanger, the first test of the engine

TABLE 1.5. **Data for Several of the Turbojet Engines that Existed in 1946 (14)**

Model	I-16	I-40	Welland I	Derwent I	Nene	Jumo 004-B4
Manufacturer	General Electric	General Electric	Rolls Royce	Rolls Royce	Rolls Royce	Junkers
Type	Turbojet	Turbojet	Turbojet	Turbojet	Turbojet	Turbojet
Compressor:						
Type	Centrifugal, double entry	Centrifugal, double entry	Centrifugal, double entry	Centrifugal, double entry	Centrifugal, double entry	Axial flow
Stages	1	1	1	1	1	8
Pressure ratio	3.8	4.1	—	3.9	—	3.0
Combustion chamber:						
Type	10 chambers, reverse flow	14 chambers, straight-through flow	10 chambers, reverse flow	10 chambers, straight-through flow	9 chambers, straight-through flow	6 chambers, straight-through flow
Turbine:						
Type	Axial	Axial	Axial	Axial	Axial	Axial
Stages	1	1	1	1	1	1
Inlet temp.	1472°F	1472°F	—	1560°F	—	1472°F
Efficiency	90%	—	—	—	—	80–85%
Exit temp.	1220°F	1170°F	1202°F	1256°F	1238°F	1150°F
Air flow (lb/s)	33	79	—	—	—	43
Weight/thrust (lb/lb t)	0.53	0.46	0.53	0.49	0.31	0.82
Frontal area (ft^2)	9.39	14.7	10.1	10.1	13.4	4.9
Fuel consumption (lb/lb t/h)	1.20	1.18	1.12	1.17	1.06	1.4
Rating, take-off thrust						
Normal (lb t)	1425	3200	1150	1550	4360	1900
Military (lb t)	1600	4000	1600	2000	5000	1980

did not include one because the heat exchanger was difficult to develop and was not ready in time for the first test.

Aviation

The gas turbine in the form of a turbojet engine was shown to be successful for propulsion of airplanes during World War II. The W1 units built by the British and the German engines that were built and used during World War II were now ready for extensive retesting, redesign, and improvement.

General Electric engineers began to make design improvements and to increase the size of the I-A engine they had been building. They built in sequence a few I-14s, many I-16s, some I-20s, and a large number of I-40s. These units were designed to develop 1400, 1600, 2000, and 4000 lb of static thrust, respectively. Engines of the first two of this series were employed to propel the twin-engine Bell P-59A aircraft, the first American jet plane. The I-20 had the best specific fuel consumption of any of these early designs, but because of the need for larger units, only a few were built. The I-40, on the other hand, was the initial design of the turbojet that has been built in larger quantities than any other unit. It was conceived early in 1943 and was run initially in December of the following year. It was first used in the Lockheed XF-80, the Shooting Star, in June 1944. In order to secure expanding facilities, the blueprints of the engine were given to the Allison Division of General Motors for mass production. The initial production model, which was designated the J33-A-21, delivered 3825 lb of static thrust with a specific fuel consumption of 1.22 lb of fuel per hour per pound of thrust. It weighed 1850 lb. Through a continual development program by Allison the engine was improved in two successive major steps. The first increased the static thrust to 4600 lb, and the second provided an even greater increase. These increases, unlike the ones in the I series, were made without increasing the diameter of the engine. Increases in thrust were obtained by increasing the air flow and the pressure ratio. Simultaneously with the thrust increase, the specific fuel consumption was reduced.

At the same time as General Electric and Allison were developing and making improvements, British engineers were also making great strides along the same lines. The DeHavilland Company produced a Goblin and the Ghost. The Rolls-Royce Company designed in order the Welland, the Derwent, and the Nene. The essential difference between the developments of the two companies was that the DeHavilland engines employed single-entry compressors, whereas Rolls-Royce concentrated on the double-entry type. The latter type of compressor has vanes on both sides of the impeller and has the advantage of increased air flow with only a small increase in weight.

Americanized versions of both the Nene I and the Nene Tay were produced by Pratt & Whitney. Other versions of the same engines were also in production in France, the Soviet Union, and Australia. Versions of the Goblin and the Ghost were produced in Sweden and Italy, and of the Derwent in Belgium and Argentina.

Although the development of axial-flow turbojet units began almost as early as the work on centrifugal type and many more axial designs have been made, production in America and England lagged behind that of the centrifugal type. The reverse was true in Germany. In spite of the fact that the first German flight was with a centrifugal-type jet and the Germans continued to make other designs, the axial-flow engines emerged the favorites and were the first to be put into production. The Jumo 004 was the only operational German turbojet. It had an eight-stage axial compressor, six combustion chambers, and a one-stage impulse turbine. It delivered a static thrust from 2200 to 2500 lb with a specific fuel consumption of about 1.36 to 1.4 lb. It was the power plant used in the ME 262, the first operational jet bomber. The construction of the Jumo 004A was begun in early 1940, and the first unit ran in December of that year. It required a period of about 6 months to eliminate the vibration troubles. Consequently the unit was not flight-tested until late in 1941.

In this country, the Westinghouse Electric Corporation and the General Electric Company pioneered in the field of axial-flow turbojets. Although some thought was given to the possibility of such engines in 1941, the fabrication of units did not begin until 1942 and the first units were tested in 1943. Westinghouse built the 19A, 19B, 9 1/2A, 9 1/2B, 19XB, 24C, and J40 in sequence. The designations of the first six of these units were based on the diameter in inches of the units, and their static thrusts delivered 1100, 1365, 270, 275, 1600, and 8000 lb of thrust, respectively.

The General Electric Company designed and built the TG180, TG190, and other new engines. The TG180, known as the J35, was placed in quantity production by Allison in 1947. The TG190, or J47, was produced by General Electric. These two engines are similar. The latter has one more compressor stage and increased air flow and thrust.

Except for the production of the British Nene, the design, development, and production of jet engines in France has concentrated on the axial-flow type. Rateau designed units with high compression ratios and large thrust values.

The Russians produced several axial-flow turbojets, which were developed from the German engines. They have improved versions of the BMW003, the Jumo 004, the Jumo 012, and the BMW018. These engines deliver thrusts of about 3750, 4000, 6600, and 700 lb, respectively. The pressure ratios vary in the range from 3.4 to 7.0, and the specific fuel consumption from 1.2 to 1.08 lb/lb-h.

One possible method of improving the static thrust and thrust at low speed is to accelerate a large mass of air. This can be done by an engine in the form of a turboprop engine.

The Armstrong-Siddeley Company used the ASX axial-flow turbojet as a basis for the development of the Python turboprop. This engine first ran in March 1945. Rolls-Royce adapted the Derwent turbojet to form an experimental turboprop, the Trent. In September 1945 it became the first turboprop to propel an aircraft.

The Bristol Company was the first to concentrate on the development of a turboprop *per se*. Their first unit was the Theseus I. It had a combined compressor consisting of eight axial stages followed by one centrifugal stage. The compressor was run by the first two stages of the turbine, and the third turbine stage drove a propeller by means of an independent shaft that extended forward through the hollow compressor rotor shaft.

The first turboprop designed in the United States was the Turbodyne of the Northrup Aircraft Company. The commitment for the production of this engine began in 1941 and the engine was wrecked in the test bed some 4 years later.

Chrysler Corporation was awarded, in the fall of 1945, a research and development contract by the Bureau of Aeronautics of the U.S. Navy to create a turboprop engine for aircraft. This program resulted in the development of a turboprop engine that achieved fuel economy approaching that of aircraft piston engines prior to being terminated in 1949.

The time between overhaul on jet engines in early 1946 was set at 50 h. In May 1947, the J33 was the first gas turbine engine to complete the 150-hour qualification (16).

A milestone was reached when a Vickers Viscount was first flown on July 16, 1948, as the world's first turboprop-driven transport. The engines were Dart R. Da. 1 turboprop engines, which developed 1380 ehp.

Other achievements were occurring in many other areas.

A milestone was reached on July 14, 1947, when the British Navy's MGB 2009 put to sea with a gas turbine engine providing part of the propulsive power. The first vessel to be propelled solely by a gas turbine was a 24-ft plane-personnel boat powered by a Boeing engine. Trials started on May 30, 1950 (17).

The second gas turbine locomotive was built by General Electric Company and first operated on November 14, 1948. Although this was an experimental unit, it was so successful that an order was placed in December 1950, with the first units delivered in January 1952.

The first gas turbine locomotive that could be called a commercial product was delivered by Brown Boveri to the Great Western Railway in England on March 10, 1950 (18).

The first gas-line unit, in the form of a gas turbine driving a centrifugal compressor, was an 1800-hp single-shaft unit built by Westinghouse. It was placed in service in May 1949 on the 22-in.-diameter line of the Mississippi Fuel Corp. at Wilman, Arkansas (19).

The first stationary gas turbine built in the United States for the purpose of generating electric power went into service on July 29, 1949 (20). It was a 3500-kW unit installed at the Belle Isle Station of the Oklahoma Gas and Electric Company in Oklahoma City. It had a pressure ratio of 6 and a turbine inlet temperature of 760°C (1400°F). The unit was retired in September 1980, moved to Schenectady, New York and dedicated as the 73rd National Historic Mechanical Engineering Landmark by the American Society of Mechanical Engineers on November 8, 1984 (21).

Brown Boveri placed a 10,000-kW, double-shaft unit with intercooling, regeneration, and reheat in operation at the Santa Rosa plant in Lima, Peru in the same year (20).

1.6 1950 TO 1960

The 1950s saw the gas turbine engine being used in practically every type of application suitable to prime movers. In several cases, it was still being used in an introductory manner; in other areas it was well entrenched.

This decade saw extensive efforts being made in materials, turbine cooling, fuels, and cycle components. It also was a time period in which some of the weak companies were forced out of the business.

The General Electric Co. was awarded a joint contract in 1951 by the Atomic Energy Commission and the U.S. Air Force to determine if an atomic-powered aircraft was feasible (5). Their responsibility was to develop sufficient data on nuclear materials and the shielding required for an atomic-powered gas turbine. Pratt & Whitney was pursuing a similar course.

In 1954, both GE and Pratt & Whitney were teamed with an airframe company for further development. All work was terminated by 1961 (5).

In 1953, GE started work on the G0L-1590, a predecessor to the J79. The G0L-1590 was a single rotor turbojet engine that used a *variable stator compressor* (5).

In 1953, the Vickers Viscount, powered by a Rolls Royce Dart turboprop engine, entered commercial service. Boeing's 707 made its maiden flight on July 15, 1954. It was powered by four Pratt & Whitney JT-3 turbojet engines.

In March 1956, a standard production model of a 1956 Plymouth powered by a gas turbine engine made the first transcontinental journey. It averaged, over the 3020-mi trip, approximately 13 mpg (22).

A modified Sikorsky S-58 powered by two General Electric T58s made the first U.S. turbine-powered helicopter flight.

In 1956, GE started development of an aft fan turbofan engine, the CJ805-23. The full engine was tested December 27, 1957. It was the first U. S. turbofan engine. It entered airline service in the 1960s, the first in the world to enter service (5). Pratt & Whitney started development of its first turbofan engine in the late 1950s. It modified a JT3 (J57) gas generator into the JT3D.

The late 1950s saw the introduction of light-weight, simple-cycle gas generators and gas turbines into the marketplaces. These simple-cycle gas turbines had pressure ratios around 12 and cycle thermal efficiencies of approximately 25%.

1.7 1960 TO 1970

The 1960s saw the gas turbine being developed into an increasingly efficient and cost-effective engine for many applications. Higher overall pressure ratios and higher turbine inlet temperatures were achieved. Improvements in materials, turbine cooling techniques, and components were achieved.

The fuel consumption per pound of thrust for U.S. commercial applications improved about 15% with the introduction of the low-bypass-ratio turbofan engines.

The General Electric Company initiated development work on a water-cooled gas turbine in the early 1960s, although the first laboratory model was not tested successfully until 1973.

Pratt & Whitney continued development of the JT3D turbofan engine. The JT3D, which made its first flight on a Boeing 707-120 on June 22, 1960, soon was selected to power some DC-8 aircraft.

The Allison Division of General Motors initiated, in September 1962, a program to assemble and test a regenerative turboprop engine (23). The regenerator was a stationary tubular-type regenerator to be built by AiResearch Manufacturing Division. The test results showed that specific fuel consumption was reduced 36% when compared with their T56 turboprop engine.

Gas turbine-powered vehicles had captured the technical community in the 1950s. Development of a gas turbine for passenger cars and trucks continued until the early 1970s. Extensive effort was devoted toward materials that would allow a gas turbine for an automobile to operate at higher turbine inlet temperatures, toward methods of reducing the production cost of an engine, and ways to improve its efficiency.

Chrysler Corporation unveiled to reporters on May 14, 1963, their plans to build 50 turbine-powered test cars to be placed in the hands of typical drivers for evaluation in everyday use. The first of these cars was delivered on October 27, 1963, the last on October 28, 1965 (22).

The world's first three-shaft turbofan engine, the Rolls-Royce Trent, was run in December 1967. Experience from this engine was used in the development of the RB.211 engine.

During the 1960s, many simple-cycle gas turbine engines were installed by utilities as peaking units. They were selected because of their short lead time and low capital costs.

1.8 1970 TO 1980

The 1970s introduced many new factors soon to influence the direction taken on gas turbine engine development. These included greater competition among the major engine manufacturers, the prospect and uncertainty of new government regulations on emissions and noise, and dramatic changes in fuel prices and availability.

Attention focused on design trade-offs among performance, weight, cost, reliability, maintainability, and manufacturability.

Research was being conducted on ways to reduce noise and emissions from aircraft engines and to increase the performance of the engines.

The fuel consumption per pound of thrust for commercial jet engines received a second major improvement when the high-bypass-ratio turbofan engines were introduced in the early 1970s. These lowered fuel consumption approximately 20% when compared to the low-bypass-ratio turbofan engines introduced in the early 1960s.

The U.S. Clean Air Act of 1970 charged the U.S. Environmental Protection Agency (EPA) with the responsibility for establishing acceptable exhaust emission levels for CO, THC, NO_x, and smoke for all types of aircraft engines. These were published on July 17, 1973 (24). The levels established in the standards and the first compliance date of January 1, 1979, have acted as a catalyst for the development of advanced technology combustors.

Two National Aeronautics and Space Administration (NASA)-sponsored low-emission-technology programs were initiated. These were the Experimental Clean Combustor Program and the Pollution Reduction Technology Program.

The 1970s saw the continued installation of combined-cycle power generating systems. The late 1960s saw the introduction of GE pre-engineered heat recovery combined-cycles for power generation. The early units were from 11 MW to 21 MW (25). Large GE power generation combined-cycles were introduced in 1971. The first GE unit, a 340 MW unit, was purchased by Jersey Central Power & Light in 1971. GE installed a total of 15 STAG systems in the 1970s (25). These ranged in size from two to eight gas turbines and one to four steam turbines with total output ranging from 105 MW to 640 MW.

After the oil embargo of 1973, it became clear that efforts to develop a more fuel-efficient air transport system needed to be accelerated. To meet this need, NASA developed an overall plan called the Aircraft Energy Efficiency (ACEE) Program, of which the Energy Efficiency Engine (E^3) project was an important part.

Noise standards were initially published in the *Federal Register* on January 11, 1969. Part 36 was adopted November 3, 1969, and became effective December 1, 1969. These standards have been modified several times and have led to extensive research into the sources of noise and ways to reduce the noise emitted from an engine.

Two areas that have received extensive attention have been acoustical treatment of the nacelle and exhaust mixers for mixed-flow turbofan engines.

In early 1974, CFM International was formed by General Electric and SNECMA. It is owned and managed on a 50/50 basis and was formed to provide management for the CFM56 engine program.

1.9 1980 TO 1990

The 1980s was a decade of concern over costs, government regulations, availability of materials, fuel efficiency, and the use of aeroderivative engines in the utility industry. It involved several joint projects, the first flights of aircraft powered by unducted fan and propfan engines, the operation of the first combined cycle with a thermal efficiency over 50%, and development work on simple-cycle gas turbines with thermal efficiencies over 40%.

Extensive research was conducted on the use of ceramics, especially for automotive applications. One benefit of using ceramics in the turbine of a gas turbine engine is a higher turbine inlet temperature since a ceramic turbine will not require cooling air. A higher turbine inlet temperature results in a more efficient engine. The major drawback to ceramics is their brittleness. They also are subject to failure due to thermal shock that can occur during start-up or rapid change in power output.

The steam-injected gas turbine (see Article 5.7) was introduced in 1984, with over 20 steam-injected gas turbines in operation by the end of this decade. The steam-injected gas turbine has a lower capital cost than a combined cycle. It is also simpler to operate and maintain and has a thermal efficiency between that of a simple-cycle gas turbine and that of a combined cycle.

International Aero Engine AG (IAE) was formed on March 11, 1983 to develop an advanced technology turbofan engine. The companies involved are Rolls-Royce, United Technologies' Pratt & Whitney, the Japanese Aero Engines Corporation, MTU, and Fiat Aviazione of Italy. The resulting engine is the V2500 turbofan engine with a baseline rating of 25,000-lb thrust.

EUROJET Turbo GmbH was formed in August 1986 to coordinate the design, development, and manufacture of the EJ-2000 augmented turbofan engine for use in a supersonic European fighter aircraft. The companies involved included Fiat Aviazione of Italy (21%), MTU (West Germany), (33%) Rolls-Royce (33%), and Sener of Spain (13%).

The last half of the 1980s saw the development of the unducted and propfan engines by General Electric, PW-Allison, and Rolls-Royce. These were engines with bypass ratios between the existing turbofan engines and turboprop engines. The one way left to make major improvements in the fuel efficiency of aircraft engines was to improve the propulsion efficiency, which involves decreasing the change in velocity between the inlet and exit from an engine.

General Electric's unducted fan (UDF™), designated GE36, has two rows of counterrotating, variable-pitch, highly swept fan blades driven directly, without a gear box, by two counterrotating turbines. It has a bypass ratio of approximately 35 and developed approximately 25,000 lb of static thrust. It was installed on a Boeing 727 with flight testing commencing on August 20, 1986. It was also flight tested on an MD-80 aircraft with flight testing beginning on May 18, 1987 (26).

On February 23, 1987, Allison and Pratt & Whitney formed PW-Allison Engines to develop and test the 578-DX engine, a geared, counterrotating propfan propulsion system. The engine was installed on a McDonnell Douglas MD-80 testbed aircraft that made its first flight on April 13, 1989.

By the end of the 1980s, installation of combined-cycle power plants with thermal efficiencies of over 50% had begun. General Electric began development work on the LM6000, which is designed around the CF6-80C2 turbofan engine. When it became operational in 1992, it was the first simple-cycle industrial gas turbine engine with an efficiency over 40% (27).

Turbine inlet temperatures and engine pressure ratios continued to increase. This required the development of improved turbine materials and highly sophisticated air-cooling techniques. Small improvements were made in component efficiencies. These improvements resulted in aircraft engines with higher thrust-to-weight ratios and shaft-power gas turbines with higher thermal efficiencies.

The GE36 unducted fan and PW-Allison propfan engines, which were flight-tested in 1986 and 1989, respectively, were being mothballed and replaced with development work on higher bypass ratio engines. The reason for low (or no) airline interest in aircraft powered by a GE36 or PW-Allison 578-DX engine was due to the low price of fuel, the noise generated by the unshrouded engines, and the fact that they had to be mounted on the back of an aircraft due to their large diameters.

2.0 1990s

Today, both the shaft-power and propulsion gas turbines are extremely reliable, efficient, and clean machines. A few of the recent developments are listed as follows.

In late 1992, Pratt & Whitney tested its Advanced Ducted Prop (ADP) engine at a thrust level exceeding 53,000-lb thrust. This is a demonstrator engine with a bypass ratio of 15:1, a 118-in.-diameter variable pitch fan and a gear box with a gear ratio of 4 to 1 (28).

On March 30, 1993, GE engineers made a preliminary test run of the GE90 powerplant taking the engine up to 85,000-lb thrust. On April 3, 1993, the engine reached a thrust level of approximately 105,000-lb thrust under sea level static conditions and an outdoor temperature of 42°F (29). This is GE's first development of a new, nonderivative commercial engine for airline passenger jets in more than 20 years.

In January 1994, Rolls-Royce's Trent 800 was operated at over 106,000 lb of thrust with one engine being operated at over 100,000 lb of thrust for over 15 minutes (30).

GE estimates that in 1990, the generating capacity of gas turbines sold worldwide totaled 34,000 MW while steam turbine generating capacity sales totaled 23,000 MW (31). This is indicative of the acceptance of the gas turbine engine in the utility industry.

Extended range twin operations (ETOPS) have verified the reliability of modern aircraft turbofan engines. The early commercial turbojet and turbofan engines, when introduced, had an in-flight shutdown rate of approximately 0.8 per 1000 engine flight hours. Today's turbofans, when introduced, had an in-flight shutdown rate of approximately 0.4 per 1000 engine flight hours (32).

The JT9D turbofan engine had an in-flight shutdown rate of 0.38 shutdowns per 1000 flight hours during its first two years of commercial service. The PW 4060 installed on Boeing 767 aircraft have had an in-flight shutdown rate below 0.02 per 1000 flight hours since their introduction in 1989 (33).

Today, many two-engine commercial aircraft are permitted to be 180 minutes from a suitable diversionary airfield with one engine inoperative. This stringent requirement resulted in 1988 when inflight shutdown rates dropped below 0.02 per 1000 engine hours.

The 1990s will continue to see improvements in the reliability, efficiency, and emissions from both shaft-power and propulsion gas turbine engines.

REFERENCES CITED

1. Potter, J. H., The Gas Turbine Cycle, ASME Paper presented at the Gas Turbine Division Forum Dinner, New York, November 27, 1972.
2. Seippel, C., Gas Turbines in Our Century, *Trans. ASME* 75:121–122 (February 1953).
3. Sawyer, R. T., Gas-Turbine Progress Report—Introduction, *Trans. ASME* 75:123–126 (February 1953).
4. Stodola, A., *Steam and Gas Turbines with a Supplement on the Prospects of the Thermal Prime Mover,* Vol. II, McGraw-Hill Book Company, New York, 1927.
5. Anon., *Eight Decades of Progress, A Heritage of Aircraft Turbine Technology,* General Electric Co., Cincinnati, Ohio, 1990.
6. Anon., A Special Design of Velox Boiler for Ships, *Brown Boveri Rev.* 29:251 (1942).
7. Pfenninger, H., Operating Experience with Brown Boveri Gas-Turbine Installations, *Brown Boveri Rev.* 40:144–166 (1953).
8. Stodola, A., Load Tests of a Combustion Gas-Turbine Built by Brown Boveri & Company, Limited, Baden, Switzerland, *Brown Boveri Rev.* 27:79–83 (1940).
9. Anon., The World's First Industrial Gas Turbine Jet at Neuchâtel (1939), An International Historic Mechanical Engineering Landmark, September 2, 1988, Neuchâtel, Switzerland.
10. Anon., Closed Cycle Gas Turbine for Efficient Power Generation, Prepared by the U.S. Department of Energy, Division of Fossil Fuels Utilization, Washington, D.C.
11. Whittle, Air Commondone F., The Early History of the Whittle Jet Propulsion Gas Turbine, Proc. Inst. Mech. Eng. (London) 152:419–435 (1945).
12. Schlaifer, R., and Heron, S. D., *Development of Aircraft Engines and Fuels,* Andover Press, Andover, Mass., 1950.
13. Meyer, A., The First Gas-Turbine Locomotive, *Brown Boveri Rev.* 29:115–126 (1942).
14. Wilkinson, P. H., *Aircraft Engines of the World,* Paul H. Wilkinson, New York, 1946.
15. Anon., *Engineering* 169 (Pt. I):305 (1950).
16. Sonnenburg, P., and Schoneberger, W., *Allison Power of Excellence 1915–1990,* Coastline Publishers, Malibu, California, 1990.
17. Dolan, W. A., Jr., and Hafer, A. A., Gas-Turbine Progress Report—Railroad, *Trans. ASME* 75:169–175 (February 1953).
18. Browne, K. A., Yellott, J. I., and Broadley, P. R., Gas-Turbine Progress Report—Railroad, *Trans. ASME* 75:161–168 (February 1953).
19. Rowley, L. N., and Skrotzki, B. G. A., Gas-Turbine Progress Report—Industrial, *Trans. ASME* 75:211–216 (February 1953).
20. Schneitter, L., Gas Turbine Progress Report—Stationary, *Trans. ASME* 75:201–209 (February 1953).
21. Anon., 3500 kW Gas Turbine at the Schenectady Plant of the General Electric Company, A National Historic Mechanical Engineering Landmark by the American Society of Mechanical Engineers, November 8, 1984.

22. Anon., *History of Chrysler Corporation Gas Turbine Vehicles March 1954–July 1974,* Chrysler Corporation Engineering Office, Technical Information, Revised July 1974.

23. Beam, P. E., and Cutler, R. E., Test of a Regenerative Turboprop Aircraft Engine, SAE Paper 807A, 1964.

24. Control of Air Pollution for Aircraft Engines—Emission Standards and Test Procedures for Aircraft, *Federal Register,* Vol. 38, No. 136, July 17, 1973.

25. Tomlinson, L. O., and Maslak, C. E., Combined-Cycle Experience, GER-3651A, *35th GE Turbine State-of-the-Art Technology Seminar,* GE Company, Schenectady, NY.

26. *Jane's All the World's Aircraft,* Jane's Publishing Inc., New York, NY, 1987–1988.

27. Oganowski, G., LM6000 Aeroderivative Industrial Gas Turbine-Development Status Update, GER-3703, *35th GE Turbine State-of-the-Art Technology Seminar,* GE Company, Schenectady, NY.

28. Anon., "105,400-LB, Thrust Reached in GE90 Test," *Aviation Week & Space Technology,* April 2, 1993.

29. Kandebo, S., Pratt & Whitney ADP Meets Test Objectives, *Aviation Week & Space Technology,* New York, NY, November 30, 1992.

30. Shifrim, C., Tent 800: Rolls' 'Best Start' Program, *Aviation Week and Space Technology,* New York, NY, March 28, 1994.

31. Valenti, M., Combined-Cycle Plants: Burning Cleaner and Saving Fuel, *Mechanical Engineering,* New York, NY, September 1991.

32. Engines Pass ETOPS Test, *Interavia Aerospace Review,* February, 1991.

33. Kandebo, S., Commonality Key to PW4084 Early ETOPS, *Aviation Week & Space Technology,* New York, NY, April 11, 1994.

2

THERMODYNAMICS

Fundamental to a study of gas turbines is an understanding of thermodynamics, mixtures, fluid flow, and combustion. Chapter 2 deals with thermodynamics and mixtures of ideal gases, Chapter 3 with fluid flow, and Chapter 4 with combustion.

It is assumed that the reader has taken at least a fundamental course in thermodynamics. Some of these principles are discussed in this chapter. The reader is referred to one or more of the references cited under "Suggested Reading" at the end of this chapter for a more detailed discussion of these topics.

2.1 FIRST LAW OF THERMODYNAMICS

The first law of thermodynamics is a statement of the conservation of energy.

A system is defined as a fixed, identifiable quantity of mass.

For a system undergoing a cycle, the first law states that the cyclic integral of the heat transfer is equal to the cyclic integral of the work, or, in equation form,

$$\oint \delta W = \oint \delta Q \tag{2.1}$$

The first law for a noncyclic system is

$$\delta Q - \delta W = dE \tag{2.2}$$

which, in integrated form, in the absence of electricity, magnetism, and surface effects, becomes for unit mass

$$ {}_1q_2 = {}_1w_2 + (u_2 - u_1) + \frac{\overline{V}_2^2 - \overline{V}_1^2}{2} + g(z_2 - z_1) \tag{2.3} $$

where

${}_1q_2$ = the heat transferred to the system in going from state 1 to state 2
${}_1w_2$ = the work done by the system in going from state 1 to state 2
u = the internal energy
$\frac{\overline{V}^2}{2}$ = the kinetic energy
gz = the potential energy

In Eq. (2.3), heat added to the system is positive, work done by the system is positive.

Many times, one is interested in an arbitrary volume in space rather than a fixed, identifiable mass.

A control volume is an arbitrary volume in space through which a fluid flows. The geometric boundary of the control volume is called the control surface.

The first law of thermodynamics for a control volume undergoing steady flow, unit mass, one stream entering and leaving, and neglecting the potential energy change, is

$$ {}_1q_2 + h_1 + \frac{\overline{V}_1^2}{2} = {}_1w_2 + h_2 + \frac{\overline{V}_2^2}{2} \tag{2.4} $$

where h, the enthalpy, is defined as

$$ h \equiv u + pv \tag{2.5} $$

Application of Eq. (2.4) to a steady-flow system when more than one stream is entering and/or leaving leads to the form

$$ {}_1\dot{q}_2 + \sum_{i=1}^{k} \left[\dot{m}_i \left(h + \frac{\overline{V}^2}{2} \right)_i \right]_{\text{in}} = {}_1\dot{w}_2 + \sum_{j=1}^{n} \left[\dot{m}_j \left(h + \frac{\overline{V}^2}{2} \right)_j \right]_{\text{out}} \tag{2.6} $$

where

${}_1\dot{q}_2$ = the rate of heat transferred to the system for the given time interval
${}_1\dot{w}_2$ = the rate of work done by the system during the given time interval
$\dot{m}$ = the mass rate of flow of each of the streams entering and leaving

2.2 MECHANICAL WORK FOR A FRICTIONLESS STEADY-FLOW PROCESS

An expression for the mechanical work of a frictionless steady-flow process is very useful.

For a frictionless, steady-flow process,

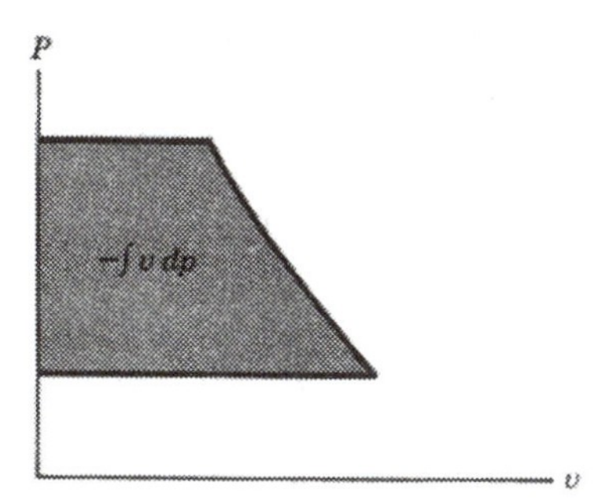

Figure 2.1. Pressure–specific volume diagram.

$$-{}_1w_2 = +\int_1^2 v\,dp + \Delta KE + \Delta PE \tag{2.7}$$

where ΔKE is the change in the kinetic energy and ΔPE is the change in the potential energy.

As is shown in Figure 2.1, the value of $-\int v\,dp$ is represented by an area on a pressure–specific volume (p–v) diagram. Therefore, the pressure–volume diagram will be a valuable tool in the analysis of the work. It must be remembered that the area has meaning only for a reversible process.

2.3 CONTINUITY EQUATION

The general form of the continuity equation will be discussed in Chapter 3. For this chapter, the one-dimensional continuity equation is needed and is

$$\dot{m} = \frac{A\overline{V}}{v} \tag{2.8}$$

where

$\dot{m}$ = the mass rate of flow
A = the cross-sectional area
$\overline{V}$ = the average velocity across the section
v = the specific volume of the fluid

2.4 SPECIFIC HEAT

Several specific heats may be defined. The two most frequently used are the constant-pressure specific heat, c_p, and the constant-volume specific heat, c_v, which are defined as

$$c_p = \left.\frac{\partial h}{\partial T}\right|_p \tag{2.9}$$

TABLE 2.1. **Table of Relative Atomic Weights of Selected Elements Based on the Atomic Mass of ^{12}C = 12.0000 (1)**

Name	Symbol	Atomic Weight
Argon	Ar	39.948
Carbon	C	12.0112
Helium	He	4.0026
Hydrogen	H	1.00797
Nitrogen	N	14.0067
Oxygen	O	15.9994
Sulfur	S	32.064

and

$$c_v = \left.\frac{\partial u}{\partial T}\right|_v \tag{2.10}$$

2.5 ATOMIC WEIGHTS

The atomic weights for some of the elements will be needed. The values of the atomic weights for a number of elements are given in Table 2.1.

A compound is composed of two or more kinds of atoms combined together in a definite ratio. The elements may exist as diatomic molecules. For example, the element nitrogen is most frequently encountered in the form of diatomic molecules and is represented by the formula N_2.

The formula weight (molecular weight, $\overline{M}$) of a compound is the sum of the atomic weights of all the elements in the compound taking into account the number of atoms of each element in the molecular formula. Molecular weights for a number of substances are given in Table 2.2.

TABLE 2.2. **Molecular Weights of Selected Compounds**

Compound	Molecular Formula	Molecular Weight
Oxygen	O_2	31.9988
Hydrogen	H_2	2.0159
Water	H_2O	18.0153
Sulfur dioxide	SO_2	64.0628
Carbon monoxide	CO	28.0106
Carbon dioxide	CO_2	44.0100
Nitrogen	N_2	28.0134
Argon	Ar	39.948

2.6 IDEAL GAS

An ideal gas is defined as a substance that has the equation of state, for unit mass,

$$pv = RT \tag{2.11}$$

where R is called the gas constant and is a different constant for each gas. R is also equal to the universal gas constant, $\overline{R}$, divided by the molecular weight of the gas. In equation form,

$$R = \frac{\overline{R}}{\overline{M}} \tag{2.12}$$

where

$$\begin{aligned} \overline{R} &= 8.314\,34 \times 10^3 \text{ J/(kmol K)} \\ &= 1.98718 \text{ Btu/(mol°R)} \end{aligned}$$

Other useful forms of the equation of state are

$$pV = mRT \tag{2.13a}$$

$$pV = n\overline{R}T \tag{2.13b}$$

where V is the total volume and n, the number of moles of a gas, is

$$n = \frac{m}{\overline{M}} \tag{2.14}$$

For an ideal gas, the enthalpy and internal energy are a function of temperature and temperature only. Therefore, for an ideal gas, Eqs. (2.9) and (2.10) become

$$c_p = \frac{dh}{dT} \tag{2.15}$$

and

$$c_v = \frac{du}{dT} \tag{2.16}$$

Combining Eqs. (2.15), (2.16), (2.5), and (2.11) yields

$$c_p - c_v = R \tag{2.17}$$

The specific heat ratio is defined as

$$k \equiv \frac{c_p}{c_v} \tag{2.18}$$

Equations (2.15), (2.16), (2.17), and (2.18), when multiplied by the molecular weight $\overline{M}$, which converts them to a mole basis, become, respectively,

$$\overline{c}_p = \frac{d\overline{h}}{dT} \tag{2.15a}$$

$$\overline{c}_v = \frac{d\overline{u}}{dT} \tag{2.16a}$$

$$\bar{c}_p - \bar{c}_v = \bar{R} \tag{2.17a}$$

$$k = \frac{\bar{c}_p}{\bar{c}_v} \tag{2.18a}$$

Now $\bar{c}_p$ and $\bar{h}$, since they are functions of temperature only, can be expressed as functions of temperature. McBride, Heimel, Ehlers, and Gordon (2) is one source of data in this form. The expressions for constant pressure specific heat and enthalpy are

$$\bar{c}_p^0 = \bar{R}\left(a_1 + a_2T + a_3T^2 + a_4T^3 + a_5T^4\right) \tag{2.19}$$

$$\bar{h}_T^0 = \bar{R}T\left(a_1 + \frac{a_2T}{2} + \frac{a_3T^2}{3} + \frac{a_4T^3}{4} + \frac{a_5T^4}{5} + \frac{a_6}{T}\right) \tag{2.20}$$

Table 2.3 lists, for several substances, the temperature coefficients needed for Eqs. (2.19) and (2.20). The coefficients listed in Table 2.3 require that the temperature in Eqs. (2.19) and (2.20) be in kelvins. Values of $\bar{c}_p^0$ are the constant-pressure specific heats at 1 atm, h_T^0 the enthalpy values at 1 atm.

Equation (2.15a) can be rearranged to

$$d\bar{h} = \bar{c}_p\, dT \tag{2.21}$$

and integrated for the change in enthalpy if the value of $\bar{c}_p$ and the initial and final temperatures are known. In processes involving an ideal gas where the temperature change is small, the change in enthalpy can be calculated by assuming a constant specific heat. Equation (2.21) then becomes

$$\bar{h}_2 - \bar{h}_1 = \bar{c}_p(T_2 - T_1) \tag{2.22}$$

When the temperature change is large, the variation of specific heat with temperature must be taken into account. This can be done by replacing the constant-pressure specific heat, $\bar{c}_p$, in Eq. (2.21) by Eq. (2.19) and integrating. This becomes

$$\bar{h}_2 - \bar{h}_1 = \int_1^2 \bar{c}_p\, dT \tag{2.23}$$

which is very time-consuming to evaluate. A simpler method is to use a table such as the *Gas Tables* (3) if they are available or to develop your own table.

Tables for several substances are included in Appendix A. The constant-pressure specific heats and enthalpy values were calculated using Eqs. (2.19) and (2.20) and the values of Table 2.3. The development and use of these tables is discussed in Section 2.13.

Values for the constant-pressure specific heat, $\bar{c}_p$, and specific heat ratio, k, at 298.15 K (536.67°R) are tabulated for several ideal gases in Table 2.4.

Figures 2.2 and 2.3 illustrate the variation of the instantaneous constant-pressure specific heat with temperature for nitrogen, N_2, and carbon dioxide, CO_2, respectively.

TABLE 2.3. **Temperature Coefficients for Thermodynamic Functions**

Substance (Chemical Symbol)	Temperature Interval (K)	a_1	a_2	a_3	a_4	a_5	a_6	a_7
Argon,	1000. 5000.	2.5000000E 00	−0.	−0.	−0.	−0.	−7.4537500E 02	4.3661076E 00
Ar(g)	300. 1000.	2.5000000E 00	−0.	−0.	−0.	−0.	−7.4537500E 02	4.3661076E 00
Carbon monoxide,	1000. 5000.	2.9511519E 00	1.5525567E-03	−6.1911411E-07	1.1350336E-10	−7.7882732E-15	−1.4231827E 04	6.5314450E 00
CO(g)	300. 1000.	3.7871332E 00	−2.1709526E-03	5.0757337E-06	−3.4737726E-09	7.7216841E-13	−1.4363508E 04	2.6335459E 00
Carbon dioxide,	1000. 5000.	4.4129266E 00	3.1922896E-03	−1.2978230E-06	2.4147446E-10	−1.6742986E-14	−4.8944043E 04	−7.2875769E-01
CO_2(g)	300. 1000.	2.1701000E 00	1.0378115E-02	−1.0733938E-05	6.3459175E-09	−1.6280701E-12	−4.8352602E 04	1.0664388E 01
Hydrogen molecule,	1000. 5000.	3.0436897E 00	6.1187110E-04	−7.3993551E-09	−2.0331907E-11	2.4593791E-15	−8.5491002E 02	−1.6481339E-00
H_2(g)	300. 1000.	2.8460849E 00	4.1932116E-03	−9.6119332E-06	9.5122662E-09	−3.3093421E-12	−9.6725372E 02	−1.4117850E 00
Hydrogen atom,	1000. 5000.	2.5000000E 00	−0.	−0.	−0.	−0.	2.5470497E 04	−4.6001096E-01
H(g)	300. 1000.	2.5000000E 00	−0.	−0.	−0.	−0.	2.5470497E 04	−4.6001096E-01
Water,	1000. 5000.	2.6707532E 00	3.0317115E-03	−8.5351570E-07	1.1790853E-10	−6.1973568E-15	−2.9888994E 04	6.8838391E 00
H_2O(g)	300. 1000.	4.1565016E 00	−1.7244334E-03	5.6982316E-06	−4.5930044E-09	1.4233654E-12	−3.0288770E 04	−6.8616246E-01
Nitrogen atom,	1000. 5000.	2.4422261E 00	1.2276187E-04	−8.4992719E-08	2.1400830E-11	−1.2511058E-15	5.6148821E 04	4.4925708E 00
N(g)	300. 1000.	2.5147937E 00	−1.1243791E-04	2.9647506E-07	−3.2464049E-10	1.2595465E-13	5.6127767E 04	4.1193032E 00
Nitrogen molecule,	1000. 5000.	2.8545761E 00	1.5976316E-03	−6.2566254E-07	1.1315849E-10	−7.6897070E-15	−8.9017445E 02	6.3902879E 00
N_2(g)	300. 1000.	3.6916148E 00	−1.3332552E-03	2.6503100E-06	−9.7688341E-10	−9.9772234E-14	−1.0628336E 03	2.2874980E 00
Nitric oxide,	1000. 5000.	3.1529360E 00	1.4059955E-03	−5.7078462E-07	1.0628209E-10	−7.3720783E-15	9.8522048E 03	6.9446465E 00
NO(g)	300. 1000.	4.1469476E 00	−4.1197237E-03	9.6922467E-06	−7.8633639E-09	2.2309512E-12	−9.7447894E 03	2.5694290E 00
Nitrogen dioxide,	1000. 5000.	4.6139219E 00	2.6386639E-03	−1.0948541E-06	2.0818425E-10	−1.4654391E-14	2.3403782E 03	1.3676372E 00
NO_2(g)	300. 1000.	3.4344563E 00	2.2234297E-03	6.7148975E-06	−9.7427719E-09	3.7212523E-12	2.8647685E 03	8.4084647E 00
Oxygen atom,	1000. 5000.	2.5372567E 00	−1.8422190E-05	−8.8017921E-09	5.9643621E-12	−5.5743608E-16	2.9230007E 04	4.9467942E 00
O(g)	300. 1000.	3.0218894E 00	−2.1737249E-03	3.7542203E-06	−2.9947200E-09	9.0777547E-13	2.9137190E 04	2.6460076E 00
Oxygen molecule,	1000. 5000.	3.5976129E 00	7.8145603E-04	−2.2386670E-07	4.2490159E-11	−3.3460204E-15	−1.1927918E 03	3.7492659E 00
O_2(g)	300. 1000.	3.7189946E 00	−2.5167288E-03	8.5837353E-06	−8.2998716E-09	2.7082180E-12	−1.0576706E 03	3.9080704E 00
Hydroxyl,	1000. 5000.	2.8895544E 00	9.9835061E-04	−2.1879904E-07	1.9802785E-11	−3.8452940E-16	3.8811792E 03	5.5597016E 00
OH(g)	300. 1000.	3.8234708E 00	−1.1187229E-03	1.2466819E-06	−2.1035896E-10	−5.2546551E-14	3.5852787E 03	5.8253029E-01

TABLE 2.4. **Ideal Gas Constants for Selected Substances**

Name	Formula	Constant-pressure Specific Heat, $\bar{c}_p$ At 298.15 K kJ/(kmol K)	Constant-pressure Specific Heat, $\bar{c}_p$ At 536.67°R Btu/(lb-mol°R)	Specific Heat Ratio, k at 298.15 K (536.67°R)
Carbon dioxide	CO_2	37.128	8.874	1.288
Nitrogen	N_2	29.125	6.961	1.399
Hydrogen	H_2	28.833	6.891	1.405
Oxygen	O_2	29.375	7.021	1.394
Water vapor	H_2O	33.577	8.025	1.329
Carbon monoxide	CO	29.143	6.965	1.399
Argon	Ar	20.786	4.968	1.666

Example Problem 2.1

Carbon dioxide, CO_2 (an ideal gas), enters a steady-flow device (see diagram) at a pressure of 1000 kPa, a temperature of 1000 K, and with negligible velocity (state 1). It expands adiabatically and leaves the steady-flow device with a velocity of 125 m/s, a temperature of 700 K, and a pressure of 100 kPa (state 2). The cross-sectional

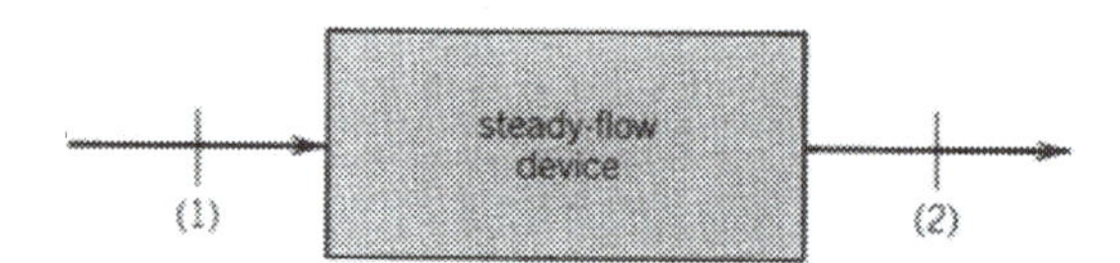

area at state 2 is 0.1 m^2. Assuming a constant specific heat based on the value given in Table 2.4, calculate:

(a) The mass rate of flow through the device
(b) The power, in watts, developed by this device

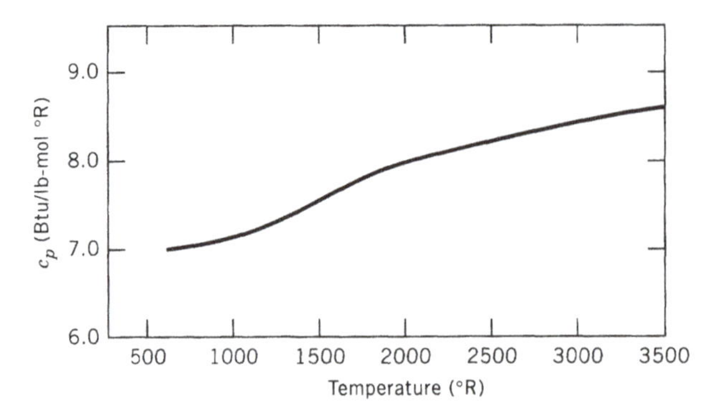

Figure 2.2. Variation of specific heat with temperature for nitrogen, N_2.

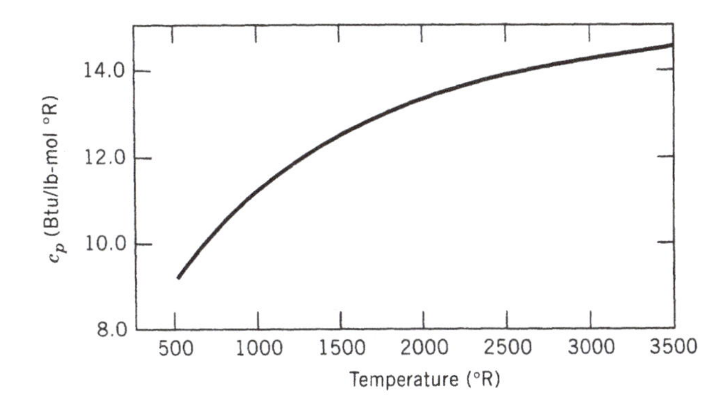

Figure 2.3. Variation of specific heat with temperature for carbon dioxide, CO_2.

Solution

Basis: 1 kg carbon dioxide

From Tables 2.2 and 2.4,

$$\overline{M} = 44.0100$$
$$\bar{c}_p = 37.128 \text{ kJ/(kmol K)}$$

The steady-flow energy equation (Eq. 2.4) becomes, for the conditions of this problem

$$_1w_2 = c_p(T_1 - T_2) - \frac{\overline{V}_2^2}{2}$$

$$= \frac{37.128 \text{ kJ}}{\text{kmol K}} \bigg| \frac{(1000 - 700) \text{ K}}{} \bigg| \frac{\text{kmol}}{44.0100 \text{ kg}}$$

$$- \frac{(125)^2 \text{ m}^2}{\text{s}^2} \bigg| \frac{}{2} \bigg| \frac{\text{kJ}}{1000 \text{ J}} \bigg| \frac{\text{N s}^2}{1 \text{ kg m}}$$

$$= 253.1 - 7.8 = 245.3 \text{ kJ/kg}$$

From the one-dimensional continuity equation,

$$\dot{m} = \frac{A_2\overline{V}_2}{v_2} = \frac{A_2\overline{V}_2 p_2}{RT_2}$$

$$= \frac{0.1 \text{ m}^2}{} \bigg| \frac{125 \text{ m}}{\text{s}} \bigg| \frac{\text{kmol K}}{8.31434 \text{ kJ}} \bigg| \frac{44.0100 \text{ kg}}{\text{kmol}} \bigg| \frac{}{700 \text{ K}} \bigg| \frac{100 \text{ kPa}}{}$$

$$= 9.452 \text{ kg/s}$$

$$\dot{w} = \dot{m}_1 w_2 = \frac{9.452 \text{ kg}}{\text{s}} \bigg| \frac{245.3 \text{ kJ}}{\text{kg}} = 2.319 \times 10^6 \text{ W}$$

Example Problem 2.1 has the units written by each term so that the reader can see the final units for each term. Several conversion factors the reader should be familiar with are

$$1 \text{ J/s} = 1 \text{ N m/s} = 1 \text{ W}$$

$$1 \text{ Pa} = 1 \text{ N/m}^2$$

In English units, several useful conversion factors are

$$1 \text{ hp} = 33{,}000 \text{ ft-lb}_f\text{/min} = 2545 \text{ Btu/h} = 0.746 \text{ kW}$$

$$1 \text{ Btu} = 778.16 \text{ ft-lb}_f$$

2.7 THERMAL EFFICIENCY

It is important at this point to introduce the concept of thermal efficiency of a heat engine. Thermal efficiency is defined as the fraction of the gross heat input to a system during a cycle that is converted into net work output, or

$$\eta_{\text{thermal}} = \frac{W_{\text{net}} \text{ (energy sought)}}{Q_{\text{in}} \text{ (energy that costs)}} \tag{2.24}$$

Combining Eqs. (2.24) and (2.1) yields another form for thermal efficiency that is often quite useful:

$$\eta_{\text{th}} = \frac{Q_{\text{in}} - Q_{\text{out}}}{Q_{\text{in}}} = 1 - \frac{Q_{\text{out}}}{Q_{\text{in}}} \tag{2.25}$$

2.8 CARNOT CYCLE

A Carnot heat engine is an *externally reversible* heat engine operating between two constant-temperature reservoirs. It consists of two constant-temperature heat transfer processes and two adiabatic and reversible processes.

For a Carnot heat engine, the thermal efficiency becomes

$$\eta_{\text{th,Carnot}} = \frac{T_H - T_L}{T_H} \tag{2.26}$$

where T_H is the absolute temperature of the high-temperature reservoir and T_L is the absolute temperature of the low-temperature reservoir.

It can be shown that no heat engine can be more efficient than an externally reversible heat engine operating between the same temperature limits and that all externally reversible engines operating between the same temperature limits have the same thermal efficiency.

Therefore, Eq. (2.26) gives the maximum thermal efficiency that any heat engine can have if the temperature extremes are known. This is a very useful value for comparison purposes.

2.9 ENTROPY

Entropy, s, is a thermodynamic property that is defined by the relation

$$ds = \left(\frac{\delta Q}{T}\right)_{\text{reversible}} \tag{2.27}$$

For an infinitesimal reversible process, for unit mass,

$$ds = \frac{\delta q}{T} \tag{2.28}$$

For an infinitesimal, irreversible (spontaneous) process, for unit mass,

$$ds > \frac{\delta q}{T} \tag{2.29}$$

Entropy is a function of the state of a system, which means that the change in its value is independent of the path between the initial and final states. However, since Eq. (2.27) applies only for a reversible process, the entropy change must be calculated from values of *heat* and *temperature* along a *reversible* path between the initial and final states.

2.10 TWO IMPORTANT RELATIONS FOR A PURE SUBSTANCE INVOLVING ENTROPY

Two very useful and important relations for a pure substance involving entropy can be derived. The first law of thermodynamics for a closed system, neglecting changes in kinetic and potential energy, is, from Eq. (2.3) written in differential form,

$$\delta q = du + \delta w \tag{2.30}$$

For a reversible process in which the only work involved is that at a moving boundary of the system,

$$\delta w = p\,dv \tag{2.31}$$

Combining Eqs. (2.30), (2.28), and (2.31) yields

$$T\,ds = du + p\,dv \tag{2.32}$$

However, once this equation has been written, it is realized that it involves only changes in properties and involves no path functions. Therefore, it must be concluded that this equation is valid for all processes, both reversible and irreversible, and that it applies to a substance undergoing a change of state as the result of flow across a boundary of a system as well as to a closed system.

Another useful form can be derived by using the definition of enthalpy,

$$h = u + pv$$

or, in differential form,

$$dh = du + p\,dv + v\,dp \tag{2.33}$$

Combining Eqs. (2.32) and (2.33) yields

$$T\,ds = dh - v\,dp \tag{2.34}$$

Equations (2.32) and (2.34), on a mole basis, become, respectively,

$$T\,d\bar{s} = d\bar{u} + p\,d\bar{v} \tag{2.32a}$$

$$T\,d\bar{s} = d\bar{h} - \bar{v}\,dp \tag{2.34a}$$

Equations (2.32a) and (2.34a) are very useful in the evaluation of the change in entropy when the substance being considered is an ideal gas. To illustrate, consider Eq. (2.34a).

For an ideal gas,

$$d\bar{h} = \bar{c}_p\,dT$$

$$p\bar{v} = \bar{R}T$$

Therefore,

$$d\bar{s} = \frac{\bar{c}_p\,dT}{T} - \frac{\bar{R}\,dp}{p} \tag{2.35}$$

Assuming constant specific heats, Eq. (2.35) becomes, when integrated between states 1 and 2,

$$\bar{s}_2 - \bar{s}_1 = \bar{c}_p \ln\frac{T_2}{T_1} - \bar{R}\ln\frac{p_2}{p_1} \tag{2.36}$$

In a similar manner, using Eq. (2.32a) yields

$$\bar{s}_2 - \bar{s}_1 = \bar{c}_v \ln\frac{T_2}{T_1} + \bar{R}\ln\frac{\bar{v}_2}{\bar{v}_1} \tag{2.37}$$

A special process of extreme interest is the adiabatic and reversible process. For an adiabatic and reversible process (isentropic process),

$$s_2 - s_1 = \bar{s}_2 - \bar{s}_1 = 0 \tag{2.38}$$

Therefore, combining Eqs. (2.36), (2.38), (2.18a), and (2.17a) yields

$$0 = \bar{c}_p \ln\frac{T_2}{T_1} - \bar{R}\ln\frac{p_2}{p_1}$$

or

$$\bar{c}_p \ln \frac{T_2}{T_1} = (\bar{c}_p - \bar{c}_v) \ln \frac{p_2}{p_1}$$

or

$$\frac{T_2}{T_1} = \left(\frac{p_2}{p_1}\right)^{(k-1)/k} \tag{2.39}$$

Combining Eq. (2.39) with Eq. (2.11) yields two other forms:

$$\frac{T_2}{T_1} = \left(\frac{v_1}{v_2}\right)^{(k-1)} \tag{2.40}$$

and

$$\frac{p_2}{p_1} = \left(\frac{v_1}{v_2}\right)^{(k)} \tag{2.41}$$

It must be remembered that Eqs. (2.39), (2.40), and (2.41) apply only to an ideal gas with constant specific heats undergoing an adiabatic and reversible process. The case with variable specific heats is discussed in Section 2.13.

Numerical values of entropy may be evaluated using the equation

$$\bar{s}_T^0 = \bar{R}\left(a_1 \ln T + a_2 T + \frac{a_3}{2}T^2 + \frac{a_4}{3}T^3 + \frac{a_5}{4}T^4 + a_7\right) \tag{2.42}$$

and the constants given in Table 2.3. Once again, the temperature used in Eq. (2.42) must be in kelvins. Here $\bar{s}_T^0$ is the entropy at the standard state, that is, at a pressure of 1 atm.

2.11 ENTROPY AS A COORDINATE

In Section 2.2 it was noted that the work for a reversible process was represented by an area on a pressure–specific volume diagram. In a similar manner, the heat transfer for a reversible process is represented by an area on a temperature–entropy (T–s) diagram. From Eq. (2.28),

$$\delta q_{\text{rev}} = (T\, ds)$$

or

$$q_{\text{rev}} = \int_1^2 T\, ds \tag{2.43}$$

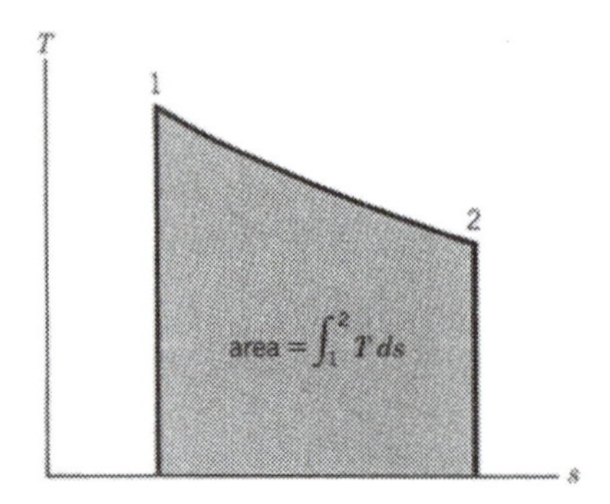

Figure 2.4. Temperature–entropy diagram.

Therefore, as shown in Figure 2.4, the heat transfer for a reversible process 1 → 2 is the area beneath the curve.

The *p*–*v* and *T*–*s* diagrams for a Carnot heat engine operating on a closed cycle are shown in Figure 2.5. From Figure 2.5a, it can be seen that heat is added during process 1 → 2 and rejected during process 3 → 4.

$$Q_{in} = {}_1Q_2 = T_h\,(s_2 - s_1)$$

$$Q_{out} = |{}_3Q_4| = T_l\,(s_3 - s_4)$$

Therefore, from Eq. (2.25),

$$\eta_{th} = \frac{Q_{in} - Q_{out}}{Q_{in}} = \frac{T_h\,(s_2 - s_1) - T_l\,(s_3 - s_4)}{T_h\,(s_2 - s_1)} = \frac{T_h - T_l}{T_h} \tag{2.44}$$

2.12 GIBBS FREE ENERGY

It is convenient to introduce another property, the Gibbs free energy. The Gibbs free energy, *G*, is defined as

$$G \equiv H - TS \tag{2.45}$$

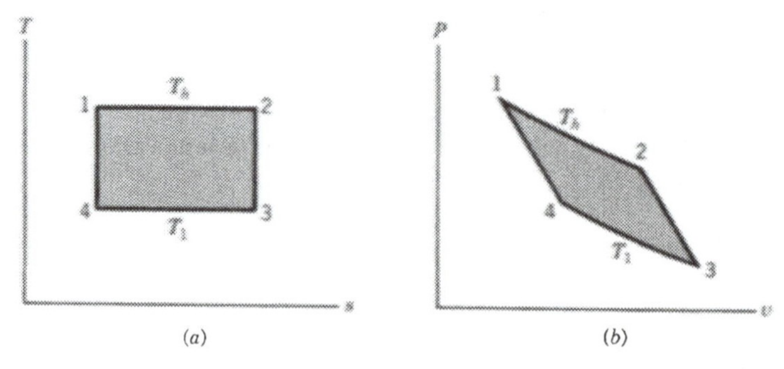

Figure 2.5. Carnot heat engine shown on (a) temperature–entropy coordinates and (b) pressure–specific volume coordinates.

or, for unit mass and per mole, respectively,

$$g = h - Ts \tag{2.45a}$$

$$\overline{g} = \overline{h} - T\overline{s} \tag{2.45b}$$

Numerical values of the Gibbs free energy may be evaluated from the defining equation. They also may be evaluated using the constants from Table 2.3 and the equation

$$\overline{g}_T^0 = \overline{R}T\left[a_1(1 - \ln T) - \frac{a_2}{2}T - \frac{a_3}{6}T^2 - \frac{a_4}{12}T^3 - \frac{a_5}{20}T^4 + \frac{a_6}{T} - a_7\right] \tag{2.46}$$

The temperature in this equation must be in kelvins. Here $\overline{g}_T^0$ is the Gibbs free energy for the standard state, that is, for a pressure of 1 atm.

2.13 GAS TABLES

The change in properties for processes involving an ideal gas where the temperature change is large should be calculated by taking into account the variation of specific heat with respect to temperature.

The change in enthalpy involves evaluation of Eq. (2.21). One could evaluate this integral between the temperature limits by using Eq. (2.19) for $\overline{c}_p$ values and calculating the change in enthalpy.

Another way is to use tables that have already been printed [such as the *Gas Tables* (3)] or by constructing your own tables.

Tables for several substances are included in Appendix A. These tables were developed using Eqs. (2.19), (2.20), (2.42), and (2.46).

Column 1 of the tables in Appendix A gives the temperature in degrees Rankine or kelvins. Column 2 lists the standard-state constant-pressure specific heat, column 3 the standard-state enthalpy, column 4 the standard-state entropy, and column 5 the standard-state Gibbs free-energy values.

The change in entropy may be determined by integrating Eq. (2.35), which, for 1 mol is

$$\int d\overline{s} = \overline{s}_2 - \overline{s}_1 = \int \frac{\overline{c}_p\, dT}{T} - \overline{R}\int \frac{dp}{p} \tag{2.47}$$

The first term on the right-hand side of Eq. (2.47) is a function of temperature only. We may use the standard state entropy values tabulated in Appendix A. Equation (2.47) then becomes

$$\overline{s}_2 - \overline{s}_1 = \overline{s}_{T_2}^0 - \overline{s}_{T_1}^0 - \overline{R}\ln\frac{p_2}{p_1} \tag{2.48}$$

For an adiabatic and reversible processes,

$$\overline{s}_2 - \overline{s}_1 = 0$$

Therefore, Eq. (2.48) becomes, for an adiabatic and reversible process,

$$\bar{s}^0_{T_2} - \bar{s}^0_{T_1} = \bar{R} \ln \frac{p_2}{p_1} \tag{2.49}$$

or

$$\left.\frac{p_2}{p_1}\right|_{s=\text{constant}} = \exp\left(\frac{\bar{s}^0_{T_2} - \bar{s}^0_{T_1}}{\bar{R}}\right) \tag{2.50}$$

Since $\bar{s}^0_{T_2}$ and $\bar{s}^0_{T_1}$ are functions of temperature only and $\bar{R}$ is a constant, the right-hand side is a function of temperature only. The relative pressure, Pr, is defined as

$$Pr \equiv \exp\left(\frac{\bar{s}^0_T}{\bar{R}}\right) \tag{2.51}$$

Therefore

$$\left.\frac{p_2}{p_1}\right|_{s=\text{constant}} = \frac{Pr_2}{Pr_1} \tag{2.52}$$

Values of Pr as a function of temperature are tabulated in column 6 of the tables of Appendix A. All of the Pr values tabulated in the tables of Appendix A have been reduced by 10 to some power. The value by which the Pr values have been reduced is the same for each table but may vary from table to table. One may determine the value by which they have been reduced by using the defining equation, that is, Eq. (2.51).

Example Problem 2.2

By what power have the Pr values been reduced for oxygen, O_2?

Solution

$$Pr = \exp\left(\frac{\bar{s}^0_T}{\bar{R}}\right)$$

At 600°R, $\bar{s}^0_T = 49.791$. Therefore,

$$Pr = \exp\left(\frac{49.791}{1.98718}\right) = 7.616 \times 10^{10}$$

This means that all values of Pr in the oxygen table have been reduced by 10^{10}.

2.14 MIXTURES OF IDEAL GASES

Many thermodynamic problems involve mixtures of different gases. Probably the most familiar example is air, which consists of a mixture of gases, mainly oxygen and nitrogen. The principles involved in determining the thermodynamic properties of a mixture of ideal gases are considered below.

The total mass of a mixture is the sum of the masses of each component, or

$$m = m_1 + m_2 + \cdots + m_k = \sum_{i=1}^{k} m_i \tag{2.53}$$

The total number of moles of the mixture is equal to the sum of the moles of the individual gases, or

$$n = n_1 + n_2 + \cdots + n_k = \sum_{i=1}^{k} n_i \tag{2.54}$$

The mole fraction of a component, x_i, is defined as

$$x_i = \frac{n_i}{n} \tag{2.55}$$

when n_i is the number of moles of the ith component and n is the total number of moles in the mixture as given by Eq. (2.54). The sum of the mole fractions is equal to 1 or,

$$\sum_{i=1}^{k} x_i = 1.0 \tag{2.56}$$

The mass of each component may be calculated by

$$m_i = n_i \overline{M}_i \tag{2.57}$$

Combining Eqs. (2.53) and (2.57) yields

$$m = \sum n_i \overline{M}_i = n\overline{M} \tag{2.58}$$

or the equivalent molecular weight may be calculated by rearranging Eq. (2.58) to

$$\overline{M} = \sum_{i=1}^{k} \frac{n_i}{n} \overline{M}_i = \sum_{i=1}^{k} x_i \overline{M}_i \tag{2.59}$$

The properties of a mixture of ideal gases may be determined from the properties of the individual gases by applying the Gibbs-Dalton law, which is usually, for clarity, stated in three parts.

1. The pressure of a mixture of ideal gases is equal to the sum of the partial pressures of the individual components, the partial pressure of any component being the pressure exerted by that component when it alone occupies the mixture volume at the mixture temperature. Therefore,

$$p = p_1 + p_1 + \cdots + p_k = \sum_{i=1}^{k} x_i p \tag{2.60}$$

The equation of state for an individual gas then becomes

$$p_i V = n_i \overline{R} T \tag{2.61}$$

Defining the part by volume (partial volume) as the volume occupied by any component if it alone were at the mixture pressure and temperature yields

$$pV_i = n_i \overline{R} T \tag{2.62}$$

Combining Eqs. (2.55), (2.60), (2.61), and (2.62) yields a useful relation,

$$\frac{n_i}{n} = \frac{p_i}{p} = \frac{V_i}{V} = x_i \tag{2.63}$$

2. The internal energy of a mixture of ideal gases is equal to the sum of the internal energies of the individual components when each component gas occupies the mixture volume at the mixture temperature, or

$$n\overline{u} = U = U_1 + U_2 + \cdots + U_k = \sum_{i=1}^{k} n_i \overline{u}_i \tag{2.64}$$

or

$$\overline{u} = \sum_{i=1}^{k} x_i \overline{u}_i \tag{2.65}$$

Some consequences are

$$n\overline{h} = H = \sum_{i=1}^{k} n_i \overline{h}_i \tag{2.66}$$

or

$$\overline{h} = \sum_{i=1}^{k} x_i \overline{h}_i \tag{2.67}$$

$$\overline{c}_p = \sum_{i=1}^{k} x_i \overline{c}_{pi} \tag{2.68}$$

$$\overline{c}_v = \sum_{i=1}^{k} x_i \overline{c}_{vi} \tag{2.69}$$

3. The entropy of a mixture of ideal gases is equal to the sum of the entropies of the individual gases taken when each component gas occupies the mixture volume at the mixture temperature, or

$$n\overline{s} = S = \sum_{i=1}^{k} n_i \overline{s}_i \tag{2.70}$$

or

$$\bar{s} = \sum_{i=1}^{k} x_i \bar{s}_i \tag{2.71}$$

The following discussion is restricted to a mixture of ideal gases for which the composition remains constant.

For a mixture of ideal gases, the change in entropy is, per mole, from Eq. (2.71),

$$\bar{s}_2 - \bar{s}_1 = \sum_{i=1}^{k} x_i (\bar{s}_2 - \bar{s}_1)_i \tag{2.72}$$

For an isentropic process, $\bar{s}_2 - \bar{s}_1 = 0$ and Eq. (2.72) becomes

$$0 = \sum_{i=1}^{k} x_i (\bar{s}_2 - \bar{s}_1)_i \tag{2.73}$$

Note that Eq. (2.73) does not say that the change in entropy of each component is zero, just the sum.

To use the tables in Appendix A to solve a problem for the isentropic expansion or compression of a mixture of ideal gas for which the composition remains constant,

$$0 = \sum_{i=1}^{k} x_i(\bar{s}^0_{T_2} - \bar{s}^0_{T_1})_i - \bar{R} \ln \frac{p_2}{p_1} \tag{2.74}$$

or

$$\sum_{i=1}^{k} x_i \bar{s}^0_{T_{2i}} = \sum_{i=1}^{k} x_i \bar{s}^0_{T_{1i}} + \bar{R} \ln \frac{p_2}{p_1}$$

Letting

$$\sum_{i=1}^{k} x_i \bar{s}^0_{T_{2i}} = \bar{s}^0_{T_{2(\text{mix})}} \tag{2.75}$$

and

$$\sum_{i=1}^{k} x_i \bar{s}^0_{T_{1i}} = \bar{s}^0_{T_{1(\text{mix})}} \tag{2.76}$$

Eq. (2.74) becomes

$$\bar{s}^0_{T_{2(\text{mix})}} = \bar{s}^0_{T_{1(\text{mix})}} + \bar{R} \ln \frac{p_2}{p_1} \tag{2.77}$$

Note that, unless the final temperature is known, a trial-and-error solution is involved. It is necessary to:

1. Assume a final temperature.
2. Calculate $\bar{s}^0_{T_{2(\text{mix})}}$.

3. See if Eq. (2.77) is satisfied.
4. Rework until an equality of Eq. (2.77) exists.

Example Problem 2.3

A gas mixture with the volumetric analysis

CO_2	17%
H_2O	8%
N_2	75%
	100%

enters a nozzle at a pressure of 48 psia, a temperature of 1700°R, and with negligible velocity. It expands adiabatically and reversibly through the nozzle to a pressure of 14 psia. Calculate, assuming variable specific heats, the velocity leaving the nozzle. Assume a constant composition.

Solution

Basis: 1 lb-mol mixture

Since the expansion through the nozzle is steady-flow, the steady-flow energy equation becomes for the conditions of this problem

$$\bar{h}_1 - \bar{h}_2 = \frac{\overline{M}V_2^2}{2}$$

or

$$\bar{V}_2 = \sqrt{\frac{(2)\,(\bar{h}_1 - \bar{h}_2)}{\overline{M}}}$$

Since the expansion is isentropic, Eq. (2.77) applies.

At the inlet condition (1700°R):

	x_i	$\overline{M}_i$	m_i	$\bar{s}_i^0$	$x_i\bar{s}_i^0$	$\bar{h}_i$	$x_i\bar{h}_i$
CO_2	0.17	44	7.48	63.596	10.811	−156,208.4	−26,555.4
H_2O	0.08	18	1.44	55.034	4.403	− 93,829.2	− 7,506.3
N_2	0.75	28	21.00	54.063	40.547	8,455.9	6,341.9
			29.92		55.761		−27,719.8

Therefore

$$\sum_{i=1}^{k} (x_i\bar{s}_i^0)_{T_2} = 55.761 + 1.987 \ln \frac{14}{48} = 53.313$$

By trial and error, it can be shown that the final temperature is 1276°R. Next, it is necessary to calculate $\bar{h}_2$ at a temperature of 1276°R. Using Tables A.3, A.7, and A.9, one can determine that

$$\sum (x_i \bar{h}_i)_{T_2} = -31{,}332.9 \text{ Btu/lb-mol mixture}$$

Therefore

$$\bar{V}_2 = \sqrt{\frac{(2)\,(32.174)\,(778)\,(-27{,}719.8 + 31{,}332.9)}{29.92}}$$

$$= 2459 \text{ ft/s}$$

Example Problem 2.4

Dry air at 1122°R (623 K) and 176.4 psia (1215.6 kPa) enters an adiabatic mixing chamber at the rate of 1 lb/s (1 kg/s) as shown below. Water at 780°R (433 K), 210 psia (1447 kPa) is sprayed into the chamber at a rate such that the air–water vapor mixture at state 3 is at 800°R (444 K) and a pressure of 176.4 psia (1215.6 kPa). Determine the rate at which water is sprayed into the chamber.

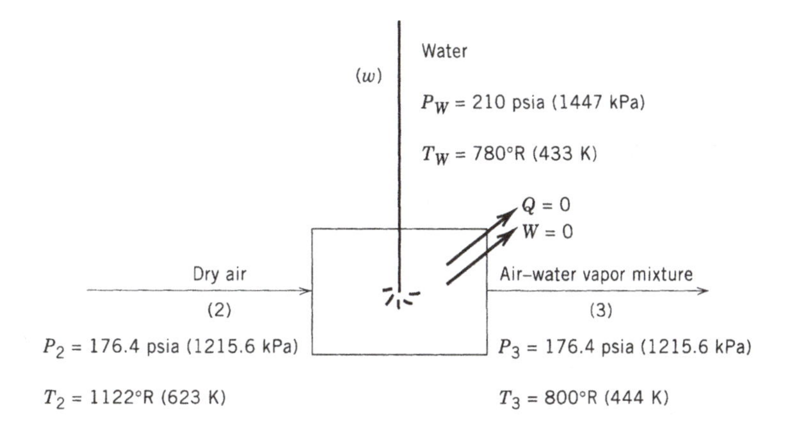

Solution

Basis: 1 lb/s (1 kg/s) dry air

$$\dot{q}_{CV} + \sum (\dot{m}h)_{in} = \dot{w}_{CV} + \sum (\dot{m}h)_{out}$$

$$\dot{q}_{CV} = 0$$

$$\dot{w}_{CV} = 0$$

Letting x represent the rate at which water is sprayed into the chamber

$$\dot{m}h_{a2} + \dot{m}_w h_{w1} = \dot{m}_a h_{a3} + \dot{m}_w h_{w3}$$

or, on a basis of 1 lb/s dry air at state 2

$$h_{a2} + xh_{wi} = h_{a3} + xh_{w3}$$

or

$$x = \frac{h_{a3} - h_{a2}}{h_{w1} - h_{w3}}$$

Using values from Keenan (4) and, assuming that the water vapor at state 3 is saturated

$$h_{w1} = h_f \text{ @ } 320°\text{F} = 290.4 \text{ Btu/lb}$$

$$h_{w3} = h_g \text{ @ } 340°\text{F} = 1190.8 \text{ Btu/lb}$$

$$x = \frac{141.7 - 61.9}{1190.8 - 290.4} = 0.089 \text{ lb/s } (0.089 \text{ kg/s})$$

One should correct the enthalpy value of the water vapor at state 3 for the fact that it is not saturated water vapor. This can be done by determining the partial pressure of the water vapor in the air–vapor mixture leaving the mixing chamber, then determining the enthalpy of the water vapor at this partial pressure and 340°F. This changes the rate at which water is being added to 0.087 lb/s (0.087 kg/s).

Example Problem 2.5

Calculate for a mixture of ideal gases with the following volumetric analysis,

	% V
Nitrogen	78.09%
Oxygen	20.95
Argon	0.93
Carbon dioxide	0.03
	100.00%

(a) The apparent molecular weight
(b) Enthalpy at 500 K
(c) Entropy at a pressure of 1 atm and 500 K
(d) Relative pressure at 500 K

Express answers on both a mole and a kilogram basis.

Solution

Basis: 1 mol mixture

	x_i	$\overline{M}_i$	$x_i\overline{M}_i$	$\overline{h}_i$	$x_i\overline{h}_i$
N_2	0.7809	28.0134	21.876	5,910.3	4615.4
O_2	0.2095	31.9988	6.704	6,087.2	1275.3
Ar	0.0093	39.948	0.372	4,195.6	39.0
CO_2	0.0003	44.0100	0.013	−385,191.3	−115.6
	1.0000		28.965		5814.1

	x_i	$\overline{s}_i^0$	$x_i\overline{s}_i^0$	ln Pr_i	x_i ln Pr_i
N_2	0.7809	206.627	161.355	24.852	19.407
O_2	0.2095	220.590	46.214	26.531	5.558
Ar	0.0093	165.477	1.539	19.903	0.185
CO_2	0.0003	234.772	0.070	28.237	0.008
			209.178		25.158

(a) $\overline{M}_{\text{app}} = \sum_{i=1}^{n} x_i\overline{M}_i = 28.965$ kg/kmol

(b) $\overline{h} = \sum_{i=1}^{n} x_i\overline{h}_i = 5814.1$ kJ/kmol

$= 200.73$ kJ/kg

(c) $\overline{s} = \sum_{i=1}^{n} (x_i\overline{s}_i^0) = 209.178$ kJ/kmol K

$= 7.222$ kJ/kg K

(d) $\text{Pr} = \exp(\overline{s}^0/\overline{R})$

$$= \exp\left(\frac{209.178}{8.31434}\right) = 8.4389 \times 10^{10}$$

Alternate Method:

$Pr = \exp(25.158) = 8.4330 \times 10^{10}$

REFERENCES CITED

1. Cameron, A. E., and Wichers, E., Report on the International Commission on Atomic Weights, *J. Am. Chem. Soc.* 84 (22:4192) (November 20, 1962).
2. McBride, B. J., Heimel, S., Ehlers, J., and Gordon, S., Thermodynamic Properties to 6000 K for 210 Substances Involving the First 18 Elements, NASA SP-3001, 1963.

3. Keenan, J. H., Chao, J., and Kaye, J., *Gas Tables*, 2nd ed., John Wiley & Sons, New York, 1980.
4. Keenan, J., Keyes, F., Hill, P., and Moore, J., *Steam Tables*, John Wiley & Sons, New York, 1969.

SUGGESTED READING

Jones, J. B., and Hawkins, G. A., *Engineering Thermodynamics*, John Wiley & Sons, New York, 1960.
Reynolds, W. C., *Thermodynamics*, 2nd ed., McGraw-Hill Book Company, New York, 1968.
Van Wylen, G. J., and Sonntag, R. E., *Fundamentals of Classical Thermodynamics*, 2nd ed., John Wiley & Sons, New York, 1973.
Wark, K., *Thermodynamics*, 2nd ed., McGraw-Hill Book Company, New York, 1971.

NOMENCLATURE

A	area
c_p	constant pressure specific heat
$\bar{c}_p$	molar constant pressure specific heat
c_v	constant volume specific heat
$\bar{c}_v$	molar constant volume specific heat
E	total stored energy
g	acceleration of gravity, Gibbs free energy
$\bar{g}$	molar Gibbs free energy
G	total Gibbs free energy
h	specific enthalpy
H	total enthalpy
$\bar{h}$	molar enthalpy
k	specific heat ratio
m	mass
$\dot{m}$	mass rate of flow
$\bar{M}$	formula (molecular) weight
n	moles
p	pressure
Pr	relative pressure
q	specific heat transfer
$\dot{q}$	rate of heat transfer
Q	total heat transfer
R	gas constant
$\bar{R}$	universal gas constant
s	specific entropy
$\bar{s}$	molar entropy
S	total entropy
T	temperature
u	specific internal energy
$\bar{u}$	molar internal energy
U	total internal energy
v	specific volume
V	total volume
$\bar{v}$	molar specific volume
$\bar{V}$	velocity
w	specific work
$\dot{w}$	power
W	total work
x	mole fraction
z	elevation
η	thermal efficiency

Superscripts

o	standard state

Subscripts

1,2	states
i, j	constituent
η	thermal efficiency
cv	control volume
h	high

PROBLEMS

2.1. Carbon monoxide, CO, an ideal gas, enters a steady-flow device at a pressure of 147 psia, a temperature of 1800°R, and with negligible velocity. It expands adiabatically and leaves the device with a velocity of 400 ft/s, at a temperature of 1200°R and at a pressure of 14.7 psia. The cross-sectional area at the outlet is 1 ft^2. Assuming constant specific heats, calculate:

(a) The mass rate of flow through the device

(b) The power developed by this device

2.2. A gas mixture with the volumetric analysis CO_2 10%, H_2O 10%, N_2 80%, enters a nozzle at a pressure of 400 kPa, a temperature of 1100 K, and with negligible velocity. It expands adiabatically and reversibly through the nozzle to a pressure of 105 kPa. Calculate, assuming variable specific heats and a constant chemical composition, the velocity leaving the nozzle.

2.3. Solve Example Problem 2.5 for a temperature of 2160°R.

2.4. Determine, assuming constant specific heats, the amount of heat transferred in cooling 100 lb of nitrogen, N_2, from 2700°R to 900°R by a steady-flow process for which the shaft work is zero and the kinetic and potential energies are negligible.

2.5. Determine, assuming constant specific heats, the amount of heat transferred in cooling 20 kg of carbon dioxide, CO_2, from 1500 K to 500 K by a steady-flow process for which the shaft work is zero and the kinetic and potential energies are negligible.

2.6. Determine, assuming variable specific heats, the amount of heat transferred in cooling 100 lb of nitrogen, N_2, from 2700°R to 900°R by a steady-flow process for which the shaft work is zero and the kinetic and potential energies are negligible.

2.7. Determine, assuming variable specific heats, the amount of heat transferred in cooling 40 kg of carbon dioxide, CO_2, from 1800 K to 600 K by a steady-flow process for which the shaft work is zero and the kinetic and potential energies are negligible.

2.8. The temperature and pressure of a mixture of gases entering a steady-flow device are at 1000 K and 680 kPa, respectively. The pressure leaving the device is 105 kPa. The composition of the gases passing through the device (assumed to be uniform and constant) are

CO_2	13.1% volume
H_2O	13.1% volume
N_2	73.8% volume
	100.0%

Assuming an adiabatic and reversible expansion through the device, that the inlet and exit velocities are negligible, and that the device develops 1000 kW, calculate, assuming variable specific heats:

(a) The temperature of the mixture leaving the turbine

(b) The mass of gas mixture passing through the device

(c) The volume of gases entering the device

2.9. A mixture of ideal gases passing through a steady-flow device has the following volumetric analysis (which remains constant):

CO_2	12.2%
N_2	73.4%
H_2O	14.4%
	100.0%

This gas mixture enters the device at a temperature of 2200°R and a pressure of 147 psia. The gas mixture leaves the device at a temperature of 1500°R and a pressure of 25 psia. Assuming adiabatic flow, steady flow with negligible kinetic energies and variable specific heats:

(a) Calculate the change in enthalpy across this device for this gas mixture, Btu/lb-mol; that is, $\bar{h}_2 - \bar{h}_1$.

(b) Calculate the change in entropy across this device for this gas mixture, Btu/lb-mol°R; that is, $\bar{s}_2 - \bar{s}_1$.

2.10. A gas turbine operates with an inlet temperature of 1540°F. The gases passing through the machine have the volumetric analysis:

N_2	40%
O_2	10%
CO_2	20%
H_2O	30%
	100%

If the turbine develops 2500 hp and the exit temperature from the turbine is 940°F, calculate the rate of flow of the gases through the turbine. Express your answer in pounds per minute. Assume adiabatic flow with negligible kinetic energies.

2.11. Dry air at 607 K (1093°R), 1114 kPa (162 psia) enters an adiabatic mixing chamber at the rate of 100 kg/s (220 lb/s). Water at 1300 kPa (189 psia), 440 K (792°R) is sprayed into the chamber at a rate such that the air–water vapor mixture at the exit is 450 K (810°R). Determine, if the pressure at state 3 is 1050 kPa (152 psia):

(a) The rate (kg/s) at which water is sprayed into the chamber.

(b) The constant pressure specific heat of the mixture at the exit. Compare to the specific heat of air at the same temperature.

2.12. Dry air at 720°R (380°C), 190 psia (1.30 MPa) enters an adiabatic mixing chamber at the rate of 95 lb/s (43 kg/s). Water at 210 psia (1.50 MPa), 358°F (180°C) is sprayed into the chamber at a rate such that the air–water vapor mixture at the exit is at its minimum temperature. Determine, if the pressure at state 3 is 185 psia (1.28 MPa):

(a) The temperature at the exit.

(b) The rate, in lb/s (kg/s), at which water is sprayed into the chamber.

3

FLUID MECHANICS

This chapter deals with the fundamental concepts of fluid mechanics. Once again, it is assumed that the reader has taken at least a fundamental course in fluid mechanics. Several suggested alternate sources of material on this subject are listed at the end of this chapter.

3.1 BASIC EQUATION FOR A SYSTEM

A system is, by definition, an arbitrary collection of matter of fixed identity. This means it is always composed of the same quantity of matter.

The conservation of mass for a system states that the total mass, M, of the system is a constant. On a rate basis, this may be written as

$$\frac{DM}{Dt} = 0 \tag{3.1}$$

Newton's second law states that, for a system moving relative to an inertial reference frame, the sum of all external forces acting on the system is equal to the time rate of change of the linear momentum of the system; that is,

$$\mathbf{F}_R = \left.\frac{D\mathbf{P}}{DT}\right|_{\text{system}} \tag{3.2}$$

where $\mathbf{F}_R$ is the resultant force and $\mathbf{P}$ is the total linear momentum of the system, given by

$$\mathbf{P}_{\text{system}} = \int_{\text{system}} \mathbf{V}\, dm \tag{3.3}$$

The moment of momentum equation for a system states that the rate of change of angular momentum is equal to the sum of all torques *acting on the system*; that is,

$$\mathbf{T}_R = \frac{D\mathbf{H}}{Dt} \tag{3.4}$$

where $\mathbf{T}_R$ is the resultant torque on the system and $\mathbf{H}$ is the moment of momentum (angular momentum) of the system. The torque may be produced by body forces, surface forces, and by shafts that cross the boundary of the system. Therefore,

$$\mathbf{T}_R = \underbrace{\int_{\text{system}} \mathbf{r} \times \mathbf{g}\, dm}_{\text{(body)}} + \underbrace{\mathbf{r} \times \mathbf{F}_S}_{\text{(surface)}} + \underbrace{\mathbf{T}_{\text{shaft}}}_{\text{(shaft)}} \tag{3.5}$$

and

$$\mathbf{H}_{\text{system}} = \int_{\text{system}} \mathbf{r} \times \mathbf{V}\, dm \tag{3.6}$$

where $\mathbf{r}$ is a position vector.

3.2 SYSTEM TO CONTROL VOLUME TRANSFORMATION

It is easier to analyze gas turbine engines using a control volume instead of a system. A control volume is a designated volume in space. The boundary of this control volume is referred to as the control surface. The amount and identity of matter within the control surface will change with time. It is usually assumed that the shape of the control surface does not vary with time; that is, it is a fixed shape.

It is now important to change the conservation of mass, linear momentum, and moment of momentum equations from a system to a control volume.

The relation between a control volume and system is

$$\frac{DN}{Dt} = \int_{\substack{\text{control}\\ \text{surface}}} \eta(\rho \mathbf{V} \cdot d\mathbf{A}) + \frac{\partial}{\partial t} \int_{\substack{\text{control}\\ \text{volume}}} \eta\rho\, dV \tag{3.7}$$

In Eq. (3.7),

$\dfrac{DN}{Dt}$ = the total rate of change of an artibrary extensive property (N) of the system

$\displaystyle\int_{\text{c.s.}} \eta\rho\mathbf{V} \cdot d\text{A}$ = the net rate of efflux of N across the control surface (c.s.) at time t

$\displaystyle\frac{\partial}{\partial t}\int_{\text{c.v.}} \eta\rho\, dV$ = the rate of change of N inside the control volume (c.v.) at time t

3.3 BASIC EQUATIONS FOR A CONTROL VOLUME

The conservation of mass for a control volume is, since $N = M$, $\eta = 1$,

$$\left.\frac{DM}{Dt}\right|_{\text{system}} = 0 = \int_{\text{c.s.}} \rho \mathbf{V} \cdot d\mathbf{A} + \frac{\partial}{\partial t}\int_{\text{c.v.}} \rho \, dV \tag{3.8}$$

The momentum equation for a nonaccelerating control volume is, since $N = \mathbf{P}$, $\eta = \mathbf{V}$,

$$\mathbf{F}_R = \int_{\text{c.s.}} \mathbf{V}(\rho \mathbf{V} \cdot d\mathbf{A}) + \frac{\partial}{\partial t}\int_{\text{c.v.}} \mathbf{V}\rho \, dV \tag{3.9}$$

where

$$\mathbf{F}_R = \mathbf{F}_{\text{surface}} + \mathbf{F}_{\text{body}}$$

The moment of momentum for a fixed control volume is, since $N = \mathbf{H}$, $\eta = \mathbf{r} \times \mathbf{V}$,

$$\mathbf{r} \times \mathbf{F}_S + \int_{\text{c.v.}} \mathbf{r} \times \mathbf{g}\rho \, dV + \mathbf{T}_{\text{shaft}}$$

$$= \int_{\text{c.s.}} \mathbf{r} \times \mathbf{V}(\rho \mathbf{V} \cdot d\mathbf{A}) + \frac{\partial}{\partial t}\int_{\text{c.v.}} \mathbf{r} \times \mathbf{V}\rho \, dV \tag{3.10}$$

Equation (3.10) is a general vector equation for the moment of momentum for an inertial control volume. The left side of the equation is an expression for all the torques that act on the control volume. The right side consists of terms that express the rate of efflux of angular momentum from the control volume and the rate of change of angular momentum within the control volume.

3.4 STAGNATION ENTHALPY AND TEMPERATURE

The stagnation (total) properties are frequently used. The stagnation (total) enthalpy is defined by

$$h_0 \equiv h + \frac{\overline{V}^2}{2} \tag{3.11}$$

where the subscript 0 refers to the total state and the unsubscripted property refers to the static value. Note that this definition places no restriction on the type of process (reversible or irreversible) or the type of fluid.

Substituting Eq. (3.11) into Eq. (2.4) yields the steady-flow energy equation in the form

$${}_1\overline{q}_2 + \overline{h}_{01} = {}_1\overline{w}_2 + \overline{h}_{02} \tag{3.12}$$

The stagnation (total) temperature of a flowing fluid is defined as the temperature that would result if the fluid were brought to rest *adiabatically* and *reversibly*.

3.5 MACH NUMBER

The sonic velocity (velocity of sound) is the velocity of a pressure wave of extremely small amplitude. Assuming the propagation of this small wave (sound wave) to be reversible and adiabatic yields the following equation for the sonic velocity:

$$c = \sqrt{\left(\frac{\partial p}{\partial \rho}\right)_s} \tag{3.13}$$

For an ideal gas for an adiabatic and reversible process,

$$\frac{p}{(\rho)^k} = \text{constant}$$

or

$$\frac{\partial p}{\partial \rho} = \frac{kp}{\rho} \tag{3.14}$$

Combining Eqs. (3.13) and (3.14) yields

$$c = \sqrt{k\,pv} \tag{3.15}$$

or, for an ideal gas,

$$c = \sqrt{k\,RT} \tag{3.16}$$

It should be noted that, as shown by Eq. (3.16), the sonic velocity of an ideal gas is a function of temperature only.

A useful parameter is the Mach number, which is the ratio of the velocity of a gas at a given point to the sonic velocity at the same point. In equation form,

$$M = \frac{\overline{V}}{c} \tag{3.17}$$

3.6 CONVERGING NOZZLES

A schematic diagram of a converging nozzle is shown in Figure 3.1. The cross-sectional area at the inlet to the nozzle is assumed to be large so that the velocity is negligibly small. This means that the pressure and temperature at the inlet are the stagnation pressure and temperature, respectively. In the following discussion, it is assumed that the inlet conditions are held constant and that the pressure in the discharge region may be varied.

The pressure at the minimum cross-sectional area (throat or mouth of the nozzle) is designated as p_{th}, the pressure in the discharge region as p_d.

If $p_d = p_{\text{inlet}}$, the pressure at all points in the nozzle is the same and there is no flow. As p_d is lowered slightly to p_a, flow occurs with $p_d = p_{\text{th}}$ and the velocity

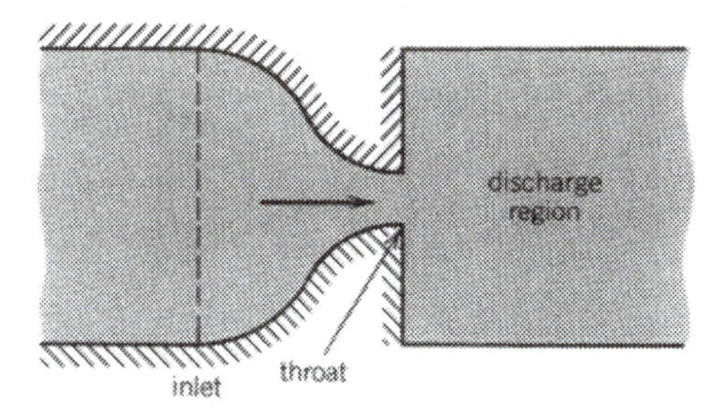

Figure 3.1. Converging nozzle.

increases from the inlet to the throat, reaching its maximum value at the throat. The variation of pressure with respect to location for this condition is shown in Figure 3.2.

When the discharge pressure is lowered further to p_b, p_d still equals p_{th} but the mass rate of flow through the nozzle increases. The pressure variation in the nozzle for this condition is also shown in Figure 3.2.

If the discharge pressure is lowered such that the velocity at the throat reaches sonic velocity, the nozzle is said to be choked. For a given nozzle size, stagnation pressure, temperature, and working fluid, the nozzle has reached its maximum mass rate of flow. For these conditions, the throat pressure is p^*, the throat temperature is T^*, and the velocity at the throat is c. The pressure variation for this condition is shown in Figure 3.2. Note that p_{th} is still equal to p_d.

Further reduction of the discharge region pressure has no effect on the throat pressure, the velocity within the converging nozzle, or the mass rate of flow through the nozzle. This is shown in Figure 3.2 by states p_e and p_f.

For the adiabatic expansion of an ideal gas with constant specific heats, the temperature at the throat when the nozzle is choked may be calculated by

$$\frac{T^*}{T_0} = \frac{2}{k+1} \tag{3.18}$$

If the expansion is adiabatic and reversible, the pressure at the throat when the nozzle is choked may be calculated by

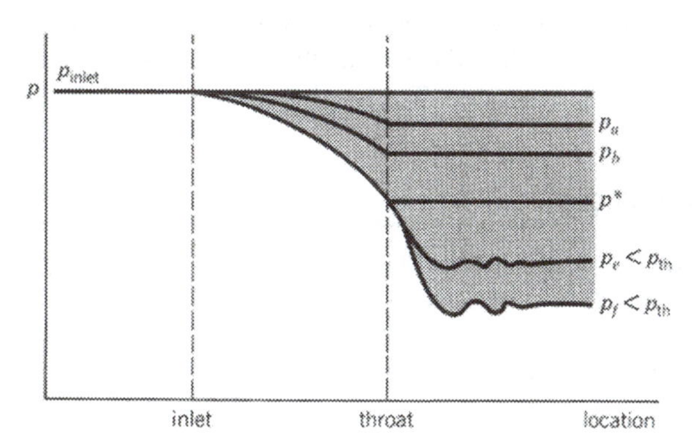

Figure 3.2. Pressure versus location within a converging nozzle for several discharge region pressures.

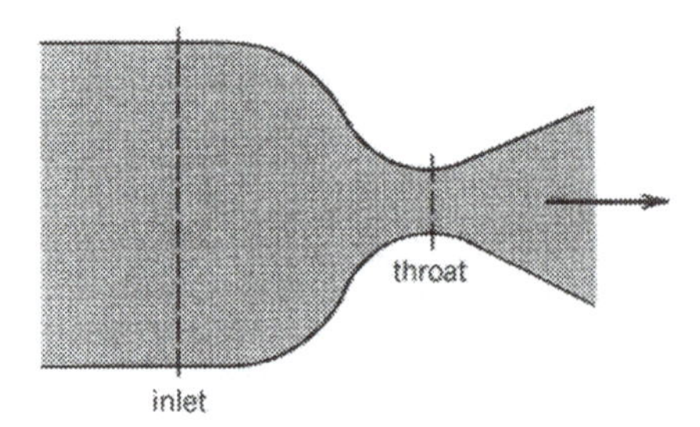

Figure 3.3. Converging-diverging nozzle.

$$\frac{p^*}{p_0} = \left(\frac{2}{k+1}\right)^{k/(k-1)} \tag{3.19}$$

3.7 CONVERGING-DIVERGING NOZZLES

In order to accelerate a fluid from a subsonic to a supersonic velocity, a diverging section must be added to the converging nozzle. This is shown in Figure 3.3.

For situations when the velocity at the throat does not reach sonic velocity, the diverging section acts as a diffuser and the velocity never reaches sonic velocity. This is shown in Figure 3.4 for the pressure p_g.

As the discharge region pressure is lowered, the velocity at the throat will reach sonic velocity. What happens in the discharge region depends on the discharge pressure.

If the discharge region pressure has a value of p_h, then the pressure in the diverging region increases as the area increases and the velocity is not supersonic at any location.

If the discharge region pressure has a value of p_k, then the pressure in the diverging section continues to decrease, the velocity continues to increase, and the velocity in the discharge region is supersonic.

If the discharge region pressure is between p_h and p_k, then the pressure decreases for a portion of the diverging section with the velocity continuing to increase. At some point, theory predicts a normal shock occurring with a resulting increase in pressure and decrease in velocity, the velocity going from supersonic to subsonic across the normal shock. After the shock, the pressure increases and the velocity decreases with an increase in area. This is shown in Figure 3.4.

Example Problem 3.1

An ideal gas expands adiabatically and reversibly through a converging nozzle. At the inlet to the nozzle, the pressure is 150 psia, the temperature is 800°R, and the velocity is negligible (state 1). The pressure in the discharge region is 80 psia. Calculate the pressure, temperature, and velocity at the throat of the nozzle if the ideal gas is

(a) Carbon dioxide
(b) Nitrogen

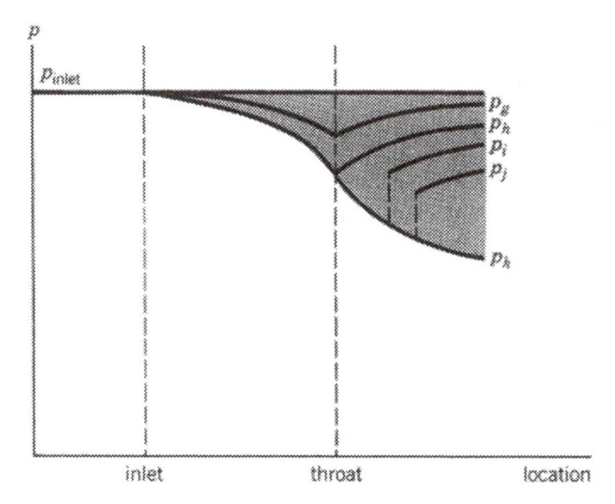

Figure 3.4. Pressure variation with location in a converging-diverging nozzle for several discharge region pressures.

Assume constant specific heats.

Solution

Basis: 1 lb of gas

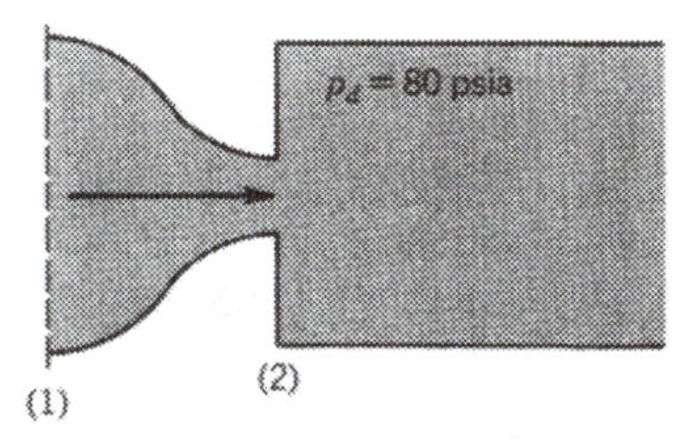

$$p_1 = 150 \text{ psia}$$

$$T_1 = 800°\text{R}$$

$$\overline{V}_1 = 0$$

(a) For carbon dioxide:

$$\overline{M} = 44.0100$$

$$R = 35.11 \text{ ft-lb}_f/\text{lb°R}$$

$$\bar{c}_p = 8.874 \text{ Btu/lb-mol°R}$$

$$k = 1.288$$

$$c_p = \frac{8.874}{44.01} = 0.2016 \text{ Btu/lb°R}$$

$$p^* = p_0 \left(\frac{2}{k+1}\right)^{k/(k-1)} = 150 \left(\frac{2}{2.288}\right)^{1.288/0.288} = 82.19 \text{ psia}$$

Therefore, the converging nozzle is choked and the pressure at the exit of the nozzle is 82.19 psia.

$$T_2 = T^* = T_0\left(\frac{2}{k+1}\right) = 800\left(\frac{2}{2.288}\right) = 699.3°\text{R}$$

$$\bar{V}_2 = c^* = \sqrt{k\,RT}$$

$$= \sqrt{(1.288)(32.174)(35.11)(699.3)} = 1009 \text{ ft/s}$$

(b) For nitrogen:

$$\bar{M} = 28.0134$$

$$R = 55.16 \text{ ft-lb}_f/\text{lb°R}$$

$$\bar{c}_p = 6.961 \text{ Btu/lb-mol°R}$$

$$k = 1.399$$

$$c_p = \frac{6.961}{28.0134} = 0.2485 \text{ Btu/lb°R}$$

$$p^* = 150\left(\frac{2}{2.399}\right)^{1.399/0.399} = 79.27 \text{ psia}$$

Therefore the converging nozzle is not choked and the pressure at the exit is 80 psia.

$$T_2 = T_0\left(\frac{p_2}{p_0}\right)^{(k-1)/k} = 800\left(\frac{80}{150}\right)^{0.399/1.399} = 668.7\ °\text{R}$$

$$T_0 = T_2 + \frac{\bar{V}_2^2}{c_p 2}$$

$$\bar{V}_2 = \sqrt{(2)(32.174)(778)(800 - 668.7)(0.2485)} = 1278 \text{ ft/s}$$

3.8 NOZZLE EFFICIENCY

Many nozzles operate adiabatically but not reversibly. These irreversibilities occur because of friction and, in the diverging sections of nozzles, because of separation.

Nozzle efficiency, η_N, is defined as the ratio of the actual kinetic energy at the nozzle exit to the kinetic energy that would have occurred if the flow through the nozzle had been adiabatic and reversible from the same inlet conditions to the same exit pressure.

In equation form,

$$\eta_N \equiv \frac{\overline{V}_{2a}^2/2}{\overline{V}_{2i}^2/2} = \frac{h_{01} - h_{2a}}{h_{01} - h_{2i}} \tag{3.20}$$

Example Problem 3.2

Oxygen, O_2, an ideal gas, with properties as given in Tables 2.2 and 2.4, enters a converging-diverging nozzle at a temperature of 335 K, a pressure of 655 kPa, and a velocity of 150 m/s (state 1). It flows through the insulated nozzle in a steady-flow manner. The flow is adiabatic and reversible in the converging section of the nozzle but irreversible in the diverging section of the nozzle. At the exit section of the nozzle (state 3), the pressure and temperature are 138 kPa and 222 K, respectively. Assuming the mass rate of flow is 9 kg/s:

(a) Determine the exit Mach number.
(b) Sketch the T-s diagram showing both the static and stagnation states at (1) and (3).
(c) Calculate the change in stagnation pressure between (1) and (3).
(d) Calculate the change in entropy between (1) and (3).
(e) Calculate the static temperature at the throat of the nozzle.
(f) Calculate the nozzle efficiency.

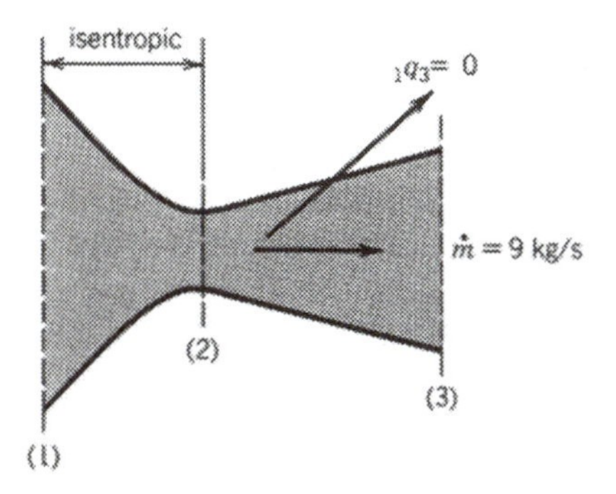

$$p_1 = 655 \text{ kPa} \qquad p_3 = 138 \text{ kPa}$$

$$T_1 = 335 \text{ K} \qquad T_3 = 222 \text{ K}$$

$$\overline{V}_1 = 150 \text{ m/s}$$

Solution

Basis: 1 lb of oxygen

From Tables 2.2 and 2.4,

$$\overline{M} = 31.9988$$

$$\overline{c}_p = 29.375 \text{ kJ/kmol K}$$

$$k = 1.394$$

$$ {}_1q_3 + h_1 + \frac{\overline{V}_1^2}{2} = {}_1w_3 + h_{3_a} + \frac{\overline{V}_{3a}^2}{2} $$

$$ \overline{V}_{3a} = \sqrt{(2)(1)\left(\frac{29.375}{32}\right)(335 - 222)(1000) + (150)^2} $$

$$ = 480 \text{ m/s} $$

$$ c_3 = \sqrt{(1.394)(1)\left(\frac{8.314 \times 10^3}{32}\right)(222)} $$

$$ = 283.6 \text{ m/s} $$

(a) $M_3 = \dfrac{480}{283.6} = 1.69$

(b)

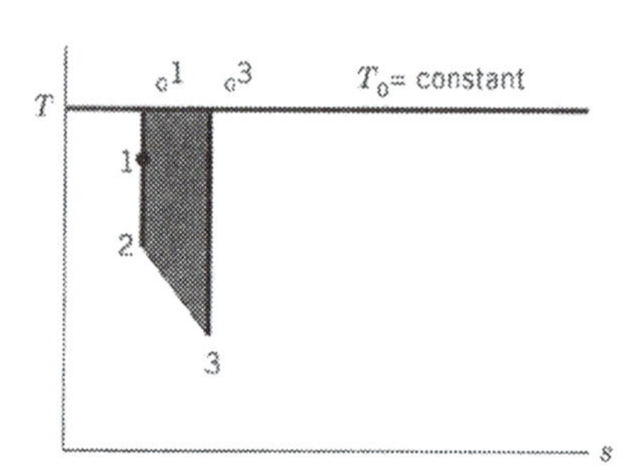

(c) $T_0 = \text{constant} = T_1 + \dfrac{\overline{V}_1^2}{2c_p}$

$$ T_0 = 335 + \frac{(150)^2(32)}{(2)(1)(1000)(29.375)} $$

$$ = 347.3 \text{ K} $$

$$ P_{01} = p_1\left(\frac{T_{01}}{T_1}\right)^{k/(k-1)} = 655\left(\frac{347.3}{335}\right)^{1.394/0.394} = 744.1 \text{ kPa} $$

$$ P_{03} = p_3\left(\frac{T_{03}}{T_3}\right)^{k/(k-1)} = 138\left(\frac{347.3}{222}\right)^{1.394/0.394} = 672.2 \text{ kPa} $$

$$ \Delta p_0 = p_{03} - p_{01} $$

$$ = 672.2 - 744.1 = -71.9 \text{ kPa} $$

(d) $s_{03} - s_{01} = c_p \ln \frac{T_{03}}{T_{01}} - R \ln \frac{p_{03}}{p_{01}}$

$$= -\left(\frac{8314}{32}\right) \ln \frac{672.2}{744.1}$$

$$= +26.403 \text{ J/kg K}$$

(e) $T^* = T_2 = T_0 \left(\frac{2}{k+1}\right) = 347.3 \left(\frac{2}{2.394}\right)$

$$= 290.1 \text{ K}$$

(f) $T_{3i} = T_o \left(\frac{p_3}{p_{01}}\right)^{(k-1)/k}$

$$= 347.3 \left(\frac{138}{744.1}\right)^{0.394/1.394}$$

$$= 215.7 \text{ K}$$

$$\overline{V}_{3i} = \sqrt{(2)(1)\left(\frac{29.375}{32}\right)(1000)(347.3 - 215.7)}$$

$$= 491.5 \text{ m/s}$$

$$\eta_N = \frac{\overline{V}_{3a}^2/2}{\overline{V}_{3i}^2/2} = \frac{(480)^2}{(491.5)^2} = 0.954$$

3.9 FLOW PARAMETER

An important relationship is that of the flow parameter for a passageway.

Consider the steady flow of an ideal gas through a converging nozzle under the following conditions:

1. Adiabatic and reversible flow
2. Inlet state is the stagnation state
3. Constant specific heats
4. The throat (mouth) area of the nozzle is A
5. One-dimensional flow

This is shown in Figure 3.5, where the properties at the inlet state are listed with a subscript o, those at the mouth of the nozzle with no subscript.

The steady-flow energy equation (Eq. 2.6) becomes, for the above conditions,

$$h_0 = h + \frac{\overline{V}^2}{2}$$

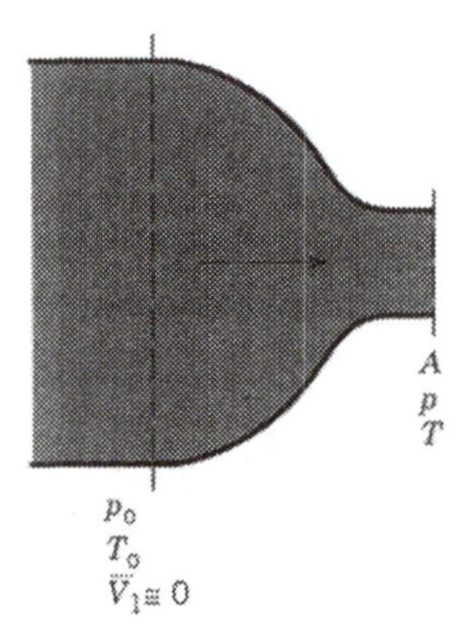

Figure 3.5. Passageway and notation used in development of flow parameter equation.

or

$$\bar{V} = \sqrt{2\, c_p\, (T_0 - T)} = \sqrt{2\, c_p T_0 \left(1 - \frac{T}{T_0}\right)} \tag{3.21}$$

Combining Eqs. (3.21), (2.17) and (2.18) yields

$$\bar{V} = \sqrt{\left(\frac{2kRT_0}{(k-1)}\right)\left(1 - \frac{T}{T_0}\right)} \tag{3.22}$$

The one-dimensional continuity equation written for values at the mouth of the nozzle is

$$\dot{m} = \frac{A\bar{V}}{v} = \frac{A\bar{V}p_0(p/p_0)}{RT_0(T/T_0)} \tag{3.23}$$

Combining Eqs. (3.22), (3.23), and (2.41) yields

$$\frac{\dot{m}\sqrt{T_0}}{p_0 A} = \sqrt{\frac{2k}{R(k-1)}}\left(\frac{p}{p_0}\right)^{1/k}\sqrt{1 - \left(\frac{p}{p_0}\right)^{(k-1)/k}} \tag{3.24}$$

The left-hand term of Eq. (3.24) is referred to as the flow parameter. Equation (3.24) shows that the flow parameter is a function of the gas (R), the specific heat ratio (k) and the expansion ratio (p/p_0). Once the nozzle is choked, the flow parameter has a fixed value.

Figure 3.6 is a plot of flow parameter versus expansion ratio for air ($R = 53.35$) for three values of specific heat ratio. Note that the expansion ratio at which the nozzle chokes and the value of the flow parameter for choked conditions is dependent on the value of k.

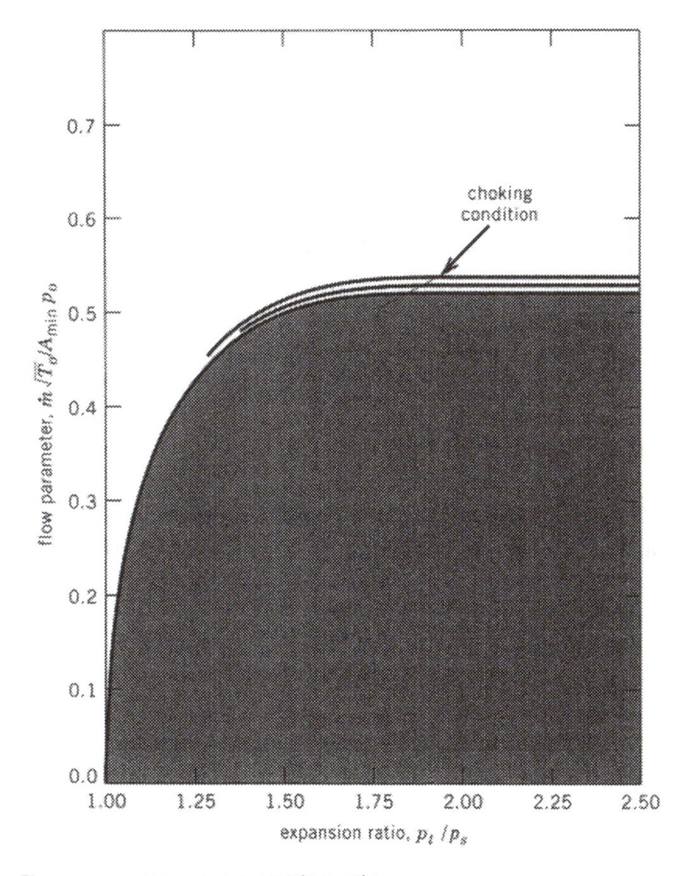

Figure 3.6. Flow parameter versus expansion ratio.

3.10 NORMAL SHOCKS

Irreversible flow discontinuities can occur in any supersonic flow field. It is important to be able to determine the change in the properties across a shock, since normal and oblique shocks are important in understanding the design of supersonic inlets.

The plane of the discontinuity in a normal shock is normal to the direction of flow. Since the distance over which this discontinuity occurs is infinitesimal, one is justified in treating the shock as being extraordinarily thin. The equations for calculating the changes in pressure, temperature, and Mach number will be developed in this section because our interest is in the change in the properties across the shock, not what happens within the shock.

Figure 3.7 shows a control volume drawn around a normal shock. The conditions on the upstream side of the normal shock are denoted by an x subscript, the conditions on the downstream side by a y subscript. The equations developed below assume steady, one-dimensional flow, no change in area across the shock because of its extremely small thickness, and adiabatic flow. Therefore, from Eq. (3.12),

$$h_{0x} = h_{0y} = h_x + \frac{\overline{V}_x^2}{2} = \text{constant} \tag{3.25}$$

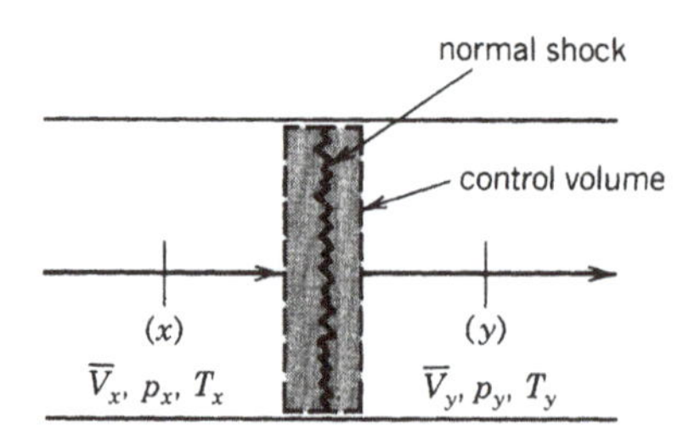

Figure 3.7. Control volume around a normal shock.

The conservation of mass, Eq. (3.8), for steady, one-dimensional flow, becomes

$$\frac{\dot{m}}{A} = \text{constant} = \rho_x \overline{V}_x = \rho_y \overline{V}_y \tag{3.26}$$

The momentum equation, Eq. (3.9), assuming negligible friction at the duct walls because the shock is so thin and assuming no body forces, becomes

$$p_x - p_y = \frac{\dot{m}}{A}(\overline{V}_y - \overline{V}_x) \tag{3.27}$$

Equations (3.25), (3.26), and (3.27) may be used to develop equations for the calculation of the fluid properties at state y if the properties at state x are known. The equations will be developed assuming that the working fluid is an ideal gas with constant specific heats.

From Eq. (3.25), it is observed that for an ideal gas with constant specific heats,

$$T_{0x} = \text{constant} \tag{3.28}$$

Also

$$c_p T_x + \frac{\overline{V}_x^2}{2} = c_p T_y + \frac{\overline{V}_y^2}{2} \tag{3.29}$$

Combining Eqs. (3.29), (3.17), (3.16), (2.18), and (2.17) yields

$$\frac{T_y}{T_x} = \frac{1 + [k - 1)/2]M_x^2}{1 + [k - 1)/2]M_y^2} \tag{3.30}$$

Combining Eqs. (3.26), (2.11), (3.17) and (3.30) yields

$$\frac{p_y}{p_x} = \frac{M_x \sqrt{1 + [(k - 1)/2]M_x^2}}{M_y \sqrt{1 + [(k - 1/2)]M_y^2}} \tag{3.31}$$

Combining Eqs. (3.27), (2.11), (3.26), and (3.27) yields

$$\frac{p_y}{p_x} = \frac{1 + kM_x^2}{1 + kM_y^2} \tag{3.32}$$

Equations (3.31) and (3.32) may be combined to eliminate p_y/p_x. This results in

$$M_y^2 = \frac{(k - 1)M_x^2 + 2}{2kM_x^2 - (k - 1)} \tag{3.33}$$

The change in the stagnation pressure across a normal shock may be evaluated as follows:

$$\frac{p_{0y}}{p_{0x}} = \left(\frac{p_{0y}}{p_y}\right)\left(\frac{p_y}{p_x}\right)\left(\frac{p_x}{p_{0x}}\right) \tag{3.34}$$

The first and third terms on the right side of Eq. (3.34) involve the relationship between the static and stagnation pressure. From Eq. (3.11) for an ideal gas with constant specific heats,

$$h_0 = c_p T_0 = c_p T + \frac{\overline{V}^2}{2} \tag{3.35}$$

Combining Eq. (3.35) with (3.17), (3.16), (2.18), and (2.17) yields

$$\frac{T_0}{T} = 1 + \left(\frac{k-1}{2}\right)M^2 \tag{3.36}$$

Therefore, the relationship between static and stagnation pressure becomes, for an ideal gas with constant specific heats,

$$\frac{p_0}{p} = \left[1 + \left(\frac{k-1}{2}\right)M^2\right]^{k/(k-1)} \tag{3.37}$$

Combining Eq. (3.34) with (3.37), (3.32) and (3.33) yields

$$\frac{p_{0y}}{p_{0x}} = \left[\frac{(k+1)M_x^2}{2+(k-1)M_x^2}\right]^{k/(k-1)} \left[\left(\frac{2k}{k+1}\right)M_x^2 - \left(\frac{k-1}{k+1}\right)\right]^{1/(1-k)} \tag{3.38}$$

It is also possible to write Eqs. (3.30) and (3.32) in terms of M_x. The results are

$$\frac{T_y}{T_x} = \left[\frac{2}{(k+1)M_x^2} + \left(\frac{k-1}{k+1}\right)\right]\left[\left(\frac{2k}{k+1}\right)M_x^2 - \left(\frac{k-1}{k+1}\right)\right] \tag{3.39}$$

$$\frac{p_y}{p_x} = \left[1 + \left(\frac{2k}{k+1}\right)(M_x^2 - 1)\right] \tag{3.40}$$

Equations (3.38), (3.39), and (3.40) allow the calculation of the fluid properties downstream from a normal shock if the conditions entering the shock are known.

Example Problem 3.3

Air enters a normal shock at a temperature of 630°R (350 K), a pressure of 20 psia (137.8 kPa), and with a velocity of 2460 ft/s (750 m/s). Determne, assuming constant specific heats with $k = 1.4$, $c_p = 0.241$ Btu/lb°R (1.008 kJ/kg K), and $\overline{M} = 28.965$:

(a) The stagnation pressure and temperature and Mach number at the inlet to the normal shock.
(b) The stagnation pressure and temperature, static pressure and temperature, Mach number, and velocity downstream from the normal shock.
(c) The change in entropy across the shock.

(d) Draw the temperature-entropy diagram showing the static and stagnation states at both inlet and exit.

Solution

(a) $c_x = \sqrt{kRT_x} = 1230$ ft/s (375 m/s)

$$M_x = \frac{\overline{V}_x}{c_x} = 2.0$$

$$T_{0x} = T_x\left[1 + \left(\frac{k-1}{2}\right)M_x^2\right] = 1134°\text{R (630 K)}$$

$$p_{0x} = p_x\left[\frac{T_{0x}}{T_x}\right]^{k/(k-1)} = 156.5 \text{ psia (1078 kPa)}$$

(b) From Eq. (3.38),

$$\frac{p_{0y}}{p_{0x}} = 0.7209$$

or

$$p_{0y} = 112.8 \text{ psia (1078 kPa)}$$

$$T_{0y} = 1134°\text{R (630 K)}$$

From Eq. (3.39),

$$\frac{T_y}{T_x} = 1.6875$$

$$T_y = 1063°\text{R (590.6 K)}$$

From Eq. (3.40),

$$\frac{p_y}{p_x} = 4.500$$

$$p_y = 90 \text{ psia (620.1 kPa)}$$

From Eq. (3.33)

$$M_y = 0.57735$$

$$\overline{V}_y = 0.57735\sqrt{(1.4)(32.174)(53.35)(1063)}$$
$$= 923 \text{ ft/s (281 m/s)}$$

(c)
$$s_y - s_x = c_p \ln\frac{T_y}{T_x} - R\ln\frac{p_y}{p_x}$$
$$= +0.0229 \text{ Btu/lb°R (0.0959 kJ/kg K)}$$

(d)

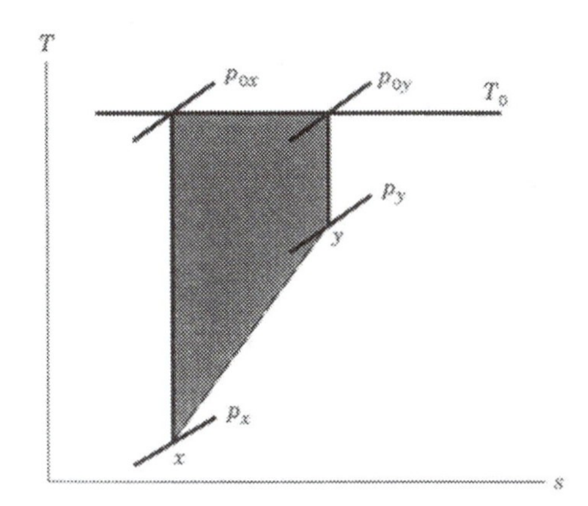

It is obvious that a very useful table could be constructed for the change in properties across a normal shock by using Eqs. (3.33), (3.38), (3.39), and (3.40). Table 3.1 is a table for an ideal gas with $k = 1.4$.

Examination of Table 3.1 shows that for the change in stagnation pressure to be small, the change in Mach number across the stock must be near unity. This means that the Mach number at the upstream station of the normal shock must be near unity.

3.11 OBLIQUE SHOCK WAVE

The preceeding section examined normal shocks, which are a special case of flow discontinuities. The general situation observed in practice is for the shock to be inclined to the direction of flow as shown in Figure 3.8. These are called oblique shocks.

The flow, as illustrated in Figure 3.8, approaches the shock wave with a velocity $\overline{V}_1$, a Mach number M_1, and at an angle θ with respect to the shock. It is turned through an angle δ as it passes through the shock, leaving with a velocity $\overline{V}_2$ and a Mach number M_2 at an angle $(\theta - \delta)$ with respect to the shock. A streamline is also shown in Figure 3.8. One of the streamlines could be replaced by a solid boundary, which would represent flow around a wedge or concave corner with an angle δ.

The inlet and exit velocities can be speарated into tangential and normal components. These components are shown in Figure 3.9.

There is no reason for the tangential component of the velocity to change across the shock. This means that

$$\overline{V}_{1,t} = \overline{V}_{2,t}$$

The normal component of velocity may be treated as flow through a normal shock. This means that $\overline{V}_{1,n}$ must be supersonic, $\overline{V}_{2,n}$ must be subsonic, and $\overline{V}_{2,n} < \overline{V}_{1,n}$. Since the normal velocity decreases across an oblique shock and the tangential velocity remains constant, the flow is always turned toward the shock as illustrated in Figure 3.8.

Equations for calculating the change in pressure, temperature, and Mach number across a normal shock were developed in the preceding section. These equations (3.33, 3.38, 3.39, and 3.40) may be used to calculate the change in pressure, temperature, and

TABLE 3.1. **Normal Shock Values for an Ideal Gas with Constant Specific Heats, $k = 1.4$**

M_x	M_y	p_y/p_x	T_y/T_x	p_{ty}/p_{tx}
3.00	0.47519	10.3333	2.6790	0.32834
2.90	0.48138	9.6450	2.5632	0.35773
2.80	0.48817	8.9800	2.4512	0.38946
2.70	0.49563	8.3383	2.3429	0.42359
2.60	0.50387	7.7200	2.2383	0.46012
2.50	0.51299	7.1250	2.1375	0.49901
2.40	0.52312	6.5533	2.0403	0.54014
2.30	0.53441	6.0050	1.9468	0.58329
2.20	0.54706	5.4800	1.8569	0.62814
2.10	0.56128	4.9783	1.7705	0.67420
2.00	0.57735	4.5000	1.6875	0.72089
1.90	0.59562	4.0450	1.6079	0.76736
1.80	0.61650	3.6133	1.5316	0.81268
1.70	0.64054	3.2050	1.4583	0.85572
1.60	0.66844	2.8200	1.3880	0.89520
1.50	0.70109	2.4583	1.3202	0.92979
1.40	0.73971	2.1200	1.2547	0.95819
1.30	0.78596	1.8050	1.1909	0.97937
1.20	0.84217	1.5133	1.1280	0.99280
1.10	0.91177	1.2450	1.0649	0.99893
1.00	1.00000	1.0000	1.0000	1.00000

Mach number across an oblique shock by using the normal components; that is, whenever the term M_x appears in Eqs. (3.33), (3.38), (3.39), and (3.40), it is replaced by the normal inlet component.

$$M_{1,n} = M_1 \sin \theta \tag{3.42}$$

One must remember that the downstream Mach number as calculated by Eq. (3.33) is the normal component $M_{2,n}$.

The tangential velocity across an oblique shock does not change but, since the

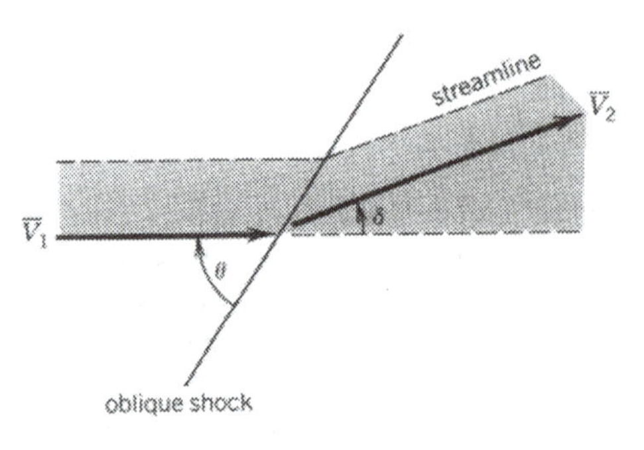

Figure 3.8. Oblique shock wave.

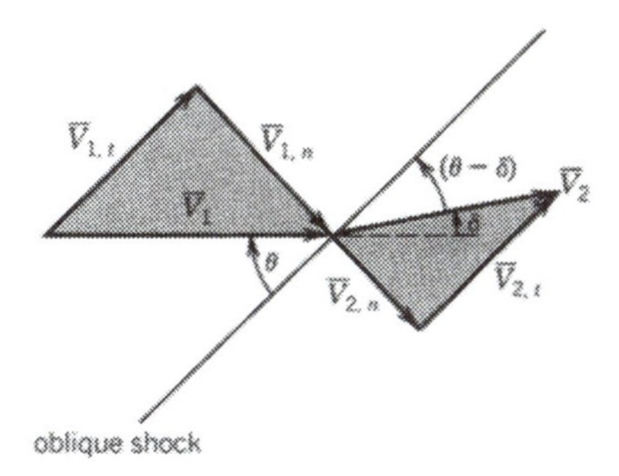

Figure 3.9. Inlet and exit velocity components across an oblique shock.

static temperature changes, the tangential Mach numbers before and after the shock are equated by the relationship

$$M_{2,t} = M_{1,t}\sqrt{\frac{T_1}{T_2}} \tag{3.43}$$

The Mach number downstream of the oblique shock can then be calculated by

$$M_2 = \sqrt{(M_{2,t})^2 + (M_{2,n})^2} \tag{3.44}$$

Example Problem 3.4

Air ($k = 1.4$) enters an oblique shock with a Mach number of 3.0. The streamline at the inlet to the oblique is at an angle of 27.8° as shown.

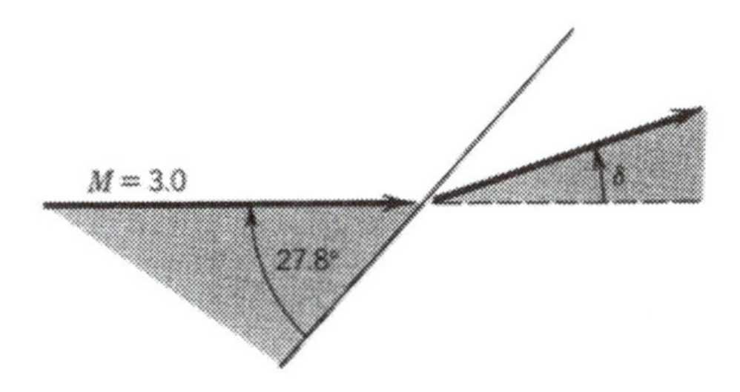

Determine:

(a) The static pressure ratio across the shock
(b) The total pressure ratio across the shock
(c) The static temperature ratio across the shock
(d) The Mach number after the oblique shock
(e) The flow deflection angle δ

Solution

The normal and tangential components of the Mach number at the inlet are

$$M_{1,t} = 3.0(\cos 27.8°) = 2.6537$$

$$M_{1,n} = 3.0(\sin 27.8°) = 1.399$$

From Table 3.1 at $M_{1,n} = 1.400$,

$$M_{2,n} = 0.73971$$

(a) $\dfrac{p_y}{p_x} = 2.1200$

(b) $\dfrac{p_{0y}}{p_{0x}} = 0.95819$

(c) $\dfrac{T_y}{T_x} = 1.2547$

$$M_{2,t} = 2.6537\sqrt{\frac{1}{1.2547}} = 2.3691$$

(d) $M_2 = \sqrt{(0.73971)^2 + (2.3691)^2}$

$= 2.482$

(e) The angle of the streamline leaving the oblique shock is shown in the drawing.

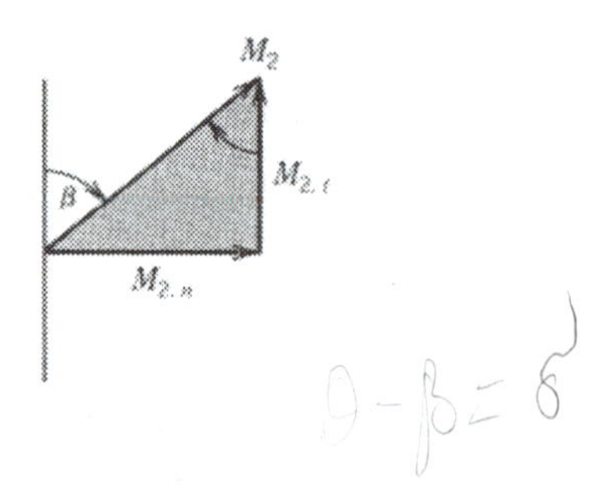

$$\tan \beta = \frac{M_{2,n}}{M_{2,t}} = \frac{0.73971}{2.3691} = 0.3122$$

$$\beta = 17.3°$$

Therefore the deflection angle $\delta = 10.5°$.

The preceding example means that if air at a Mach number of 3.0 strikes an object that is at an angle of 10.5° to the direction of flow, the oblique shock will be at an angle of 27.8° with respect to the direction of flow, the total pressure ratio across the shock will be 0.958, the Mach number leaving the shock will be 2.48 and the air will be deflected 10.5°. This is shown in Figure 3.10.

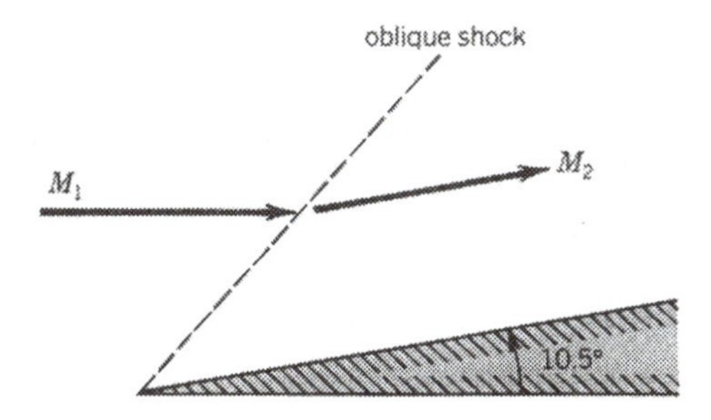

Figure 3.10. Oblique shock wave for Example Problem 3.4.

SUGGESTED READING

Hansen, A.G., *Fluid Mechanics,* John Wiley & Sons, New York, 1967.
Liepmann, H. W., and Roshko, A., *Elements of Gas Dynamics*, John Wiley & Sons, New York, 1964.
Owczarek, J. A., *Fundamentals of Gas Dynamics*, International Textbook Co., Scranton Pa., 1964.
Owczarek, J. A., *Introduction to Fluid Mechanics*, International Textbook Co., Scranton, Pa., 1968.
Potter, M. C., and Foss, J. F., *Fluid Mechanics*, The Ronald Press Co., New York, 1975.
Shames, I. H., *Mechanics of Fluids*, 2nd ed., McGraw-Hill Book Company, New York, 1982.
Shapiro, A. H., *The Dynamics and Thermodynamics of Compressible Fluid Flow*, The Ronald Press Co., New York, 1953.

NOMENCLATURE

A	area
c	sonic velocity
c_p	constant pressure specific heat
$\bar{c}_p$	molar constant pressure specific heat
$\mathbf{F}_R$	resultant force
g	acceleration of gravity
h	specific enthalpy
$\bar{h}$	molar specific enthalpy
$\mathbf{H}$	moment of momentum
k	specific heat ratio
M	total mass, Mach number
$\bar{M}$	formula (molecular) weight
$\dot{m}$	mass rate of flow
N	arbitrary extensive property
p	pressure
$\mathbf{P}$	total linear momentum
q	specific heat transfer
$\bar{q}$	molar heat transfer
$\mathbf{r}$	position vector
R	gas constant
t	time
T	temperature
$\mathbf{T}_R$	resultant torque
V	velocity
v	specific volume
w	specific work
x	state upstream normal shock
y	state downstream normal shock
η	efficiency
ρ	density

Superscript

*	sonic values

Subscript

0	stagnation state
1,2	states
N	nozzle
t	tangential

PROBLEMS

3.1. Solve Example Problem 3.1 assuming that the working fluid is air.

3.2. Carbon dioxide, CO_2, is flowing steadily in a pipe at the rate of 100 lb/s (45.3 kg/s). If the pressure is 70 psia (482.4 kPa), the temperature 700°R (389 K), and the internal diameter of the pipe is 8 in. (20 cm), calculate, assuming constant specific heats:

(a) The stagnation temperature

(b) The stagnation pressure

(c) The sonic velocity

(d) The Mach number

3.3. Oxygen, O_2, enters a passageway at a pressure of 20 psia and a temperature of 40°F with a velocity of 300 ft/s (state 1). It leaves the passageway at a pressure of 10 psia and with a velocity of 1100 ft/s (state 2). Between the entrance and exit, heat is transferred to the oxygen in the amount of 100 Btu/lb. For this process:

(a) Draw a temperature–entropy diagram showing the entrance and exit static and stagnation states.

(b) Find the change in the stagnation temperature between the entrance and exit states.

(c) Find the change in the stagnation pressure between the entrance and exit states.

3.4. Solve Problem 3.3 assuming the following values:

	State 1	*State 2*
Pressure (kPa)	140	69
Temperature (K)	280	
Velocity (m/s)	90	335
Heat transfer = 230,000 J/kg		

3.5. Solve Example Problem 3.3 assuming variable specific heats.

3.6. Nitrogen, N_2 (an ideal gas), enters a nozzle (see diagram) at a pressure of 150 psia (1034 kPa), a temperature of 1500°R (833 K), and a velocity of 600 ft/s (183 m/s). It flows adiabatically through the nozzle to a pressure of 15 psia (103.3 kPa). Calculate the following, assuming that the velocity at the nozzle

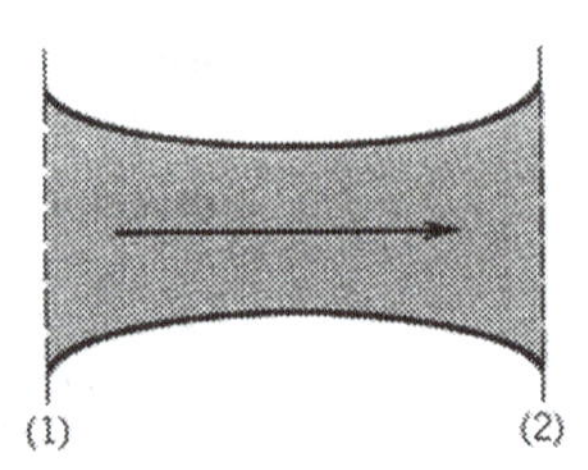

exit (state 2) is 90% of the velocity that would have existed if the flow had been adiabatic and reversible, the flow is steady, and a constant specific heat based on the value given in Table 2.4.

(a) The Mach number at the nozzle inlet (state 1).

(b) The velocity at the nozzle outlet (state 2).

(c) The temperature at the nozzle outlet (state 2).

(d) The Mach number at the nozzle exit (state 2).

(e) The change in entropy between the inlet (state 1) and the outlet (state 2).

(f) The change in total (stagnation) temperature between the inlet and the outlet.

(g) The change in total pressure between the inlet and the outlet.

(h) The mass rate of flow if the inlet area (state 1) is 10 in.^{2} (64.5 cm^2).

(i) The necessary outlet area (state 2) based on the mass rate of flow found in part (h).

(j) Draw the enthalpy—entropy diagram (also a temperature–entropy diagram for an ideal gas) showing all static and stagnation inlet and outlet states.

3.7. Verify Eq. (3.33).

3.8. Verify Eq. (3.38).

3.9. Verify Eq. (3.39).

3.10. Write a computer program that will calculate the terms contained in Table 3.1. Then construct a table when the increment of M_x is 0.01 instead of 0.1 as used in Table 3.1.

3.11. Air (k = 1.4, c_p = 1.008 kJ/kg K) enters a normal shock with a velocity of 590 m/s, a pressure of 8.120 kPa, and a temperature of 217 K. Calculate the total pressure and Mach number at the inlet to the shock, and the total pressure, static pressure, velocity, Mach number, and static temperature after the shock.

3.12. Solve Problem 3.11 assuming that the inlet pressure is 5.924 kPa, the temperature is 217 K, and the velocity is 886 m/s.

3.13. Air (k = 1.4, c_p = 0.241 Btu/lb°R) enters a normal shock with a velocity of 2420 ft/s, a pressure of 2.139 psia, and a temperature of 390°R. Calculate the total pressure, total temperature, and Mach number at the inlet to the shock, and the total pressure, static pressure, velocity, Mach number, and static temperature after the shock.

3.14. Air (k = 1.4) enters an oblique shock at a Mach number of 2.0 at an angle (θ) of 33.4°. Calculate the Mach number leaving the oblique shock, the total pressure ratio, static pressure ratio, and static temperature ratio across the shock, and the degrees the streamline is turned (δ) as it passes through the shock.

3.15. Solve Problem 3.14 if the Mach number is 2.0 and θ = 44.4°.

3.16. Solve Problem 3.14 if the Mach number is 3.0 and θ = 27.6°.

3.17. Solve Problem 3.14 if the Mach number is 2.0 and θ = 29.6°.

COMBUSTION

Combustion is the chemical combination of a substance with certain elements, usually oxygen, accompanied by the production of a high temperature or transfer of heat. The fuels used in gas turbines are usually hydrocarbons, the general formula being C_xH_y.

4.1 COMPOSITION OF DRY AIR

The oxidizer used in gas turbines is atmospheric air. Table 4.1 tabulates the U.S. Standard Atmosphere, 1962 (1), which is a dry analysis and assumed to have the given composition to altitudes of 90 km.

The composition given in Table 4.1 gives an apparent molecular weight of 28.9644 for dry air.

Table 4.2 gives the composition used in the development of a table of properties for dry air that is included as Appendix B.

The analysis given in Table 4.2 gives an apparent molecular weight for dry air of 28.965 and a gas constant for dry air of 53.35 ft-lb$_f$/lb°R. The analysis given in Table 4.2 and the tables listed in Appendix B will be used whenever possible.

4.2 COMPLETE COMBUSTION

When combustion is complete, the basic equations for combustion of carbon and hydrogen with oxygen are

$$C + O_2 \rightarrow CO_2 \tag{4.1}$$

$$H_2 + \tfrac{1}{2}O_2 \rightarrow H_2O \tag{4.2}$$

For the complete combustion of a hydrocarbon with oxygen, the general reaction is

TABLE 4.1. **U.S. Standard Atmosphere, 1962 (1)**

Constituent Gas (Formula)		Content (% by volume)	Content Variable Relative to Its Normal	Molecular Weight (^{12}C = 12.0000)
Nitrogen (N_2)		78.084	—	28.0134
Oxygen (O_2)		20.9476	—	31.9988
Argon (Ar)		0.934	—	39.948
Carbon dioxide (CO_2)		0.0314	[a]	44.00995
Neon (Ne)		0.001818	—	20.183
Helium (He)		0.000524	—	4.0026
Krypton (Kr)		0.000114	—	83.80
Xenon (Xe)		0.0000087	—	131.30
Hydrogen (H_2)		0.00005	?	2.01594
Methane (CH_4)		0.0002	[a]	16.04303
Nitrous oxide (N_2O)		0.00005	—	44.0128
Ozone (O_3)	Summer:	0 to 0.000007	[a]	47.9982
	Winter:	0 to 0.000002	[a]	47.9982
Sulfur dioxide (SO_2)		0 to 0.0001	[a]	64.0628
Nitrogen dioxide (NO_2)		0 to 0.000002	[a]	46.0055
Ammonia (NH_3)		0 to trace	[a]	17.03061
Carbon monoxide (CO)		0 to trace	[a]	28.01055
Iodine (I_2)		0 to 0.000001	[a]	253.8088

[a]The content of these gases may undergo significant variations from time to time or from place to place relative to the normal indicated for the gases.

$$C_xH_y + \left(x + \frac{y}{4}\right) O_2 \rightarrow xCO_2 + \frac{y}{2}H_2O \tag{4.3}$$

The amount of dry air theoretically required for the reaction given in Eq. (4.3) is

$$\frac{\text{moles dry air required}}{\text{moles } C_xH_y} = \frac{x + y/4}{0.2095} \tag{4.4}$$

The percent excess dry air supplied is defined as the difference between the amount supplied (S) and the amount theoretically required (R') divided by the amount theoretically required (R') or, in equation form,

$$\% \text{ excess air} = \left(\frac{S - R'}{R'}\right) \times 100 \tag{4.5}$$

Fundamental to a discussion of combustion problems are the terms enthalpy of formation, enthalpy of combustion, and enthalpy of reaction.

TABLE 4.2. **Dry Air Analysis Used for Table in Appendix B**

Constituent	Molecular Weight	Percent by Volume
Nitrogen (N_2)	28.0134	78.09
Oxygen (O_2)	31.9988	20.95
Argon (Ar)	39.948	0.93
Carbon dioxide (CO_2)	44.0100	0.03

Usually, only relative values of enthalpy ($H = U + pV$) are defined because only relative values of U are defined. With combustion calculations, it becomes necessary to tie together the U values from various tables when the reference state has been chosen arbitrarily.

Although the use of absolute energies is out of the question, it is still possible to assign zero enthalpy to suitable reference states by convention. This is very useful, because it permits the assignment and tabulation of a standard enthalpy for each substance. These standard enthalpies are useful in many calculations.

The first choice of suitable reference state might appear to be to assign a zero value of energy to the isolated atoms of all elements at absolute zero. The enthalpy of any substance would then be the change in ($U + pV$) incurred when the substance is formed from free atoms plus the heat added or removed to reach the temperature desired. This is very impractical, however, since the reference state of absolute zero is not readily accessible by experiment, therefore making it difficult (if not impossible) to measure the needed quantities with any degree of accuracy.

The convention actually chosen fulfills the experimental requirement that a reference state be realized easily and reproducibly.

A *substance* is in its *standard state* when it is under a pressure of 1 atm and at a temperature of 298.15 K (25°C, 77°F).

The *reference states* for tabulated enthalpies are the chemical *elements* in their most stable states at *standard conditions.* Under these conditions, a value of *zero enthalpy* is *assigned* to each *element.*

4.3 ENTHALPY OF FORMATION

To all *compounds* are assigned *standard molal enthalpies of formation* values, ΔH_f, which represent the enthalpy change, positive or negative, when 1 mol of the compound is formed at standard conditions from the elements in their most stable form. This is commonly called the constant-pressure heat of formation or enthalpy of formation. Values of the enthalpies of formation for the substances that will be of interest to us in our study of gas turbines are given in Table 4.3.

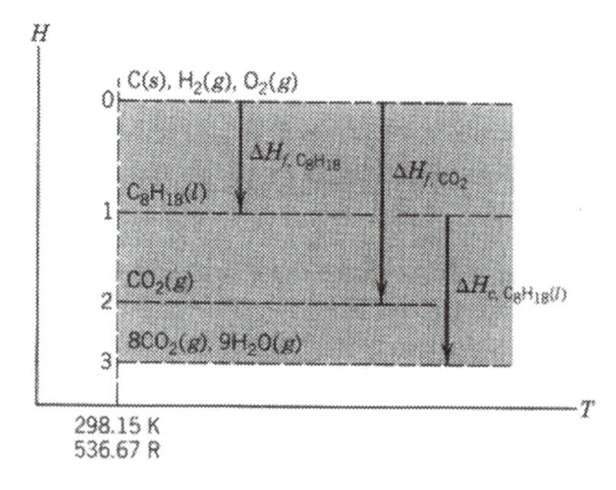

Figure 4.1. Enthalpy–temperature diagram used in the evaluation of the enthalpy of combustion.

TABLE 4.3. **Constant-Pressure Heats of Formation of Various Substances at 298.15 K/536.67°R and Low Pressure**[a]

Compound	Formula	Formula Weight	State	Constant-Pressure Heat of Formation, ΔH_f (Btu/lb-mol)	(J/mol)
Argon	Ar	39.948	Gas	0	0
Carbon (graphite)	C	12.0112	Solid	0	0
Carbon monoxide	CO	28.0106	Gas	−47,517	−110,525
Carbon dioxide	CO_2	44.0100	Gas	−169,179	−393,510
Methane	CH_4	16.0430	Gas	−32,162	−74,809
Ethane	C_2H_6	32.0701	Gas	−36,408	−84,685
Propane	C_3H_8	44.0976	Gas	−44,647	−103,849
n-Butane	C_4H_{10}	58.1248	Liquid	−63,480	−147,655
n-Heptane	C_7H_{16}	100.2064	Liquid	−96,471	−224,392
n-Octane	C_8H_{18}	114.2336	Liquid	−107,462	−249,957
Hydrogen atom	H	1.0080	Gas	93,708	217,965
Hydrogen	H_2	2.0159	Gas	0	0
Water	H_2O	18.0153	Gas	−103,963	−241,818
Helium	He	4.0026	Gas	0	0
Nitrogen atom	N	14.0067	Gas	203,226	472,704
Nitrogen	N_2	28.0134	Gas	0	0
Nitric oxide	NO	30.0061	Gas	38,800	90,249
Nitrogen dioxide	NO_2	46.0055	Gas	14,264	33,178
Ammonia	NH_3	17.0306	Gas	−19,823	−46,108
Oxygen atom	O	15.9994	Gas	107,124	249,170
Oxygen	O_2	31.9988	Gas	0	0
Hydroxyl	OH	17.0074	Gas	16,747	38,954

[a]Values from *Selected Values of Chemical Thermodynamic Properties*, National Bureau of Standards Technical Note 270-3, 1968, and F. D. Rossini, *Selected Values of Physical and Thermodynamic Properties of Hydrocarbons and Related Compounds*, Carnegie Press, Pittsburgh, Pa., 1953.

Referring to Figure 4.1, the heats of formation of C_8H_{18} (liquid) and CO_2 are, respectively (values from Table 4.3),

$$\Delta H_{f,C_8H_{18}} = H_1 - H_0 = -107{,}462 \text{ Btu/lb-mol} = -249{,}957 \text{ J/mol}$$

$$\Delta H_{f,CO_2} = H_2 - H_0 = -169{,}179 \text{ Btu/lb-mol} = -393{,}510 \text{ J/mol}$$

Note that the value for CO_2 is almost identical to the value given in Table A.3 at 298.15 K (536.67°R). This holds true because all of the values in Appendix A have as their reference state the standard state.

4.4 ENTHALPY OF COMBUSTION

The enthalpy of combustion (constant-pressure heat of combustion), ΔH_c, is defined as the heat transferred when 1 mol of a substance is completely burned at constant pressure, the products being cooled to the initial temperature of the fuel. Referring to Figure 4.1, the enthalpy of combustion of carbon, C, is (values from Table 4.3)

$$\Delta H_{c,C} = H_2 - H_0 = -169{,}179 \text{ Btu/lb-mol}$$

The enthalpy of combustion of C_8H_{18} (liquid) with all H_2O in the products a vapor is, referring to Figure 4.1 and Eq. (4.6),

$$C_8H_{18}(l) + 12\tfrac{1}{2}O_2 \rightarrow 8CO_2 + 9H_2O(g) \tag{4.6}$$

$$\Delta H_{c,C_8H_{18}}(l) = H_3 - H_1 \tag{4.7}$$

$$= (H_3 - H_0) - (H_1 - H_0) \tag{4.7a}$$

$$= \Sigma(n_i\,\Delta H_{f,i})_{pr} - \Sigma(n_i\,\Delta H_{f,i})_{re} \tag{4.7b}$$

$$= (8)(-169{,}179) + (9)(-103{,}963) - (1)(-107{,}462)$$

$$= -2{,}181{,}637 \text{ Btu/lb-mol } C_8H_{18}$$

$$= -19{,}098 \text{ Btu/lb}$$

If all H_2O in the products is a liquid, then the heat of combustion for the reaction

$$C_8H_{18}(l) + 12\tfrac{1}{2}O_2(g) \rightarrow 8CO_2(g) + 9H_2O(l)$$

is

$$\Delta H_c, C_8H_{18} = (8)(-169{,}179) + (9)(-122{,}885) - (1)(-107{,}462)$$

$$= -2{,}351{,}935 \text{ Btu/lb-mol } C_8H_{18}$$

$$= -20{,}589 \text{ Btu/lb}$$

The enthalpy of combustion can be determined for any fuel by solving Eq. (4.7b) for the fuel under consideration.

4.5 HIGHER AND LOWER HEATING VALUES

Two useful terms are the higher and lower heating values of a fuel. These values are usually expressed on a unit mass basis.

The higher heating value (HHV) of a fuel is when complete combustion occurs, the temperature of the reactants and products are the same, and all water formed from combustion is a liquid. It is the absolute value of the heat of combustion or, for $C_8H_{18}(l)$ at 77°F (25°C),

$$\text{HHV} = 20{,}589 \text{ Btu/lb } (47{,}890 \text{ kJ/kg})$$

The lower heating value (LHV) occurs when all water formed by combustion is a vapor or

$$\text{LHV} = \text{HHV} - m_{H_2O}h_{fg}$$

For $C_8H_{18}(l)$, the lower heating value at 77°F (25°C) is in English units

$$\text{LHV} = 20{,}589 - \frac{(9)(18)}{114.336}(1050.0)$$

$$= 19{,}098 \text{ Btu/lb } (44{,}422 \text{ kJ/kg})$$

TABLE 4.4. **Higher and Lower Heating Values for Various Substances at 77°F (25°C)**

			LHV		HHV	
Compound	Formula	State	Btu/lb	kJ/kg	Btu/lb	kJ/kg
Methane	CH_4	Gas	21,501	50,012	23,860	55,499
Ethane	C_2H_6	Gas	19,141	44,521	20,911	48,638
Propane	C_3H_8	Liquid	19,927	46,351	21,644	50,343
n-Butane	C_4H_{10}	Liquid	19,493	45,342	21,121	49,128
n-Heptane	C_7H_{12}	Liquid	19,155	44,555	20,666	48,069
n-Octane	C_8H_{18}	Liquid	19,098	44,422	20,589	47,890

Higher and lower heating values for several substances are tabulated in Table 4.4. It is of interest to note that on a unit mass basis, the HHV and LHV values are approximately the same for all substances listed.

4.6 ENTHALPY OF REACTION

The enthalpy of reaction, ΔH_r, is defined as the heat transferred for a specific chemical reaction when the reactants and products are at the same temperature and the pressure remains constant.

For the reaction

$$C_8H_{18}(l) + 12\tfrac{1}{2}O_2 \rightarrow 7CO_2 + CO + 9H_2O(g) + \tfrac{1}{2}O_2 \tag{4.8}$$

ΔH_r becomes, referring to Figure 4.2,

$$\Delta H_r = H_3 - H_1 = (H_3 - H_0) - (H_1 - H_0) \tag{4.9}$$

$$= \Sigma(n_i\, \Delta H_{f,i})_{\text{pr}} - \Sigma(n_i\, \Delta H_{f,i})_{\text{re}} \tag{4.10}$$

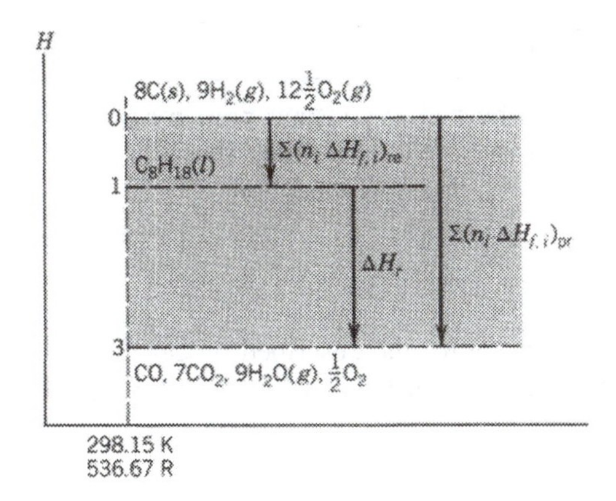

Figure 4.2. Enthalpy–temperature diagram used in the evaluation of the enthalpy of reaction.

Using Eq. (4.10), ΔH_r for the reaction in Eq. (4.8) becomes

$$H_3 - H_0 = (-110{,}525) + (7)(-393{,}510) + (9)(-241{,}818)$$

$$= 5{,}041{,}457 \text{ J/mol } C_8H_{18}$$

$$H_1 - H_0 = -249{,}957 \text{ J/mol } C_8H_{18}$$

$$\Delta H_r = -5{,}041{,}457 - (-249{,}957)$$

$$= -4{,}791{,}500 \text{ J/mol } C_8H_{18}$$

The reader should be able to solve the preceding equations using either English or SI units.

4.7 CONSTANT-PRESSURE ADIABATIC FLAME TEMPERATURE

The constant-pressure adiabatic flame temperature is the temperature that results when a fuel is burned adiabatically during a constant-pressure process. The ideal (maximum) flame temperature occurs when a fuel is burned completely. Dissociation, incomplete combustion, radiation, and excess air all lower the flame temperature.

For an adiabatic, constant-pressure, steady-flow process with no shaft work,

$$Q = 0 = H_{pr} - H_{re} \qquad (4.11)$$

where

H_{pr} = enthalpy of the products
H_{re} = enthalpy of the reactants

The solution for the adiabatic flame temperature is a trial-and-error solution of Eq. (4.11) and is illustrated in Example Problem 4.1.

Example Problem 4.1

Liquid *n*-octane, C_8H_{18}, is burned at constant pressure with 0% excess dry air. The air and fuel are supplied at 537°R. Determine the adiabatic flame temperature for these conditions assuming complete combustion.

Solution

Basis: 1 lb mol C_8H_{18}

Figure 4.3 is an *H–T* diagram for the described process.

For the complete combustion of C_8H_{18} with the theoretical amount of oxygen,

$$C_8H_{18} + 12\tfrac{1}{2}O_2 \rightarrow 8CO_2 + 9H_2O$$

$$\frac{\text{mol D.A. supplied}}{\text{mol } C_8H_{18}} = \frac{12.5}{0.2095} = 59.67$$

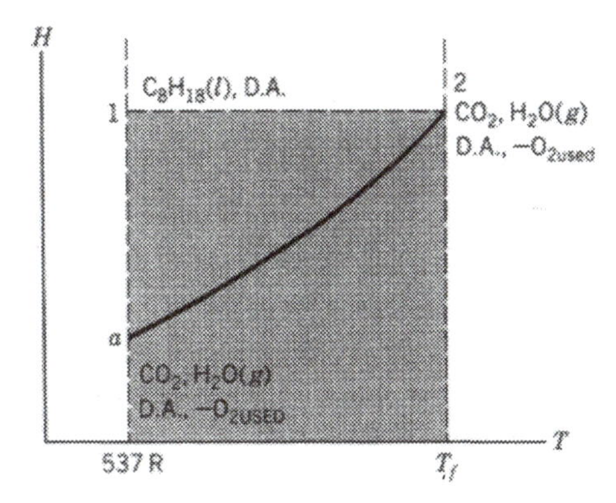

Figure 4.3. H–T diagram.

The reaction for the combustion process described is

$$C_8H_{18} + 59.67\text{D.A.} \rightarrow 8CO_2 + 9H_2O + 59.67\text{D.A.} - 12.5O_2$$

$$H_2 - H_1 = 0 = (H_2 - H_a) + (H_a - H_1) \tag{4.12}$$

where

$$H_2 - H_a = \Sigma[n_i(\bar{h}_{T_f} - \bar{h}_{537})_i]_{pr}$$

$$H_a - H_1 = [\Sigma(n_i\ \Delta H_{f,i})_{pr} - \Sigma(n_i\ \Delta H_{f,i})_{re}]$$

Since all the tables in Appendix A and B use the standard state as their reference state,

$$\Sigma(n_i\bar{h}_{537,i})_{pr} = \Sigma(n_i\ \Delta H_{f,i})_{pr}$$

or the adiabatic flame temperature is the temperature at which

$$Q = 0 = \Sigma(n_i\bar{h}_{T_{f,i}})_{pr} - \Sigma(n_i\ \Delta H_{f,i})_{re} \tag{4.13}$$

The solution to this equation is by trial and error. For an assumed final temperature of 4000°R,

$$Q = [(8)(-124{,}167.3) + (9)(-67{,}745.3) + (59.67)(27{,}772.6)$$
$$+(-12.5)(29{,}100.5)] - [(1)(-107{,}462) + (59.67)(-48.5)]$$
$$= -199{,}255 \text{ Btu/lb-mol } C_8H_{18}$$

Since Q is negative, the assumed final temperature of 4000°R is lower than the actual final temperature. By trial and error, the correct final temperature is determined to be 4308°R.

4.8 GENERAL STEADY-FLOW COMBUSTION PROBLEM

The adiabatic flame temperature problem discussed in the previous section is a special case of the general steady-flow energy problem. Another type of problem frequently encountered is a steady-flow process where work and/or heat transfer are involved.

Solution to a problem of this type involves solution of Eq. (2.6), which, for several streams entering and leaving and neglecting the kinetic energy terms, which are usually negligible, is

$$Q + H_{\text{initial}} = W + H_{\text{final}} \tag{4.14}$$

Two definitions that are very useful when working with combustion problems are the terms fuel–air ratio and percent theoretical air.

The fuel–air ratio (*f/a*) is defined as the ratio of the mass of fuel supplied to a combustion process divided by the mass of air supplied to a combustion process, both on a consistent basis, or, in equation form,

$$\frac{\text{fuel}}{\text{air}} = \frac{f}{a} = f' \left(\frac{\text{mass of fuel supplied}}{\text{mass of air supplied}} \right) \tag{4.15}$$

The percent theoretical air is equal to the percent excess air + 100 or

$$\%\ \text{theor. air} = \%\ \text{excess air} + 100 \tag{4.16}$$

Example Problem 4.2

Liquid *n*-octane, C_8H_{18}, is supplied to a combustion chamber at 298 K (see Figure 4.4). This fuel is burned completely with dry air supplied at a temperature T_2. The product gases leave the combustion chamber at 1400 K. Assuming this to be a steady-flow device, negligible heat loss from the combustion chamber, and negligible entering and leaving velocities:

(a) Derive a general expression for calculating the percent excess air supplied and the fuel–air ratio.
(b) Calculate, for a T_{2a} temperature of 700 K,
 (1) The percent excess air supplied
 (2) The composition of the gases leaving the combustion chamber
 (3) The apparent molecular weight of the product gases leaving the combustion chamber
 (4) The fuel–air ratio.

Solution

Basis: 1 mol fuel

(a) Application of Eq. (4.14) to the conditions given yields

$$Q = 0 = H_{\text{pr}} - H_{\text{re}}$$

where

H_{re} = enthalpy of the reactants
H_{pr} = enthalpy of the products

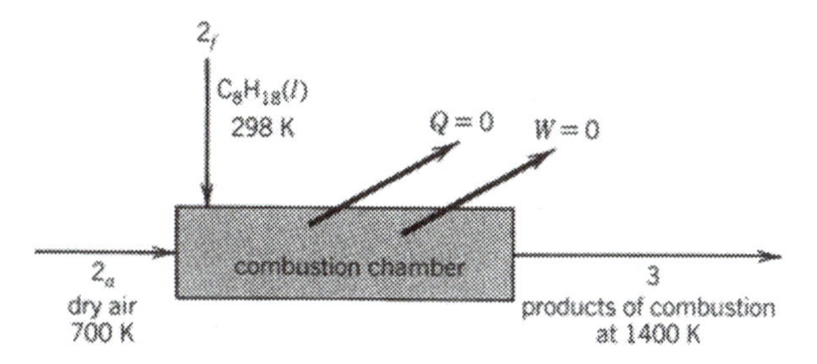

Figure 4.4. Diagram for Example Problem 4.2.

For the complete combustion of n-octane with dry air,

$$C_8H_{18} + 12\tfrac{1}{2}O_2 \rightarrow 8CO_2 + 9H_2O$$

$$\frac{\text{mol dry air required}}{\text{mol } C_8H_{18}} = \frac{12.5}{0.2095} = 59.67$$

Therefore

$$C_8H_{18} + X\text{D.A.} \rightarrow 8CO_2 + 9H_2O + X\text{D.A.} - 12.5\ O_2$$

where X is the total number of moles of dry air supplied per mole of fuel. Therefore

$$H_{pr} = \Sigma(n_i\bar{h}_{1400,i})_{pr} = -4{,}947{,}772 + (28.965)(1212.7)X \qquad (4.17a)$$

$$H_{re} = \Delta H_{f,C_8H_{18}} + X\bar{h}_{air,T_2}$$

$$= -249{,}957 + X\bar{h}_{air,T_2} \qquad (4.17b)$$

Therefore, equating (4.17a) and (4.17b) and solving for X yields

$$X = \frac{4{,}697{,}815}{35{,}118.8 - \bar{h}_{air,T_2}} \qquad (4.18)$$

$$\%\ \text{excess air} = \left(\frac{X - 59.67}{59.67}\right) \times 100 \qquad (4.19)$$

$$f' = \frac{f}{a} = \frac{114.2336}{(X)\,(28.965)} \qquad (4.20)$$

(b) For the special case where $T_2 = 700$ K,

$$X = \frac{4{,}697{,}815}{35{,}118.8 - 11{,}904.8} = 202.37$$

$$\%\ \text{excess air} = \left(\frac{202.39 - 59.67}{59.67}\right) \times 100 = 239.1\%$$

The composition of the gases leaving the combustion chamber is

	n_i	$\overline{M}_i$	$n_i\overline{M}_i$
CO_2	8	44.01	352.1
H_2O	9	18.02	162.2
D.A.	202.37	28.97	5862.7
O_2	−12.5	32.00	−400.0
	206.87		5977.0

$$\overline{M} = \frac{5977.0}{206.9} = 28.89$$

It is worthwhile to note that the apparent molecular weight of the products of combustion leaving the combustion chamber is very close to that of the dry air entering the combustion chamber.

The fuel–air ratio is

$$f' = \frac{f}{a} = \frac{114.2336}{(202.37)(28.97)} = 0.0195$$

This tells us that for every pound of air entering the combustion chamber, 1.020 lb of products leave the combustion chamber.

4.9 DISSOCIATION

The maximum temperature that could ever be considered for a turbine inlet temperature is the adiabatic flame temperature. This means, when consideration is given to the fact that the temperature of the air entering the combustion chamber may be well above 1000°R, that the theoretical adiabatic flame temperature (and therefore the turbine inlet temperature) could be on the order of 4000 to 4500°R (2200 to 2500 K).

Although no gas turbines have been designed with turbine inlet temperatures anywhere near the adiabatic flame temperature, turbine inlet temperatures have been, at least in experimental engines, over 3000°R (1650 K), with the afterburner temperature being several hundred degrees above this value.

Therefore, it is important to determine if dissociation must be considered and, if it is important, when it should be considered. The actual adiabatic flame temperature is lower than the theoretical adiabatic flame temperature because of dissociation. Dissociation has an effect similar to incomplete combustion, the amount of dissociation increasing with an increase in temperature at a given pressure and decreasing at a given temperature with an increase in pressure.

4.10 EQUILIBRIUM CRITERIA

To determine the degree of dissociation in a gas mixture that is at equilibrium, it is necessary to understand equilibrium criteria and the equilibrium constant. This section deals with the criteria of equilibrium, the next section with the equilibrium constant.

A system is said to be in equilibrium if no changes can occur in the state of the system without the aid of an external stimulus.

For a system of constant internal energy and constant volume,

$$ds\Big|_{U,V} \geq 0 \tag{4.21}$$

where the equality holds for the condition of complete equilibrium.

For a system at constant pressure and temperature,

$$dG\Big|_{p,T} \leq 0 \tag{4.22}$$

where the equality once again holds for the condition of complete equilibrium.

Thus, the Gibbs function of any system in complete equilibrium must be a minimum with regard to all states at the same pressure and temperature.

Consider an ideal gas chemical reaction where α mol of gas A react with β mol of gas B to form μ mol of gas M and ν mol of gas N. This is shown in the following equation:

$$\alpha\text{A} + \beta\text{B} \rightarrow \mu\text{M} + \nu\text{N} \tag{4.23}$$

The α, β, μ, and ν values are the stoichiometric coefficients that satisfy the reaction equation and are independent of the amount of constituents present.

By applying the equilibrium criterion of Eq. (4.22) to the ideal gas chemical reaction of Eq. (4.23), it can be shown that

$$\frac{p_{\text{M}}^{\mu}\, p_{\text{N}}^{\nu}}{p_{\text{A}}^{\alpha}\, p_{\text{B}}^{\beta}} = \text{function of temperature} = K_p \tag{4.24}$$

where p_{M}, p_{N}, p_{A}, and p_{B} are the partial pressures of constituents M, N, A, and B, respectively. K_p, as defined by Eq. (4.24), is the equilibrium constant based on partial pressures. It should be noted that the equilibrium constant has the dimension of pressure to the $(\mu + \nu - \alpha - \beta)$ power.

4.11 EQUILIBRIUM CONSTANT

The equilibrium constant, K_p, may be evaluated by the equation

$$\ln K_p = -\frac{(\Delta \bar{g}_T^0)_R}{\bar{R}T} \tag{4.25}$$

TABLE 4.5. **Logarithms of Equilibrium Constants, $\log_{10} K_p$, for Several Gas-Phase Reactions (pressure in atmospheres)**

Temperature (K)	Temperature (°R)	$H_2 + \frac{1}{2}O_2 \rightarrow H_2O$	$CO + \frac{1}{2}O_2 \rightarrow CO_2$	$CO_2 + H_2 \rightarrow CO + H_2O$	$H_2 + O_2 \rightarrow 2OH$	$N_2 + O_2 \rightarrow 2NO$	$2H \rightarrow H_2$	$2O \rightarrow O_2$	$2N \rightarrow N_2$
1500	2700	5.7252	5.3130	0.4122	-1.1226	-4.9808	9.5123	10.7899	26.4530
1600	2880	5.1802	4.7032	0.4770	-0.9603	-4.5864	8.5329	9.6829	24.3687
1700	3060	4.6986	4.1660	0.5326	-0.8177	-4.2384	7.6670	8.7052	22.5284
1800	3240	4.2701	3.6895	0.5807	-0.6914	-3.9291	6.8958	7.8354	20.8915
1900	3420	3.8864	3.2638	0.6226	-0.5789	-3.6524	6.2046	7.0564	19.4260
2000	3600	3.5407	2.8815	0.6592	-0.4779	-3.4034	5.5815	6.3548	18.1062
2100	3780	3.2277	2.5362	0.6915	-0.3870	-3.1781	5.0167	5.7196	16.9115
2200	3960	2.9430	2.2229	0.7201	-0.3046	-2.9735	4.5026	5.1417	15.8247
2300	4140	2.6829	1.9374	0.7455	-0.2297	-2.7866	4.0324	4.6137	14.8319
2400	4320	2.4443	1.6762	0.7682	-0.1613	-2.6155	3.6008	4.1294	13.9214
2500	4500	2.2247	1.4363	0.7885	-0.0987	-2.4581	3.2033	3.6836	13.0832
2600	4680	2.0220	1.2153	0.8067	-0.0411	-2.3129	2.8358	3.2719	12.3091
2700	4860	1.8341	1.0111	0.8231	0.0120	-2.1786	2.4952	2.8905	11.5920
2800	5040	1.6597	0.8218	0.8379	0.0611	-2.0539	2.1785	2.5361	10.9258
2900	5220	1.4971	0.6458	0.8513	0.1066	-1.9380	1.8833	2.2061	10.3052
3000	5400	1.3454	0.4820	0.8635	0.1488	-1.8299	1.6075	1.8979	9.7257
3100	5580	1.2034	0.3289	0.8745	0.1881	-1.7289	1.3493	1.6095	9.1833
3200	5760	1.0703	0.1857	0.8846	0.2248	-1.6343	1.1070	1.3390	8.6745
3300	5940	0.9451	0.0514	0.8937	0.2591	-1.5456	0.8792	1.0848	8.1964
3400	6120	0.8273	-0.0747	0.9021	0.2911	-1.4622	0.6646	0.8456	7.7461
3500	6300	0.7162	-0.1935	0.9097	0.3212	-1.3837	0.4621	0.6199	7.3213
3600	6480	0.6113	-0.3054	0.9167	0.3494	-1.3097	0.2707	0.4067	6.9199
3700	6660	0.5120	-0.4111	0.9231	0.3759	-1.2397	0.0896	0.2050	6.5399
3800	6840	0.4178	-0.5111	0.9289	0.4008	-1.1736	-0.0822	0.0138	6.1798
3900	7020	0.3285	-0.6058	0.9343	0.4243	-1.1110	-0.2452	-0.1676	5.8379
4000	7200	0.2436	-0.6956	0.9392	0.4464	-1.0515	-0.4002	-0.3399	5.5129
4100	7380	0.1629	-0.7808	0.9437	0.4674	-0.9951	-0.5477	-0.5039	5.2036
4200	7560	0.0859	-0.8619	0.9478	0.4871	-0.9415	-0.6882	-0.6601	4.9088
4300	7740	0.0125	-0.9390	0.9515	0.5058	-0.8905	-0.8223	-0.8090	4.6276
4400	7920	-0.0576	-1.0126	0.9550	0.5235	-0.8419	-0.9503	-0.9512	4.3589

The term $(\Delta \bar{g}_T^0)_R$ is the change in the Gibbs free energy at 1 atm for the given reaction. Equation (4.25) is sometimes written as

$$\log_{10} K_p = \frac{-(\Delta \bar{g}_T^0)_R}{(2.3025851)\bar{R}T} \tag{4.25a}$$

Note that one may calculate numerical values of $\log_{10} K_p$ for a given reaction using Eq. (4.25) or (4.25a) and the data in Appendix A.

Table 4.5 lists K_p data for several ideal gas reactions that may be of interest in gas turbine calculations. When using the data from Table 4.5, it must be remembered that the partial pressures must be in atmospheres.

The following example problem illustrates how these values were calculated.

Example Problem 4.3

Calculate, for the reaction

$$CO_2 + H_2 \rightarrow CO + H_2O$$

the value of $\log_{10} K_p$ at 2000 K.

Solution

From Eq. (4.25a),

$$\log_{10} K_p = -\frac{[(\bar{g}_T^0)_{CO} + (\bar{g}_T^0)_{H_2O} - (\bar{g}_T^0)_{CO_2} - (\bar{g}_T^0)_{H_2}]}{(2.3025851)(\bar{R})(T)}$$

$$= -\frac{[(-570976.9) + (-698305.9) - (-920351.3) - (-323690.1)]}{(2.3025851)(8.31434)(2000)}$$

$$= 0.6592$$

Example Problem 4.4

Liquid *n*-octane, C_8H_{18}, is burned with the stoichiometric amount of dry air in a steady-flow process. The *n*-octane and dry air are both supplied at 537°R. The product gases leave the steady-flow, constant-pressure device at 3960°R. Assuming that the product gases are at equilibrium and contain only Ar, CO, CO_2, H_2, H_2O, O_2, and N_2, negligible inlet and exit kinetic energies, and no shaft work, calculate:

(a) The composition of the product gases if the pressure of the reactants and products is 1 atm.
(b) The heat transfer, Btu/lb-mol C_8H_{18}.
(c) *Estimate* whether H, O, OH, N, and/or NO should have been considered.

Solution

Basis: 1 lb-mol C_8H_{18}

$$C_8H_{18} + 12.5\ O_2 \rightarrow 8\ CO_2 + 9\ H_2O$$

$$\text{lb mol } N_2 \text{ in D.A.} = \frac{(12.5)(0.7809)}{0.2095} = 46.593$$

$$\text{lb mol Ar in D.A.} = \frac{(12.5)(0.0096)}{0.2095} = 0.573$$

(a) Based on the conservation of elements, the equilibrium composition of the products may be expressed as follows:

	n_i
CO_2	$8 - x$
CO	x
H_2O	$9 - y$
H_2	y
O_2	$(x + y)/2$
Ar	0.57
N_2	46.59
	$n_{\text{total}} = \dfrac{128.32 + x + y}{2}$

Two equilibrium constants are needed. Use

(1) $CO + \frac{1}{2}O_2 \rightarrow CO_2$

$$K_{p1} = \frac{P_{CO_2}}{P_{CO}\sqrt{P_{O_2}}} = \frac{n_{CO_2}}{n_{CO}\sqrt{\dfrac{n_{O_2}}{n}P}}$$

$$= \frac{8 - x}{x[(x + y)/(128.32 + x + y)P]^{1/2}}$$

(2) $CO_2 + H_2 \rightarrow CO + H_2O$

$$K_{P_2} = \frac{P_{CO}P_{H_2O}}{P_{CO_2}P_{H_2}} = \frac{(x)(9 - y)}{(y)(8 - x)}$$

Solving the two K_p equations at 3960°R (where $K_{p1} = 167.07$ and $K_{p2} = 5.2493$) and 1 atm yields

$$x = 0.591$$

$$y = 0.135$$

(b) $Q = H_{pr} - H_{re} = \Sigma(n_i\bar{h}_{3960,i})_{pr} - \Sigma(n_i\bar{h}_{537,i})_{re}$

$= -137{,}400$ Btu/lb-mol C_8H_{18}

(c) To *estimate* the moles of N in an equilibrium mixture at 3960°R, assume the same products as in part (a) plus N and assume the composition determine in (a). Thus

$2N \rightarrow N_2$

$$K_p = 6.6788 \times 10^{15} = \frac{P_{N_2}}{(P_N)^2} = \frac{n_{N_2}n}{(n_N)^2 p}$$

TABLE 4.6. **Equilibrium Composition at 3960°R When Liquid *n*-Octane Is Burned with the Theoretical Amount of Dry Air**

	n_i		n_i		n_i
Ar	0.573	H_2	0.148	NO	0.121
CO	0.650	H_2O	8.770	O	0.012
CO_2	7.350	N	0.000	O_2	0.296
H	0.017	N_2	6.533	OH	0.148

For which

$$n_N \cong 6.7 \times 10^{-7}$$

In a similar manner, it can be shown that

$$n_H \cong 0.0165$$

$$n_O \cong 0.013$$

$$n_{OH} \cong 0.156$$

$$n_{NO} \cong 0.134$$

If the problem had been solved assuming an equilibrium composition containing Ar, CO, CO_2, H, H_2, H_2O, N, N_2, NO, O, O_2, and OH, seven equilibrium constants would have been required. The results for this equilibrium composition at 3960°R and 1 atm are shown in Table 4.6.

One should observe that

1. The equilibrium adiabatic flame temperature for the conditions of Example Problem 4.4 is higher than 3960°R.
2. Pressure has an effect on the equilibrium composition, the amount of dissociation decreasing with increasing pressure at a given temperature.
3. Increasing the number of constituents considered in the production increases the complexity of the solution.
4. OH and NO concentrations are of the same magnitude as H_2, H and O concentrations are an order of magnitude lower than H_2, N concentration is several orders of magnitude lower than that of H_2.
5. The theoretical and equilibrium adiabatic flame temperatures increase as the temperature at which the dry air is supplied increases, decrease when excess dry air is supplied.

Solution to the general equilibrium problem involves the simultaneous solution of several equilibrium constant equations and is normally done on high-speed digital computers. One technique is described in Penner (2).

Results when liquid *n*-octane is burned with dry air are tabulated in Table 4.7 and show the effect of pressure, percent excess dry air, and air supply temperature. The equilibrium composition was assumed to contain the constituents listed in Table 4.6

TABLE 4.7. **Effect of Pressure, Air Supply Temperature, and Percent Excess Dry Air When Liquid *n*-Octane Is Burned Adiabatically and at Constant Pressure with Dry Air**

Pressure (atm)	1	10	20	10	10	10	10	10
Air supply temp. (°R)	537	537	537	1000	1000	1000	1000	1000
percent excess air supplied	0	0	0	0	10	20	30	40
Adiabatic flame temperature (°R)								
Complete combustion	4308	4308	4308	4610	4354	4132	3939	3768
Equilibrium composition	4076	4171	4194	4390	4252	4072	3898	3738
Equilibrium product composition								
Total mol/mol fuel	64.795	64.530	64.468	64.755	70.402	76.241	82.141	88.086
Mol CO/mol fuel	0.853	0.525	0.448	0.825	0.315	0.130	0.057	0.026
Mol H_2/mol fuel	0.190	0.110	0.093	0.168	0.063	0.027	0.012	0.006
Mol OH/mol fuel	0.202	0.120	0.101	0.203	0.221	0.185	0.143	0.107
Mol NO/mol fuel	0.159	0.133	0.123	0.210	0.358	0.424	0.439	0.429

REFERENCES CITED

1. *U.S. Standard Atmosphere, 1962,* U.S. Government Printing Office, Washington, D.C., 1962.
2. Penner, S.S., *Thermodynamics,* Addison-Wesley Publishing Company, Reading, Mass., 1968.

SUGGESTED READING

Denbigh, K., *The Principles of Chemical Equilibrium,* 3rd ed., Cambridge University Press, Cambridge, 1971.

Jones, J. B., and Hawkins, G. A., *Engineering Thermodynamics,* John Wiley & Sons, New York, 1960.

Penner, S. S., *Chemistry Problems in Jet Propulsion,* Pergamon Press, New York, 1957.

Wark, K., *Thermodynamics,* 4th ed., McGraw-Hill Book Company, New York, 1983.

Wark, K., and Warner, C. F., *Air Pollution: Its Origin and Control,* IEP, A Dun-Donnelley Publisher, New York, 1976.

NOMENCLATURE

a air
f fuel
f' fuel–air ratio
H total enthalpy
K_P equilibrium constant
$\overline{M}$ molecular (formula) weight
p pressure
Q heat transfer
R' air theoretically required
$\overline{R}$ universal gas constant
S air actually supplied
T temperature
U total internal energy
W total work

Subscripts

c combustion
f formation
i component
pr products
R reaction
re reactants

PROBLEMS

4.1. Liquid n-octane, C_8H_{18}, is supplied to a steady-flow device at 537°R. It is burned completely with an unknown amount of excess dry air that is supplied at 147 psia, 1000°R. The products of combustion leave the steady-flow device at 3500°R, 56 psia. If the work done by the system is 115.71 Btu for every pound of air *entering* and the heat transfer from the system is 3% of the lower heating value of the fuel, calculate the percent excess dry air supplied. Assume negligible entering and leaving velocities. The lower heating value is defined as the amount of heat transferred when a unit quantity of fuel is burned completely, the products of combustion are reduced to the original temperature, and all water in the products is in the vapor phase.

4.2. Liquid n-octane, C_8H_{18}, is burned at constant pressure with 20% excess dry air. The air and fuel are supplied at 298 K. Determine the adiabatic flame temperature for these conditions assuming complete combustion.

4.3. Solve Example Problem 4.1 assuming that the dry air is supplied at 560 K.

4.4. Solve Problem 4.2 assuming that the dry air is supplied at 560 K.

4.5. Liquid n-octane, C_8H_{18}, is burned at constant pressure with 10% excess dry air supplied at 1000°R. Assuming that the fuel is supplied at 537°R, determine, the adiabatic flame temperature assuming complete combustion.

4.6. Solve Problem 4.5 assuming that the dry air is supplied at 537°R.

4.7. Solve Example Problem 4.2 for the percent excess air supplied and the fuel–air ratio for a T_2 of 300 to 1000 K in 100-degree increments and for the product gases leaving the combustion chamber at 1400 K. Then, plot air–fuel ratio and percent excess air supplied versus T_2.

4.8. Solve Example Problem 4.2 if the temperature of the product gas is 1200 K (2160°R) instead of 1400 K (2520°R).

4.9. Solve Problem 4.7 if the temperature of the product gases leaving the combustion chamber is 2160°R.

4.10. Consider the steady-flow device shown in the diagram. The fuel, ammonia, NH_3, is supplied to the device at 1 atm and 537°R (298 K). It is completely burned with an unknown amount of dry air supplied at 537°R (298 K). The products of combustion leave the device at 1 atm and 1260°R (700 K). The

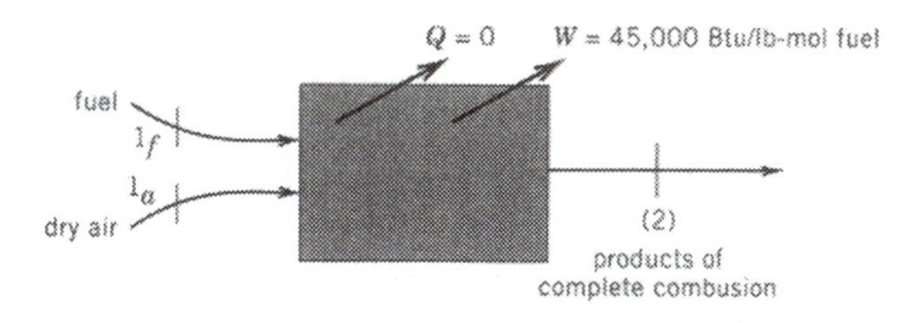

work done by the system is 45,000 Btu/lb-mol fuel. No heat is transferred to or from the system. Calculate the fuel–air ratio, that is, pounds of fuel per pound of air. For the complete combustion of ammonia with oxygen, the products are water and nitrogen or, in an *unbalanced form,* the reaction is

$$NH_3 + O_2 \rightarrow H_2O + N_2$$

4.11. Verify the value of $\log_{10} K_p$ at 4500°R as given in Table 4.6 for the reaction

$$CO + \tfrac{1}{2}O_2 \rightarrow CO_2$$

4.12. Solve Example Problem 4.4 (parts a and b) assuming that final temperature is 4140°R (2300 K). Then estimate the equilibrium adiabatic flame temperature and compare with the value listed in Table 4.7.

4.13. Liquid *n*-octane, C_8H_{18}, is burned with 20% excess dry air. The fuel is supplied at 537°R, the dry air at 1000°R. Assuming that the products are at 4072°R, that the reaction occurs at a pressure of 10 atm, and that products are in equilibrium and contain only CO_2, H_2O, O_2, N_2, Ar, and NO:

(a) Calculate the product composition.

(b) Compare your answer for NO with the appropriate value given in Table 4.7. What conclusions may be drawn from this simplified analysis when compared with the more detailed analysis used in obtaining the values listed in Table 4.7?

4.14. Table 4.7 shows, for a pressure of 10 atm and an air supply temperature of 1000°R, the variation in adiabatic flame temperature and equilibrium composition as the percent excess air supplied is increased.

(a) Plot adiabatic flame temperature versus percent excess dry air.

(b) Plot nitric oxide (NO), carbon monoxide (CO), hydrogen (H_2), and hydroxyl (OH) concentration versus percent excess dry air.

(c) Why does NO concentration increase with increasing percent excess dry air, whereas CO, H_2 and OH concentrations decrease with increasing percent excess dry air?

4.15. Solve Example Problem 4.2 if the temperature of the product gas is 1800 K (3240°R) instead of 1400 K (2520°R).

4.16. Natural gas, with the volumetric analysis

CH_4	85.6%
C_2H_6	12.3%
CO_2	0.1%
N_2	1.7%
O_2	0.3%
	100.0%

is supplied to a combustion chamber at 298 K (537°R). This fuel is burned completely with dry air supplied at a temperature T_2. The product gases leave the combustion chamber at 1400 K (2520°R). Assuming a steady-flow process, negligible heat transfer from the combustion chamber, and negligible entering and leaving velocities:

(a) Derive a general expression for calculating the percent excess air supplied and the fuel–air ratio.

(b) Calculate, for a T_2 temperature of 700 K (1260°R)

(i) The percent excess air supplied

(ii) The composition of the gases leaving the combustion chamber.

(iii) The apparent molecular weight of the product gases leaving the combustion chamber.

(iv) The fuel–air ratio.

(c) Compare with the answers for Example Problem 4.2.

4.17. Solve Problem 4.16b assuming the product gases leave the combustion chamber at 2160°R (1200 K).

4.18. Solve Problem 4.16b for air supply temperatures of 400 K (720°R), 500 K (900°R), 600 K (1080°R), 700 K (1260°R), and 800 K (1440°R). Plot fuel–air ratio and percent excess air versus air supply temperature T_2.

4.19. Solve Problem 4.16b for air supply temperatures of 720°R (400 K), 900°R (500 K), 1080°R, (600 K), 1260°R (700 K), and 1440°R (800 K). Plot fuel–air ratio and percent excess air versus air supply temperature T_2.

4.20. Air is supplied at the rate of 1 lb/s (1 kg/s) to a combustion chamber at a temperature of 1122°R (623 K) and a pressure of 176.4 psia (1215.6 kPa). Liquid *n*-octane, C_8H_{18}, is supplied to the combustion chamber at 537°R (298 K), 200 psia (1378 kPa). Steam, at a pressure of 200 psia (1378 kPa) and 1175°R (653 K), is supplied to the combustion chamber at the rate of 0.025 lb/s (0.025 kg/s). The products of combustion leave the combustion chamber at a temperature of 2520°R (1400 K) and a pressure of 174 psia (1215.6 kPa). Determine, assuming a steady-flow process, negligible heat transfer from the combustion chamber, and negligible entering and leaving velocities:

(a) The fuel–air ratio, lb fuel/lb air (kg fuel/kg air).

(b) The composition of the product gas.

(c) The apparent molecular weight of the product gas.

4.21. Solve Problem 4.20 assuming the product gas leaves the combustion chamber at 1200 K (2160°R), all other values the same.

4.21. Solve Problem 4.20 for steam flow rates of 0.01 lb/s (0.01 kg/s), 0.02 lb/s (0.02 kg/s), and 0.03 lb/s (0.03 kg/s).

5

SHAFT-POWER GAS TURBINES

A gas turbine is an engine designed to convert the energy of a fuel into some form of useful power, such as mechanical (shaft) power or the high-speed thrust of a jet. A gas turbine is basically comprised of a gas generator section and a power conversion section. The gas generator section, shown in Figure 5.1, consists of a compressor, combustion chamber, and turbine, the turbine extracting only sufficient power to drive the compressor. This results in a high-temperature, high-pressure gas at the turbine exit. The different types of gas turbines result by adding various inlet and exit components to the gas generator.

The various types of gas turbines are examined in this chapter and Chapter 6. This chapter is devoted to a study of the gas turbine cycle for gas turbines used to produce shaft work. Chapter 6 is devoted to a study of gas turbine cycles used for propulsion systems. In both cases, the same gas generator could be used.

5.1 BASIC CYCLE (AIR STANDARD)

The basic (simplest) gas turbine engine is shown in Figure 5.2. The cycle consists of a compressor where air is compressed adiabatically, a combustion chamber where the fuel is burned with the air, resulting in the maximum cycle temperature occurring at state 3. The products of combustion then expand adiabatically in the turbine (or turbines), part of the work developed in the turbine being used to drive the compressor, the remainder being delivered to equipment external to the gas turbine. This basic gas turbine engine results by adding additional turbine wheels (or a separate power turbine) behind the gas generator so that the gases can expand back to (or nearly so)

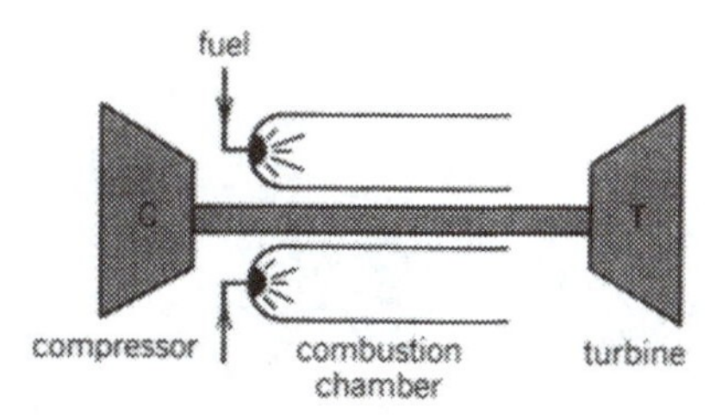

Figure 5.1. Gas generator.

the pressure at the compressor inlet. The basic gas turbine engine illustrated in Figure 5.2 has a separate power turbine (PT).

Figures 5.3 through 5.6 illustrate three different gas turbines that operate on the basic cycle. Figure 5.3 illustrates a gas turbine with an axial-flow compressor and an axial-flow turbine. Figure 5.4 shows the engine illustrated in 5.3 separated into its various components.

Figure 5.5 illustrates a gas turbine with an axial-flow compressor followed by a centrifugal-flow compressor and with an axial-flow turbine. Figure 5.6 illustrates a gas turbine with a two-stage centrifugal-flow compressor and an axial-flow turbine.

These four figures were included to illustrate the various components and to make the reader aware that there are various ways to build the components. A study of the various components is included in Chapters 7, 8, and 9, this chapter being concerned only with the cycle analysis and not with the way a certain component is constructed.

This section is concerned with the air standard cycle. In an air standard cycle, air is assumed to be the working fluid throughout, the combustion chamber being replaced by a heat addition process. The cycle, which may be considered a closed cycle, is completed by a heat rejection process.

The *ideal air standard basic cycle* assumes that the compression and expansion processes are adiabatic and reversible (isentropic), that there is no pressure drop during the heat addition process, and that the pressure leaving the turbine is equal to the pressure entering the compressor. The temperature–entropy (T–s) and pressure–specific volume (p–v) diagrams for the ideal air standard basic cycle are shown in Figure 5.7.

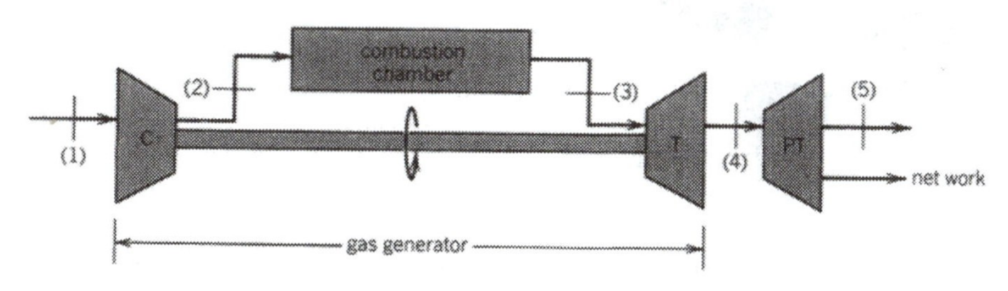

Figure 5.2. Basic gas turbine engine.

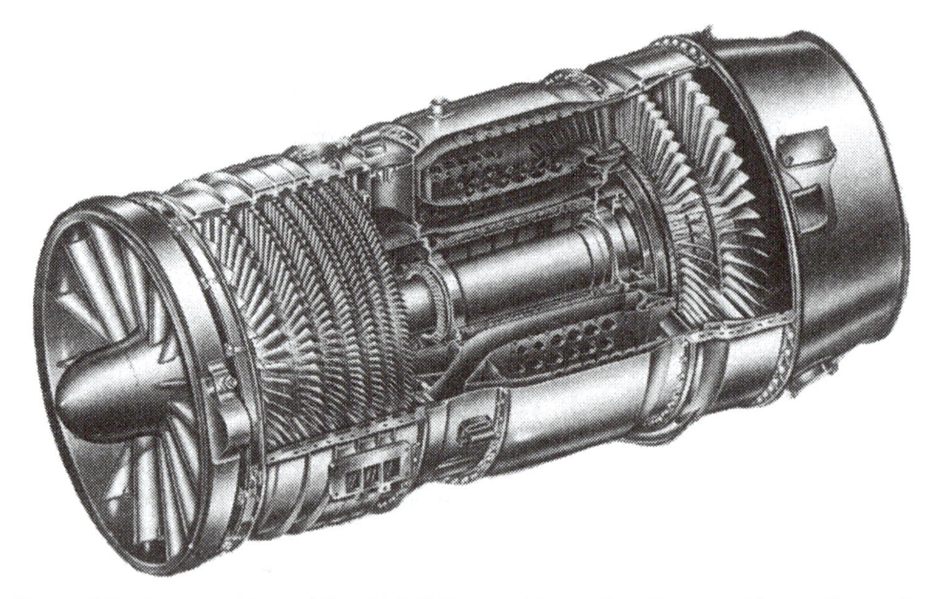

Figure 5.3. Cutaway of General Electric's LM350 gas turbine engine. (Courtesy of General Electric Co.)

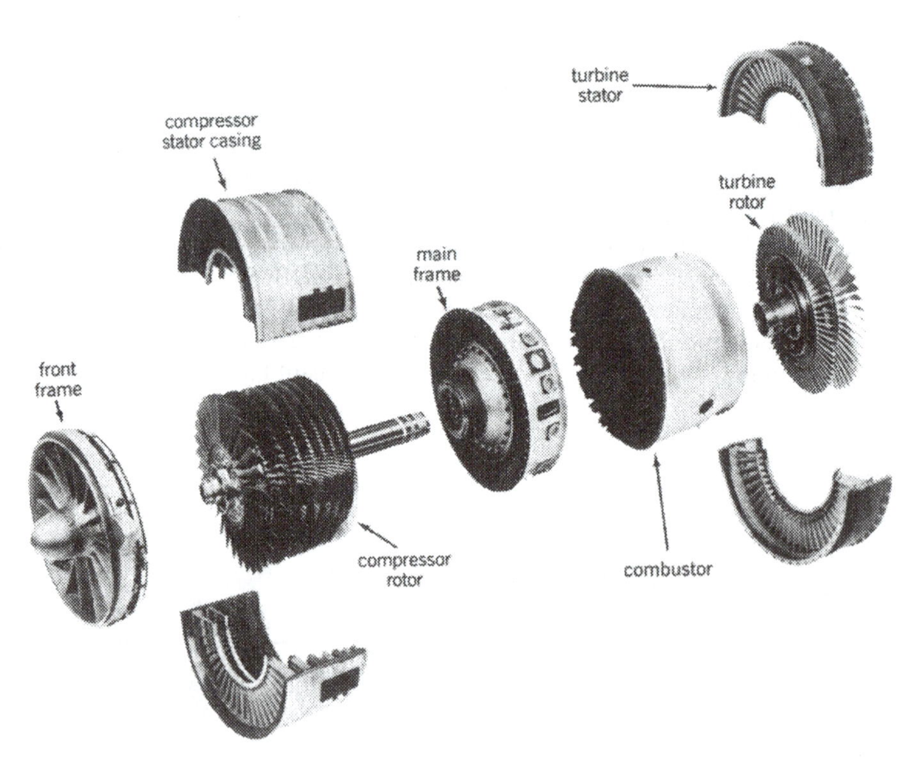

Figure 5.4. General Electric's LM350 gas turbine engine separated into components. (Courtesy of General Electric Co.)

Figure 5.5. Cutaway of AlliedSignal's T55-L-11 gas turbine engine. (Courtesy of AlliedSignal)

Each component of the gas turbine engine operates in a steady-flow manner. For a steady-flow process,

$$ {}_1q_2 + h_1 + \frac{\overline{V}_1^2}{2} = {}_1w_2 + h_2 + \frac{\overline{V}_2^2}{2} $$

Therefore, for an ideal cycle [remembering that the compressor and turbine are assumed to operate adiabatically and reversibly, that the heat addition process is one of zero work, and that the total (stagnation) enthalpy is the sum of the static enthalpy and the kinetic energy], the compressor work, the heat added, the turbine work, and the net work become, respectively,

Figure 5.6. Cutaway of AlliedSignal's IE 990 Industrial gas turbine engine. (Courtesy of AlliedSignal)

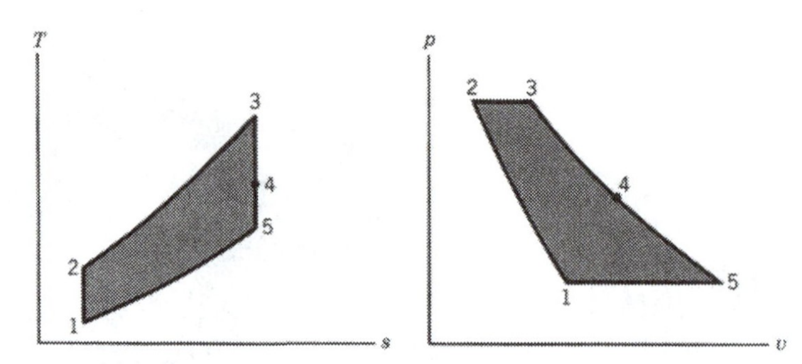

Figure 5.7. Temperature–entropy and pressure–specific volume diagrams for an ideal air standard basic cycle.

$$w_{comp} = |\,{}_1w_2| = h_{o2} - h_{o1} \tag{5.1}$$

$$q_{in} = {}_2q_3 = h_{o3} - h_{o2} \tag{5.2}$$

$$w_{turb} = {}_3w_5 = h_{o3} - h_{o5} \tag{5.3}$$

$$w_{net} = w_{turb} - w_{comp} \tag{5.4}$$

For a heat engine, the cycle thermal efficiency is useful. It was defined by Eq. (2.24) and is equal to

$$\eta_{th} = \frac{w_{net}}{q_{in}} \tag{5.5}$$

First consider the case where it is assumed that the specific heats remain constant. For constant specific heats, Eqs. (5.1) through (5.5) become

$$w_{comp} = c_p(T_{o2} - T_{o1}) \tag{5.1a}$$

$$q_{in} = c_p(T_{o3} - T_{o2}) \tag{5.2a}$$

$$w_{turb} = c_p(T_{o3} - T_{o5}) \tag{5.3a}$$

$$w_{net} = c_p(T_{o3} + T_{o1} - T_{o2} - T_{o5}) \tag{5.4a}$$

$$\eta_{th} = 1 - \frac{T_{o5} - T_{o1}}{T_{o3} - T_{o2}} \tag{5.5a}$$

For the remainder of this chapter, the velocities will be assumed to be negligible. Therefore, the static and stagnation states will be the same and the subscript *o* may be omitted. Remember: In Chapter 6 this assumption will not apply.

For an ideal gas with constant specific heats undergoing an adiabatic and reversible process,

$$\frac{p_2}{p_1} = \left(\frac{T_2}{T_1}\right)^{k/(k-1)} \tag{5.6}$$

Therefore, remembering that the compression and expansion processes are adiabatic and reversible and have the same pressure ratios,

$$\left(\frac{p_2}{p_1}\right)^{(k-1)/k} = \frac{T_2}{T_1} = \left(\frac{p_3}{p_5}\right)^{(k-1)/k} = \frac{T_3}{T_5} \tag{5.7}$$

Combining Eqs. (5.5a) and (5.7) yields

$$\eta_{th} = 1 - \frac{1}{(p_2/p_1)^{(k-1)/k}} \tag{5.8}$$

This equation illustrates the fact that for an ideal cycle, the thermal efficiency increases with increasing pressure ratio. This expression was developed for constant specific heats, but the trend also applies for variable specific heats if operating under ideal conditions, that is, isentropic compression and expansion and no pressure drop during the heat addition or heat rejection processes.

Next consider the pressure ratio that will give the maximum net work for fixed compressor and turbine inlet temperatures (or fixed ratio). Combining Eqs. (5.4a) and (5.7) yields

$$w_{net} = c_p \left[T_3 + T_1 - T_2 - \left(\frac{T_3}{T_2}\right)(T_1) \right]$$

The maximum net work occurs when the derivative of the net work with respect to T_2 is equal to zero. This gives, for fixed T_1 and T_3,

$$\frac{d(w_{net})}{d(T_2)} = 0 = c_p \left[-1 + \frac{(T_3)(T_1)}{(T_2)^2} \right]$$

or the maximum net work occurs when

$$T_2 = \sqrt{(T_1)(T_3)} \tag{5.9}$$

The pressure ratio for maximum net work is

$$\frac{p_2}{p_1} = \left(\frac{T_2}{T_1}\right)^{k/(k-1)} = \left(\frac{T_3}{T_1}\right)^{k/2(k-1)} \tag{5.10}$$

This shows that the pressure ratio for maximum net work increases with increasing turbine inlet temperature and a fixed compressor inlet temperature. This conclusion was derived for constant specific heats, but the same trend applies for variable specific heats. The compressor inlet temperature is fixed by the atmospheric conditions, whereas the turbine inlet temperature is limited by the maximum temperature the turbine materials can withstand. The turbine inlet temperature has increased steadily over the years; therefore, new engines have been designed for higher pressure ratios.

Figure 5.8 illustrates the solution to Eq. (5.10) for values of specific heat ratio of 1.4, 1.35, and 1.3. Note that the value of the optimum pressure ratio predicted by Eq. (5.10) is very dependent on the value of specific heat ratio used.

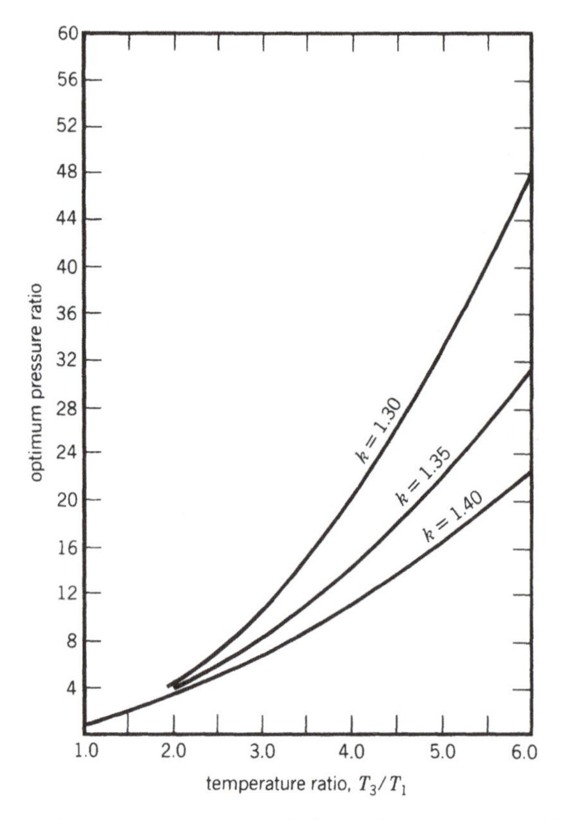

Figure 5.8. Pressure ratio versus temperature ratio for maximum net work, solution to Eq. (5.10).

Example Problem 5.1

An ideal, basic, air standard gas turbine engine (see Figure 5.2) has a compressor inlet temperature of 519°R (288 K) and a turbine inlet temperature of 2520°R (1400 K). Calculate, assuming that $c_p = 0.24$ Btu/lb°R = constant, $k = 1.4$, and that air enters the compressor at the rate of 1 lb/s,

(a) The pressure ratio that gives the maximum net work
(b) The compressor work, turbine work, heat added, and thermal efficiency for the pressure ratio determined in (a)
(c) The power, in kilowatts, developed by this engine.

Solution

(a) From Eq. (5.9)

$$T_2 = \sqrt{(519)(2520)} = 1144°\text{R}$$

From Eq. (5.10)

$$\frac{p_2}{p_1} = \left(\frac{2520}{519}\right)^{1.4/0.8} = 15.9$$

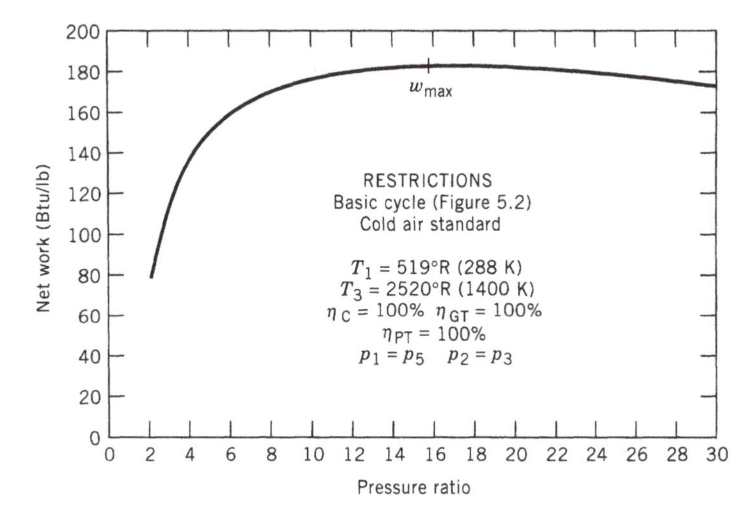

Figure 5.9. Variation of net work with pressure ratio for a cold (constant specific heats, $k = 1.4$) air standard basic cycle.

(b) $w_{comp} = c_p(T_2 - T_1) = 0.24(1144 - 519) = 149.9$ Btu/lb

$$T_5 = \frac{(T_3)(T_1)}{T_2} = T_2$$

$$w_{turb} = c_p(T_3 - T_5) = (0.24)(2520 - 1144) = 330.3 \text{ Btu/lb}$$

$$q_{in} = c_p(T_3 - T_2) = (0.24)(2520 - 1144) = 330.3 \text{ Btu/lb}$$

$$w_{net} = 330.3 - 149.9 = 180.4 \text{ Btu/lb}$$

$$\eta_{th} = \frac{180.4}{330.3} = 0.546 \text{ or } 54.6\%$$

(c) $\dot{w}_{net} = \dfrac{(180.4)(1)(3600)}{3413} = 190.3$ kW

Figure 5.9 shows how the net work varies with pressure ratio for a turbine inlet temperature of 2520°R (1400 K), compressor inlet temperature of 519°R (288 K), and specific heat ratio of 1.4.

The temperature variation in a gas turbine engine is several hundred degrees; therefore, the variation of specific heat with temperature must be considered. The variation of specific heat with temperature can be taken into account by using the tables in the appendix. Equations (5.1) through (5.5) still apply, but the method used to calculate the unknown values is different. For an isentropic process and variable specific heats,

$$Pr_2 = Pr_1\left(\frac{p_2}{p_1}\right) \tag{5.11}$$

$Pr_1 = f(T_1)$ and T_1, p_1, and p_2 would be known; therefore, Pr_2 can be calculated. This means that T_2 is known. Since for an ideal gas, $h = f(T)$ and T_1 and T_2 are known, the enthalpy values can be read from Table B, which is for dry air. The same method applies to the expansion process.

Example Problem 5.2

An ideal, basic air standard gas turbine engine has a compressor inlet temperature of 519°R (288 K)[1] and a turbine inlet temperature of 2520°R (1400 K). Assuming variable specific heats and that air enters the compressor at the rate of 1 lb/s (1 kg/s), calculate the compressor work, heat added, thermal efficiency, net work and power (in kW) developed by this engine for

(a) A compressor pressure ratio of 15.9
(b) A compressor pressure ratio of 12.0 (use SI units)
(c) A compressor pressure ratio of 4.0 (use English units)

Solution

Basis: 1 lb (kg) dry air

(a) From Appendix Table B.2 at 519°R (288 K)[1]

$$h_1 = -6.0\ (-14.3) \quad Pr_1 = 1.2093\ (1.2045)$$

$$Pr_2 = Pr_1\left(\frac{p_2}{p_1}\right) = 19.228\ (18.068)$$

By interpolation

$$T_2 = 1129°\text{R}\ (617\ \text{K}),\ h_2 = 143.5\ \text{Btu/lb}\ (322.4\ \text{kJ/kg})$$

$$w_{comp} = h_2 - h_1 = 149.5\ \text{Btu/lb}\ (336.7\ \text{kJ/kg})$$

$$T_3 = 2520°\text{R}\ (1400\ \text{K})$$

$$h_3 = 521.7\ (1212.7),\ Pr_3 = 450.9\ (450.9)$$

$$Pr_5 = Pr_3\left(\frac{p_5}{p_3}\right) = 28.359\ (28.359)$$

$$T_5 = 1254°\text{R}\ (696\ \text{K}),\ h_3 = 175.2\ (407.3)$$

$$w_{turb} = h_3 - h_5 = 346.5\ \text{Btu/lb}\ (805.4\text{kJ/kg})$$

$$w_{net} = w_{turb} - w_{comp} = 197.0\ \text{Btu/lb}\ (468.7\text{kJ/kg})$$

$$q_{in} = h_3 - h_2 = 378.2\ \text{Btu/lb}\ (890.3\ \text{kJ/kg})$$

[1]Values in parentheses are values if the temperatures had been in Kelvin. This is being done because these compressor and turbine inlet conditions will be used in several example problems, some example problems using SI units, and others in English units.

$$\eta_{th} = \frac{w_{net}}{q_{in}} = 0.521 \text{ or } 52.1\% \ (52.7\%)$$

$$\dot{W}_{net} = \frac{(197.0)(1)(3600)}{3413} = 207.8 \text{ kW} \ (468.7 \text{ kW})$$

(b) The method is the same as for part (a); therefore, only the results are given as follows.

State	T, K (°R)		h, kJ/kg (Btu/lb)		Pr	
1	288	(519)	−14.3	(−6.0)	1.2045	(1.2093)
2	580	(1046)	284.1	(122.5)	14.454	(14.512)
3	1400	(2520)	1212.7	(521.7)	450.9	(450.9)
5	751	(1351)	465.7	(200.4)	37.575	(37.575)

w_{comp} = 298.4 kJ/kg (128.5 Btu/lb)
w_{turb} = 747.0 kJ/kg (321.3 Btu/lb)
w_{net} = 448.6 kJ/kg (129.2 Btu/lb)
q_{in} = 928.6 kJ/kg (399.2 Btu/lb)
η_{th} = 0.483 or 48.3% (0.483 or 48.3%)
$\dot{w}_{net}$ = 448.6 kW/(kg/s) (203.4 kW/(lb/s))

(c) Once again, only the results are given

State	T, °R (K)		h, Btu/lb (kJ/kg)		Pr	
1	519	(288)	−6.0	(−14.3)	1.2093	(1.2045)
2	770	(427)	54.6	(126.4)	4.837	(4.818)
3	2520	(1400)	521.7	(1212.7)	450.9	(450.9)
5	1793	(996)	318.2	(739.7)	112.73	(112.73)

w_{comp} = 60.6 Btu/lb (140.7 kJ/kg)
W_{turb} = 203.5 Btu/lb (473.0 kJ/kg)
w_{net} = 142.9 Btu/lb (332.3 kJ/kg)
q_{in} = 467.1 Btu/lb (1086.3 kJ/kg)
η_{th} = 0.306 or 30.6% (0.306 or 30.6%)
$\dot{w}_{net}$ = 150.7 kW/(lb/s) (332.3 kW/(kg/s))

It is of interest to compare the results of Example Problems 5.1b and 5.2a and note the differences in the answers, especially for net work and thermal efficiency. Also, note the effect compressor pressure ratio has on compressor work, heat added, turbine work, and net work.

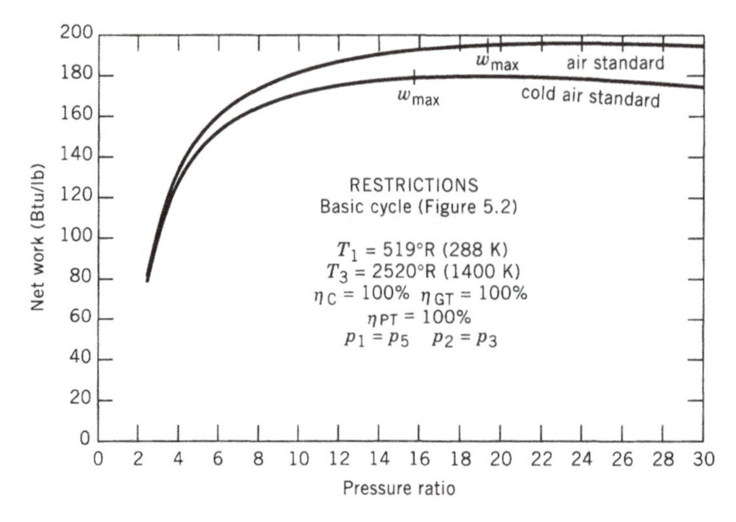

Figure 5.10. Variation, for a basic cycle, of net work with pressure ratio for a cold air standard cycle (constant specific heats, $k = 1.4$) and an air standard cycle (variable specific heats).

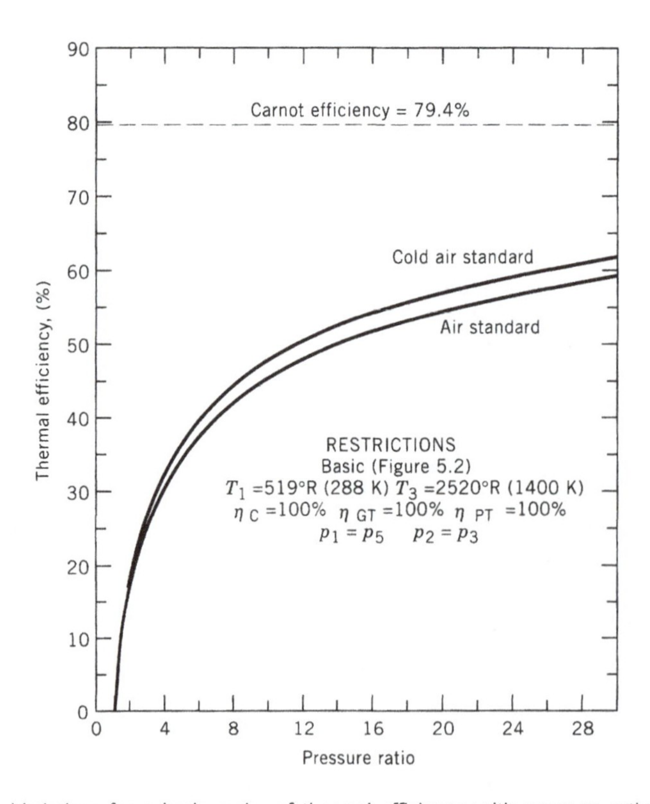

Figure 5.11. Variation, for a basic cycle, of thermal efficiency with pressure ratio for a cold air standard cycle (constant specific heats, $k = 1.4$) and an air standard cycle (variable specific heats).

Figures 5.10 and 5.11 illustrate how the net work and thermal efficiency vary with pressure ratio when the variation of specific heat with temperature is considered. Figures 5.10 and 5.11 are for an ideal basic cycle, a compressor inlet temperature of 519°R (288 K), and a turbine inlet temperature of 252°R (1400 K). Also shown in Figures 5.10 and 5.11 are the results for a cold (constant specific heat) air standard cycle operating under the same conditions. Note that the pressure ratio and numerical value for maximum net work occurs at a different value for cold air standard than for variable specific heats.

Figures 5.12 and 5.13 show, for the case of variable specific heats, the variation of net work and thermal efficiency with pressure ratio for several turbine inlet temperatures. Note on Figure 5.12 the dashed line that shows the pressure ratio at which maximum net work occurs as the turbine inlet temperature varies.

5.2 BASIC CYCLE WITH FRICTION (AIR STANDARD)

The discussion in Section 5.1 assumed that the compressor and turbine operate isentropically and that there is no pressure drop during the heat addition process. In an actual gas turbine engine, the compressor and turbine may be assumed to operate adiabatically but not isentropically. The heat addition process will have a pressure drop, and the pressure at the exit from the turbine will be above the pressure of the air entering the compressor.

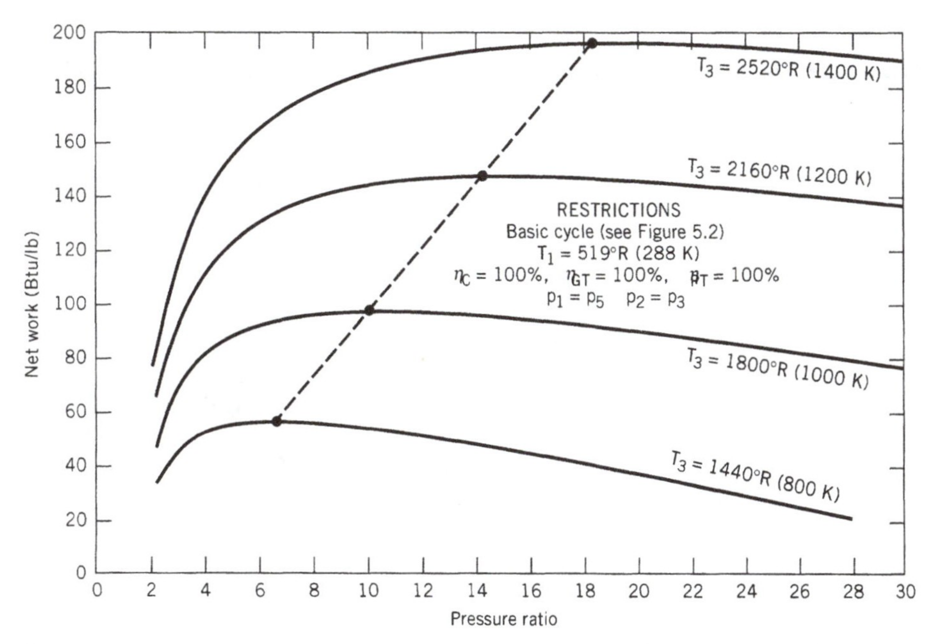

Figure 5.12. Variation, for an air standard basic cycle, or net work with pressure ratio for several turbine inlet temperatures.

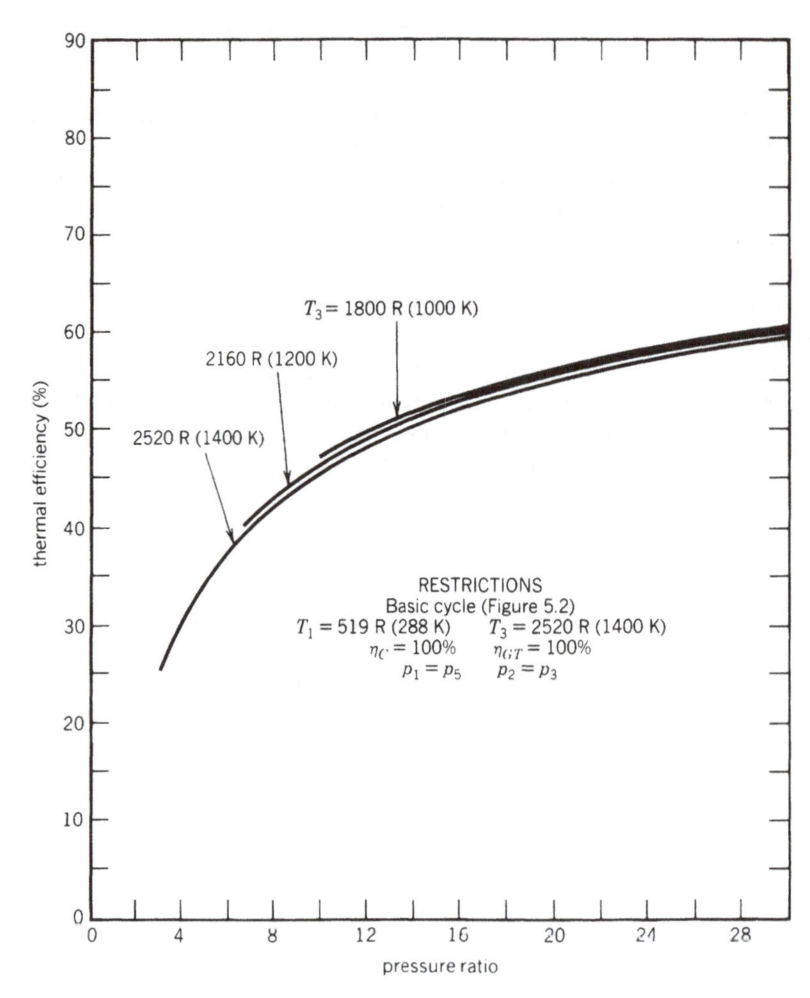

Figure 5.13. Variation, for an air standard basic cycle, of thermal efficiency with pressure ratio for several turbine inlet temperatures.

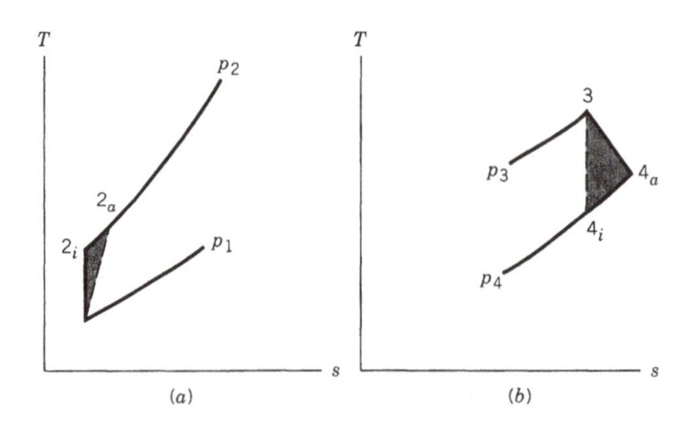

Figure 5.14. Temperature–entropy diagrams showing the ideal and actual states for a compression process (a) and an expansion process (b).

The *isentropic compressor efficiency* (sometimes referred to as the adiabatic compressor efficiency) is the ratio of the isentropic work of compression to the actual work of compression when both are compressed to the same final pressure. Care must be taken, as sometimes the final static pressure is used and sometimes the final total pressure is used. In all cases, the inlet total state is used. In this text, the isentropic compressor efficiency will be from the inlet total state to the same final total pressure, often referred to as the total *compressor efficiency.*

Referring to Figure 5.14*a,* the compressor efficiency is defined as

$$\eta_{\text{comp}} = \frac{{}_1w_{2i}}{{}_1w_{2a}} = \frac{h_{o2i} - h_{o1}}{h_{o2a} - h_{o1}} \tag{5.12}$$

The *isentropic turbine efficiency* (often called the *adiabatic turbine efficiency*) is the ratio of the actual turbine work divided by the isentropic turbine work when both expand from the same initial state to the same final pressure. Once again, the final state can be the same final static pressure or the same final total pressure. In this text, it is assumed to be the same total pressure.

Referring to Figure 5.14*b,* the turbine efficiency is defined as

$$\eta_{\text{turb}} = \frac{{}_3w_{4a}}{{}_3w_{4i}} = \frac{h_{o3} - h_{o4a}}{h_{o3} - h_{o4i}} \tag{5.13}$$

Pressure loss, in general, is expressed as the drop in total pressure divided by the inlet total pressure, or

$$\textit{pressure loss} = \frac{\Delta p}{p_{o,\text{in}}} \tag{5.14}$$

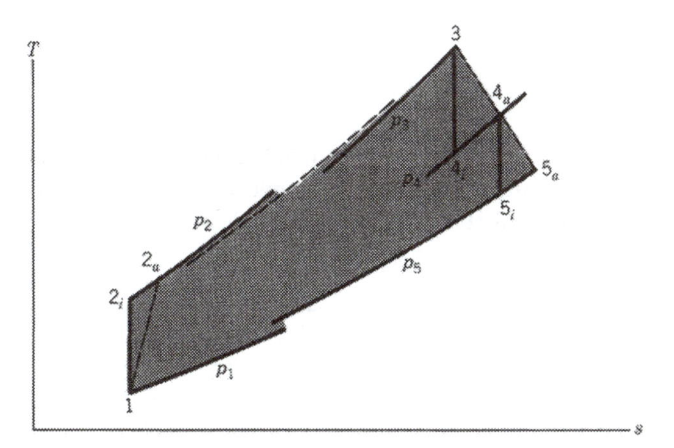

Figure 5.15. Temperature–entropy diagram for a basic gas turbine engine with friction.

Referring to Figure 5.15, which illustrates the various states for the basic gas turbine engine when friction is considered, the combustor pressure loss is

$$\left(\frac{\Delta p}{p}\right)_B = \frac{p_{o3} - p_{o2}}{p_{o2}} \tag{5.15}$$

and the exit pressure loss is

$$\left(\frac{\Delta p}{p}\right)_E = \frac{p_{o5} - p_{o1}}{p_{o5}} \tag{5.16}$$

Figure 5.15 illustrates the case where the gas turbine engine has a separate power turbine as illustrated in Figure 5.2.

Example Problem 5.3

A basic air standard gas turbine engine (see Figure 5.2) operates with a compressor inlet temperature of 288 K (519°R), a compressor efficiency of 87%, a gas generator turbine efficiency of 89%, and a power turbine efficiency of 89%. The compressor pressure ratio is 12.0. Assuming variable specific heats, a turbine inlet temperature of 1400 K (2520°R), a compressor inlet pressure of 101.3 kPa (14.70 psia), and that air enters the compressor at the rate of 1 kg/s (1 lb/s), calculate:

(a) The pressure and temperature leaving the gas generator turbine
(b) The net power, rate at which heat is added, and cycle thermal efficiency assuming no pressure drop during the heat addition process and that the pressure at the power turbine exit is 101.3 kPa (14.7 psia)
(c) The net power, rate at which heat is added, and cycle thermal efficiency if there is a 3% pressure drop in the combustion chamber and the power turbine exit pressure is 1% above the compressor inlet pressure.

Solution

Basis: 1 kg air (1 lb air)

This problem is the same as Example Problem 5.2b except for the component efficiencies and pressure loss. Information from Example Problem 5.2b will be used when possible. Values at the ideal states are tabulated below.

State	*T*, K (°R)		*p*, kPa (psia)		*h*, kJ/kg (Btu/lb)		*Pr*	
1	288	(519)	101.3	(14.70)	−14.3	(−6.0)	1.2045	(1.2093)
2	580	(1046)	1215.6	(176.4)	284.1	(122.5)	14.454	(14.512)
3	1400	(2520)	1215.6	(176.4)	1212.7	(521.7)	450.9	(450.9)

$$w_{\text{comp,i}} = 298.4 \text{ kJ/kg } (128.5 \text{ Btu/lb})$$

$$w_{\text{turb,i}} = 747.0 \text{ kJ/kg } (321.3 \text{ Btu/lb})$$

$$w_{\text{net,i}} = 448.6 \text{ kJ/kg } (192.8 \text{ Btu/lb})$$

$$q_{\text{in,i}} = 928.6 \text{ kJ/kg } (399.2 \text{ Btu/lb})$$

$$\eta_{\text{th}} = 0.483 \text{ or } 48.3\% \; (0.483 \text{ or } 48.3\%)$$

$$\dot{w}_{\text{net}} = 448.6 \text{ kW/(kg/s) } (203.4 \text{ kW/(lbs))}$$

Compressor (C)

$$w_{C,a} = \frac{w_{C,i}}{\eta_C}$$

$$= 343.0 \text{ kJ/kg } (147.7 \text{ Btu/lb})$$

$$w_{C,i} = 298.4 \text{ kJ/kg } (128.5 \text{ Btu/lb})$$

$$h_{2a} = h_1 + w_{C,a} = 328.7 \; (141.7)$$

$$T_{2a} = 623 \text{ K } (1122°\text{R})$$

(a) Gas Generator Turbine (GT)
The gas generator turbine drives the compressor. Therefore

$$w_{GT,a} = w_{c,a} = 343.0 \; (147.7)$$

$$w_{GT,i} = \frac{w_{GT,a}}{\eta_{GT}} = 385.4 \; (166.0)$$

$$h_{4i} = h_3 - w_{GT,i} = 827.3 \; (355.7)$$

$$Pr_{4i} = 151.40 \; (151.17)$$

$$p_4 = p_3 \left(\frac{Pr_{4i}}{Pr_3}\right) = 408.2 \text{ kPa } (59.1 \text{ psia})$$

(b) Power Turbine (PT)
The inlet conditions to the power turbine are the *actual* conditions leaving the gas generator turbine. These are

$$h_{4a} = h_3 - w_{GT,a} = 869.7 \; (374.0)$$

$$T_{4a} = 1109 \text{ K } (1996°\text{R})$$

$$p_4 = 408.2 \text{ kPa } (59.1 \text{ psia})$$

$$Pr_{4a} = 173.36 \; (173.20)$$

$$Pr_{5i} = Pr_{4a}\left(\frac{p_5}{p_4}\right) = 43.022\ (43.080)$$

$$h_{5i} = 495.5\ (213.3)$$

$$w_{PT,a} = \eta_{PT}(h_{4a} - h_{5i})$$

$$= 333.0\ \text{kJ/kg}\ (143.0\ \text{Btu/lb})$$

$$\dot{W}_{PT,a} = \dot{W}_{net} = 333.0\ \text{kW}\ (150.6\ \text{kW})$$

$$h_{5a} = h_{4a} - w_{PT,a}$$

$$= 536.7\ (231.0)$$

$$T_{5a} = 815\ \text{K}\ (1468°\text{R})$$

$$q_{in} = h_3 - h_{2a}$$

$$= 884.0\ \text{kJ/kg}\ (380.0\ \text{Btu/lb})$$

$$\eta_{th} = \frac{w_{PT,a}}{q_{in}} = 0.377\ \text{or}\ 37.7\%$$

(c) This part is the same as parts (a) and (b) except that the expansion ratio has been reduced. The results are

$$p_3 = (0.97)(1215.6) = 1179.1\ \text{kPa}\ (171.1\ \text{psia})$$

$$p_4 = p_3\left(\frac{Pr_{4i}}{Pr_3}\right) = 395.9\ \text{kPa}\ (57.4\ \text{psia})$$

$$h_{4a} = 869.7\ (374.0)$$

$$Pr_{4a} = 173.36\ (173.20)$$

$$p_5 = \frac{101.3}{.99} = 102.3\ \text{kPa}\ (14.85\ \text{psia})$$

$$Pr_{5i} = Pr_{4a}\left(\frac{p_5}{p_4}\right) = 44.796\ (44.837)$$

$$h_{5i} = 504.4\ (217.1)$$

$$w_{PT,a} = \eta_{PT}\ (h_{4a} - h_{5i})$$

$$= 325.1\ \text{kJ/kg}\ (139.6\ \text{Btu/lb})$$

$$\dot{W}_{PTa} = \dot{W}_{net} = 325.1\ \text{kW}\ (147.3\ \text{kW})$$

$$h_{5a} = h_{4a} - w_{PT,a}$$

$$= 544.6\ (234.4)$$

$$T_{5a} = 823\ \text{K}\ (1481°\text{R})$$

From part (b)

$$q_{in} = 884.0\ \text{kJ/kg}\ (380.0\ \text{Btu/lb})$$

$$\eta_{th} = 0.368\ \text{or}\ 36.8\%$$

TABLE 5.1. **Effect of Improving Each Component Efficiency**

η_{comp} %	η_{GT} %	η_{PT} %	Net Work kJ/kg	(Btu/lb)	Cycle Thermal Efficiency (%)	
87	89	89	333.2	(143.1)	37.68	(37.67)
88	89	89	337.1	(144.8)	37.95	(37.94)
87	90	89	336.0	(144.3)	37.99	(37.98)
87	89	90	337.0	(144.7)	38.10	(38.09)
87	90	90	339.7	(145.9)	38.42	(38.40)
88	90	90	343.6	(147.6)	38.68	(38.67)

It is of interest to determine the effect on the net work and cycle thermal efficiency if the efficiency of a component could be increased by one percentage point. The results of this type of analysis are shown in Table 5.1.

A computer program called GASTUSIM is included with this text. It allows the user to quickly determine the effect of changing one or more of the component efficiencies. The computer output for the solution to Example Problem 5.3a, b parts from this computer program are shown in Table 5.2. The answers as given in Example Problem 5.3 are not identical to those given in Table 5.2 due to interpolation of table values. It is suggested that the reader use the GASTUSIM program to determine the effect of changing one or more of the input values.

It is of interest to compare the answers calculated in Example Problems 5.2b, 5.3b, and 5.3c. In each of these examples, a pressure ratio of 12 was used. Example Problem 5.3 shows the drop in net work and thermal efficiency for realistic values of compressor efficiency, turbine efficiency, and pressure drop. It should be noted that the pressure drop in the combustion chamber had no effect on the heat added, only on the work produced by the turbine.

Next, one must consider what happens to the optimum pressure ratio for maximum net work for a basic gas turbine when variable specific heats and friction are considered. This can be answered by referring to Figures 5.16, 5.17, and 5.18. In all these figures, the pressure remains constant from compressor outlet to turbine inlet, and the pressure leaving the power turbine is assumed to be equal to the pressure entering the compressor.

Figure 5.19 illustrates the effect of compressor and turbine efficiency on thermal efficiency. Note that for a given compressor inlet temperature, turbine inlet temperature, compressor efficiency, and turbine efficiency, there is a pressure ratio at which the thermal efficiency reaches a maximum value; if the pressure ratio is increased beyond this value, the thermal efficiency will slowly decrease. This compares with the ideal basic cycle, where the thermal efficiency continues to increase with increasing pressure ratio. Table 5.3 lists the predictions of Biancardi and Peters (1) for the turbine inlet temperature, compressor pressure ratio, and turbine nominal adiabatic efficiency for baseload gas turbine engines.

Table 5.4 lists known data for several industrial gas turbines. Note the wide variation in compressor pressure ratios and flow rates. It is suggested that the reader consult

TABLE 5.2. **Output from GASTUSIM Computer Program for the Conditions in Example Problem 5.3a and b**

SI units			
Compressor inlet temperature	=	288.0	K
Low compressor efficiency	=	0.870	
Turbine inlet temperature	=	1400.0	K
Gas generator turbine efficiency	=	0.890	
Pressure drop in primary burner	=	0.0000	
Pressure drop in exhaust system	=	0.0000	
Power turbine efficiency	=	0.890	
Compressor pressure ratio	=	12.0	
Compressor work	=	342.90	kJ/kg
Temperature at compressor exit	=	622.8	K
Pressure at compressor exit (state 2)	=	1215.60	kPa
Heat added in combustion chamber	=	884.4	kJ/kg
Pressure at gas generator exit	=	408.28	kPa
Temperature at gas generator exit (state 4)	=	1109.9	K
Pressure at power turbine exit (state 5)	=	101.30	kPa
Temperature at power turbine exit (state 5)	=	815.8	K
Net work per kg into compressor	=	333.2	kJ/kg
Thermal efficiency	=	36.68	%
English units			
Compressor inlet temperature	=	519.0	°R
Low compressor efficiency	=	0.870	
Turbine inlet temperature	=	2520.0	°R
Gas generator turbine efficiency	=	0.890	
Pressure drop in primary burner	=	0.0000	
Pressure drop in exhaust system	=	0.0000	
Power turbine efficiency	=	0.890	
Compressor pressure ratio	=	12.0	
Compressor work	=	147.59	Btu/lb
Temperature at compressor exit	=	1122.3	°R
Pressure at compressor exit (state 2)	=	176.40	psia
Heat added in combustion chamber	=	379.9	Btu/lb
Pressure at gas generator exit	=	59.16	psia
Temperature at gas generator exit (state 4)	=	1997.2	°R
Pressure at power turbine exit (state 5)	=	14.70	psia
Temperature at power turbine exit (state 5)	=	1468.4	°R
Net work per pound into compressor	=	143.1	Btu/lb
Thermal efficiency	=	37.67	%

a publication such as *The 1992–1993 Handbook, Gas Turbine World* (2), for a complete listing of gas turbine specifications.

Because of the high reliability of gas turbine engines and the tremendous cost of developing a completely new engine, one will find many models of an engine. Each model will incorporate improvements in component efficiencies, increases in turbine inlet temperatures due to improvements in materials and turbine cooling methods and increases in air flow due to design changes. The change in efficiency, power, turbine inlet temperature, air flow, and pressure ratio is illustrated by the data tabulated in Table 5.5 for the Westinghouse W501 engines (3).

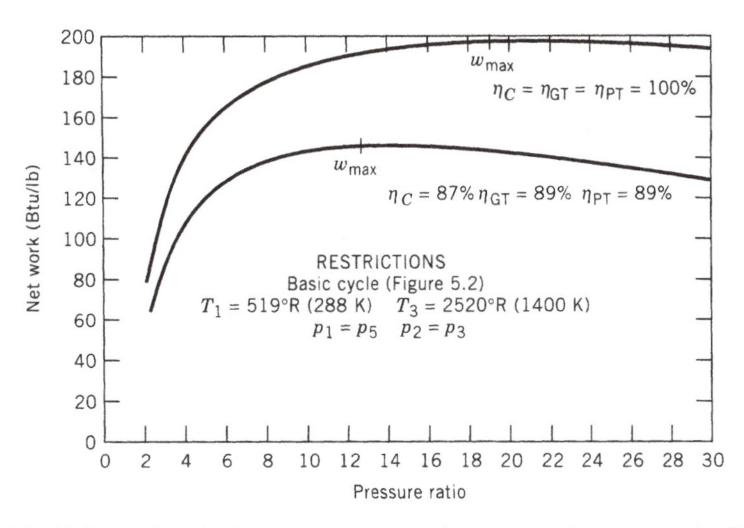

Figure 5.16. Variation, for a basic air standard cycle, of net work with pressure ratio. Comparison between actual and ideal case.

5.3 BASIC CYCLE (ACTUAL MEDIUM)

In the preceding section, air was assumed to be the working fluid for the entire cycle. This, of course, is not what occurs in an actual engine. The actual engine has air only in the compressor, with combustion occurring in the combustion chamber and the products of combustion expanding through the turbine(s).

The next problem that must be considered is the effect on the net work and thermal

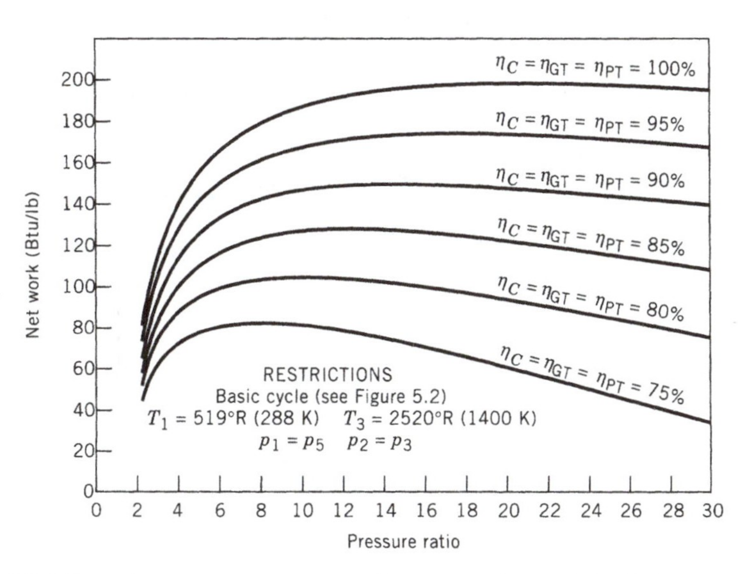

Figure 5.17. Effect of compressor and turbine efficiency on net work for a basic air standard cycle.

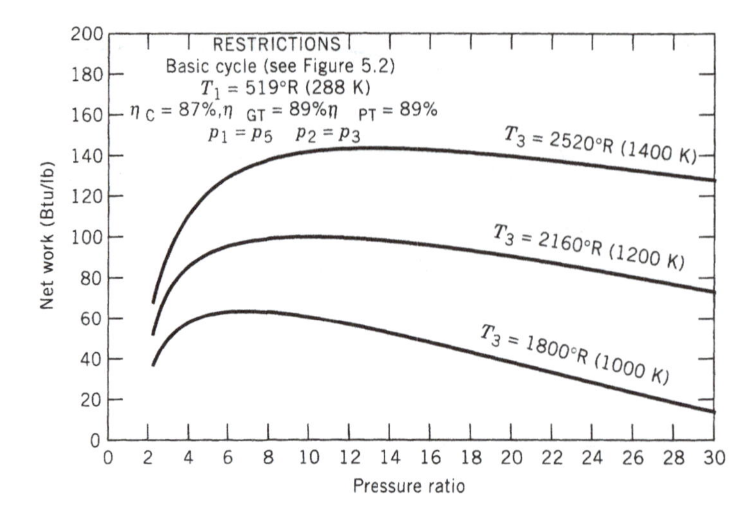

Figure 5.18. Variation, for an air standard basic cycle, of net work with pressure ratio for several turbine inlet temperatures.

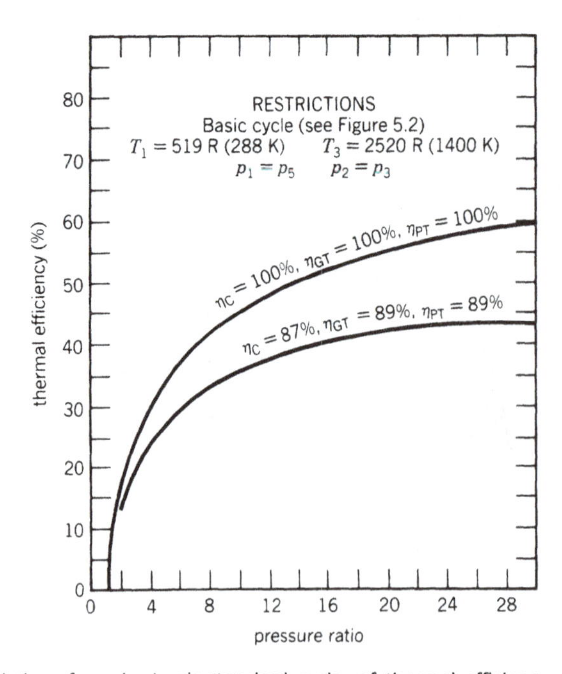

Figure 5.19. Variation, for a basic air standard cycle, of thermal efficiency with pressure ratio. Comparison between actual and ideal case.

TABLE 5.3. **Projected Technology Basis for Baseload Gas Turbine Engines (1)**

Parameter	1970 Decade	Early 1980s	Late 1980s
Turbine inlet gas	Up to 2200[a]	Up to 2400[a]	Up to 2800[a]
Temperature (°F)	Up to 2400[b]	Up to 2800[b]	Up to 3100[b]
Compressor pressure ratio	Up to 28:1	Up to 36:1	Up to 36:1
Turbine nominal adiabatic eff. (%)	90	92	93

[a]Compressor bleed air uncooled.
[b]Compressor bleed air precooled to 200°F

TABLE 5.4. **Gas Turbine Engine Specifications (in operation and operating on the basic cycle) (2)**

Manufacturer Model	Year	Nominal Rating	Heat Rate	Pressure Ratio	Flow	Turbine Inlet Temperature	Exhaust Temperature
ABB Power Generation							
GT 35	1968	23,105 hp	7,790[a]	12	202 lb/s	—	376°C
GT 8C	1994	52,800 kW	—	15.7	395 lb/s	—	517°C
GT 10	1981	34,220 hp	7,180[a]	14	173 lb/s	—	534°C
Allison Gas Turbine							
501-KC5	1983	5,278 hp	9,005[a]	9.4	34 lb/s	—	1075°F
570-K	1979	6,540 hp	8,600[a]	12.1	42 lb/s	—	1045°F
Garrett Auxiliary Power							
IM 831-800	1972	505 kW	16,400[b]	11	7.9 lb/s	1765°F	930°F
GE Marine & Industrial							
LM1600-PA	1989	13,425 kW	9,560[b]	22.3	100 lb/s	1369°F*	909°F
LM2500-PE	1973	21,925 kW	9,600[b]	18.9	153 lb/s	1490°F*	985°F
LM5000-PC	1984	33,750 kW	9,265[b]	24.8	265 lb/s	1281°F*	816°F
LM6000-PA	1992	40,400 kW	8,680[b]	29.8	275 lb/s	1555°F*	890°F
*power turbine inlet temperature							
GE Power Generation							
M3142(J)	1952	14,600 hp	9,530[a]	7.1	115 lb/s	—	979°F
PG6541(B)	1978	38,340 kW	10,880[b]	11.8	302 lb/s	2020°F	1002°F
PG6541(B)	1987	26,300 kW	11,990[b]	10.5	270 lb/s	1765°F	909°F
LM6000	1992	39,970 kW	8,790[b]	30	271 lb/s	2270°F	982°F
Pratt & Whitney Canada							
STL-T76	1975	1,440 hp	11,960[a]	7.7	13 lb/s	1793°F	1067°F
FT8	1990	25,420 kW	8,950[b]	20.3	188 lb/s	—	830°F
Solar Turbines							
Saturn	1960	1340 hp	10,933[a]	6.4	14 lb/s	—	891°F
Centaur	1970	4700 hp	9,100[a]	10.2	42 lb/s	—	844°F
Turbo Power							
FT8	1990	25,420 kW	8,950[b]	20.3	188 lb/s	2120°F	830°F

[a] Btu/shp-hr
[b] Btu/kW-hr

TABLE 5.5. **Evolution of the Westinghouse W501 Gas Turbine Engine (3)**

Engine	W501A	W501AA	W501B	W501D	W501D5
First Startup Data	1968	1971	1973	1975	1979
Power Class, MW	45	60	80	95	107
Turbine Inlet Temperature, °F	1600	1650	1800	2000	2100
Inlet Air Flow, lb/s	548	744	746	781	781
Pressure Ratio	7.5	10.5	11.2	12.6	14
Thermal Efficiency	25	27	30	32	33

efficiency when the actual working fluid is considered. The solution to this problem involves calculating the amount of excess air that must be supplied to give the specified temperature leaving the combustion chamber. Once this is known, the analysis of the product gas expanding through the turbine will be known. Equations (5.1) through (5.5) still apply, and the calculations through the compressor remain unchanged.

Calculating the percent excess air (and therefore the products leaving the combustion chamber) that must be supplied involves writing an energy balance around the combustion chamber. Since the temperature leaving the combustion chamber currently is 3000°R (1650 K) or lower, complete combustion may be assumed. The case where minute amounts of CO, NO, total hydrocarbons (THC), and so on occur will be covered in Chapter 11.

Consider the combustion of a hydrocarbon fuel with dry air. The temperature and pressure of the air entering the combustion chamber will be known from the compressor calculations, and the temperature and phase of the fuel will be known. The temperature of the products leaving the combustion chamber is a controlled (specified) quantity. From the steady-flow energy equation, assuming no heat loss from the combustion chamber,

$$H_{\text{products}} = H_{\text{reactants}}$$

Example Problem 4.2 developed general equations for the calculation of the percent excess air supplied and the fuel–air ratio when liquid *n*-octane, C_8H_{18}, is completely burned with dry air supplied at a temperature T_2, the products of combustion leaving the combustion chamber are at 1400 K (2520°R), no heat transfer or work is involved, and the entering and leaving velocities are negligible. This resulted in Eqs. (4.19) and (4.20). Solving Eqs. (4.19) and (4.20) for various values of air inlet temperatures and temperatures for the products leaving the combustion chamber yields the result shown in Figures 5.20 and 5.21. It must be remembered that Figures 5.20 and 5.21 apply to the case of complete combustion with no heat loss from the combustion chamber. These curves do not give the actual gas analysis leaving the combustion chamber, only the mass of fuel added per pound of air entering the combustion chamber.

Combustion efficiency takes into account the fact that there will be some heat loss due to radiation and conduction and that incomplete combustion might occur. *Combustion efficiency* is the actual thermal energy added to the working fluid divided by the thermal energy that should have been released had all the combustible constituents of the fuel been completely oxidized in an adiabatic combustor. This means

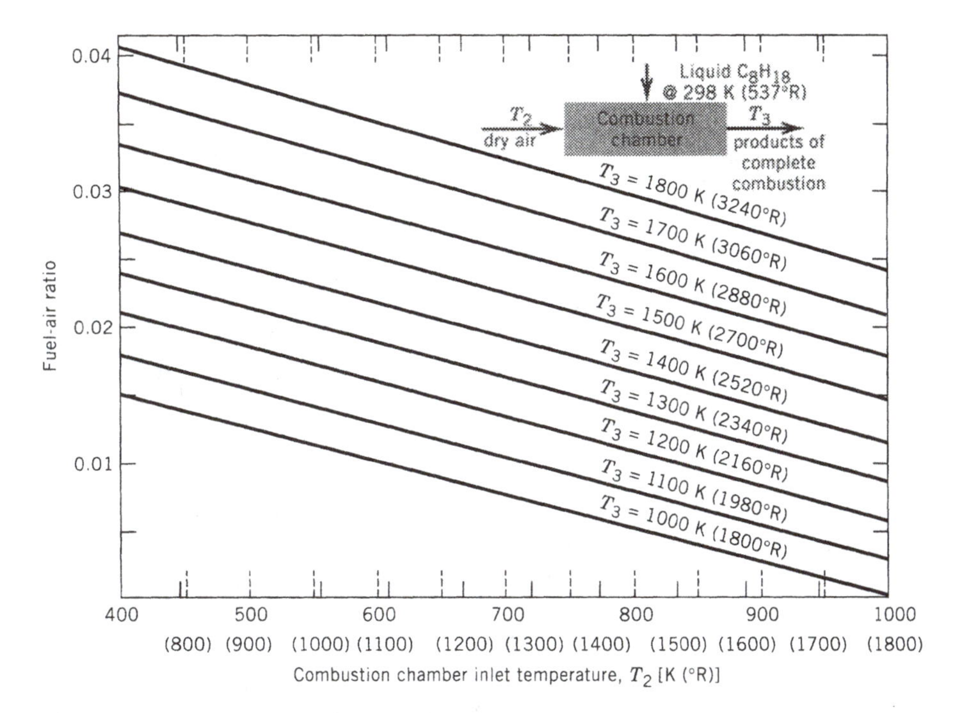

Figure 5.20. Fuel–air ratio as a function of combustion chamber inlet temperature and combustion chamber exit temperature for liquid *n*-octane, C_8H_{18}, as the fuel, complete combustion with no heat loss from the combustion chamber.

that the combustor (burner) efficiency is the ratio of the ideal fuel–air ratio (f') to the actual fuel–air ratio (f), or

$$\eta_B = \frac{f'}{f} \tag{5.17}$$

where the ideal fuel–air ratio is the ratio of the mass rate of flow of fuel divided by the mass rate of flow of dry air assuming complete combustion and no losses due to conduction and/or radiation from the combustion chamber, or

$$f' = \frac{\dot{m}'_f}{\dot{m}_a} \tag{5.18}$$

The actual fuel–air ratio is higher because of losses from conduction, radiation, and incomplete combustion, or

$$f = \frac{\dot{m}_f}{\dot{m}_a} \tag{5.19}$$

Therefore, the combustor efficiency can be written as

$$\eta_B = \frac{\dot{m}'_f}{\dot{m}_f} \tag{5.20}$$

Once the gas analysis leaving the combustion chamber is known, the work produced

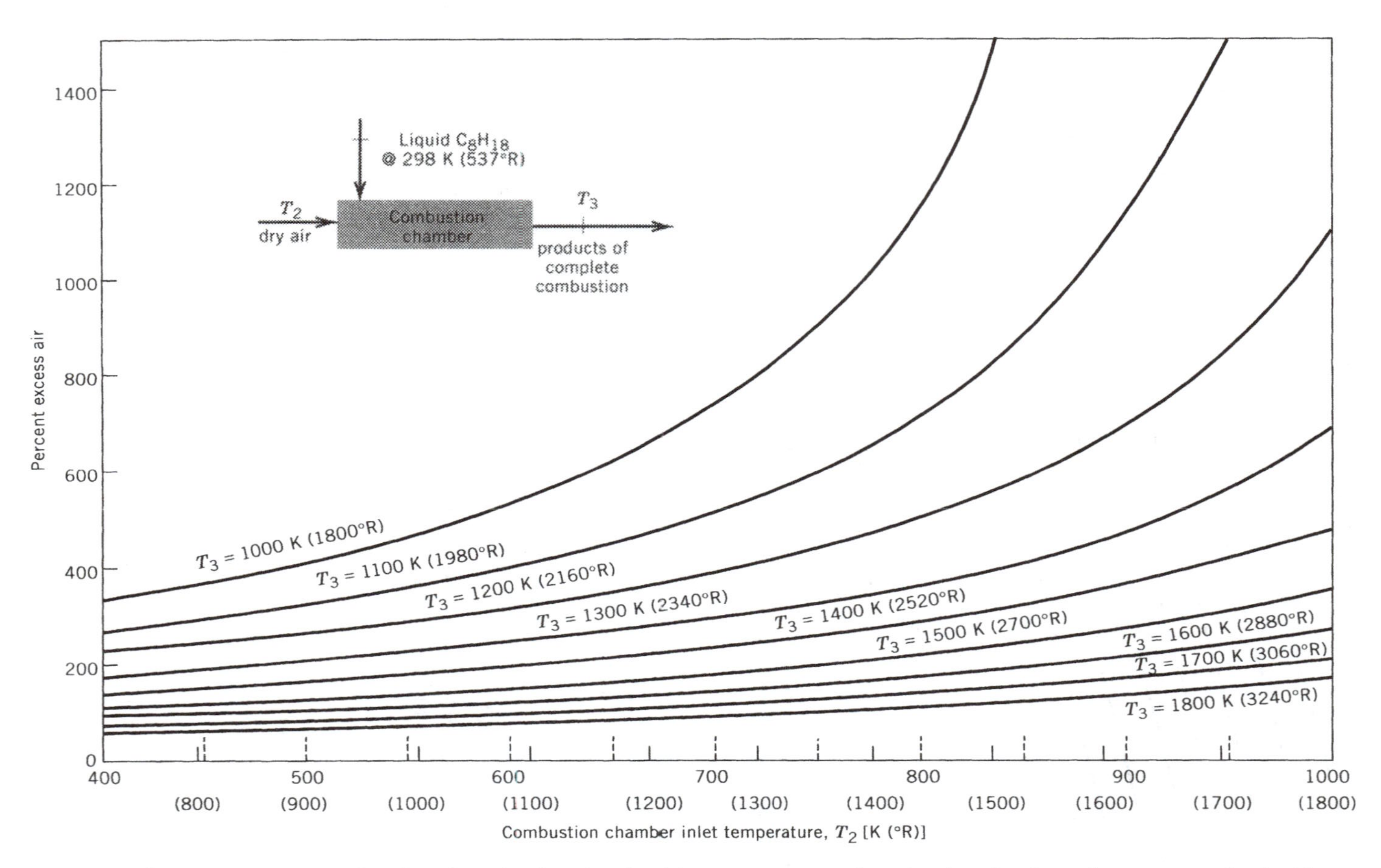

Figure 5.21. Percent excess air as a function of combustion chamber inlet temperature and combustion chamber exit temperature for liquid *n*-octane, C_8H_{18}, as the fuel, complete combustion with no heat loss from the combustion chamber.

by the turbine may be calculated. The composition of the product gas expanding through the turbine will be assumed to remain constant.

First, consider the isentropic expansion through the gas generator turbine (see Figure 5.2 for components):

$$\bar{s}_{4i} = \bar{s}_3 = \bar{s}^0_{4i} - \bar{s}^0_3 - \bar{R}\ln\frac{p_4}{p_3} \tag{5.21}$$

The pressure, temperature, and composition entering the gas generator turbine are known, along with the turbine efficiency and the actual work required. Therefore, by trial and error, the pressure and temperature at the turbine exit may be determined. This means that the conditions entering the power turbine will be known and, since the power turbine efficiency is known, the work developed by the power turbine may be determined.

Two factors that are commonly used in judging the performance of a gas turbine are the *specific fuel consumption* (SFC) and the *heat rate* (HR).

The specific fuel consumption is the mass of fuel required per hour per horsepower or kilowatt. This reduces to, in English units,

$$\text{SFC} = \frac{2545}{\eta_{\text{th}}\,|\Delta H_c|}, \frac{\text{Btu}}{\text{hph}} \tag{5.22a}$$

or

$$\text{SFC} = \frac{3413}{\eta_{\text{th}}\,|\Delta H_c|}, \frac{\text{Btu}}{\text{kWh}} \tag{5.22b}$$

where $|\Delta H_c|$ is the *lower* heating value of the fuel in Btu per pound.

The heat rate is the rate at which heat is added per unit of time divided by the power output. The heat added, in English units, is in Btu per hour. The power output can be either in horsepower or kilowatts. Therefore, in equation form

$$HR = \frac{\dot{q}_{\text{in}}}{\dot{w}_{\text{net}}}, \frac{\text{Btu}}{\text{kWh}} \text{ or } \frac{\text{Btu}}{\text{hph}} \tag{5.22c}$$

Example Problem 5.4

A gas turbine engine operating on the basic cycle (see Figure 5.2) has a compressor inlet pressure of 14.7 psia (101.3 kPa), a temperature of 519°R (288 K), a compressor exit and turbine inlet pressure of 176.4 psia (1215.6 kPa), a turbine inlet temperature of 2520°R (1400 K), a turbine exit pressure of 14.7 psia (101.3 kPa), a compressor efficiency of 87%, and a gas generator turbine and power turbine each with an efficiency of 89%. The fuel, supplied as a liquid at 537°R (298 K), is *n*-octane, C_8H_{18}. If air enters the compressor at the rate of 1.0 lb/s (1.0 kg/s), calculate, considering the actual product gas expanding through the turbine,

(a) The percent excess air supplied
(b) The net work developed per pound of air entering the compressor
(c) The thermal efficiency based on the lower heating value
(d) The power developed in horsepower and kilowatts

(e) The specific fuel consumption
(f) The heat rate

Solution
Basis: 1 lb-mol dry air

Compressor
The values for the compressor are the same as those determined in Example Problem 5.3. These values are

$$w_{C,a} = 147.7 \text{ Btu/lb } (343.0 \text{ kJ/kg})$$
$$= 4278 \text{ Btu/lb-mol } (9935 \text{ kJ/kmol})$$
$$h_{2a} = 141.7 \text{ Btu/lb } (328.7 \text{ kJ/kg})$$
$$T_{2a} = 1122°\text{R } (623 \text{ K})$$

Combustion Chamber
Equations (4.18), (4.19), and (4.20), developed in Example Problem 4.2, may be used to determine the percent excess air supplied. This results in

$$X = 183.5$$
$$f' = 0.0215$$
$$\% \text{ excess air supplied} = 207.6\%$$

The heat added in the combustion chamber is

$$q_{\text{in}} = \frac{(0.0215)(2{,}181{,}637)}{114.2336} = 410.6 \text{ Btu/lb air into compressor}$$
$$= 11{,}893 \text{ Btu/lb-mol dry air}$$

Therefore, $p_4 = 61.33$ psia
Next, the actual state leaving the gas generator turbine must be determined.

$$H_{4a} = \Sigma\,(n_i\,H_{i,4a})_{pr} = H_3 - \overline{w}_{GT,a} = -140{,}977$$

By trial and error,

$$T_{4a} = 2029°\text{R}$$

Power Turbine (PT)
In a similar manner, it can be shown that

$$S_{5i} = 10{,}064.7$$
$$T_{5i} = 1430°\text{R}$$
$$H_5 = -1{,}053{,}997$$

Therefore,

$$\overline{w}_{PT,a} = (H_{4a} - H_{5i})\,\eta_{PT} = 812{,}588 \text{ Btu/lb-mol fuel}$$
$$= 4{,}428 \text{ Btu/lb-mol dry air}$$
$$= 152.9 \text{ Btu/lb air}$$

$$T_{5a} = 1496°\text{R}$$

$$\eta_{\text{th}} = \frac{\overline{w}_{\text{PT},a}}{\overline{q}_{\text{in}}} = \frac{812{,}588}{2{,}181{,}637} = 0.372$$

$$\dot{W}_{\text{net}} = \frac{(1.0)(4428)(3600)}{(28.965)(3413)} = 161.3\ \text{kW} = 216.2\ \text{hp}$$

$$\text{SFC} = \frac{(114.2336)(1.0)(3600)}{(183.5)(28.965)(161.3)} = 0.480\ \text{lb/kWh}$$

$$= 0.358\ \text{lb/hph}$$

The products of combustion are

	n_i
CO_2	8.0
H_2O	9.0
D.A.	183.5
O_2	−12.5
	188.0

Gas Generator Turbine (GT)

$$\overline{w}_{\text{GT},a} = 4278\ \text{Btu/lb-mol dry air}$$

$$= 784{,}811\ \text{Btu/lb-mol fuel}$$

$$\overline{w}_{\text{GT},i} = \frac{\overline{w}_{\text{GT},a}}{\eta_{\text{GT}}} = 881{,}810$$

$$H_3 = \Sigma\,(\eta_i\,\overline{h}_i)_{2520} = 643{,}834$$

$$H_{4i} = H_3 - \overline{w}_{\text{GT},i} = 643{,}834 - 881{,}810 = -237{,}976$$

By trial and error,

$$T_{4i} = 1967°\text{R}$$

Next, the pressure leaving the gas generator turbine must be calculated.

$$0 = S_{4i} - S_3 - \overline{R}\,n_{pr}\,\ell\text{n}\,\frac{p_4}{p_3}$$

$$S_3 = \Sigma\,(n_i\,\overline{s}_i^0)_{2520} = 10{,}944.3$$

$$S_{4i} = \Sigma\,(n_i\,\overline{s}_i^0)_{1967} = 10{,}549.6$$

Therefore, $p_4 = 61.33$ psia

Next, the actual state leaving the gas generator turbine must be determined.

$$H_{4aa} = \Sigma\,(n_i\,H_{i,4a})_{pr} = H_3 - \overline{w}_{\mathrm{GT},a} = -140{,}977$$

By trial and error,

$$T_{4a} = 2020°\mathrm{R}$$

Power Turbine (PT)
In a similar manner, it can be shown that

$$S_{5i} = 10{,}064.7$$

$$T_{5i} = 1430°\mathrm{R}$$

$$H_5 = -1{,}053{,}997$$

Therefore,

$$\overline{w}_{\mathrm{PT},a} = (H_{4a} - H_{5i})\,\eta_{\mathrm{PT}} = 812{,}588 \text{ Btu/lb-mol fuel}$$

$$= 4{,}428 \text{ Btu/lb-mol dry air}$$

$$= 152.9 \text{ Btu/lb air}$$

$$T_{5a} = 1496°\mathrm{R}$$

$$\eta_{\mathrm{th}} = \frac{\overline{w}_{\mathrm{PT},a}}{\overline{q}_{\mathrm{in}}} = \frac{812{,}588}{2{,}181{,}637} = 0.372$$

$$\dot{W}_{\mathrm{net}} = \frac{(1.0)(4428)(3600)}{(28.965)(3413)} = 161.3 \text{ kW} = 216.2 \text{ hp}$$

$$\mathrm{SFC} = \frac{(114.2336)(1.0)(3600)}{(183.5)(28.965)(161.3)} = 0.480 \text{ lb/kWh}$$

$$= 0.358 \text{ lb/hph}$$

An alternative method for calculating the specific fuel consumption is to use Eq. (5.22). This yields

$$\mathrm{SFC} = \frac{(2545)(114.2336)}{(0.372)(2{,}181{,}637)} = 0.358 \text{ lb/hph}$$

$$= \frac{(3413)(114.2336)}{(0.372)(2{,}181{,}637)} = 0.480 \text{ lb/kWh}$$

$$\mathrm{HR} = \frac{(410.6)(1)(3600)}{161.3}$$

$$= 9164 \text{ Btu/kWh}$$

$$= 6837 \text{ Btu/hph}$$

It is of interest to compare the answers of Example Problem 5.4 with those calculated in Example Problem 5.3b. The net work when the actual products of combustion are considered is 6.8% higher than when dry air is used, the cycle thermal efficiency is

1.3% lower, and the exit temperature 29°R higher. Part of this difference results from the fact that the gases have a different composition and part because the flow through the turbines is 2.15% higher. How does the heat rate calculated compare with the heat rate values as tabulated in Table 5.4?

5.4 BASIC CYCLE (AIR EQUIVALENT)

The preceding sections examined the basic gas turbine cycle assuming in one case that air was the working fluid throughout (air standard) and, in the other case, the actual products of combustion expanding through the turbine. When the actual products are used, a number of trial-and-error solutions are involved. This can take a considerable amount of time but, if exact values are needed, is the correct way to solve the problem.

One way to simplify this problem and closely approximate the exact solution is to determine the fuel–air ratio (f') using Figure 5.20 or equations, then assume that the amount of *air* expanding through the turbine is $(1 + f')$ pounds for every pound of air entering the compressor. This has the limitation that it assumes that pure air is leaving the combustion chamber and that liquid *n*-octane, C_8H_{18}, is the fuel. This means, on a basis of 1 lb of air entering the compressor, that

$$|w_{C,a}| = (1 + f')w_{GT,a} \tag{5.23}$$

$$w_{net} = (1 + f')w_{PT,a} \tag{5.24}$$

$$q = (f')(19{,}100) \tag{5.25}$$

$$\eta_{th} = \frac{w_{net}}{q_{in}}$$

$$\text{SFC} = \frac{(f')(2545)}{w_{net}} \text{ or } = \frac{(f')(3413)}{w_{net}}$$

The solution for this type of analysis is illustrated in Example Problem 5.5.

Example Problem 5.5

Solve Example Problem 5.4 using an air-equivalent technique. Assume when determining the mass of fuel added, that the fuel is liquid C_8H_{18} supplied at 537°R (298 K) and that the fuel has a lower heating value of 19,100 Btu/lb (44,400 kJ/kg).

Solution

Basis: 1 lb air entering the compressor

From Example Problems 5.3 and 5.4,

$$f' = 0.0215$$

$$w_{C,a} = 147.7 \text{ Btu/lb } (343.0 \text{ kJ/kg})$$

The turbine portion of this problem is the same as that of Example Problem 5.3b except for the mass expanding through the turbine. Using the data from Example Problem 5.3b, where possible, yields

$$w_{GT,a} = \frac{w_{C,a}}{1+f'} = 144.6 \text{ Btu/lb } (335.8 \text{ kJ/kg})$$

$$w_{GT,i} = \frac{w_{GT,a}}{\eta_{GT}} = 162.5 \text{ Btu/lb } (377.3 \text{ kJ/kg})$$

$$T_3 = 2520°\text{R } (1400 \text{ K})$$

$$h_3 = 521.7 \text{ Btu/lb } (1212.7 \text{ kJ/kg})$$

$$Pr_3 = 450.9 \ (450.9)$$

$$p_3 = 176.4 \text{ psia } (1215.6 \text{ kPa})$$

$$h_{4i} = 359.2 \text{ Btu/lb } (835.4 \text{ kJ/kg})$$

$$Pr_{4i} = 155.2 \ (155.2)$$

$$p_4 = 60.71 \text{ psia } (418.5 \text{ kPa})$$

$$h_{4a} = 377.1 \text{ Btu/lb } (876.9 \text{ kJ/kg})$$

$$Pr_{4a} = 177.15 \ (177.29)$$

$$p_5 = 14.70 \text{ psia } (101.3 \text{ kPa})$$

$$Pr_{5i} = 42.894 \ (42.880)$$

$$h_{5i} = 212.8 \ (494.7)$$

$$w_{PT,a} = \eta_{PT} (h_{4a} - h_{5i}) (1 + f')$$

$$= 149.4 \text{ Btu/lb } (347.5 \text{ kJ/kg})$$

$$h_{5a} = 377.1 - (0.89)(377.1 - 212.8)$$

$$= 230.9 \ (536.7)$$

$$T_{5a} = 1468°\text{R } (815 \text{ K})$$

$$\dot{W}_{net} = \frac{(1)(149.4)(3600)}{3413}$$

$$= 157.6 \text{ kW } (347.5 \text{ kW})$$

$$q_{in} = f'(\text{LHV})$$

$$= 410.7 \text{ Btu/lb } (954.6 \text{ kJ/kg})$$

$$\eta_{th} = \frac{w_{PT,a}}{q_{in}} = 0.364 \text{ or } 36.4\% \ (0.364)$$

$$\text{SFC} = \frac{f'(2545)}{w_{PT,a}} = 0.366 \text{ lb/hph } (0.223 \text{ kg/kWh})$$

$$HR = \frac{(410.7)(1)(3600)}{157.6}$$

$$= 9381 \text{ Btu/kWh } (9898 \text{ kJ/kWh})$$

Example Problem 5.5 can be solved using the GASTUSIM computer program. The results are listed in Table 5.6.

TABLE 5.6. **Output from GASTUSIM Computer Program for the Conditions in Example Problem 5.5**

English Units			
Compressor inlet temperature	=	519.0	°R
Low compressor efficiency	=	0.870	
Turbine inlet temperature	=	2520.0	°R
Gas generator turbine efficiency	=	0.890	
Pressure drop in primary burner	=	0.0000	
Pressure drop in exhaust system	=	0.0000	
Power turbine efficiency	=	0.890	
Fuel is C8H18 supplied as a liquid at 537 R			
Enthalpy of CO2 at T3	=	−145261.3	Btu/lbmol
Enthalpy of H2O at T3	=	−85320.3	Btu/lbmol
Enthalpy of O2 at T3	=	15889.2	Btu/lbmol
Compressor pressure ratio	=	12.0	
Compressor work	=	147.59	Btu/lb
Temperature at compressor exit	=	1122.3	°R
Pressure at compressor exit (state 2)	=	176.40	psia
Fuel–air ratio	=	0.02147	
Percent excess air supplied	=	207.84	%
Heat added in combustion chamber	=	410.1	Btu/lb
Pressure at gas generator exit	=	60.74	psia
Temperature at gas generator exit (state 4)	=	2008.4	°R
Pressure at power turbine exit (state 5)	=	14.70	psia
Temperature at power turbine exit (state 5)	=	1468.5	°R
Net work per pound into compressor	=	149.3	Btu/lb
Thermal efficiency	=	36.41	%
Specific fuel consumption	=	0.36601	lb/(hp h)
Specific power	=	211.2	hp/(lb/s)
Heat rate	=	9374.	Btu/(kWh)
SI Units			
Compressor inlet temperature	=	288.0	K
Low compressor efficiency	=	0.870	
Turbine inlet temperature	=	1400.0	K
Gas generator turbine efficiency	=	0.890	
Pressure drop in primary burner	=	0.0000	
Pressure drop in exhaust system	=	0.0000	
Power turbine efficiency	=	0.890	
Fuel is C8H18 supplied as a liquid at 298.3 K			
Enthalpy of CO2 at T3	=	−337650.9	kJ/kmol
Enthalpy of H2O at T3	=	−85268.9	kJ/kmol
Enthalpy of O2 at T3	=	36933.4	kJ/kmol
Compressor pressure ratio	=	12.0	
Compressor work	=	342.90	kJ/kg
Temperature at compressor exit	=	622.8	K
Pressure at compressor exit (state 2)	=	1215.60	kPa
Fuel–air ratio	=	0.02150	
Percent excess air supplied	=	207.39	%
Heat added in combustion chamber	=	955.2	kJ/kg
Pressure at gas generator exit	=	419.14	kPa
Temperature at gas generator exit (state 4)	=	1116.1	K
Pressure at power turbine exit (state 5)	=	101.30	kPa
Temperature at power turbine exit (state 5)	=	815.8	K
Net work per kg into compressor	=	347.7	kJ/kg
Thermal efficiency	=	36.40	%
Specific fuel consumption	=	0.22270	kg/(kWh)
Specific power	=	347.6	kW/(kg/s)
Heat rate	=	9893.	Btu/(kW h)

TABLE 5.7. **Comparison of the Same Basic Cycle Using Various Techniques, All on a Basis of 1 lb Air Entering the Compressor**

Conditions	Example Problem	w_{net} (Btu/lb)	q_{in} (Btu/lb)	η_{th}	Fuel–Air Ratio	SPC (lb/hp-h)
Ideal air standard, variable c_p	5.2	192.8	399.2	0.483		
Actual air standard,	5.3b	143.0	380.0	0.377	—	—
variable c_p	5.3c	139.6	380.0	0.368	—	—
Actual cycle, actual medium	5.4	152.9	410.6	0.372	0.0215	0.358
Actual cycle, air equivalent	5.5	149.4	410.7	0.364	0.0215	0.366

Table 5.7 summarizes the results of solving the same basic cycle using various techniques. All of the values in Table 5.7 are on a basis of 1 lb of air entering the compressor, a turbine inlet temperature of 2520°R (1400 K), and a compressor pressure ratio of 12. Which of these techniques one should use depends on the accuracy desired.

5.5 SELECTION OF THE DESIGN POINT PRESSURE RATIO FOR A BASIC CYCLE GAS TURBINE ENGINE

Reviewing the data in Table 5.4, one observes the wide range of pressure ratios for gas turbine engines in service. Factors that must be considered include the specific power output, number of stages in the compressor and turbine, exhaust gas temperature, reliability, initial cost, and operating costs.

It is common practice to select a pressure ratio at or near the point where the specific power is a maximum. This results in a machine of minimum size since for a given power output the required air flow is inversely proportional to the specific power.

Figure 5.22 is a plot of overall thermal efficiency versus specific power output for four turbine inlet temperatures. The pressure ratios for maximum specific power for the four turbine inlet temperatures are

T_3, °R	p_2/p_1
2160	9.5
2340	11.0
2520	13.3
2700	15.4

Selecting a pressure ratio above 9.5 for a turbine inlet temperature of 2160°R will result in a higher cycle thermal efficiency but will require one or more additional compressor stages (see discussion in Chapter 7) and possibly an additional turbine stage. This adds to the initial cost of the engine but saves on operating costs. One should observe that in many cases, the pressure ratio for the engines listed in Table

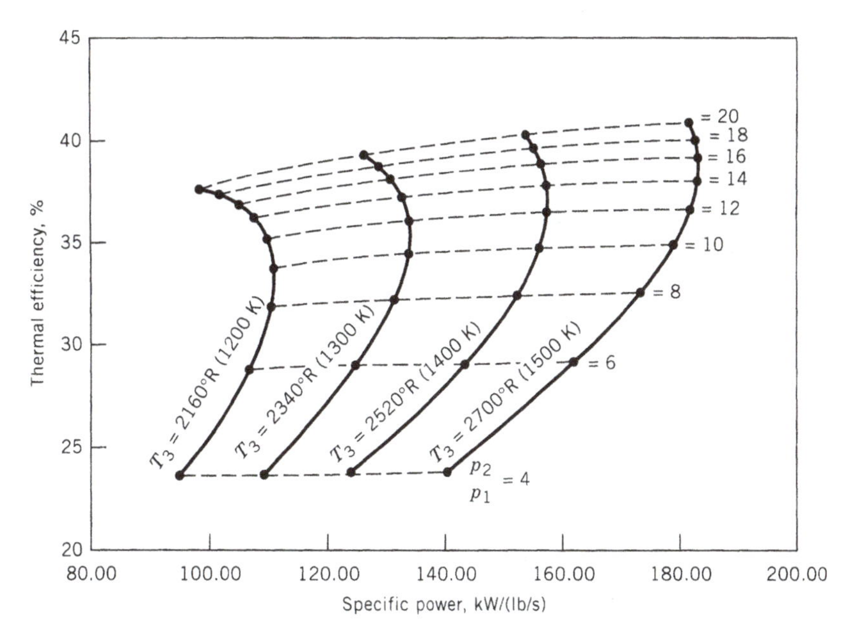

Figure 5.22. Variation of specific power with thermal efficiency for several pressure ratios and turbine inlet temperatures.

5.4 are near the pressure ratio that will give the maximum specific power output. Engines with pressure ratios near 30 are aeroderivative engines.

5.6 GAS TURBINE WITH REGENERATOR

The preceding sections examined the ideal air standard basic cycle, the air standard basic cycle with friction, and the basic cycle when the actual working medium is considered. It was noted that for an ideal basic cycle with fixed compressor and turbine inlet temperatures, there is an optimum pressure ratio that gives the maximum net work, this optimum pressure ratio increasing as the ratio of turbine inlet temperature to compressor inlet temperature is increased. For the ideal basic cycle with variable specific heats, the temperature leaving the compressor is approximately equal to the temperature of the air leaving the turbine when the compressor pressure ratio is the pressure ratio that will give maximum net work.

When friction is considered ($\eta_c < 100\%$, $\eta_t < 100\%$), pressure loss in the combustion chamber occurs, and so forth, the pressure ratio at which the maximum net work occurs is considerably below that predicted by Eq. (5.10), which assumes an ideal cold cycle (see, e.g., Figure 5.17). This means that for an actual cycle, the temperature of the air leaving the compressor is considerably below the temperature of the gas leaving the turbine.

It is next important to investigate ways of improving the thermal efficiency and/or the net work of the gas turbine. The single improvement that gives the greatest increase in the thermal efficiency is the addition of a device for transferring energy

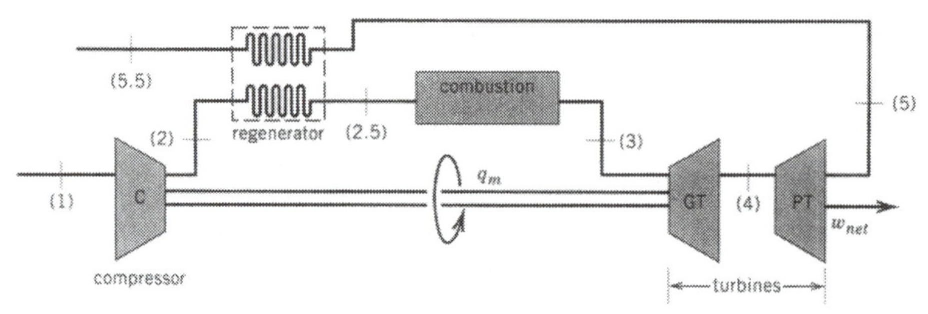

Figure 5.23. Flow diagram for the regenerative gas turbine engine.

(a heat exchanger) from the hot turbine exhaust gas to the air leaving the compressor. Figure 5.23 illustrates the flow diagram for the regenerative cycle.

There are two main types of regenerators currently being used, the recuperator and the rotating matrix regenerator. Figure 5.24 illustrates an engine with a rotating drum regenerator, Figures 5.25 and 5.26 illustrate engines with rotating disks.

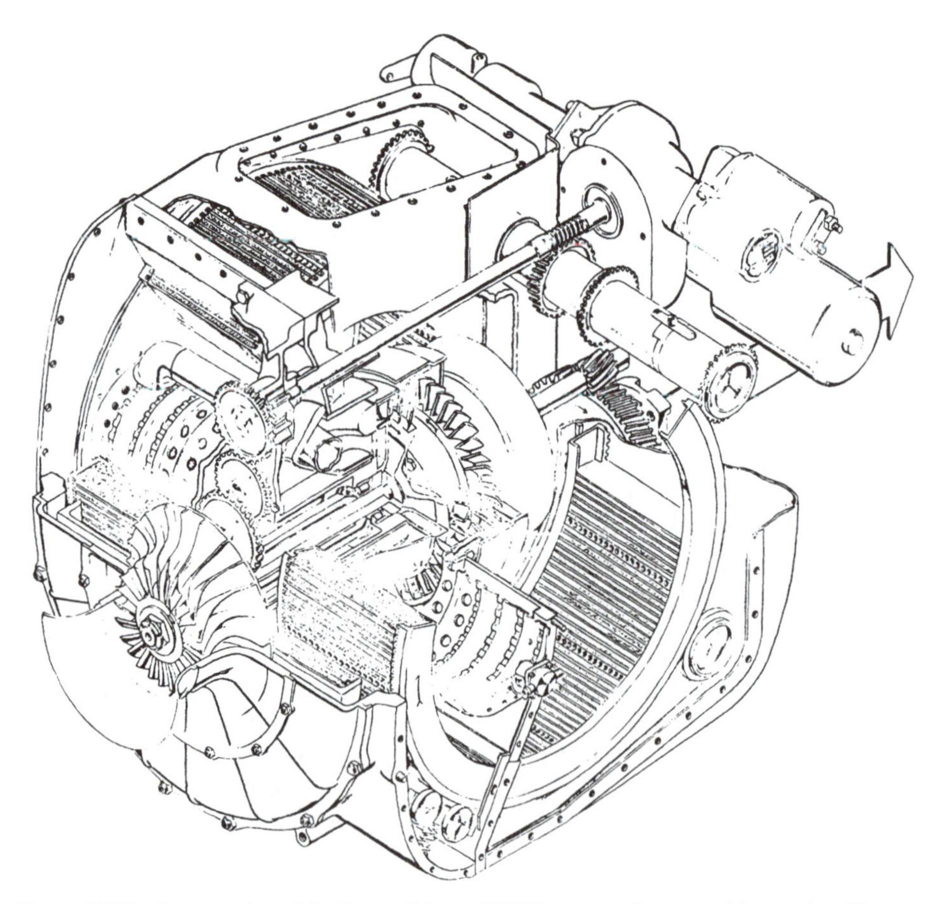

Figure 5.24. Cutaway view of the General Motors GT 305 regenerative gas turbine engine. (Courtesy of General Motors Research Laboratories)

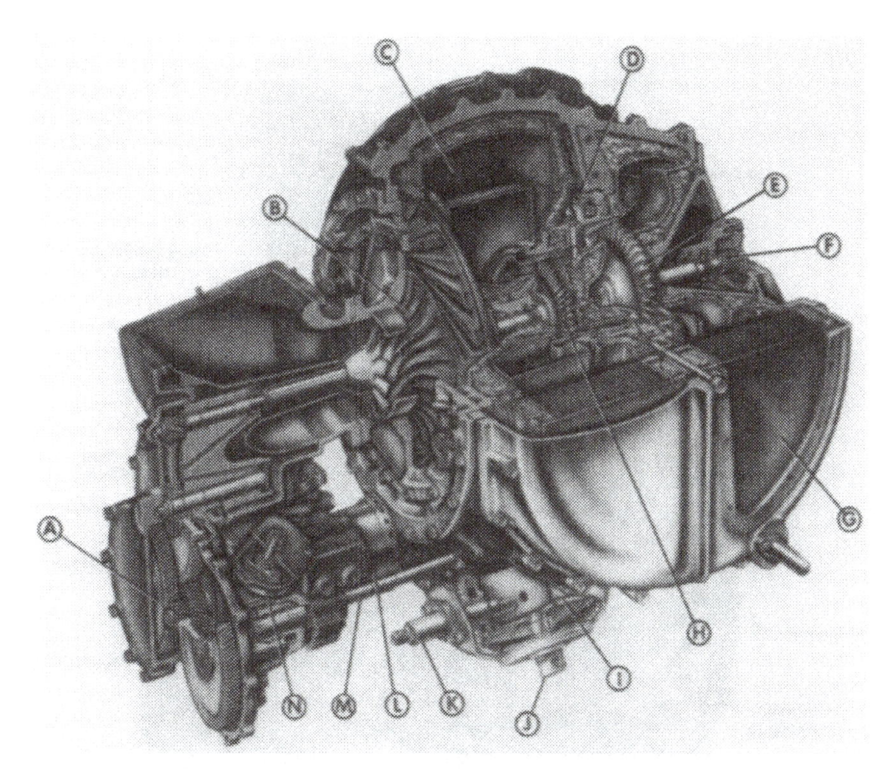

Figure 5.25. Cutaway view of the Chrysler Corporation twin-regenerative gas turbine engine. (Courtesy of Chrysler Corp.)

The two most commonly used configurations of rotating regenerators are the disk and drum types. The basic form of the disk type is simple in construction. Alternate layers of corrugated and flat thin strip material are wrapped around a central hub. This is referred to as the *matrix of the regenerator.* Gases flow perpendicular to the disk through the open spaces created by the corrugated strips. In the drum type, gases flow radially with respect to the drum. High effectiveness can be obtained since the matrix can be heated up to nearly the full exhaust gas temperature. Figure 5.27 illustrates the flow pattern for the disk and drum types of regenerators.

One of the major design problems that arises for the rotating matrix is the separation of the hot and cold gases. This sealing problem stems from the fact that the cold gases are at much higher pressure than the hot gases. The higher the engine pressure ratio, the more sealing becomes a problem since the cold gases are at the compressor discharge pressure and the hot gases are at the engine exhaust pressure, which is the compressor inlet pressure. Also, gas trapped in the matrix of the rotating regenerator is carried across the seal, causing even more leakage. Leakage because of these two reasons can amount to 3% to 4% of the air flow out of the compressor. Because the matrix is alternatively exposed to hot and cold gases, thermal expansion and contraction add to the sealing problem.

Another design problem encountered with the regenerator is that of supporting and driving the matrix at low speed, the regenerator usually rotating at a speed of approximately 20 rpm to 30 rpm.

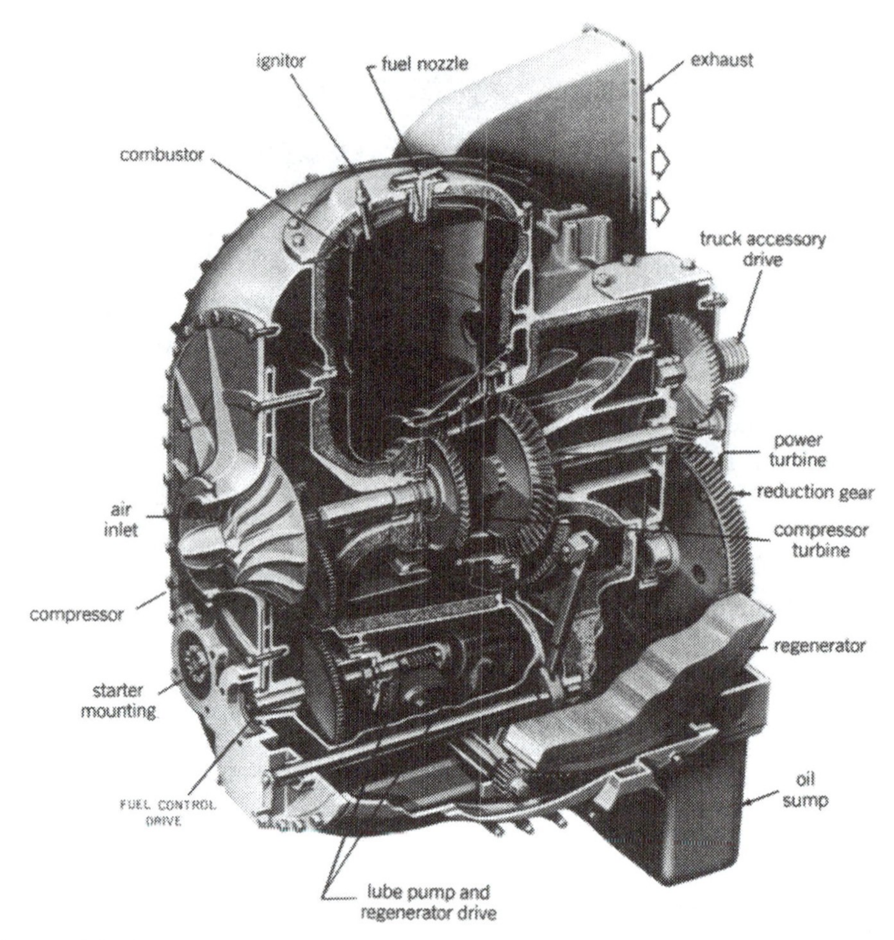

Figure 5.26. General view of Ford Motor Company prototype 707 gas turbine engine. (Courtesy of Ford Motor Company)

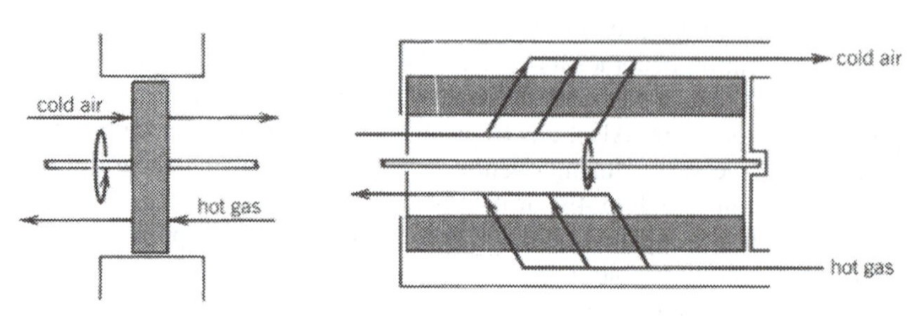

Figure 5.27. Flow pattern for disk and drum types of regenerators.

Recent developments of ceramic materials have shown good promise for use in the regenerator matrix. The most significant difference between metallic and ceramic regenerators lies in the coefficients of thermal expansion, which, in the case of ceramic material, is virtually zero and very much simplifies the sealing problem. Use of ceramic materials has enabled the matrix wall thickness to be reduced to half that of a metal matrix. Realization that glass or ceramic materials might be very suitable for regenerator matrices is generally attributed to Corning Glass Works of New York. Such materials are generally inexpensive and have good high-temperature capability. Some of them have low thermal expansion characteristics with good resistance to thermal shock. They have about one-fifth of the density of steel and have high specific heats and low conductivity, all good characteristics of matrix materials.

The recuperator type of heat exchanger generally uses common shell-and-tube heat exchangers, although recent trends have been toward plate–fin arrangements. In this type, the gases flow through closely spaced plates, usually in a counterflow fashion, exchanging heat through the metal plates. High heat transfer rates are accomplished by making flow passages small. There is a limit, however, because plugging caused by unclean exhaust gases becomes a problem.

The material requirements of a recuperator, for extended durability, must involve oxidation and corrosion resistance, stress resistance, stability under thermal cycling conditions, and be of low enough initial cost to be economically justifiable. For a given size, the recuperator is less effective in transferring energy than the regenerator because of a lower overall mean temperature difference. Because it is not rotating and is at a constant temperature, there is less thermal shock and more normal materials can be used.

When comparing the two types of heat exchangers discussed above, it must be remembered that each has its own advantages and disadvantages. The recuperative heat exchanger, though comparatively easier to build and rugged in design, must be a large, bulky piece of apparatus in order to achieve a worthwhile value of effectiveness. It must also be designed to minimize pressure drop so that power is not wasted. The regenerator, on the other hand, is smaller and, for its size, a much more efficient heat exchanger. However, it does have the problem of thermal distortion because of the cyclic thermal loading and also the problem of sealing between the hot and cold gas streams.

Thus, the specific application for a gas turbine will determine which (if any) type of heat exchanger is employed. One must remember that the increased cycle efficiency that can be gained through the use of a regenerator or recuperator must always be weighed against the disadvantages of increased service problems, cost, size, and weight.

Table 5.8 lists specifications for some gas turbine engines that have regenerators or recuperators.

Ideally, there is no pressure drop in the regenerator. Therefore, ideally, $p_2 = p_{2.5}$ and $p_5 = p_{5.5}$. Figure 5.28 shows the increase in the temperature of the high-pressure air and the resulting decrease in the temperature of the low-pressure gas. Since all the energy removed from the low-pressure gas is transferred to the high pressure (at least ideally), the area beneath the curve $5_a \rightarrow 5.5$ is equal to the area beneath the curve $2_a \rightarrow 2.5$ as shown in Figure 5.28.

TABLE 5.8. **Specifications of Several Gas Turbine Engines That Have Regenerators (4)**

Manufacturer	General Motors	Chrysler	Ford	Rover
Model no.	GT 305	A 831	705	2S/140
Power, bhp	225	130	600	150
Compressor Type	Radial	Radial	Radial	Radial
Pressure ratio	3.5	4.1	4.0	3.92
Efficiency (%)	78	80	80	79
Turbine Type	Axial	Axial	Axial	Radial
Inlet temp. (°F)	1597	1707	1748	1538
Efficiency	84	87	88	86
Regenerator Type	Drum matrix	Twin rotating disks	Recup.	Recup.
Effectiveness (%)	86	90	80	78

The regenerator effectiveness is defined as

$$\eta_{reg} = \frac{\text{actual heat transfer}}{\text{maximum heat that can be transferred}} \tag{5.26}$$

For the air standard cycle where the mass rate of flow through the turbine equals the mass rate of flow through the compressor,

$$\eta_{reg} = \frac{h_{2.5} - h_{2a}}{h_{5a} - h_{2a}} \tag{5.27}$$

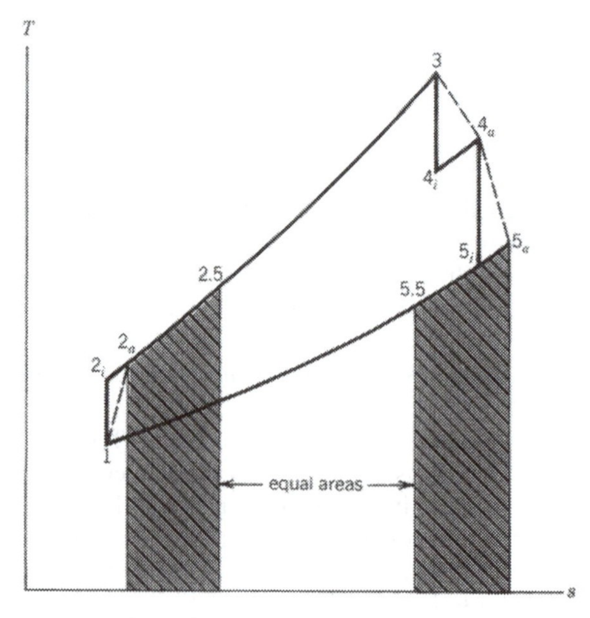

Figure 5.28. Temperature–entropy diagram for a regenerative air standard gas turbine engine.

For constant specific heats,

$$\eta_{reg} = \frac{T_{2.5} - T_{2a}}{T_{5a} - T_{2a}} \tag{5.28}$$

In an actual gas turbine with a regenerator, there will be a pressure drop in the regenerator. Effectiveness, pressure drop, and leakage are the three heat exchanger performance parameters that influence the efficiency of a gas turbine engine with exhaust heat recovery.

Collman (4) estimates that a metal recuperator core requires 2.5 times more material than a comparable metal regenerator matrix. After adding ducting and shell allowances to the metal recuperator and a cover and drive allowances for the regenerator, Collman estimates that the gas turbine engine with a regenerator maintains its weight advantage over the gas turbine engine with a recuperator.

Performance, cost, and space available must be considered when choosing among a drum design, a single-disk design, and a two-disk design. Most designers have narrowed their choices to the single- and two-disk designs. Assuming that the effects of flow maldistribution can be minimized, there should be no difference in effectiveness between the single- and two-disk arrangements. There is a difference in seal length, since these are proportional to the diameter of the disk with the total seal length of the two-disk design being 40% greater than that of the single-disk design. This suggests a potential for lower seal leakage in the single-disk arrangement. On the other hand, the larger diameter of the single-disk design results in greater thermal distortion, which complicates the seal design and can lead to lower power output.

Example Problem 5.6

An air standard gas turbine engine operates on the regenerative cycle (see Figure 5.22). The compressor inlet temperature is 519°R (288 K), compressor efficiency is 87%, turbine inlet temperature is 2520°R (1400 K), gas generator turbine efficiency is 89%, and power turbine efficiency is 89%. Calculate, assuming that $p_1 = p_5 = p_{5.5} = 14.70$ psia (101.3 kPa), $p_2 = p_{2.5} = p_3 = 58.80$ psia (405.2 kPa), and air enters the compressor at the rate of 1 lb/s (1 kg/s), the net work, heat added, thermal efficiency, and power developed (in kW) if the regenerator effectiveness is 75%. Compare with the results for a cycle without a regenerator. Neglect the mass of fuel added.

Solution

Basis: 1 lb dry air

From Table B.2 (B.1)

Compressor (C)

$$h_1 = -6.0\ (-14.3),\ Pr_1 = 1.2093\ (1.2045)$$

$$Pr_2 = Pr_1\left(\frac{p_2}{p_1}\right) = 4.837\ (4.818)$$

$$h_{2i} = 54.6\ (126.4)$$

$$w_{C,a} = \frac{h_{2i} - h_1}{\eta_C} = 69.7 \text{ Btu/lb } (161.7 \text{ kJ/kg})$$

$$h_{2a} = 69.7 - 6.0 = 63.7 \ (147.4)$$

$$T_{2a} = 807°\text{R} \ (448 \text{ K})$$

Gas Generator Turbine (GT)

$$T_3 = 2520°\text{R} \ (1400 \text{ K}),\ h_3 = 521.7 \ (1212.7),\ Pr_3 = 450.9 \ (450.9)$$

$$p_3 = 58.8 \text{ psia } (405.2 \text{ kPa})$$

$$w_{GT,a} = w_{C,a} = 69.7 \ (161.7)$$

$$w_{GT,i} = \frac{w_{GT,a}}{\eta_{GT}} = 78.3 \text{ Btu/lb } (181.7 \text{ kJ/kg})$$

$$h_{4i} = h_3 - w_{GT,i} = 443.4 \ (1031.0)$$

$$Pr_{4i} = 279.2 \ (279.4)$$

$$p_4 = p_3 \left(\frac{Pr_{4i}}{Pr_3}\right) = 36.41 \text{ psia } (251.1 \text{ kPa})$$

Power Turbine (PT)

$$p_4 = 36.41 \text{ psia } (251.1 \text{ kPa})$$

$$h_{4a} = h_3 - w_{PT,a} = 452.0 \ (1051.0)$$

$$Pr_{4a} = 295.0 \ (295.3)$$

$$p_5 = 14.70 \text{ psia } (101.3 \text{ kPa})$$

$$Pr_{5i} = Pr_{4a}\left(\frac{p_5}{p_4}\right) = 119.10 \ (119.13)$$

$$h_{5i} = 325.1 \ (756.8)$$

$$w_{PT,a} = (h_{4a} - h_{5i})\eta_{PT} = 112.9 \text{ Btu/lb } (261.8 \text{ kJ/kg})$$

$$h_{5a} = h_{4a} - w_{PT,a} = 339.1 \ (789.2)$$

$$T_{5a} = 1870°\text{R} \ (1040 \text{ K})$$

(a) For basic (simple) cycle

$$q_{in} = h_3 - h_{2a} = 458.0 \text{ Btu/lb } (1065.3 \text{ kJ/kg})$$

$$w_{net} = w_{PT,a} = 112.9 \text{ Btu/lb } (261.8 \text{ kJ/kg})$$

$$\dot{W}_{net} = \frac{(112.9)(1)(3600)}{3413} = 119.1 \text{ kW } (261.8 \text{ kW})$$

$$\eta_{th} = \frac{w_{net}}{q_{in}} = 0.247 \text{ or } 24.7\% \ (0.246 \text{ or } 24.6\%)$$

(b) For engine with a regenerator

$$\eta_{\text{reg}} = 0.75 = \frac{h_{2.5} - h_{2a}}{h_{5a} - h_{2a}}$$

$$h_{2.5} = 270.3 \ (628.8)$$

$$T_{2.5} = 1616°\text{R} \ (898 \text{ K})$$

$$q_{\text{in}} = h_3 - h_{2.5} = 251.4 \text{ Btu/lb} \ (583.9 \text{ kJ/kg})$$

$$\eta_{\text{th}} = \frac{112.9}{251.4} = 0.449 \text{ or } 44.9\% \left(\frac{261.8}{583.9} = 0.448\right)$$

Example Problem 5.6 illustrates the change in thermal efficiency with no change in net work since no pressure losses were considered, thereby illustrating that the addition of a regenerator can greatly increase the thermal efficiency of the gas turbine. It must be remembered that in an actual gas turbine with a regenerator, there will be a pressure drop in the regenerator and possible leakage, thereby lowering the net work from the cycle. Also, addition of a regenerator increases the initial cost of the gas turbine, partially offsetting the savings resulting from the higher thermal efficiency.

Figure 5.29 illustrates the effect of regenerator effectiveness on cycle thermal efficiency as a function of pressure ratio for a fixed turbine inlet temperature. Figure

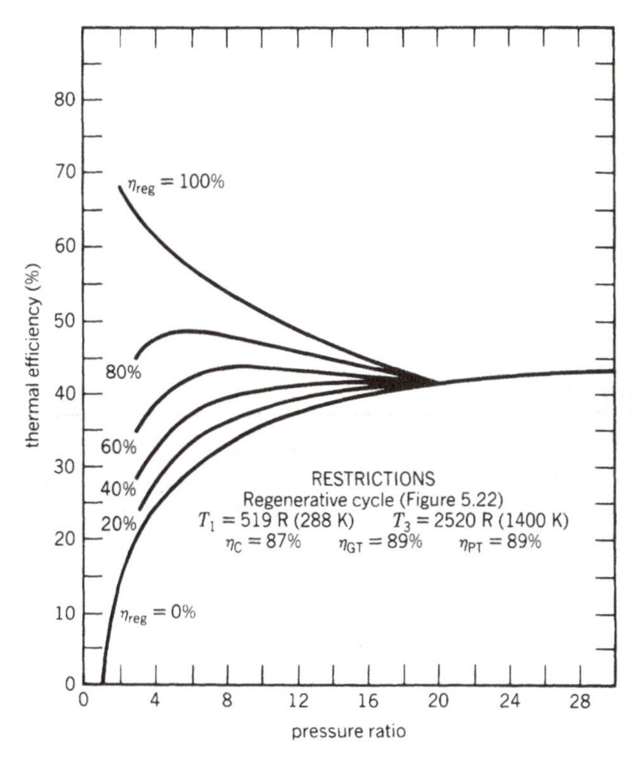

Figure 5.29. Effect of regenerator effectiveness on cycle thermal efficiency.

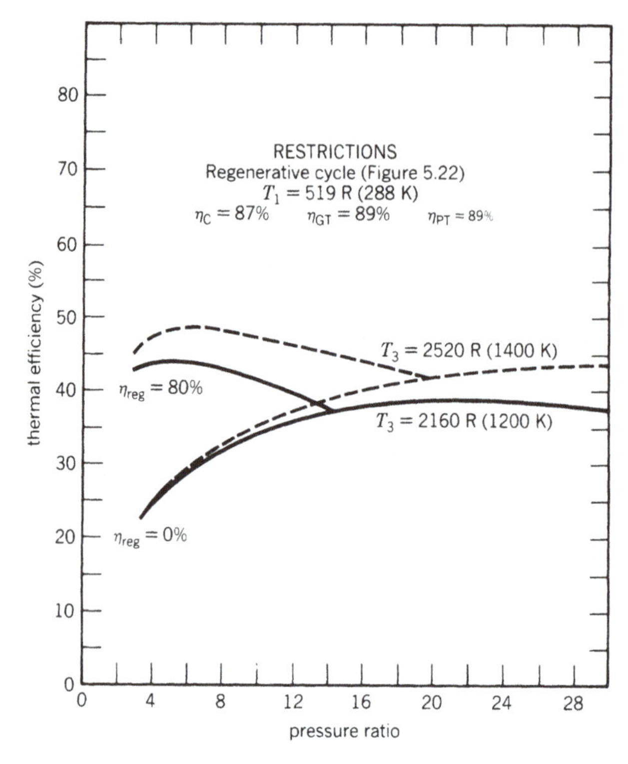

Figure 5.30. Effect of turbine inlet temperature and regenerator effectiveness on cycle thermal efficiency.

5.30 illustrates the effect of turbine inlet temperature on cycle thermal efficiency as a function of pressure ratio for a fixed regenerator effectiveness.

Example Problem 5.7

An air standard gas turbine engine operates on the regenerative cycle shown below. The compressor inlet temperature is 288 K (519°R), compressor efficiency is 87%, turbine inlet temperature 1400 K (2520°R), gas generator turbine efficiency 89% and power turbine efficiency is 89%. Assume that 2.0% of the air leaving the compressor leaks to state 5.5 from state 2. Assume that no energy is transferred to this air; that is, it leaks from state 2 to the pressure at state 5.5 with no change in temperature. Calculate, assuming that $p_1 = p_5 = p_{5.5} = 101.3$ kPa (14.70 psia), that $p_2 = p_{2.5} = p_3 = 405.2$ kPa (58.80 psia), and that air enters the compressor at the rate of 1 kg/s (1 lb/s), the net work, heat added, thermal efficiency and power developed by the power turbine (in kW) if the regenerator effectiveness is 85%. Neglect the mass of fuel added. Compare the answers with those of Example Problem 5.6.

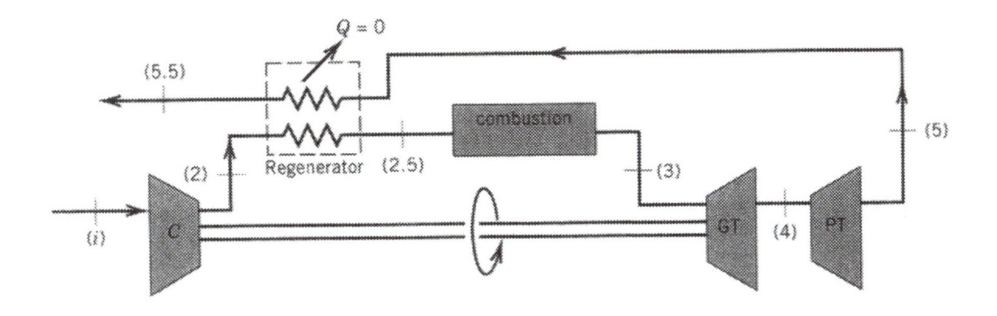

Solution

Basis: 1 kg air into the compressor

The compressor portion of this problem was solved in Example Problem 5.6. Values are summarized as follows:

State	T, K (°R)		p, kPa (psia)		h, kJ/kg (Btu/lb)		Pr	
1	288	(519)	101.3	(14.70)	−14.3	(−6.0)	1.2045	(1.2093)
2a	448	(807)	405.2	(58.80)	147.4	(63.7)	4.818	(4.837)
3	1400	(2520)	405.2	(58.80)	1212.7	(521.7)	450.9	(450.9)

$$w_{C,a} = 161.7 \text{ kJ/kg } (69.7 \text{ Btu/lb})$$

Gas Generator Turbine

The power developed by the turbine must be equal to the power required to drive the compressor (neglecting bearing losses, etc.) or on a basis of one kg into the compressor.

Gas Generator Turbine (GT)

$$|(1)(w_{C,a})| = |(0.98)(w_{GT,a})|$$

$$w_{GT,a} = \frac{161.7}{0.980} = 165.0 \text{ kJ/kg } (71.1 \text{ Btu/lb})$$

$$w_{GT,i} = \frac{W_{GT,a}}{\eta_{GT}} = 185.4 \text{ kJ/kg } (79.9 \text{ Btu/lb})$$

$$h_{4i} = h_3 - w_{GT,i} = 1027.3 \ (441.8)$$

$$Pr_{4i} = 276.5 \ (276.3)$$

$$p_4 = p_3\left(\frac{Pr_{4i}}{Pr_3}\right) = 248.5 \text{ kPa } (36.03 \text{ psia})$$

$$h_{4a} = h_3 - w_{GT,a} = 1047.7 \ (450.6)$$

$$Pr_{4a} = 292.7 \ (292.4)$$

Power Turbine (PT)

$$P_4 = 248.5 \text{ kPa } (36.03 \text{ psia})$$

$$p_5 = 101.3 \text{ kPa } (14.70 \text{ psia})$$

$$h_{4a} = 1047.7 \text{ kJ/kg } (450.6 \text{ Btu/lb})$$

$$Pr_{4a} = 292.7\ (292.4)$$

$$Pr_{5i} = Pr_{4a}\left(\frac{p_5}{p_4}\right) = 119.32\ (119.30)$$

$$h_{5i} = 756.1\ (325.3)$$

$$w_{PT,a} = \eta_{PT}(h_{4a} - h_{5i}) = 259.5 \text{ kJ/kg } (111.5 \text{ Btu/lb})$$

$$h_{5a} = h_{4a} - w_{PT,a} = 788.2 \text{ kJ/kg } (339.1 \text{ Btu/lb})$$

$$T_{5a} = 1039 \text{ K } (1870°\text{R})$$

$$\eta_{reg} = 0.85 = \frac{h_{2.5} - h_{2a}}{h_{5a} - h_{2a}}$$

$$h_{2.5} = 692.1 \text{ kJ/kg } (297.8 \text{ Btu/lb})$$

$$T_{2.5} = 954 \text{ K } (1718°\text{R})$$

$$q_{in} = h_3 - h_{2.5} = 520.6 \text{ kJ/kg } (223.9 \text{ Btu/lb})$$

$$w_{net} = w_{PT,a} = 259.5 \text{ kJ/kg through power turbine } (111.5 \text{ Btu/lb})$$

On a basis of 1 kg/s of air into the compressor (1 lb/s)

$$\dot{W}_{net} = \dot{W}_{PT,a} = (w_{PT,a})(1 - 0.02)$$

$$= 254.3 \text{ kW/(kg/s into compressor) } (115.3 \text{ kW} = 109.3 \text{ Btu/s})$$

$$\dot{q}_{in} = (q_{in})(1 - 0.02)$$

$$= 510.2 \text{ kJ/s } (219.4 \text{ Btu/s})$$

$$\eta_{th} = \frac{\dot{W}_{net}}{\dot{q}_{in}} = 0.498 \text{ or } 49.8\%\ (0.498 \text{ or } 49.8\%)$$

One should observe that when the two regenerative cycles are compared, the cycle with the recuperator has a higher specific power, but the cycle with the rotating regenerator with 2% leakage has a higher thermal efficiency, therefore a lower specific fuel consumption.

Solving Example Problem 5.7 for air leakage rates of 4%, 6%, and 8% yields the results shown in Table 5.9.

5.7 STEAM-INJECTED GAS TURBINE CYCLE

A modification to the basic cycle, which has been used for power generation since the mid-1980s, is the steam-injected gas turbine. The flow diagram for the steam-injected gas turbine is shown in Figure 5.31. The exhaust gases from the power

TABLE 5.9. **Effect of Regenerator Leakage on Thermal Efficiency and Specific Power**

% Leakage	Thermal Efficiency, %	Specific Power, kW/(kg/s) [kW/(lb/s)]
2.0	49.8	254.3 (115.3)
4.0	49.2	245.9 (111.5)
6.0	48.5	237.8 (107.8)
8.0	47.8	229.4 (104.0)

turbine are used as the energy source in a heat recovery steam generator (HRSG) where energy is transferred from the exhaust gases to boiler feedwater. The water at the exit from the HRSG is high-pressure steam, which is injected along with fuel in the combustion chamber. It is also possible to inject a portion of the high pressure steam into the gas turbine between the gas generator turbine and the power turbine.

The advantages of the steam-injected gas turbine are:

1. Able to place into service quickly.
2. Helps reduce NO_x emissions.
3. Increases the net power output compared with a comparable simple cycle gas turbine.
4. Has a lower heat rate (and a higher thermal efficiency) than a comparable simple cycle gas turbine.
5. Lower capital cost than for a combined cycle which is discussed in Section 5.12.
6. Can be converted to a combined cycle.

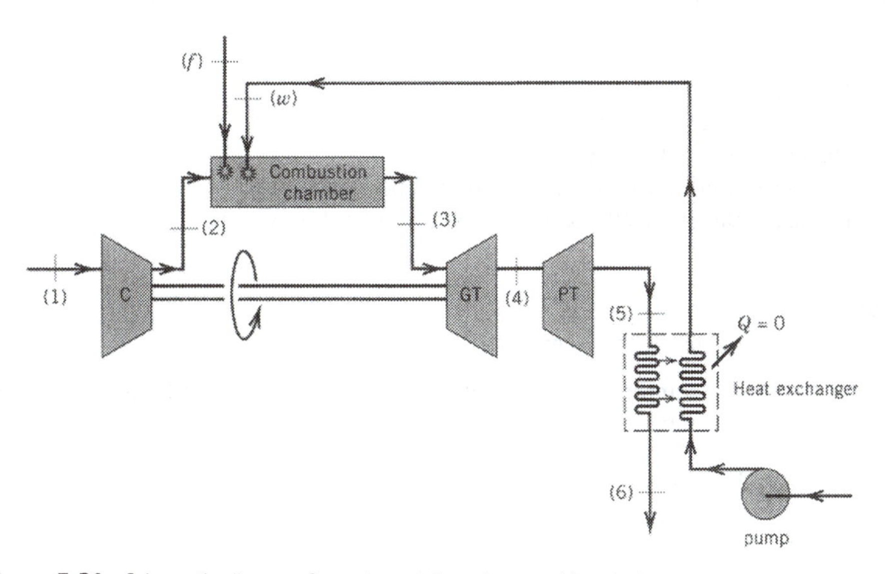

Figure 5.31. Schematic diagram for a steam-injected gas turbine cycle.

Disadvantages of the steam injected gas turbine are:

1. A constant supply of purified water is required since the steam after being injected into the gas turbine cycle is exhausted to the atmosphere.
2. A heat recovery steam generator must be added.
3. Care must be taken to avoid injecting compressed liquid (water slugs) into the combustion chamber as this will cause a partial or complete extinction of the flame in the combustion chamber.

The amount of water injected into the discharge area between the compressor discharge and combustion chamber is usually between 2% and 3% of the compressor discharge air, although flow can be as high as 5% of the compressor discharge air flow.

The injected steam must be superheated under all conditions and at a pressure higher than the compressor discharge pressure.

Example Problem 5.8

A gas turbine engine operating on the steam-injected cycle (see Figure 5.31) has a compressor inlet pressure of 14.7 psia (101.3 kPa), a temperature of 519°R (288 K), a compressor exit and turbine inlet pressure of 176.4 psia (1215.6 kPa), a turbine inlet temperature of 2520°R (1400 K), a power turbine exit pressure of 14.7 psia (101.3 kPa), a compressor efficiency of 87% and a gas generator turbine and power turbine each with an efficiency of 89%. The fuel, supplied as a liquid at 537°R (298 K), is *n*-octane, C_8H_{18}. Steam at a pressure of 200 psia (1378.2 kPa) and 1175°R (653 K) and equal to 2.5% of the compressor discharge flow is injected into the combustion chamber. If air enters the compressor at the rate of 1.0 lb/s, calculate, assuming the products leaving the combustion chamber are dry air and water vapor but taking into account the mass of fuel added

(a) The fuel–air ratio, lb fuel/lb air
(b) The pressure at the exit from the gas generator turbine
(c) The cycle thermal efficiency
(d) The net power, in kW, developed by the power turbine
(e) The heat rate, Btu/kWh, based on the lower heating value

Solution

Basis: 1 lb/s air entering the compressor

The values for the compressor are the same as those determined for Example Problem 5.3. These values are:

$$w_{C,a} = 147.7 \text{ Btu/lb } (343.0 \text{ kJ/kg})$$

$$h_{2a} = 141.7 \text{ Btu/lb } (328.7 \text{ kJ/kg})$$

$$T_{2a} = 1122°\text{R } (623 \text{ K})$$

(a) Combustion Chamber (CC)

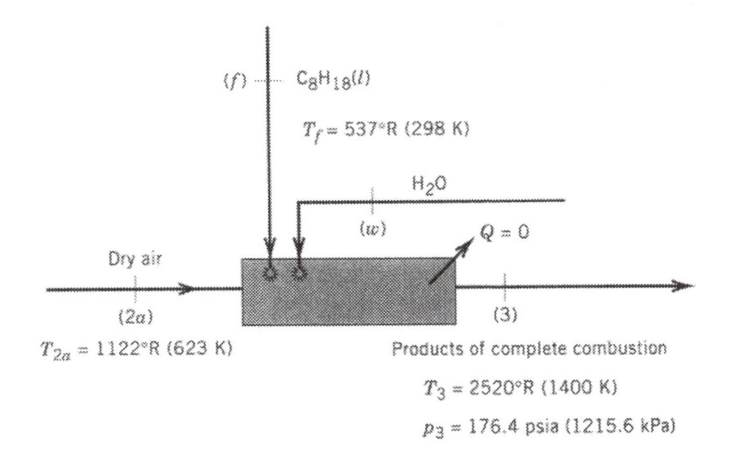

Writing a conservation of mass on the basis of one pound mole of C_8H_{18} around the combustion chamber yields the following equation

$$C_8H_{18} + X\ \text{D.A.} + \frac{(28.965X)(0.025)}{18}\ H_2O$$

$$\rightarrow 8\ CO_2 + (9 + 0.040X)H_2O + X\ \text{D.A.} - 12.5\ O_2$$

Writing an energy balance around the combustion chamber yields

$$X = 175.1$$

$$f' = \frac{114.2336}{(175.1)(28.965)} = 0.02252\ \frac{\text{lb fuel}}{\text{lb dry air}}$$

It will now be assumed that 1.02252 lb of dry air and 0.025 lb of H_2O enter the gas generator turbine for every pound of air entering the compressor.

(b) Gas Generator Turbine (GT)

$$\dot{w}_{C,a} = \dot{w}_{GT,a}$$

$$(1)(147.7) = (1.0225)(521.7 - h_{4a})_{air} + \frac{0.025}{18}[-85320.3 - h_{4a}]_{H_2O}$$

By trial and error, the actual temperature at state 4 is 2033°R (1129 K). Taking into account the gas generator turbine efficiency of 87%, the resulting ideal temperature and pressure at the exit from the gas generator turbine are 1972°R (1096 K) and 63.68 psia (438.9 kPa), respectively.

Power Turbine (PT)
The conditions at the inlet to the power turbine are

$$p_4 = 63.68 \text{ psia } (438.9 \text{ kPa})$$

$$T_{4a} = 2033°\text{R } (1129 \text{ K})$$

$$\dot{m}_{air} = 1.0225 \text{ lb/s}$$

$$\dot{m}_{H_2O} = 0.025 \text{ lb/s}$$

The pressure at the exit from the power turbine is 14.70 psia (101.3 kPa). The ideal temperature at the exit from the power turbine can be solved by trial and error. The determined value is 1405°R (781 K). The details of the trial and error solutions for the gas generator turbine and power turbine are left as an exercise for the reader.

It is next important to determine the specific power developed by the power turbine on the basis of 1 lb/s of air entering the compressor. The results are

$$\dot{w}_{PT,a} = \left[(1.0225)(383.6 - 214.5) + \frac{0.025}{18}(-90507.0 + 96613.2)\right](0.89)$$

$$= 161.5 \text{ Btu/s per lb/s air into compressor}$$

The resulting actual temperature at the exit from the power turbine is 1476°R (820 K).

(c) $\eta_{th} = \dfrac{(161.5)}{(1.0)(0.02252)(19100)} = 0.375 \text{ or } 37.5\%$

(d) $\dot{w}_{PT,a} = \dfrac{(161.5)(3600)}{3413} = 170.3 \text{ kW}$

(e) $HR = \dfrac{(0.02252)(19100)(3600)}{170.3} = 9093 \text{ Btu/kWh}$

Comparing the results of Example Problem 5.8 with the results for Example Problem 5.5, which is a simple cycle for the same pressure ratio, component efficiencies, and turbine inlet temperature, shows that the specific power increased by 8.1% and the heat rate was lowered by 3.1%.

Table 5.10 lists known data for several gas turbines operating on the steam-injected gas turbine cycle. One should note the wide range of compressor pressure ratios, nominal outputs, and steam flow rates as a percent of compressor air flows for the engines listed. It is also of interest to compare the heat rates for the steam-injected gas turbines listed in Table 5.10 to the comparable simple cycle gas turbines listed in Table 5.4.

TABLE 5.10. **Steam-injected Gas Turbine Engine Specifications (2)**

Manufacturer Model	Year	Nominal Rating	Heat Rate	Pressure Ratio	Flow	Exhaust Temper-ature	Comments
Allison Gas Turbine							
501-KH Steam	1986	5,940 kW	8670 Btu/kWh	11.8	40.6 lb/s	915°F	6.0 lb/s steam
GE Marine & Industrial							
LM1600-PB STIG	1991	16,900 kW	8,607 Btu/kWh	25.1	116 lb/s	878°F	20,000 lb/h steam
LM2500-PH STIG	1986	26,300 kW	8,800 Btu/kWh	19.7	164 lb/s	937°F	40,000 lb/h steam
LM5000-PD STIG	1986	50,425 kW	7,950 Btu/kWh	31.7	347 lb/s	752°F	78,000 lb/h HP & 56,000 lb/h LP steam

5.8 EVAPORATIVE–REGENERATIVE GAS TURBINE CYCLES

Another way to increase the specific power output of the simple gas turbine cycle is the evaporative–regenerative gas turbine cycle, which is shown schematically in Figure 5.32. In this cycle, water is sprayed into the air as it leaves the compressor. As the water evaporates, it lowers the temperature of the air, resulting in a large temperature difference between the temperature of the air leaving the power turbine (state 5) and the air leaving the evaporator (state 2.3), thereby justifying the addition of a regenerator/recuperator between the evaporator exit (state 2.3) and the combustion chamber inlet (state 2.5).

The mass of water added and the fact water has a higher specific heat than air results in a gas turbine with a higher specific power and a lower heat rate when compared to a gas turbine operating on a simple cycle.

Care must be taken to be certain all of the water sprayed into the air in the evaporator is evaporated. The maximum amount is limited in theory to that which results in the air being saturated at the evaporator exit (state 2.3) or when the temperature at the evaporator exit is equal to the temperature of the water being sprayed into the evaporator.

Example Problem 5.9

A gas turbine engine operates on the evaporative-regenerative cycle as shown in Figure 5.32. Known values are

$T_1 = 288$ K (519°R) $\quad p_1 = p_5 = p_{5.5} = 101.3$ kPa (14.70 psia)

$T_{2.3} = 444$ K (800°R) $\quad P_2 = P_{2.3} = P_{2.5} = P_3 = 1216.2$ kPa (176.4 psia)

$T_3 = 1400$ K (2520°R) $\quad p_8 = 1447$ kPa (210 psia)

$T_8 = 433$ K (780°R)

$\eta_C = 87\%$, $\eta_{GT} = 89\%$, $\eta_{PT} = 89\%$, $\eta_{reg} = 85\%$,

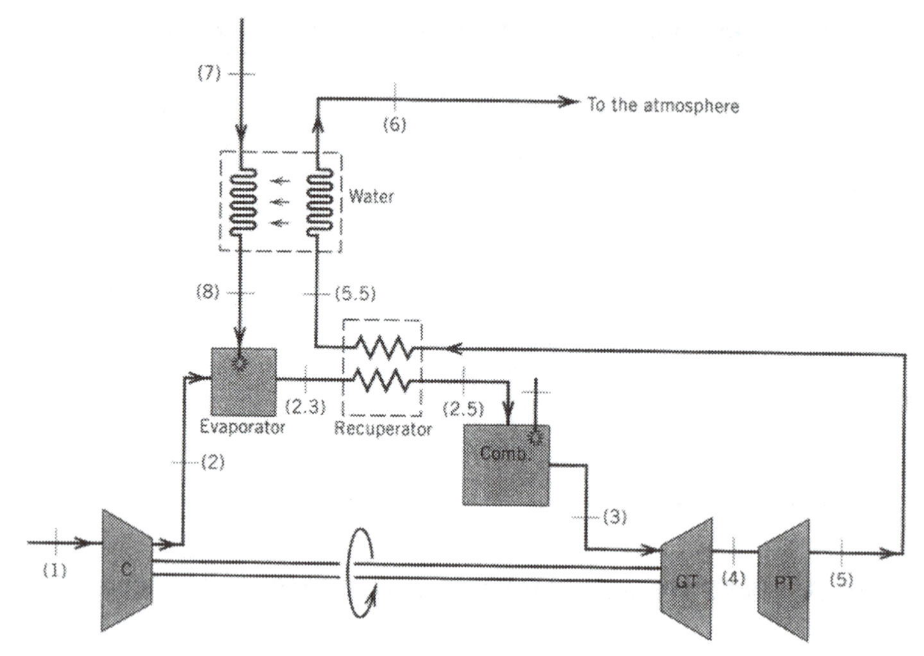

Figure 5.32. Evaporative–regenerative gas turbine cycle.

Calculate, neglecting the mass of fuel added, assuming air enters the compressor at the rate of 1 kg/s (1 lb/s), taking into account the mass of water added and assuming an air–water vapor mixture after the evaporator:

(a) The pressure and temperature at the exit from the gas generator turbine
(b) The net power, in kW, developed by the power turbine
(c) The cycle thermal efficiency
(d) The heat rate, kJ/kWh (Btu/kWh)

Solution

Basis: 1 kg/s (1 lb/s)

The values for the compressor and the evaporator were determined in Example Problems 5.3 and 2.4, respectively. These values are

$$w_{C,a} = 343.0 \text{ kJ/kg } (147.7 \text{ Btu/lb})$$

$$T_{2a} = 623 \text{ K } (1122°\text{R})$$

$$h_{2a} = 328.7 \text{ kJ/kg } (141.7 \text{ Btu/lb})$$

$$\dot{m}_w = 0.089 \text{ kg/s } (0.089 \text{ lb/s})$$

Gas Generator Turbine (GT)

$$\dot{w}_{C,a} = \dot{w}_{GT,a}$$

$$(1)(343.0) = (1)(1212.7 - h_{4a})_a + \frac{0.089}{18}(-198322 - \bar{h}_{4a})_w$$

(a) By trial and error, the actual temperature at the exit from the gas generator turbine is determined to be

$$T_{4a} = 1156 \text{ K } (2080°\text{R})$$

$$w_{GT,i} = \frac{343.0}{0.89} = 385.4 \text{ kJ/s}$$

Therefore

$$385.4 = \Sigma\,(m_i h_i)_3 - \Sigma(m_i h_i)_4$$

By trial and error

$$T_{4i} = 1125 \text{ K } (2025°\text{R})$$

The pressure at the exit from the gas generator turbine may be determined by the equation

$$\Sigma(\dot{m}_i s_i)_3 - \Sigma(m_i s_i)_4 = 0$$

or

$$\Sigma\eta_i\left(\bar{s}_3^0 - \bar{s}_4^0 - \overline{\text{R}}\,\ell\text{n}\,\frac{p_3}{p_4}\right) = 0$$

By trial and error

$$p_4 = 478.2 \text{ kPa } (69.4 \text{ psia})$$

Power Turbine (PT)

$$p_4 = 478.2 \text{ kPa } (69.4 \text{ psia})$$

$$T_{4a} = 1156 \text{ K } (2080°\text{R})$$

$$p_5 = 101.3 \text{ kPa } (14.70 \text{ psia})$$

The ideal temperature at the exit from the power turbine may be determined by solving the equation

$$\Sigma n_i\left(\bar{s}_4^0 - \bar{s}_{5i}^0 - \overline{\text{R}}\,\ell\text{n}\,\frac{478.2}{101.3}\right)_i = 0$$

By trial and error

$$T_{5i} = 789 \text{ K } (1420°\text{R})$$

(b) $\dot{w}_{PT,a} = 0.89\left[(1)(924.0) - 507.7) + .00494(-209233 + 224246)\right]$

$$= 436.6 \text{ kJ/s } (187.6 \text{ Btu/s})$$

$$= 436.6 \text{ kW } (197.9 \text{ kW})$$

$$T_{5a} = 831 \text{ K } (1496°\text{R})$$

(c) It is necessary to determine the temperature at the exit from the recuperator to determine the heat added.

For the regenerator

$$0.85 = \frac{h_{2.5} - h_{2.3}}{h_{5a} - h_{2.3}}$$

$$h_{2.3} = (1)(143.4) + 0.00494\,(-236848)$$

$$= -1027.7$$

$$h_{5a} = (1)(554.0) + (0.00494)(-222616)$$

$$= -546.7$$

$$h_{2.5} = 0.85(-546.7 + 1027.7) - 1027.7$$

$$= -618.9$$

$$T_{2.5} = 769 \text{ K } (1383°\text{R})$$

$$\dot{q}_{\text{in}} = [(1)(1212.7) + 0.00494(-198321.7)] - (-618.9)$$

$$= 851.0 \text{ kJ/s } (369.2)$$

$$\eta_{\text{th}} = \frac{436.6}{851.0} = 0.513\ (0.508)$$

(d) $HR = \dfrac{(851.0)(3600)}{436.6} = 7017 \text{ kJ/kWh } (6716 \text{ Btu/kWh})$

5.9 GAS TURBINE WITH INTERCOOLING

The preceding section discussed one way of improving the performance of the basic gas turbine engine. This involved adding a regenerator to the gas turbine, which increased the cycle thermal efficiency without, at least ideally, changing the net work for the cycle.

Two other improvements can be made in the gas turbine cycle. These are intercooling and reheat. Both of these improvements increase the net work obtained from the cycle, intercooling decreasing the compressor work without changing the turbine work, reheat increasing the turbine work without changing the compressor work. The following discussion of intercooling and reheat assumes the same compressor inlet temperature, turbine inlet temperature, and same overall pressure ratio as the basic gas turbine cycle. This section discusses intercooling; reheat is discussed in the next section.

Compressors, as used in gas turbines, operate adiabatically. Ideally, the compression process is reversible. For a reversible, steady-flow process, the work is

$$-{}_1w_2 = \int_1^2 v\,dp + \Delta\text{KE} + \Delta\text{PE} \tag{2.7}$$

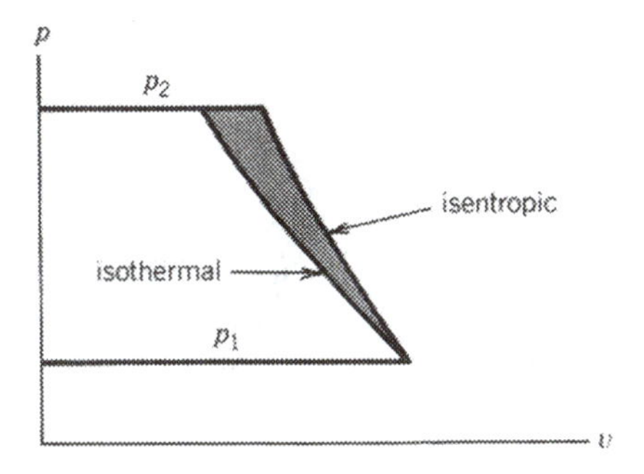

Figure 5.33. Pressure–specific volume diagram for an isothermal and isentropic compression process.

Since the initial and final states usually measured are the total states, Eq. (2.7) reduces to

$$-{}_1w_2 = \int_1^2 v\,dp$$

Examination of the pressure-volume diagram in Figure 5.33 shows that the work for an isothermal, steady-flow compression process from p_1 to p_2 is less than that for an isentropic compression process between the same two pressures, the decrease in work being represented by the shaded area.

It is impractical to construct a gas turbine compressor that would operate isothermally. The compression process can be made to approximate the isothermal compression process by intercooling, which involves the use of two or more compressors.

Figure 5.34 illustrates the schematic diagram for a gas turbine engine with two compressors and intercooling between the compressors. In the engine illustrated in Fig. 5.34, it is assumed that the gas generator turbine (GT) drives both compressors, the power turbine delivering power to an external device (load).

The first compressor (C_1) compresses air from ambient pressure to some intermediate pressure. The air leaving the first compressor enters the intercooler, where heat is removed. Ideally, no pressure drop occurs in the intercooler and the temperature of the air as it leaves the intercooler (state 1.5) is the same as the temperature at the inlet to the first compressor (state 1). From a practical standpoint, there is a pressure

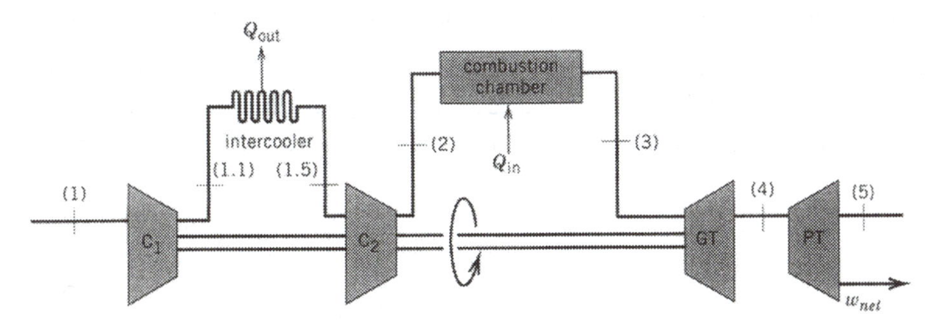

Figure 5.34. Gas turbine engine with intercooling.

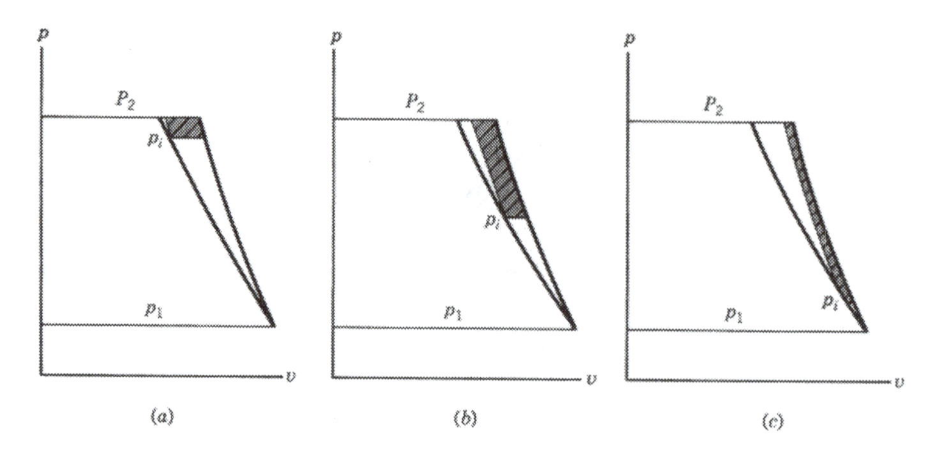

Figure 5.35. Pressure–specific volume diagrams for compression process with intercooling. (a) High intercooler pressure. (b) Moderate intercooler pressure. (c) Low intercooler pressure.

drop in the intercooler and the temperature of the air leaving the intercooler will be considerably above the temperature entering the compressor C_1, since the intercooler is normally an air-to-air heat exchanger and it is impossible to cool back to the compressor inlet temperature. The second compressor (C_2) completes the compression process to the desired final pressure. Figure 5.35 illustrates three p–v diagrams for different intermediate pressures. In all three of the diagrams, it is assumed that the temperature leaving the intercooler is the same as that entering the first compressor.

The three diagrams illustrate the work saved by using intercooling, this work saved being represented by the shaded area. Examination of Figure 5.35 suggests that there should be an optimum intercooler pressure. It can be shown that this optimum intercooler pressure is (assuming no pressure drop in the intercooler and that the temperature leaving the intercooler is the same as the temperature entering the first compressor)

$$p_i = \sqrt{p_1 p_2} \tag{5.29}$$

Because of the cost and complexity of adding an intercooler, no more than one intercooler is ever used in the design of a gas turbine engine. Since adding an intercooler lowers the work required to drive the compressor and changes the total turbine work only slightly (how much depends on the component arrangement), the net work for the cycle will increase. This is illustrated in the following example problem.

Example Problem 5.10

An air standard gas turbine operates on the cycle shown in Figure 5.34. Known values are

$T_1 = 519°R\ (288\ K)$	$\eta_{C_1} = 87\%$	$p_1 = p_5 = 14.7$ psia (101.3 kPa)
$T_{1.5} = 519°R\ (288\ K)$	$\eta_{C_2} = 87\%$	$p_2 = p_3 = 176.4$ psia (1215.6 kPa)
$T_3 = 2520°R\ (1400\ K)$	$\eta_{GT} = 89\%$	$p_{1.1} = p_{1.5} = 50.9$ psia (350.9 kPa)
	$\eta_{PT} = 89\%$	

Calculate, on a basis of 1 lb/s (1 kg/s) of air entering the first compressor and neglecting the fuel added:

(a) The optimum intercooler pressure.
(b) The total compressor work, Btu/lb (kJ/kg)
(c) The net power, kW.
(d) The cycle thermal efficiency.
(e) Compare the results with a basic gas turbine operating with the same overall pressure ratio, compressor and turbine inlet temperatures, and efficiencies.

Solution

Basis: 1 lb dry air

(a) The optimum intercooler pressure is, from Eq. (5.29),

$$p_{1.1} = p_{1.5} = \sqrt{p_1\, p_2} = 50.92 \text{ psia } (350.9 \text{ kPa})$$

Therefore, the intercooler pressure is the optimum intercooler pressure.

Compressor C_1

(b) From Table B

$$h_1 = -6.0\,(-14.3)$$

$$Pr_1 = 1.2093\,(1.2045)$$

$$Pr_{1.1i} = 1.2093\left[\frac{50.9}{14.7}\right] = 4.187\,(4.174)$$

$$h_{1.1i} = 47.0\,(109.1)$$

$$w_{C1,a} = \frac{47.0 - (-6.0)}{0.87}$$

$$= 60.9 \text{ Btu/lb } (141.8 \text{ kJ/kg})$$

$$h_{1.1a} = -6.0 + 6.09 = 54.9\,(127.5)$$

$$T_{1.1a} = 771°\text{R } (428 \text{ K})$$

Compressor C_2

Since the pressure ratio, inlet temperature, and efficiency of compressor C_2 are the same as for C_1:

$$w_{C2,a} = 60.9 \text{ Btu/lb } (141.8)$$

$$T_{2a} = 771°\text{R } (428 \text{ K})$$

$$h_{2a} = 54.9\,(127.5)$$

$$w_{C,\text{total}} = w_{C1} + w_{C2}$$

$$= 121.8 \text{ Btu/lb } (283.6 \text{ kJ/kg})$$

Gas Generator Turbine (GT)

$$h_3 = 521.7\ (1212.7)$$

$$Pr_3 = 450.9\ (450.9)$$

$$w_{GT,a} = w_{C,total} = 121.8\ (283.6)$$

$$w_{GT,i} = \frac{121.8}{0.89} = 136.9\ (318.7)$$

$$h_{4i} = 521.7 - 136.9$$

$$= 384.8\ (894.0)$$

$$Pr_{4i} = 187.17\ (186.97)$$

$$p_4 = 176.4 \left[\frac{187.17}{450.9}\right]$$

$$= 73.2 \text{ psia } (504.1 \text{ kPa})$$

Power Turbine

$$h_{4a} = 521.7 - 121.8$$

$$= 399.9\ (929.1)$$

$$Pr_{4a} = 208.3\ (208.0)$$

$$Pr_{5i} = 208.3 \left[\frac{14.7}{73.2}\right]$$

$$= 41.831\ (41.782)$$

$$h_{5i} = 210.4\ (488.9)$$

(c) $w_{net} = w_{PT,a} = \eta_{PT}(h_{4a} - h_{5i})$

$$= 0.89\ (399.9 - 210.4)$$

$$= 168.7 \text{ Btu/lb } (391.8 \text{ kJ/kg})$$

$$\dot{W}_{net} = \frac{(168.7)(1)(3600)}{3413}$$

$$= 177.9 \text{ kW } (391.8 \text{ kW})$$

(d) $q_{in} = h_3 - h_{2a} = 521.7 - 54.9$

$$= 466.8 \text{ Btu/lb } (1085.2 \text{ kJ/kg})$$

$$\eta_{th} = \frac{168.7}{466.8} = 0.361\ (0.361)$$

(e) An engine with no intercooler for the same conditions was solved in Example Problem 5.3b. Comparing the results of Example Problem 5.3b with this Example Problem shows that by adding only an intercooler, the compressor work is decreased by approximately 18%, the net work increases by approximately 18%, the heat added increases by approximately 23%, and the thermal efficiency *decreases* by approximately 4%.

5.10 GAS TURBINE WITH REHEAT

Turbines in gas turbine engines operate approximately adiabatically. Ideally, the expansion process is reversible. A line of reasoning similar to that used with intercooling shows that the work obtainable from a turbine operating between fixed inlet and discharge pressures can be increased by constructing a turbine that operates isothermally. In practice, this is approximated by allowing the gases to expand from the maximum cycle pressure to some intermediate pressure, reheating at constant pressure to the maximum cycle temperature, and then expanding the gases in a second turbine to the minimum cycle pressure. Figure 5.36 is a schematic diagram for a gas turbine engine operating on the reheat cycle. The engine illustrated in Figure 5.36 has turbine GT driving the compressor, the net work developed by the power turbine (PT). Figure 5.37 illustrates the temperature-entropy diagram for a gas turbine with reheat.

The optimum intermediate pressure with reheat is the same as with intercooling or

$$p_{re} = \sqrt{p_3 p_5} \tag{5.30}$$

Adding reheat to a gas turbine engine increases the turbine work (therefore the net work) without changing the compressor work. This is illustrated in the following example problem.

Example Problem 5.11

An air standard gas turbine engine operates on the cycle shown in Figure 5.36. Known values are

$T_1 = 288$ K (519°R)	$\eta_C = 87\%$	$p_1 = p_5 = 101.3$ kPa (14.7 psia)
$T_3 = 1400$ K (2520°R)	$\eta_{GT} = 89\%$	$p_2 = p_3 = 1215.6$ kPa (176.4 psia)
$T_{4.5} = 1400$ K (2520°R)	$\eta_{PT} = 89\%$	

Calculate, on a basis of 1 kg/s (1 lb/s) of air entering the compressor, neglecting the fuel added and assuming the optimum reheat pressure

(a) The optimum reheat pressure
(b) The net power, kW
(c) The cycle thermal efficiency
(d) Compare the results with a basic gas turbine operating with the same pressure ratio, compressor and turbine inlet temperatures, and efficiencies

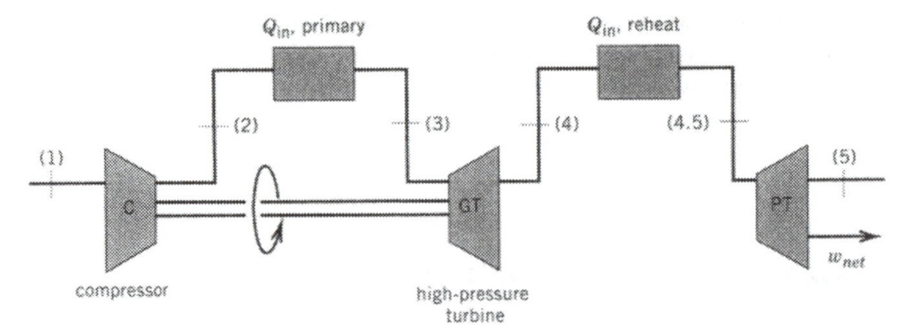

Figure 5.36. *Schematic diagram of a gas turbine with reheat where the high-pressure turbine drives the compressor.*

Solution

(a) The optimum reheat pressure is, from Eq. (5.30),

$$p_{re} = p_{4.5} = p_4 = \sqrt{p_3 p_5}$$
$$= 350.9 \text{ kPa } (50.92 \text{ psia})$$

Since the turbine GT drives the compressor, the pressure at the exit from this turbine is fixed by an energy balance. These are the same conditions as used in Example Problem 5.3b, which resulted in

$$p_4 = 408.2 \text{ kPa } (59.1 \text{ psia})$$

This pressure is above the optimum reheat pressure but is the one that would result because of the mechanical arrangement.

The solution to this problem is the same as that of Example Problem 5.3b except for the inlet temperature to the power turbine. Values at the states through the exit from the gas generator turbine are summarized below.

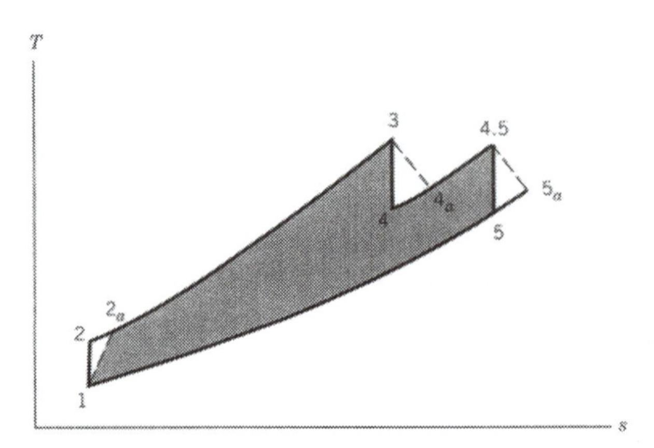

Figure 5.37. Temperature–entropy diagram for a gas turbine with reheat.

State	T, K (°R)		p, kPa (psia)		h, kJ/kg (Btu/lb)		Pr	
1_a	288	(519)	101.3	(14.70)	−14.3	(−6.0)	1.2045	(1.2093)
2_a	623	(1122)	1215.6	(176.4)	328.7	(141.7)	18.719	(18.789)
3	1400	(2520)	1215.6	(176.4)	1212.7	(521.7)	450.9	(450.9)
4_a	1109	(1996)	408.2	(59.1)	869.7	(374.0)	173.36	(173.20)

$$w_{C,a} = 343.0 \text{ kJ/kg } (147.7 \text{ Btu/lb})$$

(b) The conditions at the inlet to the power turbine are

$$p_{4.5} = 408.2 \text{ kPa } (59.1 \text{ psia})$$

$$T_{4.5} = 1400 \text{ K } (2520°\text{R})$$

$$h_{4.5} = 1212.7\ (521.7)$$

$$Pr_{4.5} = 450.9\ (450.9)$$

$$Pr_{5i} = 450.9 \left[\frac{101.3}{408.2}\right]$$

$$= 111.90\ (112.15)$$

$$h_{5i} = 737.6\ (317.6)$$

$$w_{\text{net}} = w_{\text{PT},a} = 0.89\ (1212.7 - 737.6)$$

$$= 422.8 \text{ kJ/kg } (181.6 \text{ Btu/lb})$$

$$\dot{W}_{\text{net}} = 422.8 \text{ kW } (191.6 \text{ kW})$$

$$q_{\text{in}} = (h_3 - h_{2a}) + (h_{4.5} - h_{4a})$$

$$= (1212.7 - 328.7) + (1212.7 - 869.7)$$

$$= 1227.0 \text{ kJ/kg } (527.7 \text{ Btu/lb})$$

(c) $\eta_{\text{th}} = \dfrac{422.8}{1227.0} = 0.345\ (0.344)$

(d) Comparing the results with those obtained in Example Problem 5.3b, one observes that the net work has increased by approximately 27% and the thermal efficiency has *decreased* by approximately 9%.

5.11 GAS TURBINE WITH INTERCOOLING, REHEAT, AND REGENERATION

Intercooling and reheat, when used alone, decrease the cycle thermal efficiency; therefore, they are seldom, if ever, used alone.

Figure 5.38 illustrates a schematic diagram of a reheat–regenerative–intercooled

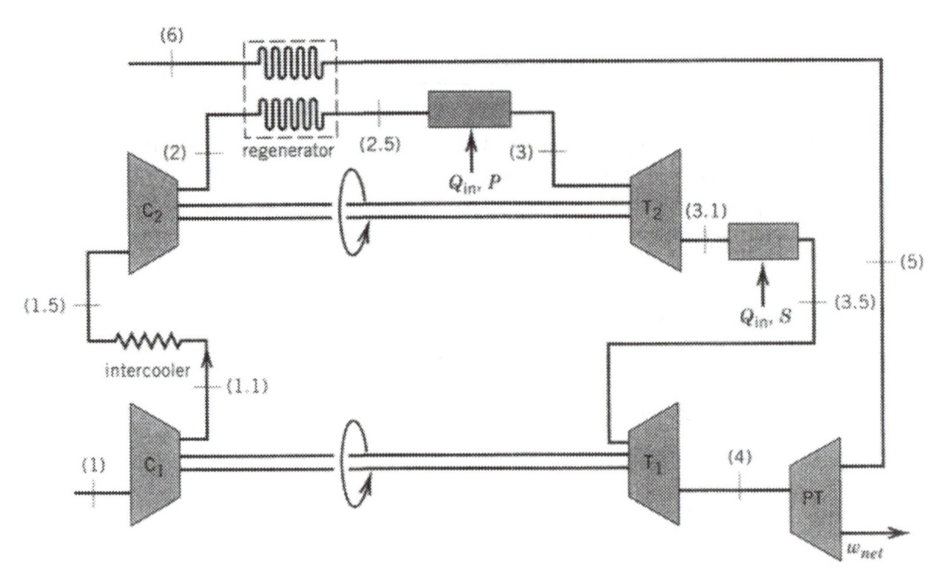

Figure 5.38. Schematic diagram of an intercooled-reheat-regenerative gas turbine cycle.

cycle that is usually used whenever reheat and/or intercooling are used. The engine illustrated in Figure 5.38 uses turbine T_2 (HPT) to drive compressor C_2 (HPC) and turbine T_1 (LPT) to drive the compressor C_1 (LPC). The power turbine (PT) delivers work to external equipment.

Figure 5.39 illustrates the temperature-entropy diagram for a gas turbine with intercooling–reheat–regeneration, the diagram corresponding to the schematic diagram shown in Figure 5.38.

Example Problem 5.12

An air standard gas turbine operates on the cycle shown in Figure 5.38. Known values are

T_1 = 519°R (288 K) $\quad \eta_{C1} = 87\% \quad p_1 = p_5 = p_6 = 14.7$ psia (101.3 kPa)

$T_{1.5}$ = 519°R (288 K) $\quad \eta_{C2} = 87\% \quad p_{1.1} = p_{1.5} = 50.9$ psia (350.9 kPa)

T_3 = 2520°R (1400 K) $\quad \eta_{T2} = 89\% \quad p_2 = p_{2.5} = p_3 = 176.4$ psia (1215.6 kPa)

$T_{3.5}$ = 2520°R (1400 K) $\quad \eta_{T2} = 89\% \quad p_{3.1} = p_{3.5}$

$\eta_{PT} = 89\%$

Recuperator Effectiveness = 80%

Calculate, on a basis of 1 lb/s (1 kg/s) of air entering compressor C_1 and neglecting the mass of fuel added:

(a) The total compressor work, Btu/lb (kJ/kg)
(b) The net work, Btu/lb (kJ/kg)
(c) The specific power, kW, developed by this engine

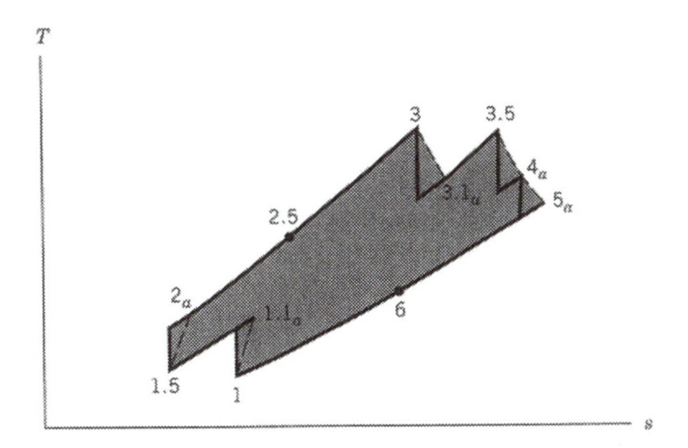

Figure 5.39. Temperature-entropy diagram of an intercooled-reheat-regenerative gas turbine cycle as shown in Figure 5.38.

(d) The cycle thermal efficiency

(e) Compare the results with a basic gas turbine operating with the same pressure ratio, compressor and turbine inlet temperatures, and efficiencies

Solution

Basis: 1 lb dry air

The solution to this problem through compressor C_2 is the same as that of Example Problem 5.10. Temperature, pressure and enthalpy values through state 2 are summarized below. See Figure 5.38 for state locations.

State	T, °R (K)		p, psia (kPa)		h, Btu/lb (kJ/kg)	
1	519	(288)	14.7	(101.3)	−6.0	(−14.3)
1.1_a	771	(428)	50.9	(350.9)	54.9	(127.5)
1.5	519	(288)	50.9	(350.9)	−6.0	(−14.3)
2_a	771	(428)	176.4	(1215.6)	54.9	(127.5)

$$w_{C,1a} = 60.9 \text{ Btu/lb } (141.8 \text{ kJ/kg})$$

$$w_{C,2a} = 60.9 \text{ Btu/lb } (141.8 \text{ kJ/kg})$$

Since turbine T_2 drives compressor C_2

$$w_{T,2a} = w_{C,2a} = 60.9\ (141.8)$$

$$w_{T,2i} = \frac{60.9}{0.89}$$

$$= 68.4\ (159.3)$$

$$68.4 = 521.7 - h_{3.1i}$$

$$h_{3.1i} = 453.3\ (1053.4)$$

$$Pr_{3.1i} = 297.5\ (297.3)$$

$$p_{3.1} = 176.4\left[\frac{297.5}{450.9}\right]$$

$$= 116.4 \text{ psia } (801.5)$$

$$h_{3.1a} = 521.7 - 60.9$$

$$= 460.8\ (1070.9)$$

Turbine T_1 drives compressor C_1. The known values are

$$p_{3.5} = 116.4 \text{ psia } (801.5)$$

$$T_{3.5} = 2520°\text{R } (1400 \text{ K})$$

$$h_{3.5} = 521.7\ (1212.7)$$

$$Pr_{3.5} = 450.9\ (450.9)$$

$$w_{T,1a} = w_{C,1a} = 60.9\ (141.8)$$

$$w_{T,1i} = \frac{60.9}{0.89} = 68.4\ (159.3)$$

$$h_{4i} = 453.3\ (1053.4)$$

$$Pr_{4i} = 297.5\ (297.3)$$

$$P_4 = 116.4\left[\frac{297.5}{450.9}\right]$$

$$= 76.8 \text{ psia } (528.5 \text{ kPa})$$

$$h_{4a} = 460.8\ (1070.9)$$

$$Pr_{4a} = 312.0\ (311.9)$$

Power Turbine (PT)

$$Pr_{5i} = 312.0\left[\frac{14.7}{76.8}\right] = 59.72\ (59.78)$$

$$h_{5i} = 246.0\ (572.2)$$

$$w_{\text{net}} = w_{\text{PT,a}} = 0.89(460.8 - 246.0)$$

$$= 191.2 \text{ Btu/lb } (443.8 \text{ kJ/kg})$$

$$\dot{W}_{\text{net}} = \frac{(191.2)(1)(3600)}{3413}$$

$$= 201.7 \text{ kW } (443.8 \text{ kW})$$

$$q_{\text{in}} = (h_3 - h_{2.5}) + (h_{3.5} - h_{3.1})$$

The enthalpy at state 2.5 can be determined since the regenerator effectiveness is known.

$$0.80 = \frac{h_{2.5} - h_{2a}}{h_{5a} - h_{2a}}$$

$$h_{5a} = 460.8 - 191.2 = 269.6\ (627.1)$$

$$h_{2.5} = 0.80(269.6 - 54.9) + 54.9$$

$$= 226.7\ (527.2)$$

$$q_{\text{in}} = (521.7 - 226.7) + (521.7 - 460.8)$$

$$= 355.9 \text{ Btu/lb } (827.3 \text{ kJ/kg})$$

$$\eta_{\text{th}} = \frac{191.2}{355.9} = 0.537\ (0.536)$$

(d) One should note that the reheat pressure of 116.4 psia (801.5 kPa) is considerably above the optimum reheat pressure of 50.9 psia (350.9 kPa). This results because the high-pressure turbine exit pressure is fixed by the fact that this turbine extracts only sufficient power to drive the high-pressure compressor. Possibly the secondary burner should be placed between the low-pressure turbine and the power turbine.

Figure 5.40 illustrates the effect on net work as a function of pressure ratio when a gas turbine operates on the cycle shown in Figure 5.38. Also shown on the same diagram is the net work of a basic cycle gas turbine with the same compressor and turbine inlet temperatures and compressor and turbine efficiencies.

Figure 5.41 illustrates the effect on cycle thermal efficiency as a function of pressure ratio when a gas turbine operates on the cycle as shown in Figure 5.38. Also shown on the same diagram is the thermal efficiency of a basic cycle with the same compressor and turbine inlet temperatures and compressor and turbine efficiencies.

One engine that has been constructed using an intercooler, regenerator, and reheat is the Ford Motor Company Model 705 engine (6).

5.12 COMBINED-CYCLE POWER PLANT

There are three distinct types of loads that a utility must plan to meet. These are

1. Baseload demands
2. Intermediate or midrange demands
3. Peak load demands

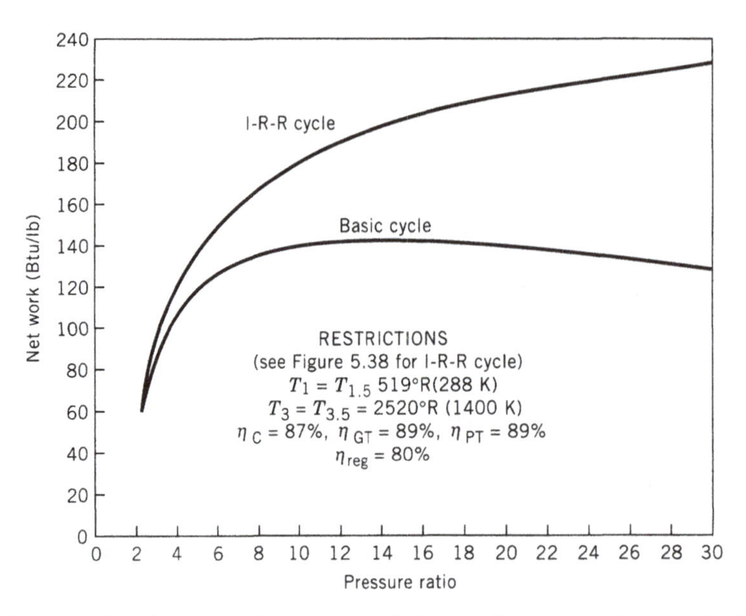

Figure 5.40. Variation, for a gas turbine engine with intercooling–regenerator–reheat, of net work with pressure ratio. Basic cycle variation is shown for comparison.

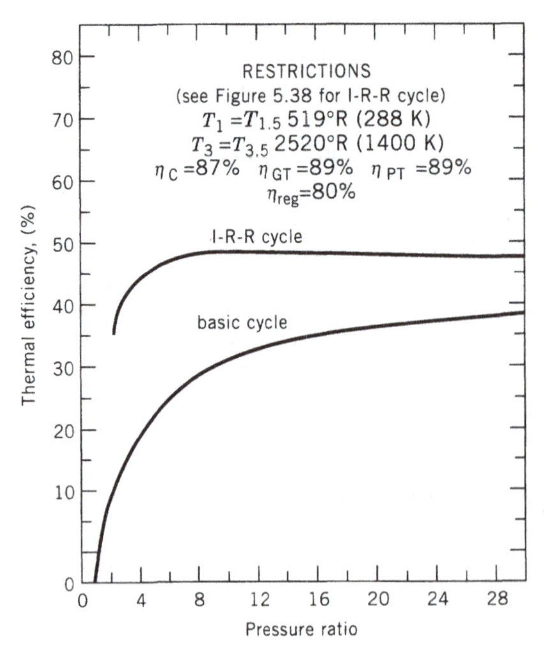

Figure 5.41. Variation, for a gas turbine engine with intercooling–regeneration–reheat, of thermal efficiency with pressure ratio. Basic cycle variation shown for comparison.

There is no clear distinction as to where one type of load ends and the next one begins. Baseload units operate essentially all year long. These units are usually the most efficient and reliable units, with a major design consideration being units that give the lowest cost in mills per kilowatt.

Peaking units usually operate for no more than 5% to 6% of the time during the year. Units in this category must provide reliable standby reserve. Equipment used in this category is characterized with the lowest possible capital cost with little concern for operating efficiency.

The intermediate load is between the baseload units and the peaking units. It must be able to take the swing load, operates up to 75% of the time, and must have a short startup time.

Steam turbines have traditionally been used as the major source of power for the baseload demand. Since the eastern U.S. blackout of November 1965, gas turbines operating on the basic cycle have made considerable headway in meeting peak load demands. This has resulted because of low initial cost and fast startup. One drawback of the early simple gas turbines used for peaking load demands was their high heat rates. Some of the current aircraft derivative gas turbines, because of their high turbine inlet temperatures and component efficiencies, are overcoming this drawback. Gas turbine engines operating on the simple cycle that are currently being placed in service have cycle thermal efficiencies of approximately 40%.

A review of the literature suggests many ways to combine two or more thermal cycles into a power plant. The only one that has been widely accepted is the combination of the gas turbine and a steam turbine.

Reasons for considering a combined cycle are:

1. High cycle thermal efficiency or low heat rate
2. Reduces need for cooling water since only the steam turbine portion of the cycle needs cooling water.
3. Able to build in stages. The gas turbine can be installed initially and operated as a simple cycle gas turbine. At a later date, the steam turbine can be added and then operated as a combined-cycle power plant.
4. Able to operate the gas turbine with the steam turbine idle if a diverter valve was installed so the gas turbine exhaust gas can bypass the heat recovery steam generator.
5. Smaller unit size.

Studies have shown that a combined-cycle system using gas and steam turbines could provide an optimum plant for the intermediate and base load areas with thermal efficiencies well above 50%.

The combined-cycle plant usually is composed of one or more gas turbines exhausting into a heat recovery steam generator (HRSG). Some of the units have supplementary fuel added to the gas turbine exhaust gases, others do not use any additional fuel. Figures 5.42, 5.43, and 5.44 illustrate three possible configurations for a combined-cycle power plant.

In Figure 5.42, no additional fuel is supplied between the exhaust of the gas turbine and the boiler. In Figure 5.43, additional fuel is supplied between the gas turbine and the boiler. When supplementary firing is used, the HRSG inlet temperatures are most often in the range of 800 to 1500°F.

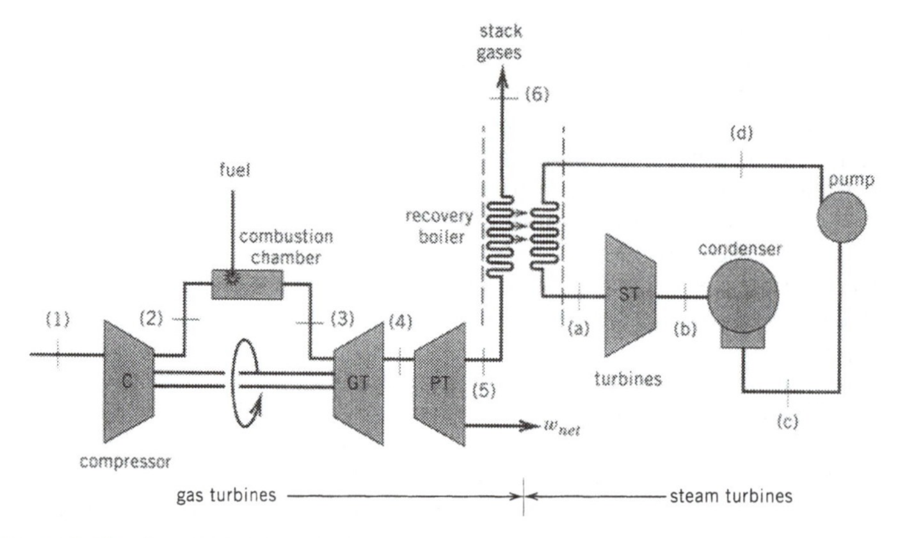

Figure 5.42. Combined cycle with no supplementary burning.

In Figure 5.44, the turbine of the gas turbine is situated downstream of the boiler. This means that the boiler will be supercharged (operating at a high pressure), which could present problems from leakage, and so on. It also lowers the temperature of the gases entering the gas generator turbine, which is undesirable.

In designing a combined-cycle power plant, one must be certain to make optimum use of the energy in the gases leaving the gas turbine in the heat recovery steam generator (HRSG). A typical temperature-heat transfer diagram for a single-pressure steam turbine is shown in Figure 5.45.

The gases leaving the gas turbine at state 5, as illustrated in Figure 5.42, enter the HRSG and leave it at state 6. The water enters the economizer as a subcooled liquid

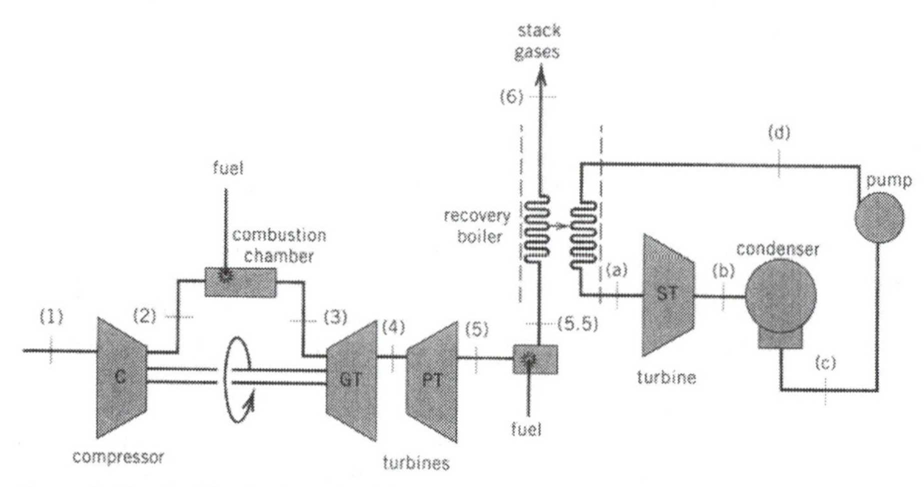

Figure 5.43. Combined cycle with additional combustion chamber.

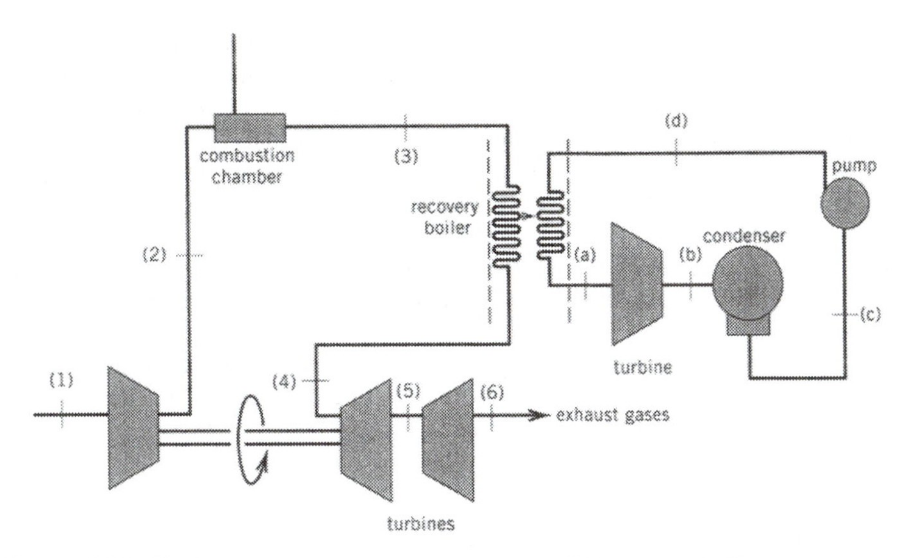

Figure 5.44. Combined cycle with heat recovery boiler between gas turbine compressor and turbines.

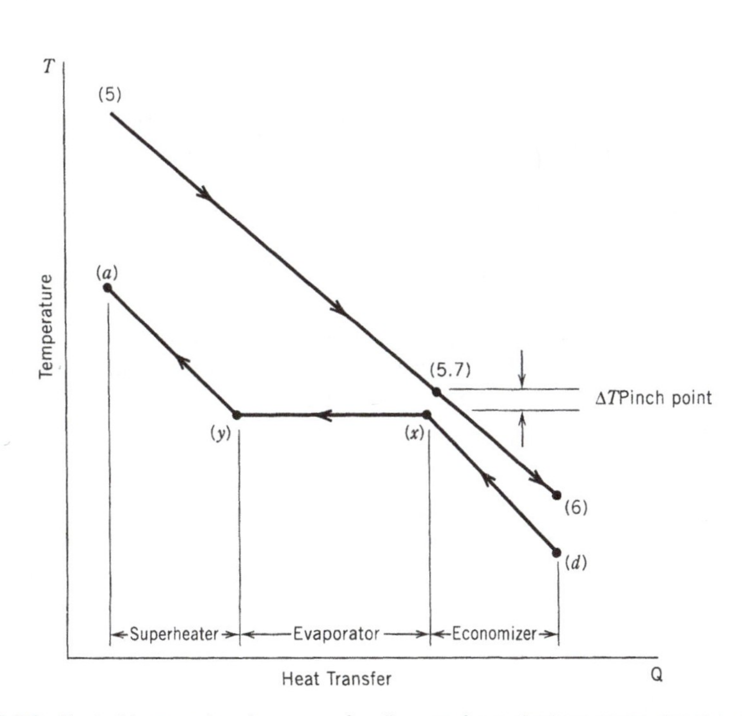

Figure 5.45. Typical temperature-heat transfer diagram for a single-pressure steam turbine combined cycle.

at state d. At state x, the exit from the economizer, as shown in Figure 5.45, the water is a saturated liquid. At this point, the minimum temperature difference between the gas stream and the water stream occurs and is called the *pinch point.* Typical pinch point temperature differences are 30°–50°F (15°–30°C).

From state x to state y, in the evaporator, the water is converted from saturated liquid (state x) to saturated vapor (state y). In the superheater (state y to state a), the water is superheated.

Care must be taken to be certain that low temperature corrosion does not occur in the HRSG. This means that the temperature at state 6 must be 300°F (150°C) or higher.

The amount of energy transferred from the exhaust gases to the water can be increased by decreasing the pinch point temperature difference. This requires an increase in the heat exchanger surface area.

Example Problem 5.13

A combine cycle operates on the cycle illustrated in Figure 5.42. Known values are

$T_1 = 519°\text{R}$ (288 K) $\quad \eta_C = 87\%$ $\quad p_1 = p_5 = 14.70$ psia (101.3 kPa)

$T_3 = 2520°\text{R}$ (1400 K) $\quad \eta_{GT} = 89\%$ $\quad p_2 = p_3 = 176.4$ psia (1215.6 kPa)

$T_6 = 800°\text{R}$ (445 K) $\quad \eta_{PT} = 89\%$

Steam Power Plant

$t_a = 900°\text{F}$ (500°C) $\quad \eta_{\text{turbine}} = 90\%$ $\quad p_a = p_d = 900$ psia (6.0 MPa)

$\eta_{\text{pump}} = 60\%$ $\quad p_b = 1$ psia (6.0 kPa)

$p_c = 1$ psia, saturated liquid (6.0 kPa)

Calculate, on the basis of 1 lb of air per second (1 kg/s) entering the gas turbine and neglecting the mass of fuel added, the

(a) Thermal efficiency and heat rate assuming no steam power plant after the gas turbine, Btu/kWh
(b) Heat rate for the combined-cycle power plant, Btu/kWh
(c) Overall cycle thermal efficiency
(d) Fraction of the total power output generated by the gas turbine
(e) Minimum temperature difference between the exhaust gases and the water in the HRSG; that is, pinch point temperature difference
(f) Quality of the steam at the steam turbine exit

Solution

(a) The gas turbine portion of this problem was solved in Example Problem 5.3b. Results from Example Problem 5.3 were, on a basis of 1 lb of air entering the compressor

$$w_{GT,\text{net}} = 143.0 \text{ Btu/lb } (333.0 \text{ kJ/kg})$$

$$\dot{w}_{GT} = 150.6 \text{ kW } (333.0 \text{ kW})$$

$$q_{\text{in}} = 380.0 \text{ Btu/lb } (884.0 \text{ kJ/kg})$$

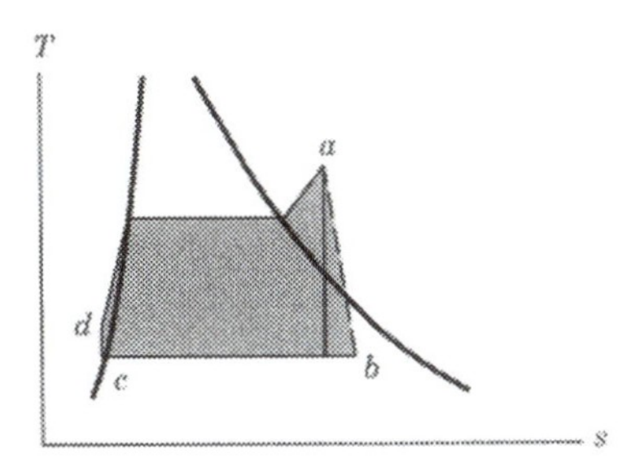

Figure 5.46. Temperature–entropy diagram for the steam cycle.

$$T_{5a} = 1468°\text{R} \ (815 \text{ K})$$

$$h_{5a} = 231.0 \text{ Btu/lb} \ (536.7 \text{ kJ/kg})$$

$$\eta_{th} = 0.377 \text{ or } 37.7\%$$

$$HR = \frac{(380)(1)(3600)}{150.6} = 9084 \text{ Btu/kWh} \ (9557 \text{ kJ/kWh})$$

(b) Figure 5.46 illustrates the temperature-entropy diagram for the steam cycle. At state c, using values from Keenan (7)

$$p_c = 1 \text{ psia, saturated liquid} \ (6 \text{ kPa})$$

$$t_c = 101.70°\text{F} \ (36.16°\text{C})$$

$$v_c = 0.01614 \text{ ft}^3\text{/lb} \ (0.0010064 \text{ m}^3\text{/kg})$$

$$h_c = h_f = 69.74 \text{ Btu/lb} \ (151.53 \text{ kJ/kg})$$

For an ideal pump, constant specific volume, and negligible kinetic energies,

$$w_{p,a} = \frac{w_i}{\eta_p} = -\frac{\int v\,dp}{\eta_p} = -\frac{v\,\Delta p}{\eta_p}$$

$$= -\frac{(0.01614)(899)(144)}{(0.6)(778)} = -4.475 \text{ Btu/lb} \ (-10.05 \text{ kJ/kg})$$

$$h_d = 69.74 + 4.48 = 74.22 \text{ Btu/lb} \ (161.58 \text{ kJ/kg})$$

$$p_a = 900 \text{ psia} \ (6.0 \text{ MPa})$$

$$t_a = 900°\text{F} \ (500°\text{C})$$

$$h_a = 1451.9 \text{ Btu/lb} \ (3422.2 \text{ kJ/kg})$$

$$s_a = 1.6257 \text{ Btu/lb°R} \ (6.8803 \text{ kJ/kg K})$$

The energy available to produce steam is the energy removed from the gas turbine exhaust gases. Writing an energy balance between states 5 and 6 yields

$$\dot{m}_w(h_a - h_d) = \dot{m}_a(h_{5a} - h_6)$$

Therefore, the mass rate of water flow is

$$\dot{m}_w = \frac{(1)(231.0 - 61.9)}{1451.9 - 74.2}$$

$$= 0.1227 \text{ lb/s } (0.1203 \text{ kg/s})$$

For the steam turbine for an ideal expansion

$$p_b = 1 \text{ psia } (6.0 \text{ kPa})$$

$$s_b = s_a = 1.6257 \text{ Btu/lb°R } (6.8803)$$

$$1 - x_b = \frac{1.9779 - 1.6257}{1.8453} = 0.191 \ (0.186)$$

$$h_{bi} = 1105.8 - (0.191)(1036.0)$$

$$= 907.9 \ (2118.0)$$

$${}_aw_{ba} = w_{ST,a} = (0.90)(1451.9 - 907.9)$$

$$= 489.6 \text{ Btu/lb } (1173.8 \text{ kJ/kg})$$

$$\dot{w}_{ST,net} = \frac{0.1227 \ (489.6 - 4.5)(3600)}{3413}$$

$$= 62.8 \text{ kW } (140.0 \text{ kW})$$

$$\dot{w}_{total} = \dot{w}_{GT} + \dot{w}_{ST}$$

$$= 150.6 + 62.8$$

$$= 213.4 \text{ kW } (473.0 \text{ kW})$$

$$\text{Overall } HR = \frac{(380.0)(1)(3600)}{213.4}$$

$$= 6410 \text{ Btu/kWh } (6728 \text{ kJ/kWh})$$

(c) $$\eta_{th} = \frac{w_{total}}{q_{in}} = \frac{143.0 + 59.5}{380.0}$$

$$= 0.533 \text{ or } 53.3\% \ (0.535)$$

(d) The fraction of the output generated by the gas turbine is

$$\text{fraction} = \frac{150.6}{213.4} = 0.706 \ (0.704)$$

(e) The minimum temperature difference occurs between the exhaust gases as they exit the evaporator and the saturation steam temperature as shown in Figure 5.45. The temperature at state 5.7 can be determined by the following energy balance.

$$\dot{m}_a(h_{5a} - h_{5.7}) = \dot{m}_w(h_a - h_x)$$

where

$$h_x = h_f \text{ @ } 900 \text{ psia} = 526.6\ (1213.4)$$

$$T_x = 532.12^\circ\text{F}\ (275.6^\circ\text{C})$$

$$(1)(231.0 - h_{5.7}) = (0.1227)(1451.9 - 526.6)$$

$$h_{5.7} = 117.5 \text{ Btu/lb } (271.0 \text{ kJ/kg})$$

$$\Delta T_{pp} = T_{5.7} - T_x$$

$$= 1025 - (532 + 460)$$

$$= 33^\circ\text{F}\ (19^\circ\text{C})$$

(f) $$h_{ba} = 1451.9 - 489.6$$

$$= 962.3$$

$$1 - x_{ba} = \frac{1105.8 - 962.3}{1036.0}$$

$$= 0.139\ (0.132)$$

It is of interest to determine the effect of the maximum steam power plant pressure on the amount of energy transferred in the HRSG, the gas temperature at the exit from the HRSG (state 6), the overall heat rate, and the quality of the steam at the turbine exit. The results of solving Example Problem 5.13 assuming T_6 is 800°R (445 K) in both cases, that the pinch point temperature is unknown, and that $p_a = p_d = 600$ psia (4 MPa), all other values unchanged, are shown in Figure 5.47. The results for Example Problem 5.13 are also shown on the same figure.

One should observe that solving Example Problem 5.13 with T_6 unknown, a pinch point temperature difference of 33°F (18.3°C), and lowering the pressure at the inlet to the steam turbine from 900 psia (6 MPa) to 600 psia (4 MPa) results in

(a) Lowering the overall heat rate from 6410 Btu/kWh to 6387 Btu/kWh
(b) Changing the fraction of the total power from the gas turbine from 70.6% to 70.3%
(c) Lowering the temperature of the gases at the exit from the HRSG from 800°R (340°F) to 763°R (303°F)
(d) The quality of the steam leaving the turbine increasing from 86.1% to 88.8%

One should now realize that for a given gas turbine engine, the steam power plant operating conditions and the pinch point temperature greatly influence the overall performance of the combined cycle. The results of solving Example Problem 5.13 for several different steam turbine inlet pressures and temperatures are shown in Table 5.11.

The preceding example problem, figures, and tables illustrate the improvement that may be achieved by converting a basic (simple) gas turbine engine into a single-pressure combined cycle.

Table 5.12 lists data for several simple cycle gas turbine engines and several

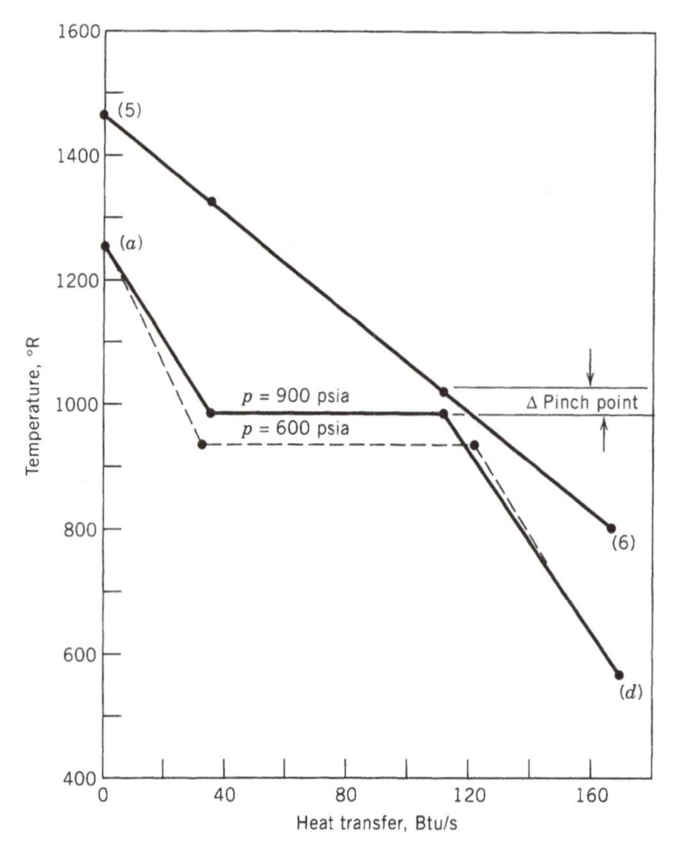

Figure 5.47. Temperature-heat transfer diagram for a single-pressure combined cycle operating at steam turbine supply conditions of 900 psia, 900°F and 600 psia, 900°F and a pinch point temperature difference of 33°F.

TABLE 5.11. **Effect of Steam Turbine Inlet Temperature and Pressure on Combined-Cycle Performance***

Steam Temperature °F	Steam Pressure psia	Steam Flow Rate lb/s	Quality Steam Turbine Exit	Gas Temperature HSRG Exit °F	Fraction Total Output From Gas Turbine	Overall Heat Rate Btu/kWh
900	900	0.1227	86.1%	340	70.6%	6410
900	600	0.1281	88.8%	303	70.3%	6387
900	800	0.1318	84.2%	318	71.7%	6410
800	800	0.1238	86.9%	332	71.6%	6410
600	600	0.1530	80.0%	270	70.6%	6117
400	600	0.1648	79.4%	240	71.1%	6465

* Solution of Example Problem 5.13 for various steam turbine inlet pressures and temperatures. Pinch point temperature difference in all cases is 33°F, and the condenser pressure is 1 psia.

TABLE 5.12. **Typical Combined Cycles Compared with Comparable Simple Gas Turbine Cycles (2)**

Manufactuer Model	Year	Net Plant Output kW	Heat Rate Btu/kWh	LHV Efficiency	Gas Turbine Power kW	Steam Turbine Power kW	Number and Type Gas Turbine
ABB Power Generation							
GT 10	1981	24,630	9,770	—	24,630	—	—
KA 10-2	1989	69,100	7,035	48.5%	47,600	21,500	2-GT 10
GT 13E2	1993	164,300	9,558	—	164,300	—	—
KA 13E2-4	1993	978,700	6,305	54.1%	637,200	341,500	4-GT13E2
GE Power Generation							
MS9331 (FA)	1991	226,500	9,570	—	226,500	—	—
S-207FA	1991	483,200	6,250	54.6%	311,200	178,100	1-MS 7001FA
LM6000	1992	39,970	8,790	—	39,970	—	—
S-260	1992	109,100	6,790	50.3%	79,000	32,000	2-LM 6000
Siemens KWU							
V84.2	1985	103,200	10,120	—	103,200	—	—
GUD 3.84.2	1987	488,000	6,535	52.2%	306,000	190,000	3-V84.2
V64.3	1990	60,500	9,705	—	60,500	—	—
GUD 1.64.3	1990	88,200	6,640	51.5%	59,000	31,000	1-64.3
V94.3	1993	219,000	9,450	—	219,000	—	—
GUD 2.94.3	1994	649,000	6,320	54.0%	426,000	236,000	2-V94.3
Turbo Power							
FT8	1990	25,420	8,950	—	25,420	—	—
FT8	1990	32,280	7,010	48.7%	24,700	7,580	1-FT8
Westinghouse							
251B12	1982	49,100	10,450	—	49,100	—	—
PACE70	1982	69,700	7,200	47.4%	47,400	24,050	1-251B12
501D5	1975	109,800	10,040	—	109,800	—	—
PACE150	1981	158,000	6,900	49.5%	106,100	56,800	1-501D5
501F	1989	161,300	9,450	—	161,300	—	—
PACE220	1989	236,000	6,350	53.7%	155,500	85,500	1-501F

combined cycles. One should note once again the improvement in the heat rate by converting from a simple cycle gas turbine to a combined cycle.

5.13 COMBINED-CYCLE POWER PLANT WITH SUPPLEMENTARY FIRING

The preceding section examined the fundamentals of the single-pressure combined-cycle power plant with no additional energy being added to the gas turbine exhaust gases between the exit from the gas turbine power turbine and the inlet to the HRSG.

When additional heat is added between the gas turbine exit and the inlet to the heat recovery steam generator, the fraction of the total output produced by the gas turbine decreases. Advantages of adding supplementary firing to the combined cycle, as illustrated in Figure 5.43, include the ability to

1. Increase the total combined cycle output.
2. Control the temperature at the inlet to the HRSG. The gas temperature and air

flow rate at the exit from the gas turbine are very dependent on the temperature at the inlet to the gas turbine compressor.

The effect of adding supplementary firing to the combined cycle solved in Example Problem 5.13 is illustrated by the following example problem.

Example Problem 5.14

A combined cycle operates with supplementary firing as illustrated in Figure 5.43. Solve Example Problem 5.13 assuming the temperature at state 5.5 is 1040 K (1872°R), all other given values unchanged.

Solution

The main change in the solution is the mass rate of flow of water in the steam power plant. Writing an energy balance for the HRSG yields

$$\dot{m}_w = \frac{(1)(789.7 - 144.4)}{3422.2 - 161.6}$$

$$= 0.1979\ (0.2017)$$

(a) $\dot{w}_{GT} = 333.0 \text{ kW } (150.6 \text{ kW})$

$$HR_{GT} = 9557 \text{ kJ/kWh } (9084 \text{ Btu/kWh})$$

(b) $\dot{w}_{ST} = (0.1979)(1173.8 - 10.1)$

$$= 230.3 \text{ kW } (103.2 \text{ kW})$$

$$\dot{w}_{combined} = 333.0 + 230.3$$

$$= 563.3 \text{ kW } (253.8 \text{ kW})$$

$$q_{in} = (1)[(h_3 - h_{2a}) + (h_{5.5} - h_{5a})]$$

$$= (1)[884.0 + (789.7 - 536.7)]$$

$$= 1137 \text{ kJ/s } (488.8 \text{ Btu/s})$$

$$HR_{combined} = \frac{(1137)(3600)}{563.3}$$

$$= 7266 \text{ kJ/kWh } (6933 \text{ Btu/kWh})$$

(c) $\eta_{th} = \frac{563.3}{1137}$

$$= 0.495\ (0.492)$$

(d) $\text{fraction} = \frac{333.0}{563.3}$

$$= 0.591\ (0593)$$

(e) $(1)(789.7 - h_{5.7}) = (0.1979)(3422.2 - 1213.4)$

$$h_{5.7} = 352.6\ (153.2)$$

$$T_{5.7} = 645\ \text{K}\ (1167°\text{R})$$

$$\Delta T_{PP} = 645 - (276 + 273)$$

$$= 96°\text{C}\ (175°\text{F})$$

(f) $x_{ba} = 0.132\ (0.139)$

Comparing the results from Example Problems 5.13 and 5.14, one observes that

1. A higher fraction of the output results from the steam turbine portion of the combined cycle when supplementary firing is used.
2. The pinch point temperature difference is much higher for the combined-cycle with supplementary firing since the inlet and exit temperatures were assumed to be the same for both cycles.
3. The unfired combined-cycle power plant had a lower heat rate.
4. Decreasing the pinch point temperature difference for the cycle with supplementary firing to that of the unfired cycle would decrease the HRSG exit temperature for the exhaust gases and increase the fraction of power developed by the steam turbine.

Figure 5.48 illustrates the temperature-heat transfer diagram for an unfired single-pressure combined cycle and a supplementary fired single-pressure combined cycles. The results plotted are for Example Problems 5.13 and 5.14.

One should recognize that there is a significant difference between a conventional steam power plant and the combined-cycle power plant. A conventional steam power plant achieves a higher efficiency if the feedwater is increased to a higher temperature by means of one or more open and/or closed feedwater heaters. The combined-cycle power plant achieves a higher efficiency with a low feedwater temperature.

For a combined cycle, a high steam pressure does not necessarily mean a high thermal efficiency (or low overall heat rate). A high steam pressure does increase the steam turbine cycle efficiency but decreases the rate of energy transferred from the exhaust gases.

5.14 MULTIPRESSURE COMBINED-CYCLE POWER PLANTS

The preceding two sections assumed a single-pressure steam turbine cycle for the combined cycle. Designing combined-cycle steam turbines with steam supplied to the steam turbines at two different pressures increases the cycle thermal efficiency over that of a one-pressure steam turbine cycle, allows for better utilization of the energy in the exhaust gases and, when reheat is used, allows for good control of the quality of the steam at the exit from the turbines. A simplified temperature-heat transfer diagram for the HRSG of a two-pressure combined cycle is shown in Figure

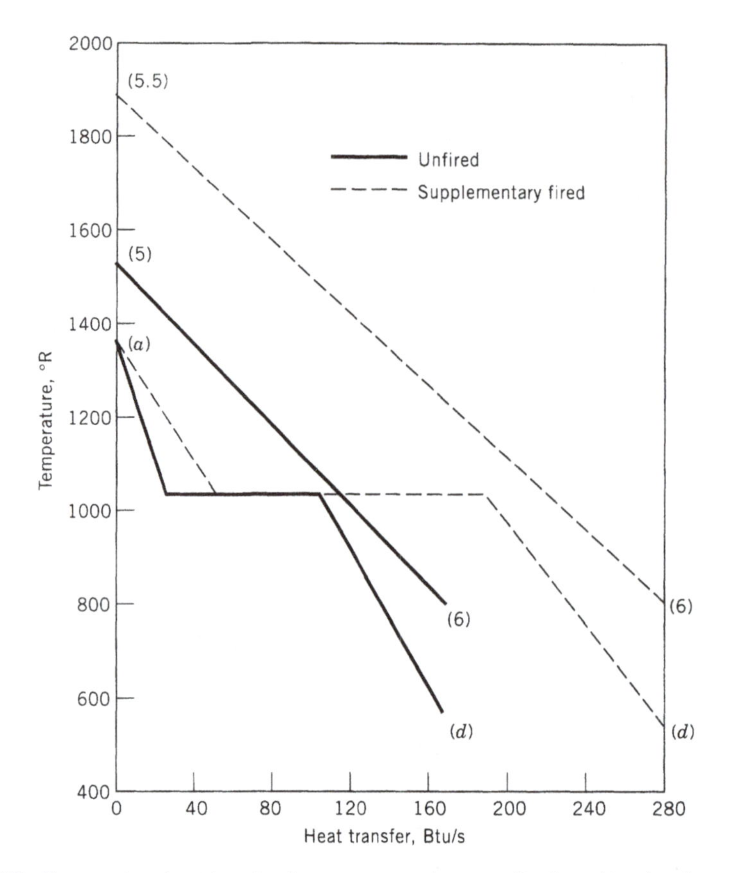

Figure 5.48. Temperature-heat transfer diagram comparing an unfired combined cycle and a supplementary combined cycle for the same HSRG exit gas temperature.

5.49. For the steam cycle illustrated in Figure 5.50, the water leaves the low pressure pump at state d. From state d to state m, the subcooled water is heated in the economizer at pressure p_x until it is a saturated liquid. The temperature difference, $T_{5.7} - T_m$, is referred to as the low pressure pinch point, ΔT_{LPPP}.

From state m to state n, a portion of the water is converted from saturated liquid to saturated vapor. At state n, the saturated water enters a second pump where its pressure is increased to pressure p_y. This fraction of the water is then heated to state a, first to state p where it is a saturated liquid, then evaporated from state p to state q, and finally superheated to its final state.

The temperature difference, $T_{5.5} - T_p$, is referred to as the high pressure pinch point, ΔT_{HPPP}. The saturated vapor at state n can either be taken directly to a low pressure turbine or can be superheated before entering the low pressure turbine. A simplified flow diagram for a cycle of this type where the saturated water vapor portion of the steam at pressure P_x is superheated before entering the low pressure turbine is illustrated in Figure 5.50. Two other multipressure combined cycle flow diagrams are illustrated in Figures 5.51 and 5.52.

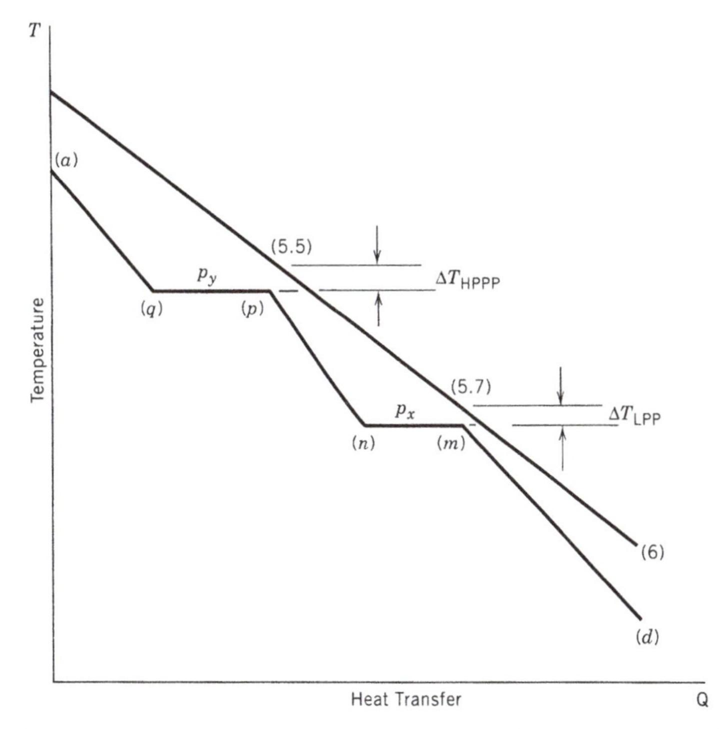

Figure 5.49. Temperature-heat transfer diagram for the HRSG of a two-pressure combined cycle.

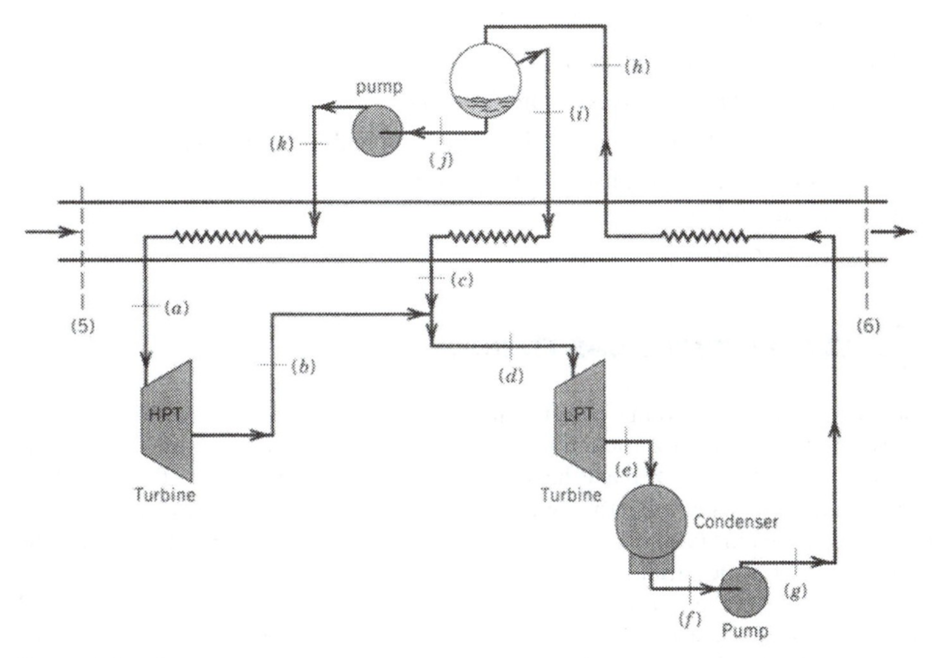

Figure 5.50. Schematic diagram for a two-pressure combined cycle with superheating of the low pressure steam.

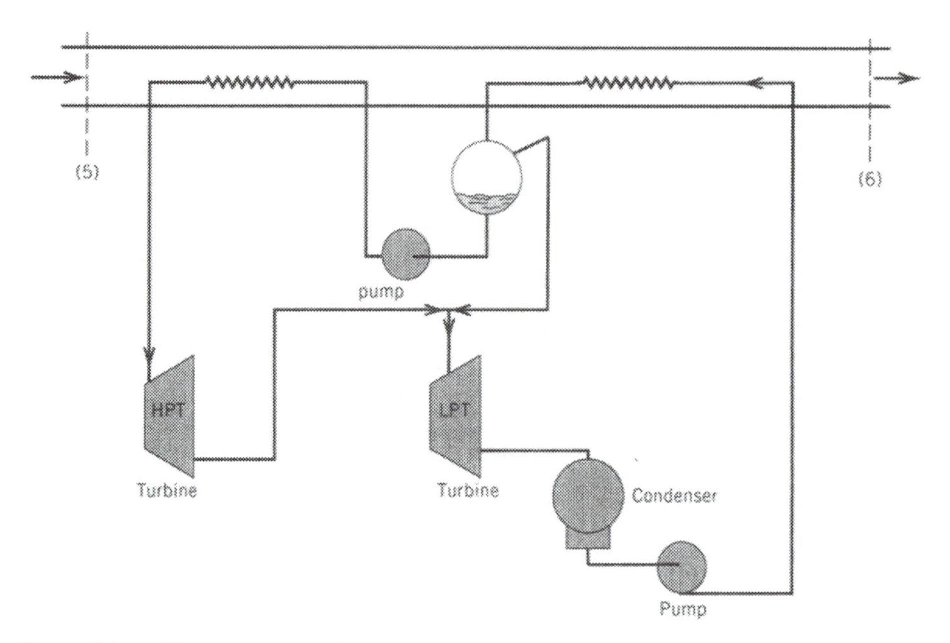

Figure 5.51. Schematic diagram for a two-pressure combined cycle with no superheat of the low pressure steam.

The major design parameters that influence the performance of the steam portion of the combined-cycle power plant are the

1. Steam pressure and temperature at the inlet to the high pressure turbine
2. Steam pressure and temperature at the inlet to the low pressure turbine
3. High pressure and low pressure pinch points
4. Amount of superheat (if any) of the steam supplied to the low pressure turbine
5. Fraction of the total steam flow that passes through both the high pressure and low pressure turbine
6. Condenser pressure

5.15 CLOSED-CYCLE GAS TURBINES

The closed-cycle gas turbine power plant has been under development by the firm of Escher-Wyss, Ltd., of Zurich, Switzerland, since 1940. Since that time, more than a dozen central station-type plants have been put into operation in the 12,000-kW class.

In Europe, particularly Germany and Switzerland, the merits of the closed-cycle gas turbine have long been recognized for combined electrical power and the utilization of waste heat for industrial uses and urban district heating.

Closed-cycle gas turbines did not receive much attention in the United States until the early 1970s, when fuel conservation became increasingly important due to the lack of the availability of abundant, low-cost, clean gaseous and liquid fuels.

The advantages of the closed-cycle gas turbine include:

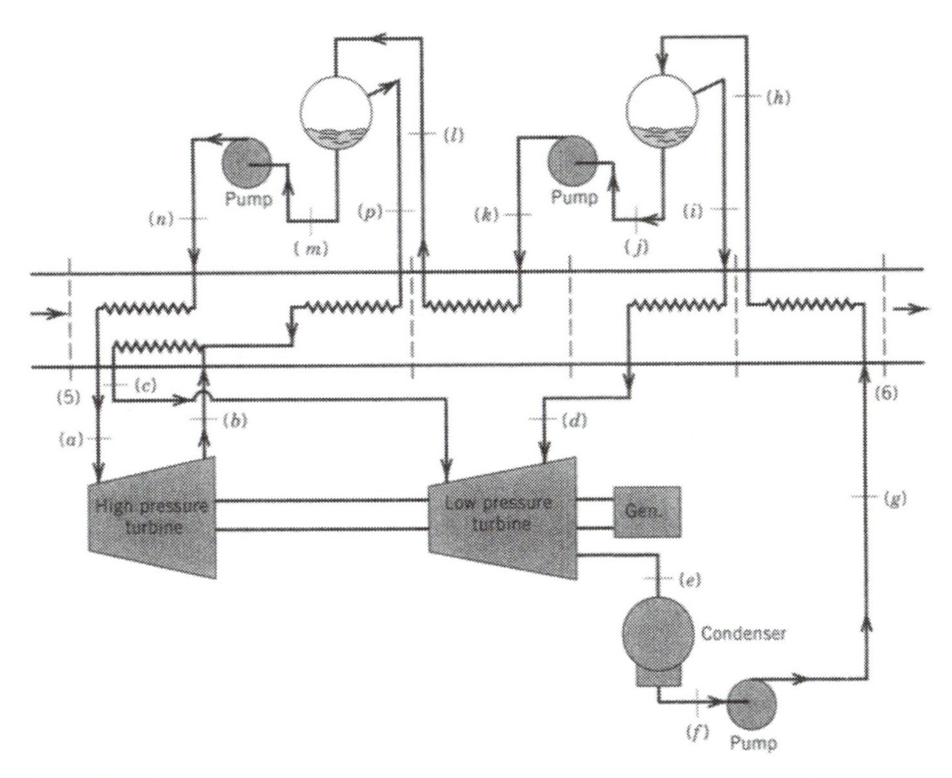

Figure 5.52. Schematic diagram for a three-pressure combined cycle with superheating and reheat between the high pressure and intermediate pressure steam turbines.

1. Possibility of using a wide variety of medium gases such as air, carbon dioxide, helium, hydrogen, or nitrogen.
2. Possibility of using a wide variety of fuels such as coal, oil, gas, nuclear, solar, or stored energy.
3. Ability to operate with a much higher pressure. Increasing the pressure decreases the specific volume, thereby permitting more work per unit area. A decrease in specific volume also improves the heat transfer characteristics of the gas turbine medium.
4. Opportunity to vary the total pressure of the system. This offers an opportunity to control part-load output that is independent of the turbine inlet temperature. The total pressure may be varied by adding or withdrawing fluid from the gas turbine. This is accomplished by the use of pressurized storage tanks.
5. Working fluid (and components) are not contaminated by the combustion process.
6. Heat rejection characteristics are well suited to either dry cooling, wet-dry cooling, district heating, or utilization of a binary cycle. The cooling of the closed-cycle gas turbine exhaust gas may be achieved by a dry air-cooling tower because of the large difference between the exhaust gas temperature and the ambient air temperature.

TABLE 5.13. **Operating Closed-cycle Gas Turbine Systems (7)**

Plant	Pioneer CCGT	RAVENSBURG	OBERHAUSEN I	COBURG	HAUS ADEN	GELSENKIRCHEN	OBERHAUSEN 2
Country	Switzerland	Germany	Germany	Germany	Germany	Germany	Germany
Manufacturer	ESCHER WYSS	GHH	GHH	GHH	GHH	GHH	GHH
Fuel	Oil	Coal	Coal/coke oven gas	Coal	Coal/mine gas	Furnace gas	Coke oven gas
Working fluid	Air	Air	Air	Air	Air	Air	Helium
Electrical output (MWe)	2.0	2.3	13.75	6.6	6.4	17.25	50[a]
Heat output, (MWt)	—	2.3–4.1	18.5–28	8–16	7.8	20–29	53.5
Efficiency at terminal (%)	32.6	25.0	29.5	28.9	29.5	30.8	31.3
Fuel utilization (%)	—	—	65.6	64.5	65.4	65.9	65.0
Turbine inlet temp., °F (°C)	1202 (650)	1220 (660)	1310 (710)	1256 (680)	1256 (680)	1312 (711)	1382 (750)
Turbine inlet press., psia (MPa)	363 (2.5)	392 (2.7)	464 (3.2)	399 (2.75)	450 (3.1)	551 (3.8)	406 (2.8)
Commissioning date	1939	1956	1960	1961	1963	1967	Started 1975
Running time, hours to June 1976	6,000	120,000	100,000	100,000	100,000	75,000	[b]
Plant availability, %	Plant operated during World War II	87	73	83	91	75	—

[a] Physical dimensions of turbomachinery and heat exchangers equivalent to a nuclear helium gas turbine of about 300 MWe.

[b] Plant not yet commissioned. More than 5000 h of operation up to Spring 1977 at electrical power outputs up to 30 MWe. Full heat load of 53.5 MWt delivered during Winter 1976–1977 to city of Oberhausen district heat system.

TABLE 5.14. **Results for Various Gas Turbine Engine Configurations**

Number	Cycle	Example Problem	Pressure Ratio	Specific Power kW/(lb/s)(kW/(kg/s))	Thermal Efficiency %
1	Basic (ideal)	5.2b	12.0	203.4 (448.6)	48.3
2	Basic (actual)	5.3b	12.0	150.6 (333.0)	37.7
3	Basic (actual)	5.6a	4.0	119.1 (261.8)	24.7
4	Regenerative	5.6b	4.0	119.1 (261.8)	44.9
5	Steam-injected	5.8	12.0	170.3 (375.4)	37.5
6	Evaporative-regenerative	5.9	12.0	197.9 (436.6)	51.3
7	Intercooled-regenerative-reheat	5.12	12.0	201.7 (443.8)	53.7
8	Single-pressure combined cycle	5.13	12.0	213.4 (473.0)	53.5
9	Combined cycle with supplementary firing	5.14	12.0	253.8 (563.3)	49.5

Table 5.13 lists the major closed-cycle gas turbines in Europe.

5.16 COMPARISON OF VARIOUS CYCLES

Various configurations of shaft-power gas turbine engines have been discussed in this chapter. A number of example problems were included to illustrate the effect on specific power and cycle thermal efficiency with each change.

Table 5.14 and Figure 5.53 summarize the results for the various configurations. All cycles listed are for a turbine inlet temperature of 2520°R (1400 K) and assume air is the working fluid throughout the entire cycle.

One must keep in mind that this was a very simple analysis. Assumptions made

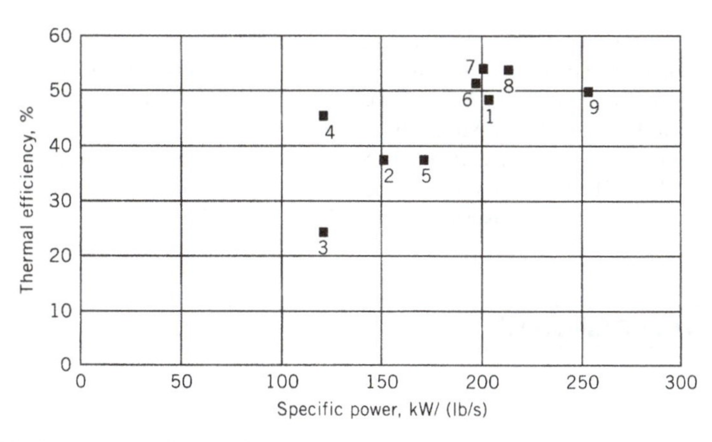

Figure 5.53. Comparison of thermal efficiency and specific power for several gas turbine configurations. See Table 5.14 for engine configurations.

in solving the example problems included

1. All were for a turbine inlet temperature of 2520°R (1400 K).
2. No pressure loss in the combustion chamber.
3. A compressor efficiency of 87% and turbine efficiencies of 89%.
4. A pressure ratio of 12 in most cases.

Questions that need to be considered include:

1. What is the best turbine inlet temperature for current materials and turbine cooling techniques?
2. What pressure ratio should be used?
3. What are the component efficiencies for current designs?
4. What are realistic pressure losses for a combustion chamber and exhaust system?
5. The analysis used is at one operating point. What happens to the net power, cycle thermal efficiency, and component efficiencies when the engine is operated at part power; that is, at a lower turbine inlet temperature?

Some of these questions are answered in Chapter 10. Others require a more detailed analysis than has been used in this chapter.

REFERENCES CITED

1. Biancardi, F. R., and Peters, G. T., Utility Applications for Advanced Gas Turbines to Eliminate Thermal Pollution, ASME paper No. 70-WA/GT-9, 1970.
2. *The 1992-93 Handbook,* Gas Turbine World, Southport, Conn., 1992.
3. Maurer, R., Destec's Successes and Plans for Coal Gasification Combined Cycle (CGCVC) Power Systems, in *1992 ASME Cogen-Turbo, 6th International Conference in Cogeneration and Utility,* edited by Cooke, D. H., et al, pp 75–85, ASME, 1992.
4. Sawyer, J. W. (ed.), *Gas Turbine Engineering Handbook,* Gas Turbine Publications, Stamford, Conn., 1966.
5. Collman, J. S., Amann, C. A., Matthews, C. C., Stettler, R. J., and Verkamp, F. J., *The GT-225—An Engine for Passenger-Car Gas-Turbine Research,* SAE Paper 750167, 1975.
6. Peitsch, G., and Swatman, I. M., A Gas Turbine Super Transport Truck Power Package, Society of Automotive Engineers Paper No. 991B, January 1965.
7. Keenan, J., Keyes, F., Hill, P., and Moore, J., *Steam Tables,* John Wiley & Sons, New York, 1969.
8. McDonald, C. F., The Closed-Cycle Turbine—Present and Future Prospectives for Fossil and Nuclear Heat Sources, ASME paper no. 78-GT-102, April 1978.

SUGGESTED READING

Cohen, H., Rogers, G. F. C., and Saravanamuttoo, H. I. H., *Gas Turbine Theory,* John Wiley & Sons, New York, 1972.

Dusinberre, G. M., and Lester, J. C., *Gas Turbine Power,* International Textbook Co., Scranton, Pa., 1958.

Hill, P. G., and Peterson, C. R., *Mechanics and Thermodynamics of Propulsion,* Addison-Wesley Publishing Co., Reading, Mass., 1965.

Lewis, A. D., *Gas Power Dynamics,* D. Van Nostrand Company, Princeton, N.J., 1962.

NOMENCLATURE

c_p	constant pressure specific heat
f	actual fuel–air ratio
f'	ideal fuel–air ratio
h	specific enthalpy
H	total enthalpy
HR	heat rate
k	specific heat ratio
$\dot{m}$	mass rate of flow
p	pressure
Pr	relative pressure
q	specific heat transfer
$\dot{q}$	rate of heat transfer
$\overline{R}$	universal gas constant
s	specific entropy
$\overline{s}$	molar specific entropy
S	total entropy
SFC	specific fuel consumption
T	temperature
v	specific volume
$\overline{V}$	velocity
w	specific work
$\dot{w}$	power
ΔH_c	enthalpy of combustion
ΔT	pinch point temperature difference
η	efficiency

Superscript

0	standard state

Subscripts

1,2,3,4,5	states
a	actual
B	burner
C	compressor
f	fuel
GT	gas generator turbine, gas turbine
i	ideal
o	stagnation (total)
PP	pinch point
pr	products
PT	power turbinere
reg	generator
ST	steam turbine
T	turbine
th	thermal
W	water

PROBLEMS

5.1. An ideal, basic air standard gas turbine engine has a compressor inlet temperature of 519°R (288 K) and a turbine inlet temperature of 2160°R (1200 K). Calculate, assuming constant specific heats:

(a) The pressure ratio that gives the maximum net work

(b) The compressor work, turbine work, heat added, and thermal efficiency for the pressure ratio of part (a)

5.2. Solve Problem 5.1 assuming variable specific heats and a pressure ratio of 8. Compare the net work and thermal efficiency with the values determined in Problem 5.1.

5.3. Solve Problem 5.1 assuming variable specific heats and a pressure ratio of 4. Compare the net work and thermal efficiency with the values determined in Problems 5.1 and 5.2.

5.4. Solve Example Problem 5.2 assuming that the compressor inlet temperature is 560°R (311 K).

5.5. An ideal, basic, air standard gas turbine engine has a compressor inlet temperature of 519°R (288 K) and a turbine inlet temperature of 2160°R (1200 K). Air enters the compressor at the rate of 100 lb/s (45.4 kg/s). Calculate, assuming variable specific heats and a pressure ratio of 11.6:

(a) The power required to drive the compressor

(b) The fraction of the work developed by the turbine that is required to drive the compressor

(c) The power the turbine can deliver to an external generator

5.6. One of the earliest uses for a gas turbine was to produce compressed air. Shown in the sketch is one technique by which this could be accomplished. Assuming that turbine T_1 develops only sufficient power to drive the compressor, calculate the amount of compressed air delivered at state 2_d for every pound of air entering the compressor. Known values are shown below the diagram.

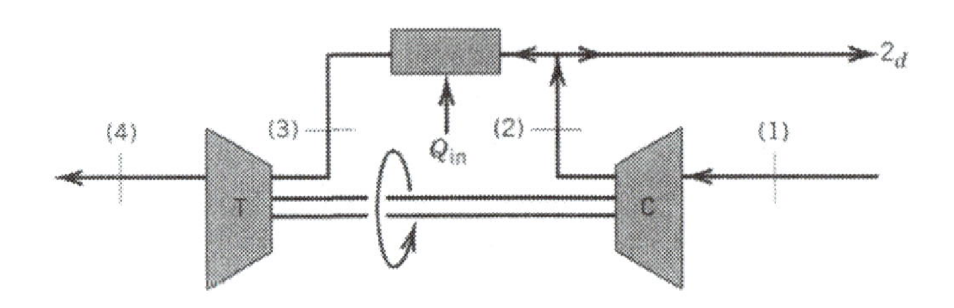

State	Pressure (psia)	Pressure (kPa)	Temperature (°R)	Temperature (K)	
1	14.7	101.3	519	288	$\eta_c = 100\%$
2	73.5	506.5			
3	73.5	506.5	2160	1200	$\eta_t = 100\%$
4	14.7	101.3			

5.7. Solve Example Problem 5.3 for a compressor inlet temperature of 560°R (311 K), all other values being the same. What is the percent decrease in net work per pound of air?

5.8. A basic, air standard gas turbine has a compressor inlet temperature and pressure of 519°R and 14.7 psia, respectively, a compressor exit and turbine inlet pressure of 231.5 psia, a turbine inlet temperature of 2520°R, a turbine exit pressure of 14.7 psia, a compressor efficiency of 82%, and a turbine efficiency of 85%.

Air enters the compressor at the rate of 100 lb/s. Calculate, assuming variable specific heats:

(a) The power required to drive the compressor

(b) The fraction of the work developed by the turbine that is required to drive the compressor

(c) The power the turbine can deliver to an external generator

5.9. Solve Example Problem 5.3 assuming that the compressor efficiency is 82%, gas generator efficiency is 89%, power turbine efficiency is 89%, all other values the same.

5.10. A basic air-standard gas turbine engine operates with a compressor inlet temperature of 288 K (519°R), a compressor efficiency of 87%, a gas generator efficiency of 89%, and a power turbine efficiency of 89%. The compressor pressure ratio is 4. Assuming variable specific heats and turbine inlet temperature of 1200 K (2160°R), calculate the net work, heat added, and cycle thermal efficiency assuming no pressure drop during the heat addition process and that the pressure at the power turbine exit is equal to the compressor inlet pressure.

5.11. Solve Problem 5.10 assuming a compressor efficiency of 82%, gas generator efficiency of 85%, power turbine efficiency of 85%, all other values being the same.

5.12. Solve Problem 5.6 assuming a compressor efficiency of 82% and a turbine efficiency of 85%.

5.13. A basic, air standard gas turbine engine has a compressor inlet temperature and pressure of 519°R and 14.7 psia, respectively, a compressor exit and turbine inlet pressure of 298.4 psia, a turbine inlet temperature of 2160°R, a turbine exit pressure of 14.7 psia, a compressor efficiency of 85%, and a turbine efficiency of 88%. Air enters the compressor at the rate of 188 lb/s. Calculate, assuming variable specific heats:

(a) The power required to drive the compressor

(b) The fraction of the work developed by the turbine that is required to drive the compressor

(c) The power the turbine can deliver to an external generator

Compare with the values in Table 5.4 for the FT8 turbine.

5.14. Estimate, for the Allison Gas Turbine 501-KC5 and 570-K gas turbine engines, the turbine inlet temperature. Use the data given in Table 5.4.

5.15. A basic gas turbine engine operates under the conditions given.

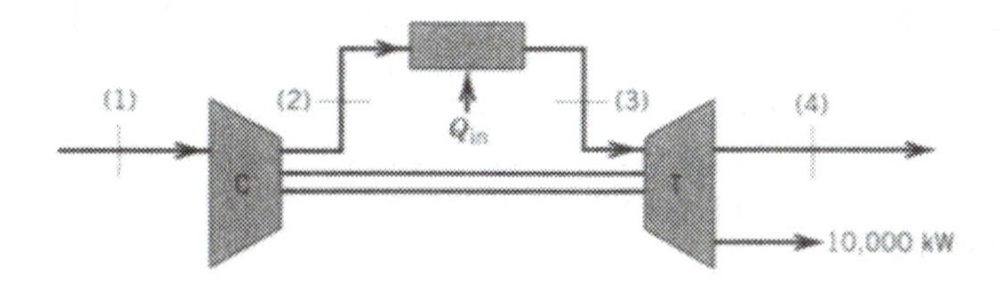

$p_1 = 14.7$ psia $\quad T_1 = 519°R$

$p_2 = 176.0$ psia

$p_3 = 172.0$ psia $\quad T_3 = 2160°R$

$p_4 = 15.0$ psia

$\eta_c = 82\%$ $\quad \eta_t = 85\%$

The fuel used is liquid C_8H_{18} supplied at 537°R.

(a) Determine the fuel–air ratio for this engine.

(b) Determine the percent excess air supplied.

(c) Determine how many pounds of air must be supplied to the compressor, pounds per second, for this engine to produce 10,000 kW of power (net power).

(d) Determine, for this engine, the heat rate, which is the heat added in Btu per hour divided by the power output in kilowatts. Base the heat input on the lower heating value of the fuel.

5.16. An article in the *SAE Journal* (vol. 75, number 6, pp. 71–77) states that "ammonia is superior to hydrocarbons as a fuel for gas turbines on the basis of power output and thermal efficiency but poorer on the basis of fuel economy." To verify this, solve Example Problem 5.5 assuming ammonia, NH_3, to be the fuel that is supplied as a gas. The reaction for complete combustion of ammonia is

$$2NH_3 + 1\tfrac{1}{2}O_2 \rightarrow N_2 + 3H_2O$$

5.17. A gas turbine operating on the basic cycle has a compressor inlet pressure and temperature of 14.7 psia and 519°R, respectively, a compressor exit pressure of 231.5 psia, a turbine inlet pressure and temperature of 222 psia and 2520°R, respectively, a turbine exit pressure of 15 psia, a compressor efficiency of 82%, and a turbine efficiency of 85%. The fuel, supplied as a liquid at 537°R, is *n*-octane, C_8H_{18}. If 5.0 lb of air are supplied to the compressor per second, calculate:

(a) The percent excess air supplied

(b) The net work developed per pound of air entering the compressor

(c) The thermal efficiency based on the lower heating value of the fuel

(d) The horsepower developed by the engine

(e) The specific fuel consumption

5.18. A basic, air-standard gas turbine engine has a compressor inlet temperature and pressure of 288 K (519°R) and 101.3 kPa (14.70 psia), respectively, a compressor exit and turbine inlet pressure of 3039 kPa (441 psia), a turbine inlet temperature of 1517 K (2730°R), a turbine exit pressure of 101.3 kPa (14.70 psia), a compressor efficiency of 88%, and a turbine efficiency of 90%. Air enters the compressor at the rate of 271 lb/s. Calculate, assuming variable specific heats and that the fuel is liquid C_8H_{18} supplied at 298 K (537°R), taking into account the mass of fuel supplied but assuming air is the fluid expanding through the turbines:

(a) The temperature of the air at the exit from the gas generator turbine

(b) The temperature of the air leaving the power turbine

(c) The fuel–air ratio for this engine

(d) The heat rate for this engine, kJ/kWh

(e) The power developed by the power turbine

Solve first using values from the tables, then using the GASTUSIM computer program.

5.19. Solve Problem 5.18 assuming a 6% pressure loss in the combustion chamber and a 1% loss in the exhaust system; that is, the pressure at state 5 is 1% above the inlet pressure to the compressor.

5.20. Solve Example Problem 5.5 for combustion chamber pressure losses of 1%, 2%, 3%, 4%, 5%, 6%, and 7%. It is suggested that the GASTUSIM computer program be used.

5.21. Determine, for the pressure ratios that give the maximum specific power for the four turbine inlet temperatures in Figure 5.22, the temperature at the exit from the power turbine. Then solve for the temperature at the exit from the power turbine for turbine inlet temperatures of 2520°R (1400 K) and 2700°R (1500 K) and a pressure ratio of 30, all other values (η_C, η_{GT}, etc.) the same as those used in developing Figure 5.22. It is suggested that the GASTUSIM computer program be used in solving this problem.

5.22. A gas turbine operates on the basic cycle shown in the diagram. Known values are as follows:

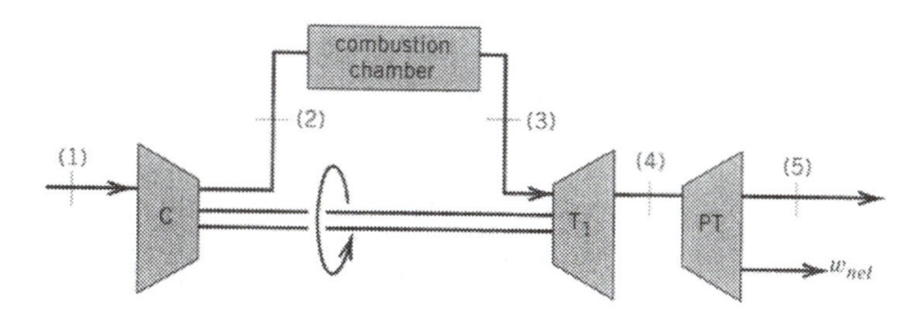

Compressor inlet pressure: 14.45 psia (99.6 kPa)
Compressor inlet temperature: 560°R (311 K)
Pressure leaving compressor: 223.98 psia (1543.5 kPa)
Turbine T_1 inlet pressure: 212.13 psia (1461.8 kPa)
Turbine T_1 inlet temperature: 2160°R (1200 K)
Power turbine exit pressure: 15.05 psia (103.7 kPa)
Compressor efficiency: 0.80
Turbine T_1 efficiency: 0.84
Power turbine efficiency: 0.85
Mass flow into compressor: 2.71 lb/s (1.23 kg/s)

Assuming that turbine T_1 drives the compressor and taking into account the mass of fuel added but assuming air to be the working fluid throughout, calculate:

(a) The horsepower developed by this gas turbine (net power)

(b) The specific fuel consumption for this engine

5.23. It is desirable to have a comparison between two regenerative engines, one using a rotating matrix, one using a stationary heat exchanger. To make this comparison, calculate the net work, thermal efficiency, and specific fuel consumption for the two cycles described.

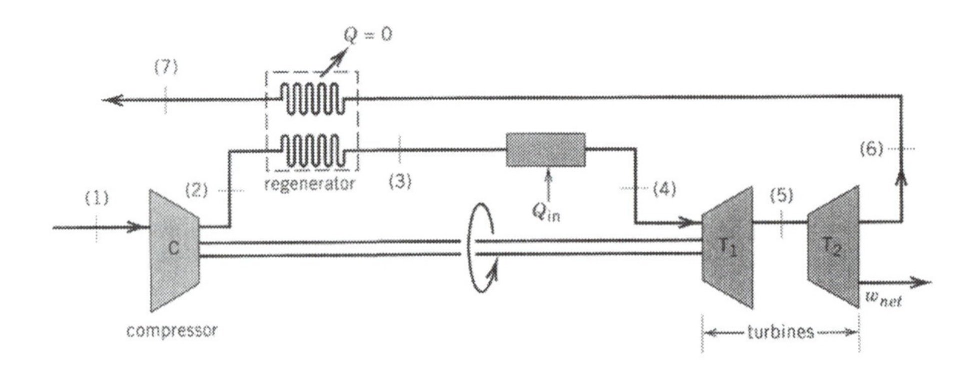

Known Values

	Rotating Regenerator	Stationary Regenerator
p_1	14.7 psia (101.3 kPa)	14.7 psia (101.3 kPa)
p_2	73.5 psia (505.6 kPa)	73.5 psia (505.6 kPa)
p_3	73.1 psia (503.7 kPa)	72.5 psia (499.6 kPa)
p_4	72.6 psia (500.3 kPa)	72.0 psia (496.2 kPa)
p_6	15.3 psia (105.4 kPa)	15.8 psia (108.9 kPa)
p_7	14.7 psia (101.3 kPa)	14.7 psia (101.3 kPa)
T_1	519°R (288 K)	519°R (288 K)
T_4	2160°R (1200 K)	2160°R (1200 K)
Regenerator Effectiveness	84%	75%
Compressor Efficiency	85%	85%
Turbine T_1 Efficiency	90%	90%
Turbine T_2 Efficiency	90%	90%
Leakage[a]	2.5%	0.00%

Turbine T_1 drives the compressor, turbine T_2 delivers power to an external system.

[a]Discussion on air leakage: For the rotating regenerator, a leakage of 2.5% of the air leaving the compressor has been assumed. Assume for your calculations that 2.5% of the air leaving the compressor leaks through to state 7 from state 2. Assume that no energy is transferred to this gas; that is, it leaks from state 2 to the pressure at state 7 with no change in temperature.

5.24. A regenerative gas turbine engine has a full-power pressure ratio of 4 and a regenerator effectiveness of approximately 85%. Does the pressure ratio selected for this engine seem reasonable? Why?

5.25. A regenerative gas turbine engine operates on the cycle shown in the diagram. Turbine T_1 drives the compressor, turbine T_2 delivers power to an external system. Known design point data are as follows:

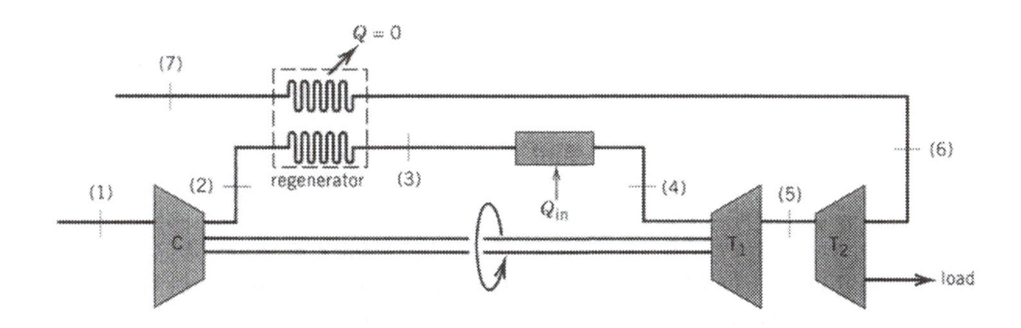

$p_1 = p_7 = 14.7$ psia (101.3 kPa)	Compressor efficiency = 80%
$p_2 = 58.8$ psia (405.2 kPa)	Turbine T_1 efficiency = 87%
$p_3 = 57.5$ psia (396.2 kPa)	Turbine T_2 efficiency = 84%
$p_4 = 56.0$ psia (385.9 kPa)	Air flow entering compressor = 2.2 lb/s (998 g/s)
$p_6 = 15.5$ psia (106.8 kPa)	Assume no leakage through the regenerator
$T_1 = 545°R$ (303 K)	
$T_3 = 1550°R$ (861 K)	
$T_4 = 2160°R$ (1200 K)	

Assuming an air standard cycle but taking into account the mass of fuel added, calculate:

(a) The temperature at state 5
(b) The pressure at state 5
(c) The horsepower developed by this engine
(d) The temperature at state 6
(e) The temperature at state 7
(f) The regenerator effectiveness
(g) The specific fuel consumption

5.26. Solve Problem 5.25 assuming that 4% of the air leaving the compressor leaks through to state 7 from state 2. Assume that no energy is transferred to this air; that is, it leaks from state 2 to the pressure at state 7 with no change in temperature. Calculate the temperature at state 7 both before this air is mixed with the air leaving the regenerator and after it is mixed.

5.27. A recuperator is added to the gas turbine described in Problem 5.22. All values are the same with the following exceptions or additional information:

Temperature leaving recuperator: 1426°R (792 K)
Pressure leaving power turbine: 15.92 psia (109.7 kPa)
Calculate, taking into account the mass of fuel added but assuming air to be the working fluid throughout:

(a) The horsepower developed by this gas turbine (net power)

(b) The specific fuel consumption for this engine

(c) The recuperator effectiveness

Compare your answers with those calculated in Problem 5.22.

5.28. A gas turbine operating on the steam-injected cycle (see Figure 5.31) has a compressor inlet pressure of 101.3 kPa (14.70 psia), a temperture of 288 K (519°R), a compressor exit and turbine inlet pressure of 1013 kPa (147.0 psia), a turbine inlet temperature of 1200 K (2160°R), a power turbine exit pressure of 101.3 kPa (14.70 psia), a compressor efficiency of 87% and a gas generator and power turbine each with an efficiency of 89%. The fuel, supplied as a liquid at 298 K (537°R), is *n*-octane, C_8H_{18}. Steam at a pressure of 1378.2 kPa (200 psia) and 653 K (1175°R) and equal to 2% of the compressor discharge flow is injected into the combustion chamber. If air enters the compressor at the rate of 1 kg/s (1 lb/s), calculate, assuming the products leaving the combustion chamber are dry air and water vapor but taking into account the mass of fuel added

(a) The fuel-air ratio, kg fuel/kg air

(b) The pressure at the exit from the gas generator exit

(c) The cycle thermal efficiency

(d) The net power, in kW, developed by the power turbine

(e) The heat rate, kJ/kWh, based on the lower heating value

5.29. Solve Example Problem 5.8 assuming the mass of steam added is 4.0% of the compressor discharge flow, all other values the same.

5.30. Solve Example Problem 5.8 assuming that an additional 2.5% of steam is added between the gas generator turbine and power turbine, all other values the same.

5.31. Solve Problem 5.28 assuming the mass of steam added is 4.0% of the compressor discharge flow, all other values the same.

5.32. Solve Problem 5.28 assuming the compressor pressure ratio is 8.0, all other values the same.

5.33. A gas turbine engine operates on the evaporative-regenerative cycle as shown in Figure 5.32. Known values are

$T_1 = 519°R\ (288\ K)$ $\quad p_1 = p_5 = p_{5.5} = 14.70$ psia (101.3 kPa)
$T_{2.3} = 756°R\ (420\ K)$ $\quad p_2 = p_{2.3} = p_{2.5} = p_3 = 147.0$ psia (1013 kPa)
$T_3 = 2160°R\ (1200\ K)$ $\quad p_8 = 160$ psia (1103 kPa)
$T_8 = 711°R\ (395\ K)$
$\eta_C = 87\%$, $\eta_{GT} = 89\%$, $\eta_{PT} = 89\%$, $\eta_{reg} = 85\%$

Calculate, neglecting the mass of fuel added, assuming air enters the rate of 1 lb/s (1 kg/s), taking into account the mass of water added and assuming an air–water vapor mixture after the evaporator

(a) The pressure and temperature at the exit from the gas generator turbine

(b) The net power, in kW, developed by the power turbine

(c) The cycle thermal efficiency

(d) The heat rate, Btu/kWh (kJ/kWh)

5.34. Solve Example Problem 5.9 assuming air is the working fluid after the evaporator, all other conditions the same.

5.35. Solve Example Problem 5.9 assuming the fuel is *n*-octane, C_8H_{18}, supplied as a liquid at 298 K (537°R). Take into account the mass of fuel added but assume an air–water vapor mixture after the combustion chamber.

5.36. Prove that the optimum intercooler pressure for a gas turbine engine with intercooling is

$$p_i = \sqrt{p_1 p_2}$$

as stated in Eq. (5.29).

5.37. Solve Example Problem 5.10 assuming the temperature at state 1.5 is 330 K (594°R), all other values the same.

5.38. An air standard gas turbine engine operates on the cycle shown in Figure 5.34. Known values are

$T_1 = 288$ K (519°R) $\quad \eta_{C1} = 87\%$ $\quad p_1 = p_5 = 101.3$ kPa (14.70 psia)

$T_{1.5} = 288$ K (519°R) $\quad \eta_{C2} = 87\%$ $\quad p_2 = p_3 = 1013$ kPa (147.0 psia)

$T_3 = 1200$ K (2160°R) $\quad \eta_{GT} = 89\%$ $\quad p_{1.1} = p_{1.5} = 320.3$ kPa (46.50 psia)

$\eta_{PT} = 89\%$

Calculate, on the basis of 1 kg/s (1 lb/s) of air entering the first compressor and neglecting the fuel added:

(a) The optimum intercooler pressure

(b) The total compressor work, kJ/kg (Btu/lb)

(c) The net power, kW

(d) The cycle thermal efficiency

5.39. An air standard gas turbine engine operates on the reheat cycle as shown in Figure 5.36. Known values are

$T_1 = 519$°R (288 K) $\quad \eta_C = 87\%$ $\quad p_1 = p_5 = 14.70$ psia (101.3 kPa)

$T_3 = 2160$°R (1200 K) $\quad \eta_{GT} = 89\%$ $\quad p_2 = p_3 = 147.0$ psia (1013 kPa)

$T_{4.5} = 2160$°R (1200 K) $\quad \eta_{PT} = 89\%$

Calculate, on a basis of 1 lb/s (1 kg/s) of air entering the compressor, neglecting the mass of fuel added and assuming the optimum reheat pressure

(a) The optimum reheat pressure

(b) The net power, kW

(c) The cycle thermal efficiency

5.40. Solve Example Problem 5.12 assuming the temperature at state 1.5 is 330 K (594°R), all other values the same.

5.41. An air standard gas turbine engine operates on the intercooled-regenerative-reheat cycle illustrated in Figure 5.38. Known values are

T_1 = 519°R (288 K)	η_{C1} = 80%	p_1 = 14.70 psia (101.3 kPa)
$T_{1.5}$ = 612°R (340 K)	η_{C2} = 80%	$p_{1.1}$ = 58.8 psia (405.2 kPa)
T_3 = 2160°R (1200 K)	η_{T2} = 87%	$p_{1.5}$ = 57.9 psia (399.1 kPa)
$T_{3.5}$ = 2160°R (1200 K)	η_{T1} = 87%	p_2 = 231.6 psia (1596.4 kPa)
	η_{PT} = 88%	$p_{2.5}$ = 228.1 psia (1572.5 kPa)
Recuperator effectiveness = 80%		p_3 = 219.0 psia (1509.6 kPa)
		$p_{3.5}$ = 0.97 $p_{3.1}$ (0.97 $p_{3.1}$)
		p_5 = 1.06 p_6 (1.06 p_6)
		p_6 = 14.70 psia (101.3 kPa)

Calculate, on a basis of 1 lb/s (1 kg/s) of air entering compressor C_1 and neglecting the mass of fuel added

(a) The total compressor work, Btu/lb (kJ/kg)

(b) The net work, Btu/lb (kJ/kg)

(c) The specific power, kW, developed by this engine

(d) The cycle thermal efficiency

5.42. Solve Problem 5.41 assuming

$$p_{1.1} = p_{1.5}, p_2 = p_{2.5} = p_3, p_{3.1} = p_{3.5} \text{ and } p_5 = p_6$$

all other values the same. Compare the results with those calculated in Problem 5.41.

5.43. Solve Example Problem 5.13 assuming the pinch point temperature difference, ΔT_{PP}, is 10°C (18°F), all other values the same.

5.44. Solve Example Problem 5.13, all known values the same except T_3 = 2160°R (1200 K). Compare your answers with those for Example Problem 5.13.

5.45. A combined-cycle power plant operates on the cycle illustrated in Figure 5.42. Known values are as follows:

Gas Turbine (Air Standard)

T_1 = 288 K (519°R)	η_C = 88%	$p_1 = p_5 = p_6$ = 101.3 kPa (14.70 psia)
T_3 = 1420 K (2560°R)	η_{GT} = 90%	$p_2 = p_3$ = 1347 kPa (195.5 psia)
	η_{PT} = 90%	
$\Delta T_{\text{pinch point}}$ = 15°C (27°F)		

Steam Power Plant

t_a = 505°C (940°F)	η_{turbine} = 90%	$p_a = p_d$ = 6.9 MPa (1000 psia)
	η_{pump} = 60%	p_b = 6.0 kPa (1.0 psia)
		p_c = 6.0 kPa, saturated liquid (1.0 psia)

Calculate, on a basis of 150 kg/s (330 lb/s) entering the gas turbine compressor and neglecting the mass of fuel added:

(a) The thermal efficiency and heat rate assuming no steam power plant after the gas turbine, kJ/kWh (Btu/kWh)

(b) The heat rate for the combined-cycle power plant, kJ/kWh (Btu/kWh)

(c) The overall thermal efficiency

(d) The fraction of the total power output generated by the gas turbine

5.46. Solve Problem 5.45 assuming supplementary firing (see Figure 5.43) that raises the temperature at the inlet to the HRSG to 1890°R (1050 K), all other values the same. Compare your answers with those determined for Problem 5.45.

5.47. Solve Problem 5.45 assuming the pressure at the inlet to the steam turbine is 500 psia (3.45 MPa), all other values the same.

5.48. A combined-cycle power plant operates on the single-pressure cycle illustrated in Figure 5.42. Known values are as follows:

Gas Turbine (Air Standard)

T_1 = 519°R (288 K) $\quad$ η_C = 87% $\quad$ $p_1 = p_6$ = 14.70 psia (101.3 kPa)

T_3 = 2700°R (1500 K) $\quad$ η_{GT} = 89% $\quad$ p_2 = 195.5 psia (1347 kPa)

η_{PT} = 89% $\quad$ p_3 = 186.0 psia (1281 kPa)

p_5 = 15.60 psia (107.5 kPa)

$\Delta T_{\text{pinch point}}$ = 18°F (10°C)

Steam Power Plant

t_a = 950°F (510°C) $\quad$ $\eta_{turbine}$ = 89% $\quad$ $p_a = p_d$ = 950 psia (6.5 MPa)

η_{pump} = 65% $\quad$ p_c = 1.0 psia (6.0 kPa)

p_d = 1.0 psia, saturated liquid (6.0 kPa)

Calculate, on the basis of 700 lb/s (320 kg/s) of air entering the compressor of the gas turbine and neglecting the mass of fuel added:

(a) The heat rate assuming no steam turbine after the gas turbine, Btu/kWh (kJ/kwh)

(b) The heat rate for the combined cycle

5.49. A two-pressure steam turbine combined-cycle power plant operates on the cycle illustrated in Figure 5.50. Known values are as follows:

Gas Turbine (Air Standard)

T_1 = 519°R (288 K) $\quad$ η_C = 87% $\quad$ p_1 = 14.70 psia (101.3 kPa)

T_3 = 2700°R (1500 K) $\quad$ η_{GT} = 89% $\quad$ p_2 = 198.5 psia (1370 kPa)

T_6 = 760°R (420 K) $\quad$ η_{PT} = 89% $\quad$ p_3 = 190.0 psia (1313 kPa)

p_5 = 15.60 psia (107.0 kPa)

p_6 = 14.70 psia (101.3 kPa)

Steam Power Plant

t_a = 950°F (510°C)	η_{ST} = 89%	p_a = 920 psia (6.3 MPa)
t_c = 330°F (165°C)	η_P = 65%	$p_c = p_g = p_j$ = 52.0 psia (360 kPa)
		p_f = 1.0 psia (6.0 kPa)
$\dot{m}_a = 0.83\ \dot{m}_f$		

Calculate, on the basis of 540 lb/s (300 kg/s) of air entering the compressor of the gas turbine and neglecting the mass of fuel added:

(a) The minimum temperature difference between the air and H_2O in the HRSG and where it occurs

(b) The mass rate of flow, in lb/s (kg/s), of the H_2O in the steam cycle

(c) The net power developed by the gas turbine, kW

(d) The heat rate for the gas turbine, Btu/kWh (kJ/kWh)

(e) The net power developed by the steam turbine, kW

(f) The heat rate for the combined cycle, Btu/kWh (kJ/kWh)

5.50. Solve Problem 5.49 assuming the mass rate of flow at state a is 77% of the mass rate of flow at the condenser exit. Compare your answers with those calculated for Problem 5.49.

5.51. A two-pressure steam turbine combined-cycle power plant operates on the cycle illustrated in Figure 5.50. Known values are as follows:

Gas Turbine (Air Standard)

T_1 = 288 K (519°R)	η_C = 87%	$p_1 = p_6$ = 101.3 kPa (14.70 psia)
T_3 = 1500 K (2700°R)	η_{GT} = 89%	p_2 = 3039 kPa (441.0 psia)
	η_{PT} = 89%	p_3 = 2917 kPa (423.4 psia)
		p_5 = 103.4 kPa (15.00 psia)

Steam Power Plant

t_a = 425°C (800°F)	η_{ST} = 90%	$p_a = p_k$ = 6.2 MPa (900 psia)
t_c = 165°C (330°F)	η_p = 60%	$p_c = p_g = p_j$ = 350 kPa (50 psia)
		p_f = 6.0 kPa (1.0 psia)

Minimum temperature difference between the air and H_2O in the HRSG is 15°C (27°F).

Calculate, on the basis of 350 kg/s (630 lb/s) of air entering the compressor of the gas turbine and neglecting the mass of fuel added:

(a) The net power developed by the gas turbine, kW

(b) The fraction of the flow in the steam turbine that passes through the high pressure turbine

6

GAS TURBINES FOR AIRCRAFT PROPULSION

Before an analysis can be made of the various types of gas turbines used as airplane propulsion systems, the general thrust equation and performance parameters used in evaluating propulsion systems must be examined.

6.1 GENERAL THRUST EQUATION

A fairly general equation for evaluating thrust from an air-breathing jet engine can be derived from the conservation of momentum and mass equations. Figure 6.1 will be used in the derivation of a general equation from the position of an observer riding with the thrust-producing device.

In Figure 6.1, the control surface is shown by the dashed line. Surface A is far upstream, where the pressure and velocity may be assumed uniform over the entire surface. Surface B is at the exit from the thrust-producing device. The side control surfaces are parallel to the upstream velocity and are far removed from the thrust-producing device. The derivation of the thrust equation will be for the time-invariant (steady-flow) condition.

Values that are known are shown in Figure 6.1. The fuel is assumed to be added at a right angle to the direction of flow. The velocity and pressure at the inlet are assumed to be uniform, and the mass of air entering the thrust-producing device, $\dot{m}_a$, enters the device through an area A_i. The products of combustion leave the thrust-producing device through an area A_e with a velocity $\overline{V}_e$, a mass rate of flow $\dot{m}_e$, and at a pressure p_e. The velocity and pressure of the air that passes around the thrust-producing device are $\overline{V}$ and p_a, respectively. A rate of air, $\dot{m}_s$, leaves through the side control surface.

The thrust developed by the generalized thrust producer may be derived from

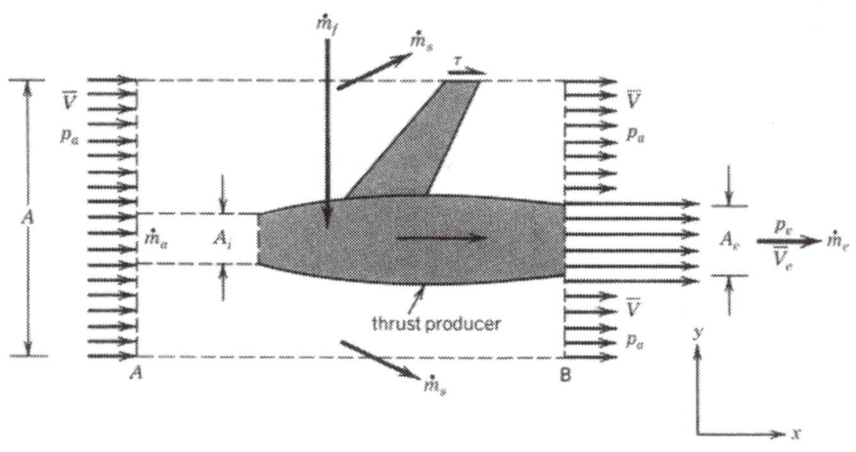

Figure 6.1. Generalized thrust-producing device.

Newton's second law (Eq. 3.9). For the time-invariant (steady-flow) condition and neglecting the body forces, Eq. (3.9) becomes

$$\mathbf{F}_{\mathrm{R}} = \int_{\mathrm{C.S.}} \mathbf{V}\rho\mathbf{V} \cdot d\mathbf{A} \tag{6.1}$$

Considering the components of force and momentum flux in the x direction (direction of flight) yields

$$\Sigma F_x = \int_{\mathrm{C.S.}} \rho \overline{V}_x (\mathbf{V} \cdot d\mathbf{A}) \tag{6.2}$$

For the case illustrated in Figure 6.1, the left-hand side of Eq. (6.2) becomes

$$\Sigma F_x = \tau + p_a A - p_a(A - A_e) - p_e A_e \tag{6.3}$$

which reduces to

$$\Sigma F_x = \tau + (p_a - p_e)A_e \tag{6.4}$$

Equation (6.4) is for the general case where the exhaust (exit) pressure may be different than the inlet pressure. This will occur when the expansion pressure ratio across the exhaust nozzle is sufficient to produce supersonic flow and expansion in the nozzle is not back to ambient pressure. The net pressure force resulting because of this pressure difference is

$$(p_a - p_e)A_e \tag{6.5}$$

The conservation of mass (Eq. 3.8) for a steady-flow process, assuming that the fuel is added from outside the control volume, is

$$\int_{\mathrm{C.S.}} \rho\mathbf{V} \cdot d\mathbf{A} = 0 \tag{6.6}$$

Therefore,

$$\dot{m}_A + \dot{m}_f = \dot{m}_s + \dot{m}_B \tag{6.7}$$

or

$$\rho\overline{V}A_i + \rho\overline{V}(A - A_i) + \dot{m}_f = \dot{m}_s + \rho_e\overline{V}_eA_e + \rho\overline{V}(A - A_e) \tag{6.8}$$

which simplifies to

$$\dot{m}_f = \dot{m}_s + A_e(\rho_e\overline{V}_e - \rho\overline{V}) \tag{6.9}$$

For the engine (thrust producer),

$$\dot{m}_a + \dot{m}_f = \dot{m}_e \tag{6.10}$$

Combining Eqs. (6.9) and (6.10) yields

$$\dot{m}_s = \rho\overline{V}(A_e - A_i) \tag{6.11}$$

Since the side control surface is far removed from the thrust-producing device, the velocity of $\dot{m}_s$ in the y direction will be approximately zero and in the x direction approximately $\overline{V}$. Therefore, the right-hand side of Eq. (6.2) becomes

$$\int_{\text{C.S.}} \overline{V}_x\,(\rho\mathbf{V}\cdot d\mathbf{A}) = \dot{m}_e\overline{V}_e + \rho\overline{V}(A - A_e)\overline{V} + \dot{m}_s\overline{V}$$

$$- m_a\overline{V} - \rho\overline{V}(A - A_i)\overline{V} \tag{6.12}$$

$$= \dot{m}_e\overline{V}_e - \dot{m}_a\overline{V} \tag{6.13}$$

Substituting Eqs. (6.4), (6.13), and (5.19) into Eq. (6.2) yields

$$\tau = (p_e - p_a)A_e + \dot{m}_a[(1 + f)\overline{V}_e - \overline{V}] \tag{6.14}$$

Equation (6.14) was derived for one stream passing through the thrust-producing device. In some cases, two separate streams pass through the thrust-producing device. Assuming that fuel is added only to the hot stream and identifying the hot and cold streams by h and c subscripts yields, for the general case,

$$\tau = (\dot{m}_{a_h} + \dot{m}_f)\overline{V}_{e_h} - m_{a_h}\overline{V}$$

$$+ \dot{m}_{a_c}(\overline{V}_{e_c} - \overline{V}) + (p_{e_h} - p_a)A_{e_h} + (p_{e_c} - p_a)A_{e_c} \tag{6.15}$$

The specific thrust, I, is defined as the thrust produced when a unit mass of air per second enters the device, or

$$I = \frac{\tau}{\dot{m}_a} \tag{6.16}$$

6.2 ENGINE PERFORMANCE PARAMETERS

Three performance parameters are of interest for thrust-producing devices. These are the propulsion efficiency, the thermal efficiency, and the overall efficiency.

Propulsion Efficiency

One measure of the performance of a thrust-producing device is the ratio of the thrust power to the jet power. The thrust power is defined as the product of the thrust and the flight velocity. The jet power is the change in kinetic energy of the gases passing through the device. Therefore, referring to Figure 6.1, the propulsion efficiency of a thrust-producing device with a single jet stream is

$$\eta_p = \frac{\text{thrust power}}{\text{jet power}} = \frac{\tau \overline{V}}{\dot{m}_a[(1 + f)(\overline{V}_e^2/2) - \overline{V}^2/2]} \tag{6.17}$$

Neglecting the pressure term and assuming that the fuel–air ratio (f) is zero yields

$$\eta_p = \frac{\overline{V}(\overline{V}_e - \overline{V})}{(\overline{V}_e^2)/2 - \overline{V}^2/2} = \frac{2\overline{V}/\overline{V}_e}{1 + \overline{V}/\overline{V}_e} \tag{6.18}$$

Examination of Eq. (6.18) indicates that the maximum propulsion efficiency occurs when the velocity leaving the thrust-producing device equals the velocity of the device. This, however, is not practical, since when $\overline{V} = \overline{V}_e$, the thrust is zero.

Thermal Efficiency

The thermal efficiency for the thrust-producing devices of interest to us is defined as the ratio of the change in kinetic energy of the gases passing through the device to the rate at which energy is added to the device. Therefore,

$$\eta_{\text{th}} = \frac{\dot{m}_a[(1 + f)(\overline{V}_e^2/2) - \overline{V}^2/2]}{\dot{m}_f|\Delta H_c|} \tag{6.19}$$

Here $|\Delta H_c|$ is the lower heating value of the fuel.

Overall Efficiency

The overall efficiency is defined as the product of the thermal efficiency and the propulsion efficiency, or

$$\eta_o = \eta_{\text{th}}\, \eta_p \tag{6.20}$$

6.3 ALTITUDE TABLES

Ambient pressure and temperature vary with altitude. One source of data is the *U.S. Standard Atmosphere, 1962* (1). An abridged version of this table is in the appendix as Table C.

One should note that the temperature decreases from sea level to 36,089 ft (11,000 m), remains constant from 36,089 ft (11,000 m) to 65,617 ft (20,000 m), then increases

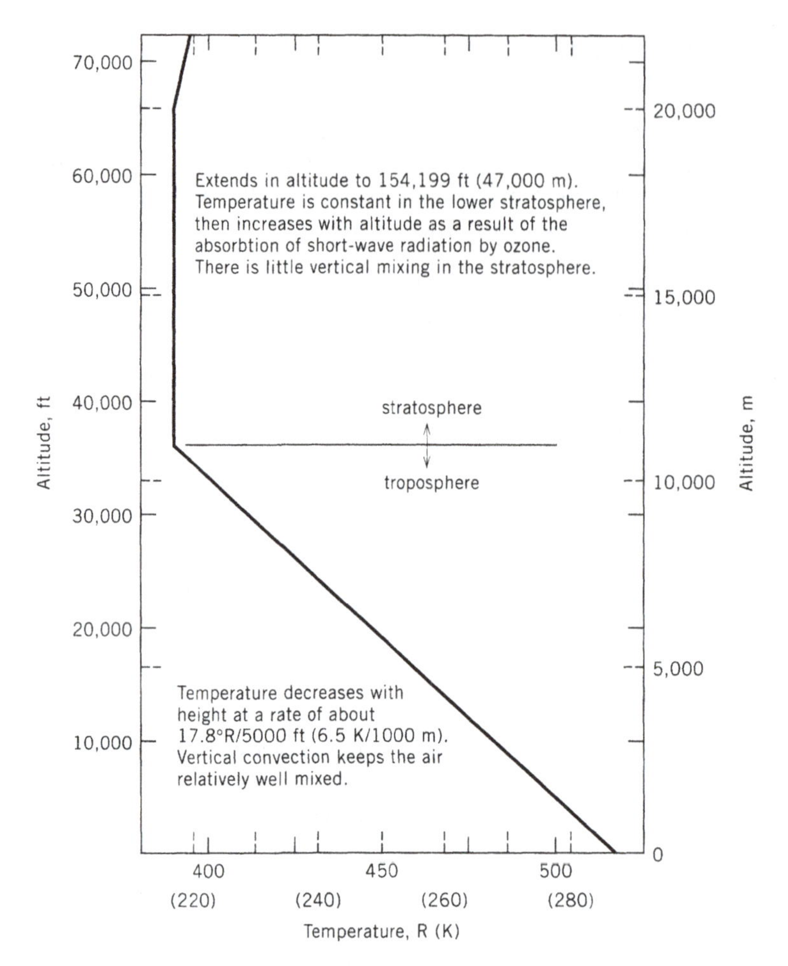

Figure 6.2. Atmospheric temperature profile.

from 65,617 ft (20,000 m) to 154,199 ft (47,000 m). These temperature regions are shown in Figure 6.2 with the zones identified.

6.4 TURBOJET (AIR STANDARD)

There are various types of gas turbine engines used to propel aircraft. These usually fall into the categories of the turbojet, turbofan, and turboprop engines. All could be constructed from the same gas generator. This section deals with the analysis of the air-standard nonafterburning turbojet engine. Variations of the turbojet engine along with the turbofan and turboprop engines will be discussed in later sections of this chapter.

Figure 6.3 illustrates a schematic diagram for a nonafterburning turbojet engine. Ambient air enters the diffuser where the velocity decreases and the static pressure increases. Figure 6.4 illustrates the temperature–entropy diagram for the entire cycle. Figure 6.5 illustrates the temperature–entropy diagram for the diffuser.

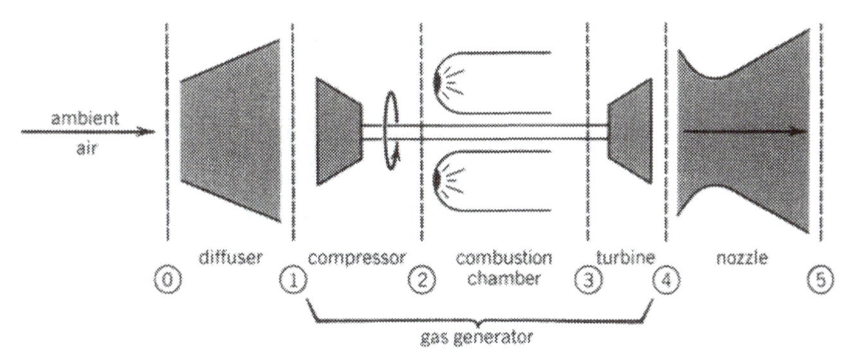

Figure 6.3. *Schematic diagram for the general nonafterburning turbojet engine.*

The adiabatic diffuser efficiency is defined (referring to Figure 6.5) as the ratio of the actual isentropic enthalpy change to the ideal isentropic enthalpy change. Therefore,

$$\eta_d = \frac{h_1' - h_0}{h_{o0} - h_0} \tag{6.21a}$$

In terms of the static and total pressures, the diffuser efficiency becomes

$$\eta_d = \frac{p_{o1} - p_0}{p_{o0} - p_0} \tag{6.21b}$$

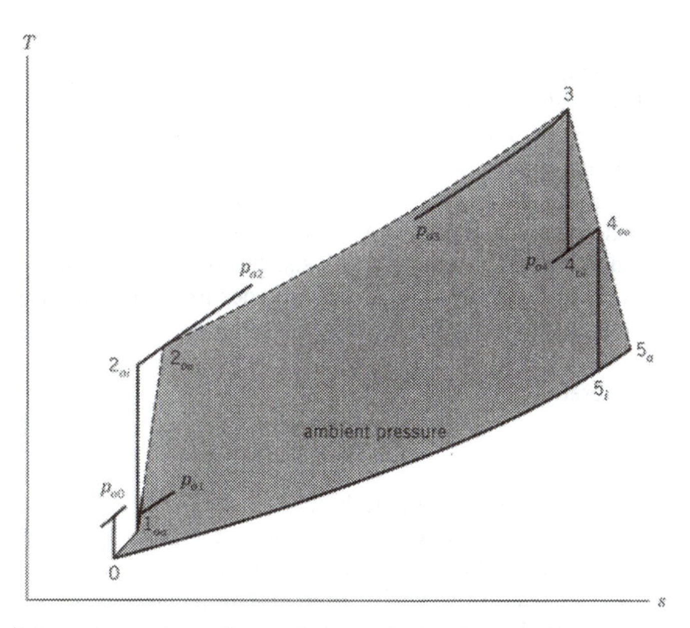

Figure 6.4. Temperature–entropy diagram for a nonafterburning turbojet engine with nozzle expansion back to ambient pressure.

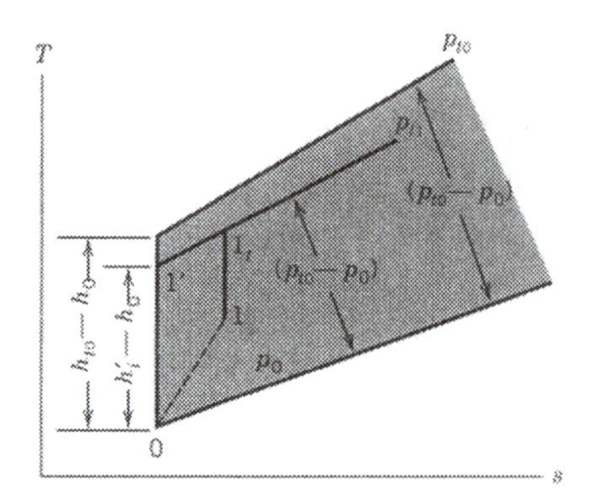

Figure 6.5. Temperature–entropy diagram for a diffuser.

The diffuser efficiency is often measured as the recovery ratio, which is defined as

$$\eta_r = \frac{p_{o1}}{p_{o0}} \tag{6.22}$$

MIL-E-5007D inlet recovery is specified as (2)

$$\frac{p_{o1}}{p_{o0}} = 1.00 \qquad \text{from } M = 0 \text{ to } 1.0 \tag{6.23a}$$

$$\frac{p_{o1}}{p_{o0}} = 1.00 - 0.076\,(M - 1)^{1.35} \qquad \text{from } M > 1.0 \text{ to } 5.0 \tag{6.23b}$$

The air, upon leaving the diffuser, enters the compressor. The compressor, combustion chamber (heat addition process in the air standard cycle), and turbine were discussed in Section 5.2. Compressor efficiency, turbine efficiency, and pressure drop in the combustion chamber were all defined and discussed in Section 5.2.

The inlet and outlet states to the compressor, combustion chamber, and turbine that are used in defining the efficiencies and pressure drop are usually the total (stagnation) pressures. In the air standard analysis, the combustion chamber is replaced by a heat addition process. Ideally, no pressure drop occurs in the combustion chamber (heat addition process). In the actual case, a pressure drop does occur.

The actual work delivered by the turbine is equal to the work required to drive the compressor. Therefore,

$$\left|{}_1w_{2a}\right| = \left|{}_3w_{4a}\right| \tag{6.24}$$

Since the work (power) done by the turbine is equal to the work (power) required to drive the compressor, the pressure at the turbine exit is high. The air next expands through a nozzle, where the velocity is increased with a resulting decrease in pressure. Ideally, the nozzle operates isentropically. Actually, it operates adiabatically but irreversibly. The nozzle efficiency was defined in Section 3.8 and is the ratio of the actual to the ideal kinetic energy when expanding from the same inlet total (stagnation)

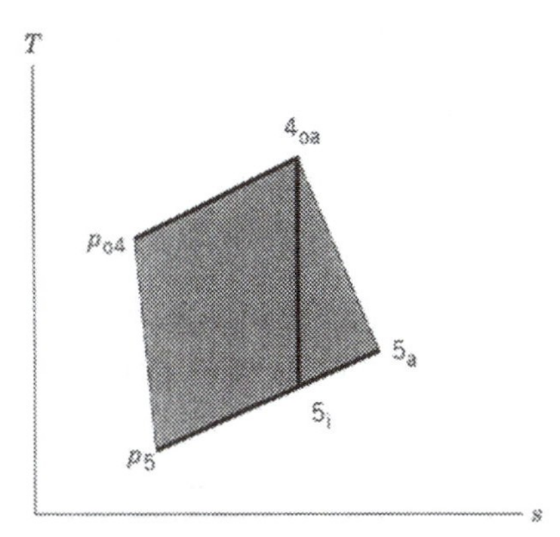

Figure 6.6. Temperature–entropy diagram for a nozzle.

state to the same *static* pressure. Therefore, referring to Figures 6.4 and 6.6, the nozzle efficiency is

$$\eta_N = \frac{\overline{V}_{5a}^2/2}{\overline{V}_{5i}^2/2} \tag{6.25a}$$

$$= \frac{h_{o4} - h_{5a}}{h_{o4} - h_{5i}} \tag{6.25b}$$

In the analysis of a turbojet engine, one item of interest is the thrust. This can be evaluated by use of Eq. (6.14). Example Problems 6.1 and 6.2 illustrate the calculation of the thrust of a turbojet engine at sea level static and for a flight condition, respectively. Both problems assume an air standard analysis and neglect the mass of fuel added.

Example Problem 6.1

A turbojet engine is operating under the following conditions:

Flight speed at sea level, standard day	0
Air flow entering compressor	1 lb/s (1 kg/s)
Compressor pressure ratio (total to total)	12
Efficiencies	
Diffuser	100%
Compressor	87%
Turbine	89%
Jet nozzle	100%
Turbine inlet temperature (stagnation)	2520°R (1400 K)

Assuming an air standard cycle, that the stagnation pressure remains constant from compressor outlet to turbine inlet, and neglecting the mass of fuel added:

(a) Calculate the thrust developed by and the heat added to this engine assuming a converging nozzle

(b) Calculate the thrust developed by, the heat added to, and the thermal efficiency of this engine assuming a converging-diverging nozzle, that is, expansion in the nozzle back to ambient pressure

Solution

From Table C:

Ambient temperature	518.670°R (288.150 K)
Ambient pressure	14.696 psia (101.325 kPa)

Diffuser

Since velocity is zero, the conditions at the compressor inlet are ambient conditions. The values for the gas generator (through state 4) are the same as those determined in Example Problem 5.3. These values are

State	T,°R (K)		p, psia (kPa)		h, Btu/lb (kJ/kg)	
1	519	(288)	14.70	(101.3)	−6.0	(−14.3)
2_a	1122	(623)	176.4	(1215.6)	141.7	(328.7)
3	2520	(1400)	176.4	(1215.6)	521.7	(1212.7)
4_a	1996	(1109)	59.1	(408.2)	374.0	(869.7)

$$w_{C,a} = 147.7 \text{ Btu/lb } (343.0 \text{ kJ/kg})$$

$$q_{in} = 380.0 \text{ Btu/lb } (884.0 \text{ kJ/kg})$$

Gas Generator

The conditions leaving the gas generator turbine and entering the exhaust nozzle for either the engine with a converging nozzle or the engine with the converging–diverging nozzle are as follows:

$$p_{o4} = 59.1 \text{ psia } (408.2 \text{ kPa})$$

$$T_{o4} = 1996°\text{R } (1109 \text{ K})$$

$$h_{o4} = 374 \text{ Btu/lb } (869.7 \text{ kJ/kg})$$

$$Pr_{o4} = 173.20\ (173.36)$$

(a) For the converging nozzle, the velocity at the nozzle exit will be sonic. To approximate the exit temperature and therefore the exit velocity,

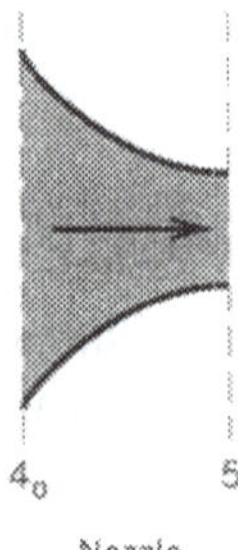

$$\frac{T}{T_o} = \frac{2}{k+1} = \frac{T_5}{T_{o4}}$$

From Table B.3,

$$k \cong 1.34$$

Therefore,

$$T_5 \cong T_{o4} \frac{2}{k+1} = 1996 \left(\frac{2}{2.34}\right) = 1706°\text{R (948 K)}$$

Based on a $T_5 = 1706°\text{R}$ (948 K),

$$h_{o4} = h_5 + \frac{\overline{V}_5^2}{2}$$

$$\overline{V}_5 = \sqrt{(2)(32.174)(778)(374.0 - 294.5)}$$
$$= 1{,}995 \text{ ft/s (607.8 m/s)}$$

The sonic velocity at a temperature of 1706°R is

$$c_5 = \sqrt{kg_cRT} = \sqrt{(1.340)(32.174)(53.35)(1{,}706)}$$
$$= 1{,}981 \text{ ft/s (603.9 m/s)}$$

T_5	h_5	$\overline{V}_5$	c_5
1,707	294.8	1,991	1,981
1,708	295.1	1,987	1,982
1,709	295.4	1,984	1,983

Therefore,

$$T_5 = 1709°\text{R (950 K)}$$
$$\overline{V}_5 = 1984 \text{ ft/s (604 m/s)}$$
$$Pr_5 = 93.16\ (93.38)$$

Next, the pressure at state 5 must be calculated.

$$p_5 = p_{o4}\left(\frac{Pr_5}{Pr_4}\right)$$
$$= 59.1\left(\frac{93.16}{173.20}\right)$$
$$= 31.79 \text{ psia (219.9 kPa)}$$

Before the thrust can be calculated, it is necessary to determine the exit area.

$$\dot{m} = \frac{A_5\overline{V}_5}{v_5} = \frac{A_5\overline{V}_5 p_5}{RT_5}$$

or, for a mass rate of flow of 1 lb/s,

$$A_5 = \frac{(1)(53.35)(1{,}709)}{(31.79)(144)(1{,}984)}$$

$$= 1.004 \times 10^{-2}\ \text{ft}^2\ (2.053 \times 10^{-3}\ \text{m}^2)$$

From Eq. (6.14), neglecting the mass of fuel added and for $\overline{V}_1 = 0$,

$$\tau = (31.79 - 14.70)(144)(0.01004) + \frac{1{,}984}{32.174}$$

$$= 24.71 + 61.66 = 86.4\ \text{lb}_f\ (848\ \text{N})$$

(b) For the case where the engine has a converging-diverging nozzle, expansion in the nozzle will be assumed to be back to ambient pressure. Therefore,

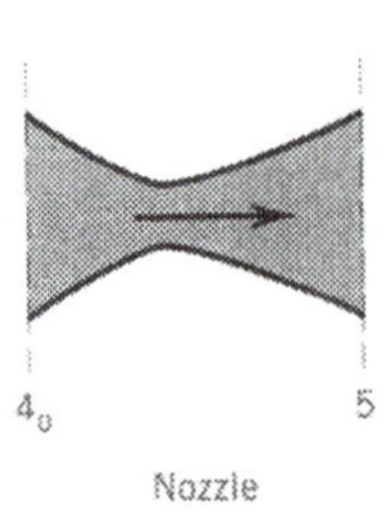

$$Pr_{5i} = Pr_4\left(\frac{p_5}{p_{o4}}\right)$$

$$= 173.20\left(\frac{14.70}{59.1}\right)$$

$$= 43.080\ (43.021)$$

$$h_{5i} = 214.3\ (495.5)$$

$$T_{5i} = 1404°\text{R}\ (778\ \text{K})$$

$$\overline{V}_{5i} = \sqrt{(374.0 - 214.3)(2)(32.174)(778)}$$

$$= 2828\ \text{ft/s}\ (865.0\ \text{m/s})$$

$$\tau = \frac{(1)(2828)}{32.174} = 87.9\ \text{lb}_f\ (865.0\ \text{N})$$

Since $q_{in} = 380.0$ Btu/lb (884.0 kJ/kg)

$$\eta_{th} = \frac{(2828)^2}{(2)(380.0)(32.174)(778)}$$

$$= 0.420\ (0.423)$$

Example Problem 6.2

A turbojet engine is operating under the following conditions:

Altitude (standard atm)	11,000 m (36,089 ft)
Flight Mach number	0.85
Air flow entering compressor	1 kg/s (1 lb/s)
Compressor pressure ratio (total to total)	12
Efficiencies	
Diffuser	100%
Compressor	87%
Turbine	89%
Nozzle	100%
Turbine inlet temperature (stagnation)	1400 K (2520°R)

Assuming an air standard cycle, that the stagnation pressure remains constant from compressor outlet to turbine inlet, and neglecting the mass of fuel added:

(a) Calculate the thrust developed by and the heat added to this engine assuming a converging nozzle
(b) Calculate the thrust developed by, the heat added to, and the thermal efficiency of this engine assuming a converging-diverging nozzle with expansion back to ambient pressure in the nozzle

Solution

From Table C2:

ambient temperature = 217 K (390°R)

ambient pressure = 22.63 kPa (3.283 psia)

Diffuser

$$h_1 = -85.6\ (-37.0)$$

$$Pr_1 = 0.4481\ (0.4449)$$

$$\overline{V}_{plane} = \overline{V}_1 = 0.85\sqrt{(1.400)(1)\left(\frac{8314}{28.965}\right)(217)}$$

$$= 251\ \text{m/s}\ (823\ \text{ft/s})$$

$$h_{o1} = h_1 + \frac{\overline{V}_1^2}{2}$$

$$= -54.1 \text{ kJ/kg } (-23.4 \text{ Btu/lb})$$

The remainder of the solution to this problem is similar to that of Example Problem 6.1 except for the values at the various states. Therefore, the values at the various states for the gas generator are summarized below with some of the calculations shown.

State	p kPa (psia)		h kJ/kg (Btu/lb)		Pr	
1_o	36.26	(5.267)	−54.1	(−23.4)	0.7180	(0.7138)
2_{oi}	435.1	(63.20)	203.7	(87.3)	8.616	(8.566)
2_{oa}	435.1	(63.20)	242.2	(103.8)	—	
3_o	435.1	(63.20)	1212.7	(521.7)	450.9	(450.9)
4_{oi}	172.6	(25.12)	879.6	(347.6)	178.9	(179.2)
4_{oa}	172.6	(25.12)	916.2	(394.4)	200.2	(200.5)

The conditions entering the nozzle for either engine are those at state 4_{oa}.

(a) For the converging nozzle,

$$T_5 = 986 \text{ K } (1775°\text{R})$$

$$h_5 = 728.0 \ (313.3)$$

$$\overline{V}_5 = \sqrt{(2)(1)(916.2 - 728.0)(1{,}000)}$$

$$= 614 \text{ m/s } (2{,}016 \text{ ft/s})$$

$$p_5 = p_{o4}\left(\frac{Pr_5}{Pr_{O4a}}\right)$$

$$= 93.27 \text{ kpa } (13.56 \text{ psia})$$

$$A_5 = 4.942 \times 10^{-3} \text{ m}^2 \ (2.406 \times 10^{-2} \text{ ft}^2)$$

$$\tau = (p_5 - p_1)A_5 + \overline{V}_5 - \overline{V}_1$$

$$= 349 + 363$$

$$= 712 \text{ N } (72.7 \text{ lb}_f)$$

(b) For the engine with the converging-diverging nozzle,

$$Pr_5 = Pr_{o4}\left(\frac{p_5}{p_{o4}}\right)$$

$$= 26.25 \ (26.20)$$

$$h_{5i} = 392.0 \ (168.4)$$

$$\overline{V}_{5i} = 1{,}024 \text{ m/s } (3{,}363 \text{ ft/s})$$

$$\tau = 1{,}024 - 251$$

$$= 773 \text{ N } (78.9 \text{ lb}_f)$$

$$\eta_{th} = \frac{\bar{V}_5^2 - \bar{V}_1^2}{2q_{in}}$$

$$= 0.508$$

Analyzing the results of Example Problems 6.1 and 6.2 leaves unanswered many questions.

1. Would the pressure ratio be the same at both the sea level statis (SLS) and altitude points? If not, why and how would one determine the change in the pressure ratio?
2. Does the thrust decrease by only 10.5% at the altitude point when compared with the SLS point for the engine with the converging-diverging nozzle, 16% for the engine with the converging nozzle?
3. Is it reasonable to assume the same component efficiencies and turbine inlet temperature at both operating conditions?
4. Is it reasonable to assume the same air flow for both flight conditions? The answer is obviously no, but how does one determine the variation with flight velocity and/or altitude?

Answers to these questions must wait until engine components and component matching have been discussed.

Figures 6.7 and 6.8 illustrate two turbojet engines with axial-flow compressors and turbines. One should note the large number of compressor stages compared to the number of turbine stages.

Figures 6.9, 6.10, 6.11, and 6.12 illustrate the variation of thrust with compressor pressure ratio for various altitudes, flight velocities, component efficiencies, and

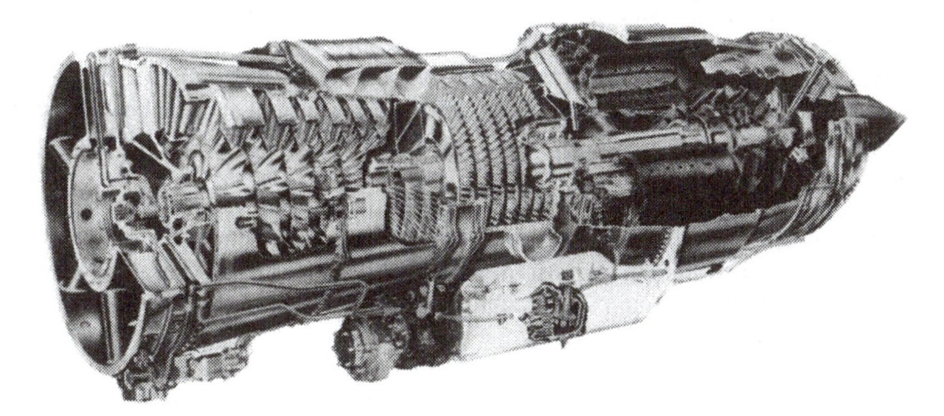

Figure 6.7. Pratt & Whitney's JT3 (J57) dual-spool axial-flow compressor turbojet engine. (Courtesy of Pratt & Whitney Aircraft Group)

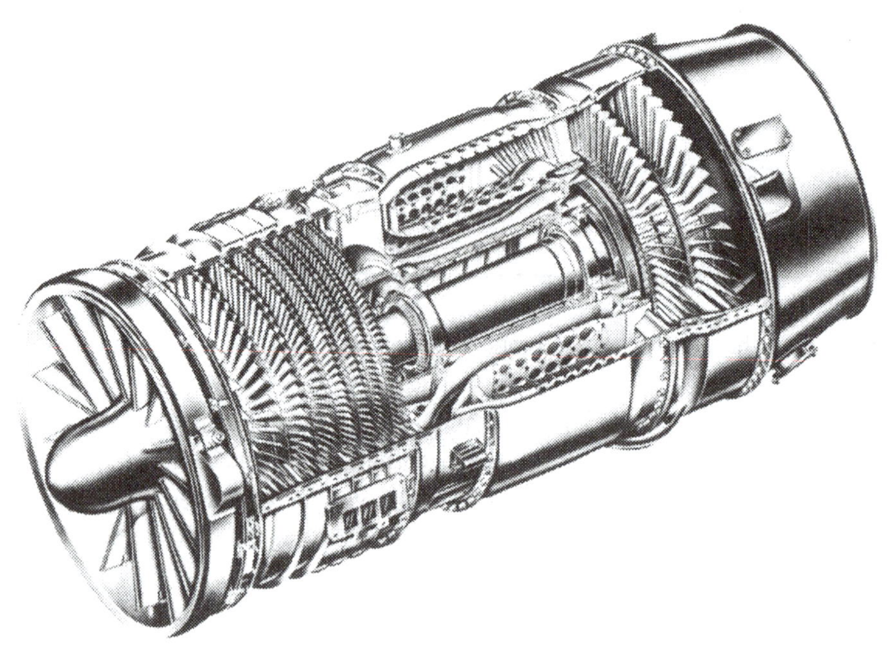

Figure 6.8. General Electric's CJ610 turbojet engine. (Courtesy of General Electric Co.)

turbine inlet temperatures. A converging-diverging nozzle has been assumed in all cases, with expansion back to ambient pressure.

Figure 6.9 shows the variation in specific thrust with compressor pressure ratio for sea level static conditions and a turbine inlet temperature of 2520°R. Both the ideal case (compressor and turbine efficiencies of 100%) and one with realistic compressor and turbine efficiencies are shown. One should observe the loss in thrust when realistic efficiencies are used. Note that the maximum thrust occurs at a pressure ratio of 19.3 for the ideal case, at a pressure ratio of 13.1 for the actual case.

Figure 6.10 is for the same conditions as in Figure 6.9 except that it is for a Mach number of 0.85 at 11,000 m (36,089 ft).

Once again, note the pressure ratio at which the maximum thrust occurs. One should be careful in comparing Figures 6.9 and 6.10 because in an actual engine, the mass rate of flow into an engine will be considerably lower at the altitude point than at sea level static.

Figure 6.11 illustrates the variation in specific thrust with compressor pressure ratio for sea level static conditions and a turbine inlet temperature of 2520°R. This figure illustrates the variation in thrust for various component efficiencies. One should note that decrease in maximum specific thrust and the pressure ratio at which it occurs as component efficiency decreases.

Figure 6.12 illustrates the effect of turbine inlet temperature on specific thrust and the compressor pressure ratio at which the maximum specific thrust occurs. Once again, observe the decrease in maximum specific thrust and pressure ratio where

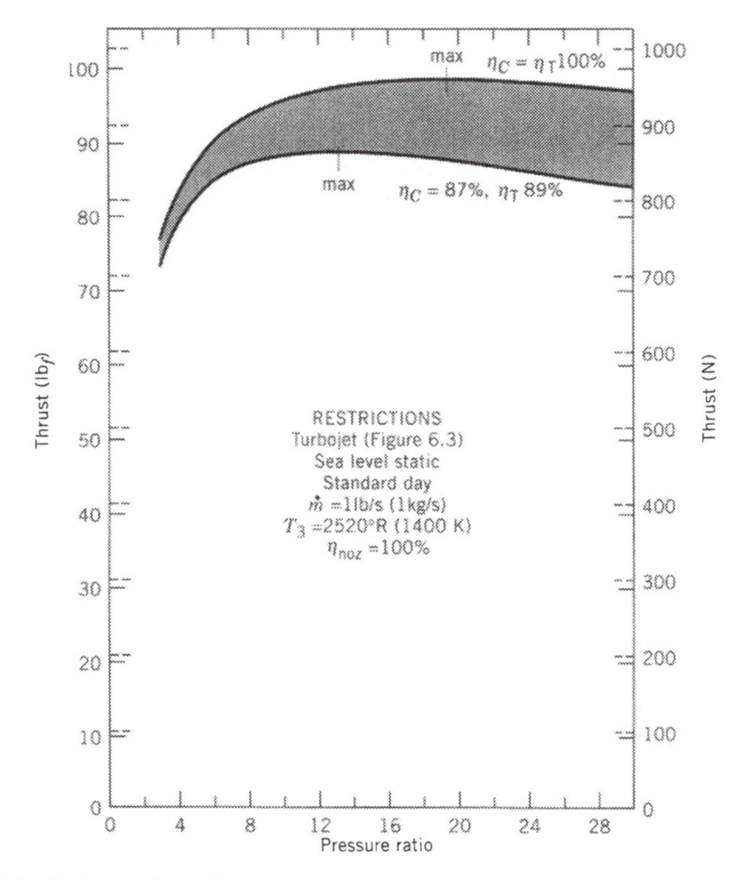

Figure 6.9. Variation of specific thrust with pressure ratio for an air standard turbojet engine for both an ideal engine and one with realistic efficiencies.

maximum thrust occurs as the turbine inlet temperature is decreased for fixed component efficiencies.

6.5 TURBOJET (ACTUAL MEDIUM)

The preceding section discussed the air standard turbojet engine where air is the working fluid throughout. In the actual engine, fuel is added in the combustion chamber and burned with the air leaving the compressor. Since the maximum cycle temperature is 3000°R or lower, the excess air supplied is large; therefore, combustion is complete and the products of combustion have properties very close to those of air.

A solution considering the actual medium would be to use air as the working fluid through the compressor, then the products of combustion after the combustion chamber. The effect on the calculated values when the actual working fluid is considered instead of an air standard cycle was illustrated in Section 5.3. This involves a considerable

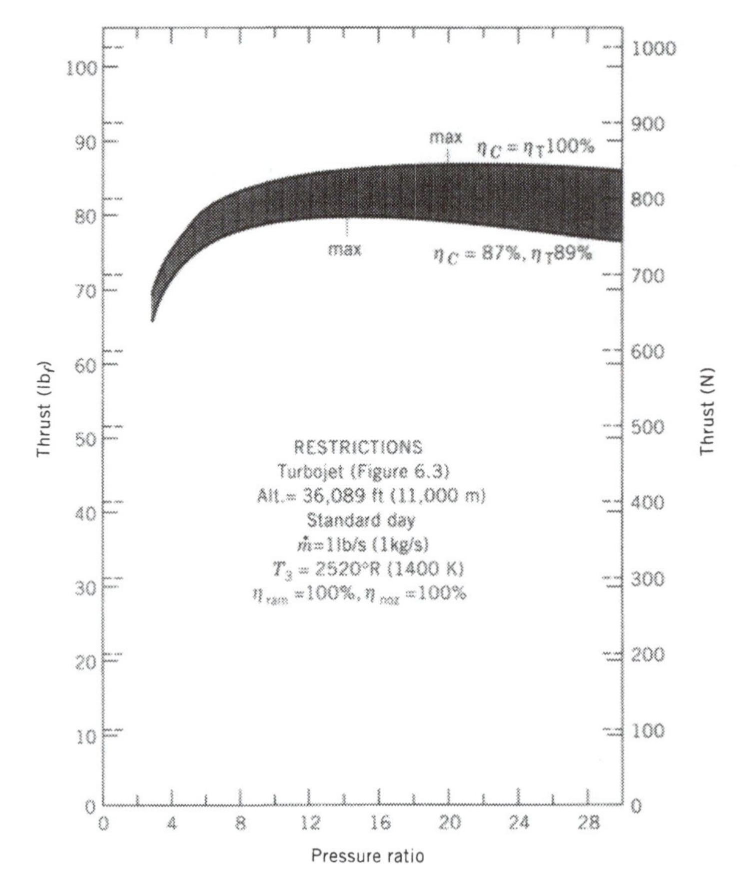

Figure 6.10. Variation of specific thrust with pressure ratio for an air standard turbojet engine for both an ideal engine and one with realistic efficiencies.

number of trial-and-error solutions, especially when the turbine and/or nozzle efficiencies are less than 100%.

A much simpler method that closely approximates the values calculated when the actual working fluid is used is to calculate the actual (true) fuel–air ratio but assume air to be the working fluid throughout. This was illustrated in Section 5.4.

When the mass of fuel added in the combustion chamber is taken into account, it means that for every pound (or kilogram) of air entering the compressor, $(1 + f)$ pounds (or kilograms) of air leave the combustion chamber and expand through the turbine and nozzle. Therefore,

$$\dot{m}_{\text{air}} \text{ into comp. } \left|w_{\text{comp. act}}\right| = \dot{m}_{\text{air}} \text{ into turbine } \left|w_{\text{turb. act}}\right| \tag{6.26}$$

and the thrust equation is as given in Eq. (6.14).

One factor used in judging the performance of turbojet engines is the thrust-specific fuel consumption (TSFC), which is the mass of fuel added per unit of time divided by the thrust developed by the engine, or

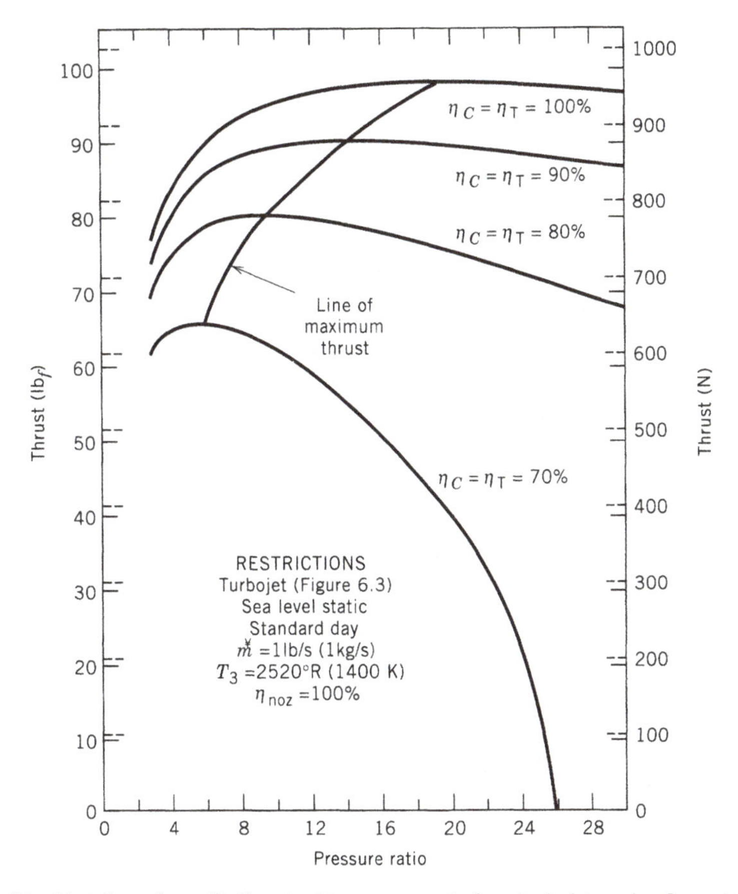

Figure 6.11. Variation of specific thrust with pressure ratio for a turbojet engine for various component efficiencies.

$$\text{TSFC} = \frac{\dot{m}_f}{\tau} \tag{6.27}$$

The solution of Example Problems 6.1 and 6.2 when the mass of fuel added is considered but air is assumed to be the working fluid throughout is illustrated in Example Problems 6.3 and 6.4, respectively. Only the case where the engine has a converging-diverging nozzle is illustrated.

Example Problem 6.3

Solve Example Problem 6.1b assuming that *n*-octane, C_8H_{18}, is supplied to the combustion chamber as a liquid at 537°R (298 K). Calculate, taking into account the mass of fuel added but assuming that air is the fluid expanding through the turbine and nozzle:

(a) The thrust developed by the engine
(b) The thrust-specific fuel consumption

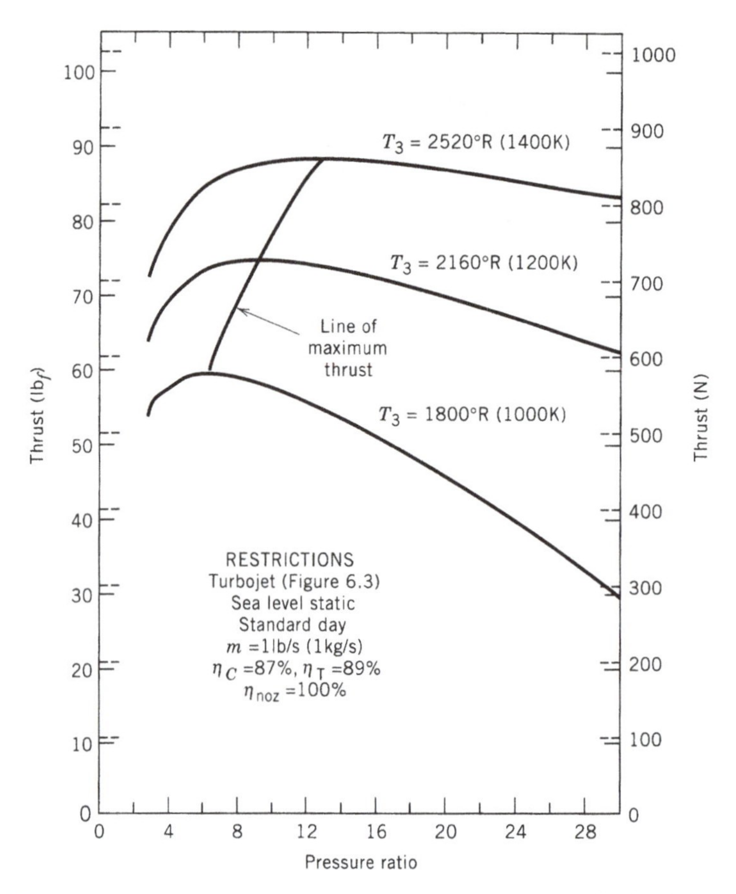

Figure 6.12. Variation of specific thrust with pressure ratio for an air standard turbojet engine with fixed component efficiencies for various turbine inlet temperatures.

Solution

This problem is the same as Example Problem 5.5 through the gas generator turbine. Therefore, referring to Example Problem 5.5, known values are

$$f' = 0.0215$$

$$q_{\text{in}} = 410.7 \text{ Btu/lb } (954.6 \text{ kJ/kg})$$

$$p_{o4} = 60.71 \text{ psia } (418.5 \text{ kPa})$$

$$h_{o4a} = 377.1 \text{ Btu/lb } (876.9 \text{ kJ/kg})$$

$$Pr_{o4a} = 177.15 \ (177.29)$$

For a converging-diverging nozzle with expansion to ambient pressure in the nozzle,

$$Pr_{5i} = Pr_{4a}\left(\frac{p_5}{p_4}\right)$$

$$= 177.17\left(\frac{14.70}{60.71}\right)$$

$$= 42.90\ (42.85)$$

$$h_{5i} = 212.8\ (494.4)$$

$$\overline{V}_{5i} = \sqrt{(2)(32.174)(778)(377.1 - 212.8)}$$

$$= 2{,}868 \text{ ft/s}$$

$$\tau = \frac{(2{,}868)(1.0215) - 0}{32.174}$$

$$= 91.1 \text{ lb}_f\ (892 \text{ N})$$

In English units,

$$\text{TSFC} = \frac{0.0215 \text{lb}}{\text{s}}\left|\frac{3600 \text{ s}}{\text{h}}\right|\frac{}{91.1 \text{ lb}_f}$$

$$= 0.850 \text{ lb/lb}_f\text{-h}$$

In SI units,

$$\text{TSFC} = \frac{0.0215 \text{ kg}}{\text{s}}\left|\frac{}{892 \text{ N}}\right|\frac{10^6 \text{ mg}}{\text{kg}}$$

$$= 24.1 \text{ mg/N s}$$

Example Problem 6.4

Solve Example Problem 6.2b assuming that *n*-octane, C_8H_{18}, is supplied to the combustion chamber as a liquid at 298 K (537°R). Calculate, taking into account the mass of fuel added but assuming that air is the fluid expanding through the turbine and nozzle:

(a) The thrust developed by the engine
(b) The thrust-specific fuel consumption

Solution

From Example Problem 6.2,

$$h_{2a} = 242.0\ (103.8)$$

Using Eqs. (4.18) and (4.20) (or Figure 5.21),

$$X = \frac{4{,}697{,}815}{35{,}118 - (242.0)(28.965)} = 167.2$$

$$f' = \frac{114}{(28.965)(167.2)} = 0.0235$$

$$w_{T,a} = \frac{w_{C,a}}{1 + f'}$$

$$= 289.5 \text{ kJ/kg } (124.3 \text{ Btu/lb})$$

The remainder of the solution is the same as that of Example Problem 6.2b except for the values. Therefore, the values at the various states are summarized in the following table.

State	p kPa	(psia)	h kJ/kg	(Btu/lb)	M Pr	
4_{oi}	176.8	(25.72)	887.2	(397.3)	183.2	(183.5)
4_{oa}	176.8	(25.72)	923.0	(831.9)	204.3	(204.6)
5_i	22.63	(3.283)	391.2	(168.2)	26.15	(26.12)

$$\overline{V}_{5i} = 1{,}031 \text{ m/s } (3{,}387 \text{ ft/s})$$

$$\tau = (1.0251)(1{,}031) - 251$$

$$= 804 \text{ N } (82.2 \text{ lb}_f)$$

$$\text{TSFC} = \frac{(0.0235)(10^6)}{804}$$

$$= 29.2 \text{ mg/N s } (1.03 \text{ lb/lb}_f\text{-h})$$

Table 6.1 summarizes the results for a turbojet engine operating at two flight conditions. Values tabulated are for ideal components, taking into account realistic compressor and turbine efficiencies and then taking into account the mass of fuel added but assuming air to be the working fluid throughout.

Table 6.2 lists known data for several operational turbojet engines. One should observe the wide range of engine pressure ratios and maximum thrust levels and that the thrust-specific fuel consumption levels are all near 1.00 lb/lb$_f$-h. It is of interest to compare the pressure ratios and thrust-specific fuel consumption values given in Table 6.2 with the values listed in Table 6.1.

Figure 6.13 illustrates the variation of specific thrust with altitude and flight velocity for a turbojet engine with a turbine inlet temperature of 2520°R (1400 K), a compressor

TABLE 6.1. **Comparison of the Same Turbojet Engine Operating Two Flight Conditions**[a]

Conditions	Example Problem	Thrust N(lb$_f$)	TSFC mg/N s(lb/lb$_f$-h)
		Sea Level Static	
Ideal, no fuel			
conv. nozzle		914 (93.1)	
c-d nozzle		947 (96.6)	
Actual, no fuel			
conv. nozzle	6.1a	848 (86.4)	
c-d nozzle	6.1b	865 (87.9)	
Actual, mass of fuel considered,			
c-d nozzle	6.3	892 (91.1)	24.1 (0.850)
		11,000 m (36,089 ft), M = 0.85	
Actual, no fuel			
conv. nozzle	6.2a	712 (72.7)	
c-d nozzle	6.2b	773 (78.9)	
Actual, mass of fuel considered,			
c-d nozzle	6.4	804 (82.2)	29.2 (1.03)

[a] All on a basis of 1 lb/s (kg/s) of air entering the compressor. All calculations are for variable specific heats.

pressure ratio of 12, a compressor efficiency of 87%, and a turbine efficiency of 89%. This figure is misleading, as the specific thrust in the figure increases with altitude.

6.6 TURBOFAN ENGINE

Sections 6.4 and 6.5 discussed the nonafterburning turbojet engine. Example Problems 6.3 and 6.4 illustrated the specific thrust and thrust-specific fuel consumption for two flight conditions. Typical U.S. turbojet specifications were listed in Table 6.2.

The propulsion efficiency of the turbojet engine is quite low except at high flight speeds. This is evident by the large velocity of the stream of gases leaving the engine.

To increase the propulsion efficiency, the nozzle exit velocity must be decreased. This can be accomplished by extracting more power in the turbine without increasing the power required to drive the compressor. This increase in power can be used to compress additional air, which is diverted around the combustion chamber, thereby increasing the mass of air compressed without increasing the amount of fuel supplied to the engine. The type of engine that does this is called a turbofan engine.

A turbofan engine is basically a turbojet engine in which some front-end compressor stages have been removed and replaced by larger-diameter stages that are usually called fans. This is illustrated in Figure 6.14. More turbine capability is needed to drive the combination fan and compressor. The added capability depends on the fan pressure ratio and the amount of air bypassing the basic gas generator (turbojet or core engine).

The fan pressure ratio (pressure ratio across the fan) varies from slightly above

TABLE 6.2. **Typical Turbojet Specifications (1)**

Manufacturer Model No. or Designation	Number of Compressor Stages	Pressure Ratio at Maximum rpm	Maximum Power at Sea Level, lb thrust	Specific Fuel Consumption at Maximum Power lb/lb$_f$-hr	Dry Weight Less Tailpipe, lb
General Electric Co.,					
GE Aircraft Engines					
J85-GE-17 A, B, CAN40[a]	8	6.9	2,850	0.99	400
J85-GE-21[a]	9	8.3	3,670	0.98	465
CJ610-6, 8 A[a]	8	6.8	2,950	0.97	411
Teledyne CAE					
304[c]	1	5.5	55	1.2	6
312-2[b]	1,1	8.1	235	1.06	40
320-1[c]	1	5.7	240	1.11	45
United Technologies					
Pratt & Whitney					
Government Engine &					
Space Propulsion					
J52-P-8A, B[a]	12	6.4	9,300	0.86	2,129
J60-P-3, 5, 6[a]	9	11.9	3,000	0.96	5,875
Williams International Corp.					
WR 24-7[b]	1,1	5.4	176	1.20	44
WR 2-6[c]	1	4.2	125	1.21	28
Turbomeca					
Arbizon 3B[b]	1,1	5.85	907	1.12	253
Bristol					
Viper 11[a]	7	4.3	2,500	1.07	625
Viper 531[a]	8	5.2	3,120	1.0	760
Viper 680[a]	8	6.8	4,370	0.98	836

[a] Axial-flow compressor.
[b] Axial-centrifugal flow compressor.
[c] Centrifugal-flow compressor.

1.0 to about 3.0. The bypass ratio (BPR) is defined as the ratio of the air flow passing through the fan tips and the duct to the air flow passing through the gas generator (core) engine. Therefore,

$$1 + \text{BPR} = \frac{\dot{m}_{\text{total}}}{\dot{m}_{\text{gas generator}}} \tag{6.28}$$

It is usually true that the higher fan pressure ratios are associated with low bypass ratio engines and the high-bypass-ratio engines are associated with low fan pressure ratios.

The turbofan has the advantage that a large increase in thrust may be achieved by taking an existing turbojet and adding a fan to it. This lowers the exit velocity and increases the propulsion efficiency. The turbofan engine, since it has no increase in

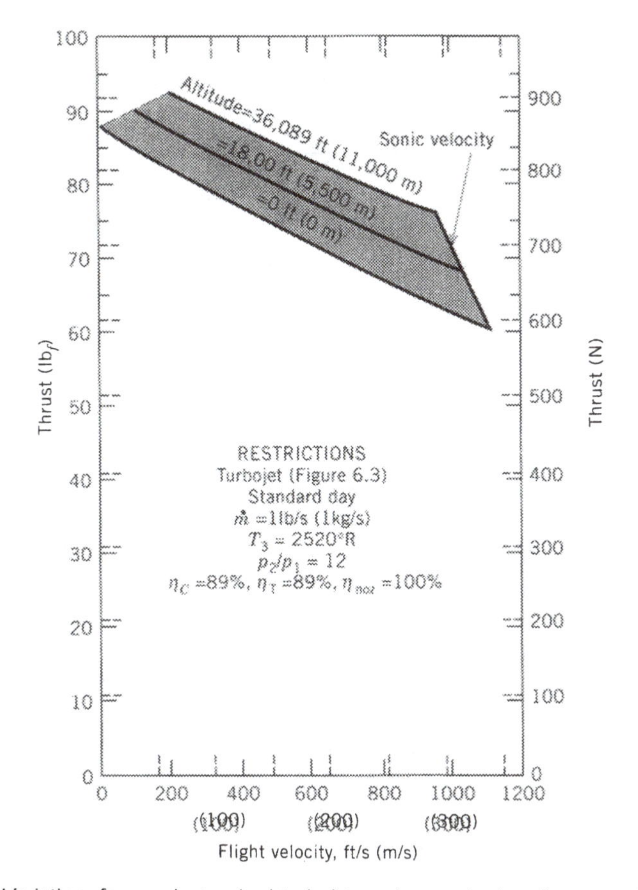

Figure 6.13. Variation, for an air standard turbojet engine neglecting the mass of fuel added, of specific thrust with flight velocity for various altitudes.

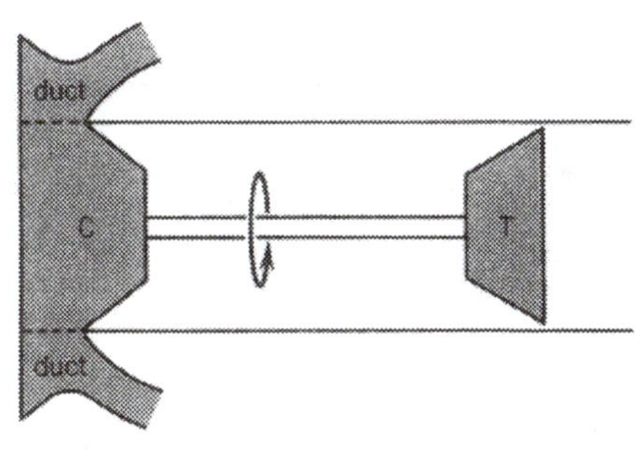

Figure 6.14. General layout of a turbofan engine.

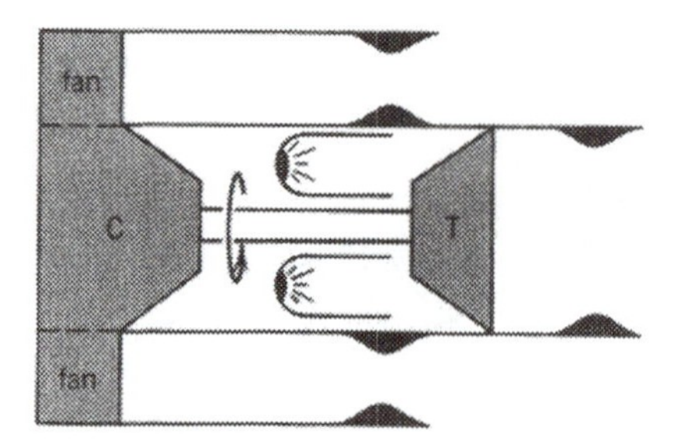

Figure 6.15. Nonmixed turbofan engine.

fuel flow for this added thrust, has more thrust per mass of air entering the gas generator and therefore a lower thrust-specific fuel consumption.

There are two general types of turbofan engines. These are the nonmixed turbofan engine (separate streams) and the mixed-flow (static pressure balanced) turbofan engine. These are shown in Figures 6.15 and 6.16.

In the nonmixed turbofan engine (Figure 6.15), the air that passes through the fan enters a duct, then passes through its own nozzle. In the mixed-flow turbofan engine (Figure 6.16), the air that passes through the fan is ducted around the combustion chamber and mixed with the gas generator (core) gases behind the turbine. The static pressure at the point of mixing must be the same for both streams. The mixed streams then pass through the common nozzle. Each type of engine is currently being built.

Turbofan engines built in the United States over the years have taken on many different configurations. Some of these are illustrated in Figures 6.17 through 6.19.

Figure 6.17 illustrates the General Electric CF700 turbofan engine. This engine had the fan blades as an extension of the turbine blades. It eliminated the need to beef up the shaft between the turbine and the compressor fan to handle the increased power that would be required to drive the fan compressor unit but did require a seal between the hot, high-pressure turbine blades and the cold, low-pressure fan blades and could present high thermal stresses in the turbine fan blades because of the temperature difference of the two streams.

Figure 6.18 illustrates the General Electric CF6 turbofan engine, a high-bypass-ratio engine.

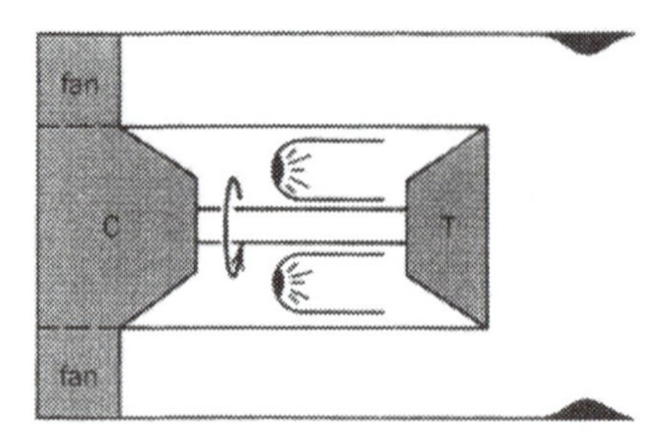

Figure 6.16. Mixed turbine engine.

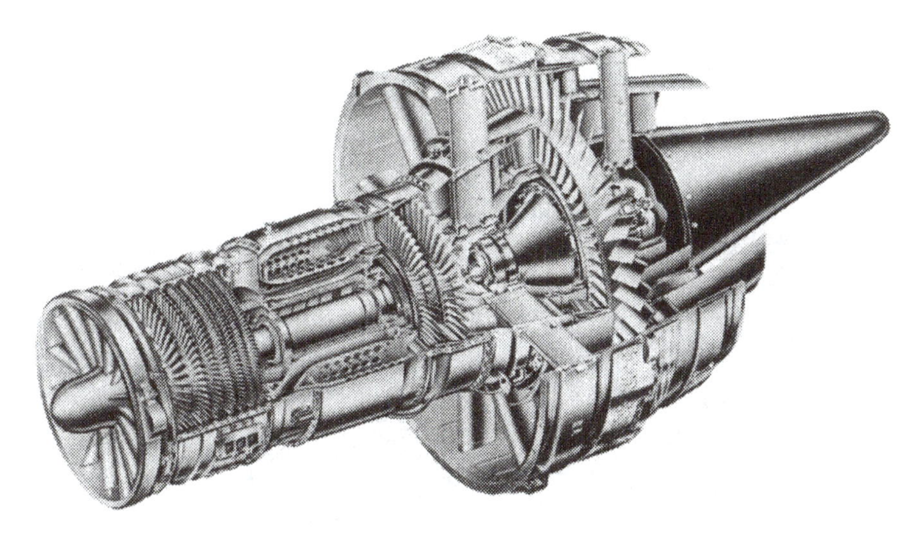

Figure 6.17. General Electric's CF700 turbofan engine. (Courtesy of General Electric Co.)

Observe the relative size of the fan-low compressor unit compared to the high-pressure compressor spool and the number of compressor stages compared to the number of turbine stages.

Figure 6.19 illustrates AlliedSignal's TFE 731 turbofan engine. This is a two-spool turbofan engine with a single-stage gear-driven fan. The fan is driven by the low-pressure spool through a planetary gear system to provide an approximate 2-to-1 speed reduction.

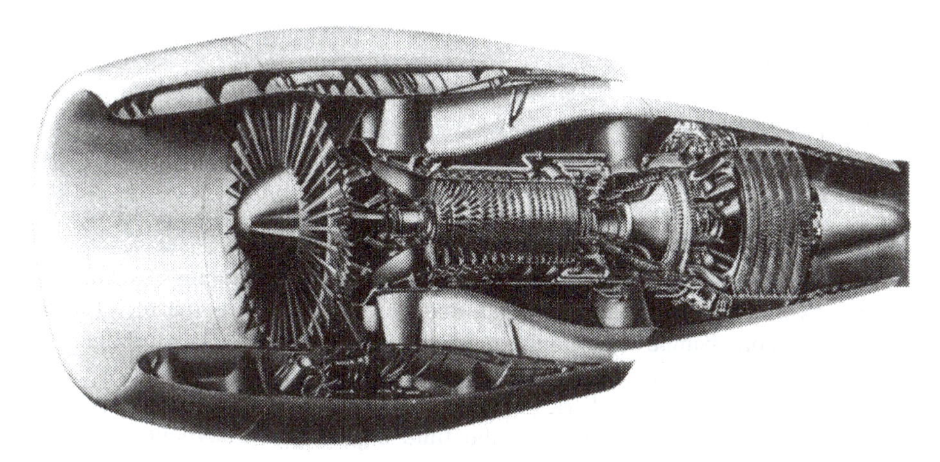

Figure 6.18. General Electric's CF6-6 turbofan engine. (Courtesy of General Electric Co.)

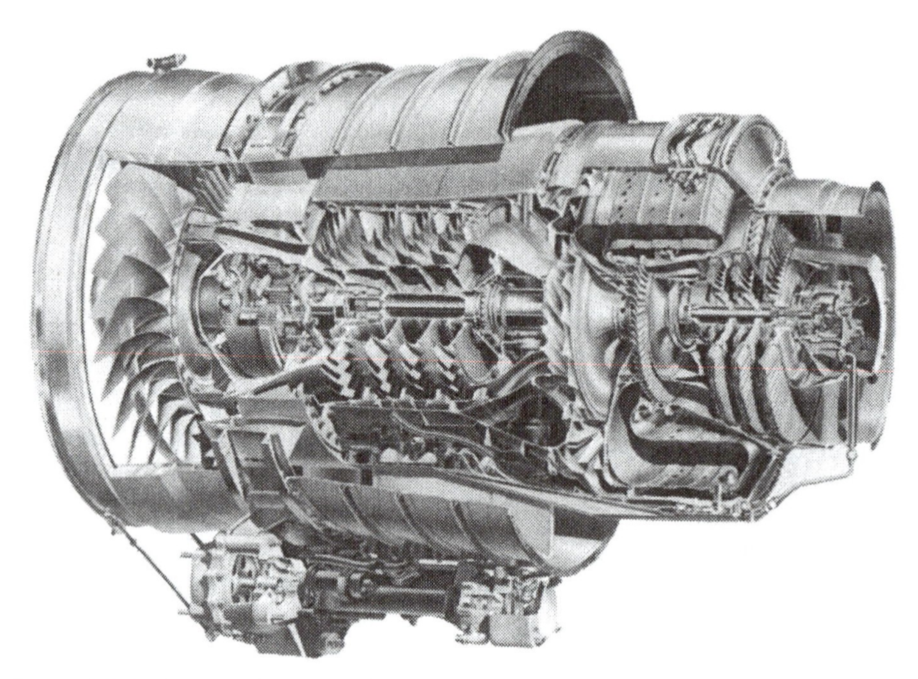

Figure 6.19. Cutaway view of AlliedSignal's TFE 731 turbofan engine. (Courtesy of AlliedSignal.)

Example Problem 6.5

A turbofan engine is operating under the following conditions:

Flight speed at sea level, standard day	0
Air flow entering the gas generator (core)	1 lb/s (1 kg/s)
Compressor pressure ratio (total to total)	12
Fan pressure ratio (total to total)	1.35
Efficiencies	
Diffuser	100%
Compressor	87%
Fan	85%
Turbine	89%
Gas generator nozzle	100%
Fan nozzle	100%
Turbine inlet temperature (stagnation)	2520°R (1400 K)

Assume an air standard cycle, that the stagnation pressure remains constant from the compressor outlet to the turbine inlet, that *n*-octane, C_8H_{18}, is the fuel supplied to the combustion chamber as a liquid at 537°R (298 K), and that the turbofan is of the nonmixed type as shown in the diagram, with expansion in both the gas generator and fan nozzles back to atmospheric pressure. Calculate, taking into account the mass of fuel added but assuming air to be the fluid expanding throughout the turbine and nozzle:

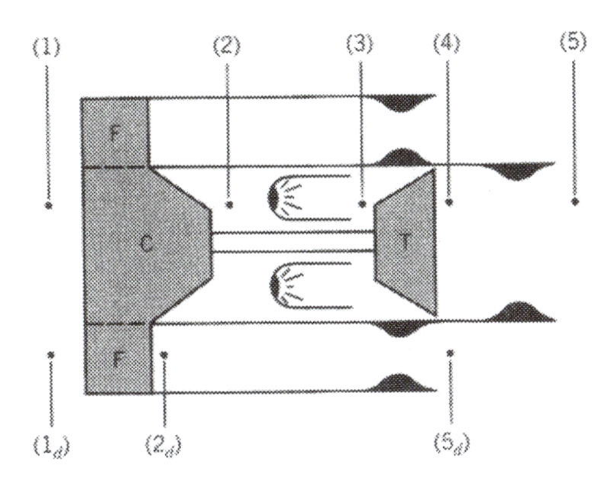

(a) The thrust, thrust-specific fuel consumption, and thermal efficiency if the BPR is 2.0
(b) The thrust, thrust-specific fuel consumption, and thermal efficiency if the BPR is 5.0

Solution

Basis: 1 lb/s (1 kg/s) of air entering the gas generator (core).

The solution of the gas generator portion of this problem is the same as for the engine of Example Problem 5.5 up to the entrance to the turbine. From Example Problem 5.5,

$$w_{C,a} = 147.7 \text{ Btu/lb } (343.0 \text{ kJ/kg})$$

$$f' = 0.0215$$

$$p_{o3} = 176.4 \text{ psia } (1215.6 \text{ kPa})$$

$$q_{\text{in}} = 410.7 \text{ Btu/lb } (954.6 \text{ kJ/kg})$$

For the fan (per pound of air entering)

$$T_1 = 519^\circ\text{R } (288 \text{ K})$$

$$p_1 = 14.70 \text{ psia } (101.3 \text{ kPa})$$

$$h_1 = -6.0 \text{ Btu/lb } (-14.0 \text{ kJ/kg})$$

$$Pr_1 = 1.2093 \ (1.2045)$$

$$Pr_{2d} = 1.35(\text{Pr}_1)$$

$$= 1.6327 \ (1.6274)$$

$$h_{2\text{di}} = 5.2 \ (11.6)$$

$$w_{\text{fan},a} = \frac{h_{2\text{di}} - h_1}{\eta_{\text{f}}}$$

$$= 13.2 \text{ Btu/lb } (30.5 \text{ kJ/kg})$$

$$h_{2\text{da}} = 7.2 \text{ Btu/lb } (16.2 \text{ kJ/kg})$$

$$Pr_{2da} = 1.7207\ (1.7096)$$

$$Pr_{5d} = Pr_{2da}\left(\frac{1}{1.35}\right)$$

$$= 1.2746\ (1.2664)$$

$$h_{o5d} = -4.1\ (-10.1)$$

$$\overline{V}_{5d} = \sqrt{(7.2 - (-4.1))(2)(32.174)(778)}$$

$$= 752 \text{ ft/s } (229 \text{ m/s})$$

For a BRP = 2.0,

$$w_{T,a} = \frac{w_{C,a} + (2)w_{f,a}}{1 + f'}$$

$$= 170.4 \text{ Btu/lb } (395.4 \text{ kJ/kg})$$

Since the remainder of the solution is quite similar to Example Problem 6.3, only the values are tabulated.

State	p psia (kPa)		h Btu/lb (kJ/kg)		Pr	
3_o	176.4	(1215.6)	521.7	(1215.1)	450.9	(450.9)
4_{oi}	48.63	(335.4)	330.3	(768.1)	124.30	(124.42)
4_{oa}	48.63	(335.4)	351.4	(817.0)	146.38	(146.51)
5	14.70	(101.3)	215.8	(501.6)	44.25	(44.25)

$$\overline{V}_{5i} = 2605 \text{ ft/s } (794 \text{ m/s})$$

$$\tau = \frac{[(1.0215)(2{,}605) + (2)(752)] - 0}{32.174}$$

$$= 129.5 \text{ lb}_f\ (1{,}269 \text{ N})$$

$$\% \text{ thrust for gas generator} = \frac{(1.0215)(2{,}605)}{(32.174)(129.5)} \times 100$$

$$= 63.9\%$$

$$\text{TSFC} = \frac{(0.0215)(3{,}600)}{129.5}$$

$$= 0.598 \text{ lb/lb}_f\text{-h } (16.9 \text{ mg/N s})$$

$$\eta_{th} = \frac{[(2{,}605)^2(1.0215) - (2)(752)^2]/(2)(32.174)(778)}{11{,}005/28.965}$$

$$= 0.416$$

(b) Solution for a BPR of 5.0 is similar to that for a BPR = 2.0. Therefore, only the values are tabulated below

$$w_{T,a} = \frac{343.0 + (5.0)(30.4)}{1.0215}$$

$$= 485.1 \text{ kJ/kg (209.2 Btu/lb)}$$

State	p kPa (psia)		h kJ/kg (Btu/lb)		Pr	
3_o	1,215.6	(176.4)	1,215.1	(521.7)	450.9	(450.9)
4_{oi}	234.4	(33.96)	667.7	(187.1)	86.94	(86.81)
4_{oa}	234.4	(33.96)	727.6	(312.9)	108.08	(107.95)
5	101.3	(14.70)	513.9	(221.1)	46.71	(46.73)

$$\overline{V}_{5i} = 654 \text{ m/s (2,143 ft/s)}$$

$$\tau = 1{,}813 \text{ N (185 lb}_f)$$

$$\% \text{ thrust for G.G.} = 36.8\%$$

$$\text{TSFC} = 11.9 \text{ mg/N s (0.418 lb/lb}_f\text{-h)}$$

$$\eta_{th} = 0.390$$

It must be remembered that the turbofan engine has some disadvantages, including being considerably heavier and larger in diameter.

Table 6.3 compares the turbojet engine with two turbofan engines. In all cases, the same gas generator (turbojet engine) could be used. Note that if the turbojet engine was converted to a turbofan engine with a bypass ratio of 5.0, the thrust would approximately double with the thrust-specific fuel consumption decreasing by 50%. Remember that the turbofan engine would have a much larger frontal area, would be much heavier, and, if one had started with an engine that had been designed as a turbojet, the shaft would have to be modified to transmit the additional power required

TABLE 6.3. **Comparison of Turbojet Engine and Two Turbofan Engines that Use the Core of the Turbojet Engine as their Gas Generator**[a]

Type	BPR	Thrust lb_f (N)	TSFC $\frac{lb}{lb_f\text{-h}} \left(\frac{mg}{N\ s}\right)$	percent Thrust from Gas Generator	Velocity ft/s (m/s) Core	Velocity ft/s (m/s) Fan
Turbojet	0.0	91.1 (892)	0.850 (24.1)	100%		—
Turbofan	2.0	130 (1269)	0.598 (16.9)	63.9%	2605 (794)	753 (229)
Turbofan	5.0	185 (1813)	0.418 (11.9)	36.8%	2143 (654)	753 (229)

[a] All values calculated on a basis of 1 lb/s (1 kg/s) entering the gas generator.

TABLE 6.4. **Typical U.S. Turbofan Specifications (3)**

Manufacturer Model No. or Designation	Number of Fan-Compressor Stages	Maximum Power at Sea Level, lb. thrust	Specific Fuel Consumption at Maximum Power	Pressure Ratio at Maximum rpm
Allied Signal Garrett Engine Div.				
TFE 731-3[a]	1,4,1	3,700	0.506	14.6
F-109-GA-100[a]	1,2	1,300	0.396	16.5
General Electric Co., GE Aircraft Engines				
CF6-45A2[b]	4,14	46,500	0.354	26.3
CF6-50C[b]	4,14	51,000	0.390	29.3
CF6-80A[b]	4,14	48,000	0.344	27.3
CF6-80C2B6[b]	5,14	60,800	0.334	31.1
TF34-GE-400A[b]	1,14	9,275	0.363	21.0
CF700-2D-2[b]	1,8	4,500	0.65	6.8
General Motors Allison Gas Turbine Division				
TF41-A-400, A-2B[b]	3,2,11	15,000	0.67	21.4
United Technologies Pratt & Whitney				
TF30-P-408[b]	16	13,400	0.64	18.8
TF33-P-3[b]	15	17,000	0.52	13.0
F117-PW-100[b]	17	41,700	0.34	30.8
JT8D-9, 9A[b]	13	14,500	0.60	15.9
JT9D-7[b]	15	45,600	0.362	22.3
JT9D-7R4H1	16	56,000	0.364	26.7
PW4056	16	56,750	0.319	30.5

[a] Axial-centrifugal-flow compressor.
[b] Axial-flow compressor.

to drive the fan and compressor, and additional turbine capacity would have to be added.

Table 6.4 lists known data for several operational U.S. turbofan engines. It is of interest to note how the compressor pressure ratio has increased and the thrust-specific fuel consumption decreased as later engines have been built. One example is the progression JT3D → JT8D → JT9D.

Table 6.5 lists performance and design characteristics for two of the CF6 family of high-bypass-ratio engines (4).

Figures 6.20 and 6.21 illustrate how thrust and thrust-specific fuel consumption vary with compressor pressure ratio for a fixed fan pressure ratio, turbine inlet temperature, and component efficiencies. These figures are for sea level static conditions. Note that, for a fan pressure ratio of 1.70 and a bypass ratio of 5.0, the thrust drops off sharply at both low and high compressor pressure ratios; in fact, the turbine for a pressure ratio of 3, or 30, cannot extract sufficient power to drive the fan.

Figure 6.22 illustrates the variation in percent thrust from the gas generator (core) of the engine for varying bypass ratio for a fixed compressor pressure ratio, turbine inlet temperature, fan pressure ratio, and component efficiency. Figure 6.23 is for the same conditions as Figure 6.22 but illustrates how the velocity for the gas generator decreases with increasing bypass ratio. Note that, for a fan pressure ratio of 1.70 and

TABLE 6.5. **Performance and Design Characteristics of Two of the CF6 Family of High-Bypass-Ratio Turbofans (4)**

Engine	CF6-6	CF6-50A
Take-off thrust		
Ideal nozzle	40,000 lb	49,000 lb
Real nozzle	39,000 lb	48,000 lb
Maximum cruise thrust (35,000 ft, Mach 0.85)		
Ideal nozzle	9,125 lb	10,840 lb
Real nozzle	8,800 lb	10,500 lb
Cruise-specific fuel consumption:		
Ideal nozzle	0.612	0.623
Real nozzle	0.635	0.651
Basic engine weight	7,350 lb	8,100 lb
Fan diameter	86.4 in.	86.4 in.
Compressor stages		
Low pressure	2	4
High pressure	16	14
Turbine stages		
Low pressure	5	4
High pressure	2	2
Bypass ratio	6.25	4.3
Take-off airflow (lb/s)		
Fan	1,160	1,178
Core	183	270
Cruise airflow (lb/s)		
Fan	511	582
Core	82	110
Pressure ratio		
Fan	1.64	1.69
Overall	26.6	29.9

a bypass ratio of 5.0, the exit velocity from the gas generator is lower than that of the fan stream.

Various techniques have been used to increase the thrust from an existing engine. These include;

1. Increasing the fan pressure ratio. This may involve adding an additional fan stage, which will add weight to the engine and require that additional power be extracted in the turbine, and may require an additional turbine stage.
2. Increasing the core (overall) pressure ratio. This can be accomplished by either increasing the rotational speed of the compressor and/or by adding another compressor stage.
3. Increasing the bypass ratio. This can be accomplished by increasing the diameter of the fan blades and therefore the diameter of the engine. In addition to requiring that additional power be extracted in the turbine, this will increase the weight of the engine and will reduce the ground clearance between the engine and the ground if the engine is mounted beneath the wing of the aircraft.
4. Increasing the turbine inlet temperature. This may require a redesign of the blades at the inlet to the turbine, might require additional turbine cooling and/or a different blade cooling technique.

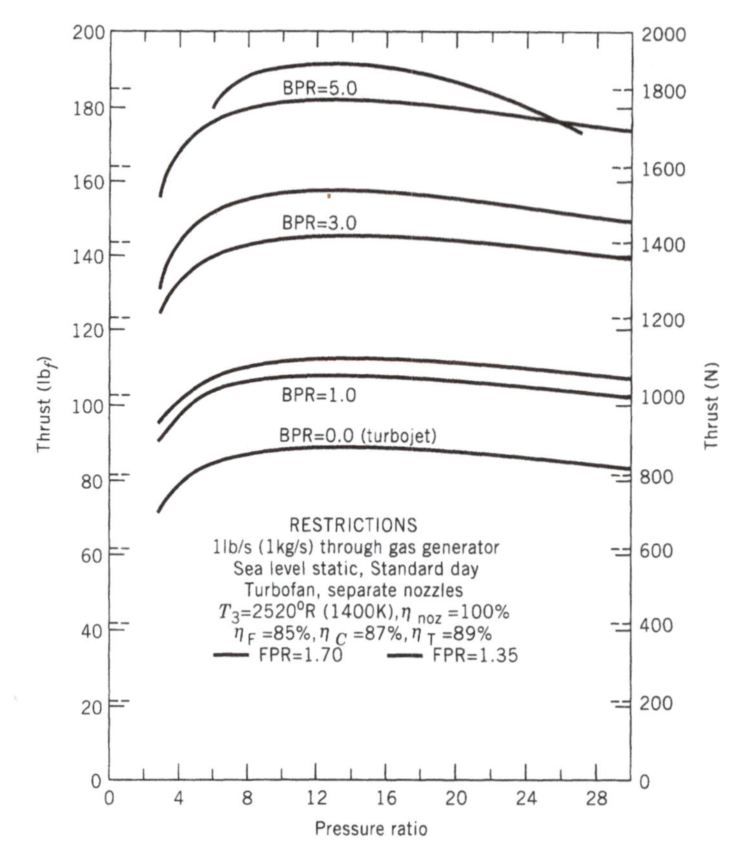

Figure 6.20. Variation of thrust with pressure ratio for an air standard turbofan engine with fixed component efficiencies for various bypass and fan pressure ratios.

The reason each of the techniques will increase the thrust from an engine is apparent if one examines the various plots in this chapter. Additional reasons why and problems associated with each technique will become apparent once components and component matching have been discussed.

Table 6.6 tabulates approximate temperatures and pressures at sea level static conditions for the Pratt & Whitney Aircraft JT9D-7R4 series turbofan engines. Note how both the turbine inlet temperature and overall pressure ratio change. Is this important? Why? Note that the pressures leaving the low-pressure turbine and leaving the fan are approximately equal.

6.7 TURBOPROP (TURBOSHAFT) ENGINE

A third type of gas turbine used for propulsion of an airplane is the turboprop engine. Propulsion by a turboprop engine is accomplished through the combined action of a propeller at the front of the engine and the thrust produced by the exhaust gases from

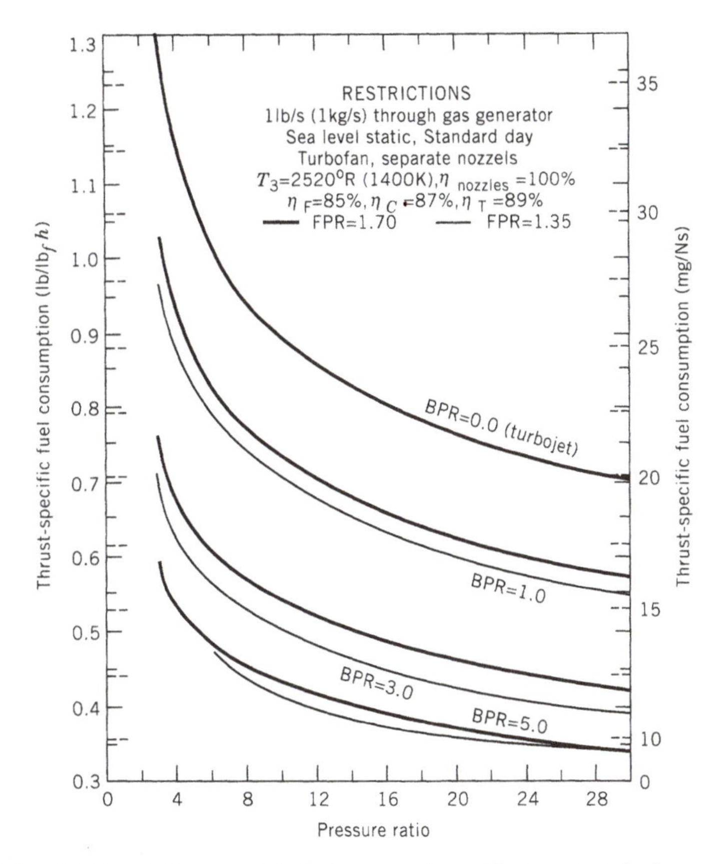

Figure 6.21. Variation of thrust-specific fuel consumption with pressure ratio for an air standard turbofan engine with fixed component efficiencies for various bypass and fan pressure ratios.

the gas turbine. A turbojet engine may be converted to a turboprop engine by adding an additional turbine to drive a propeller through a speed-reducing gear system. This type of engine is shown schematically in Figure 6.24.

A turboprop engine combines the advantages of a turbojet engine with the propulsion efficiency of a propeller. The turbojet engine derives its thrust from a large momentum change of a relatively small mass of air, whereas the turboprop engine develops its propulsive force by imparting a small momentum change to a relatively large mass of air.

The turbine of a turbojet engine extracts only the power necessary to drive the compressor and accessories, whereas the turbine of a turboprop engine is designed not only to absorb the power required to drive the compressor and accessories but also to deliver to the propeller shaft the maximum torque possible.

The propeller of a typical turboprop engine is responsible for roughly 90% of the total thrust at sea level static conditions. This percentage varies with airspeed, altitude, and other engine parameters.

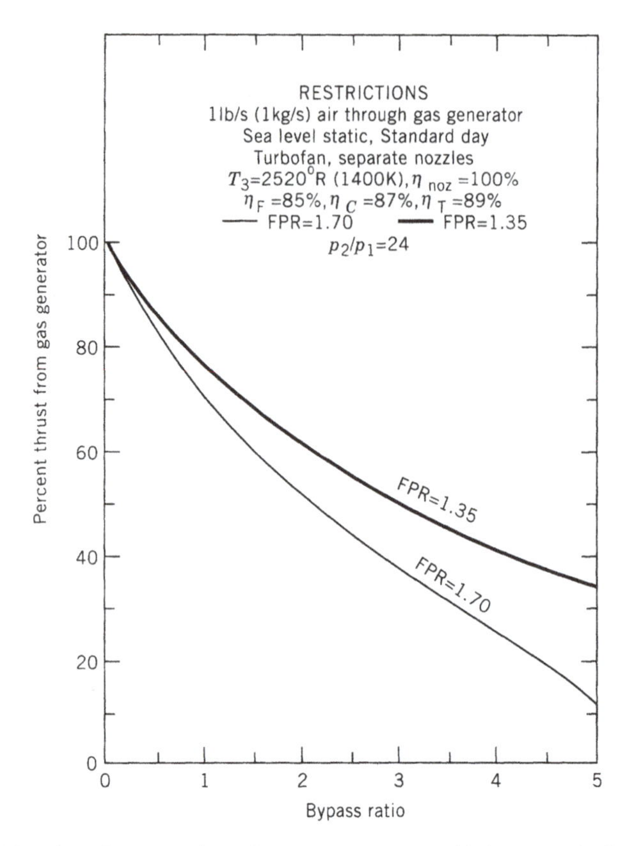

Figure 6.22. Variation of percent thrust from gas generator with bypass ratio for an air standard turbofan engine with fixed component efficiencies for two fan pressure ratios.

The turboprop engine has a lower thrust-specific fuel consumption during take-off and at low to moderate subsonic flight speeds than do the turbojet and turbofan engines, this advantage decreasing as the altitude and airspeed increase. The propeller efficiency remains fairly constant up to a Mach number of approximately 0.5, then drops rapidly. This means that aircraft propelled by turboprop engines are usually limited, by current technology, to about 400 mph.

The choice between the turbojet/turbofan (jet thrust) and the turboprop (shaft power and jet thrust) engines involves many considerations. For instance, the higher the operating speed, the larger may be the proportion of total output in the form of jet thrust. Also, an extra turbine stage may be required if more than a certain proportion of the total power is to be provided by the propeller.

Since the turbojet is rated in thrust and the turboprop engine in shaft horsepower plus thrust, no direct comparison may be made between the two. A comparison can be made by converting the horsepower developed by the turboprop engine to thrust or the thrust developed by the turbojet to horsepower.

When a turboprop engine operates under static conditions, one shaft horsepower delivered to the propeller is assumed to produce 2.5 lb of thrust. The power produced

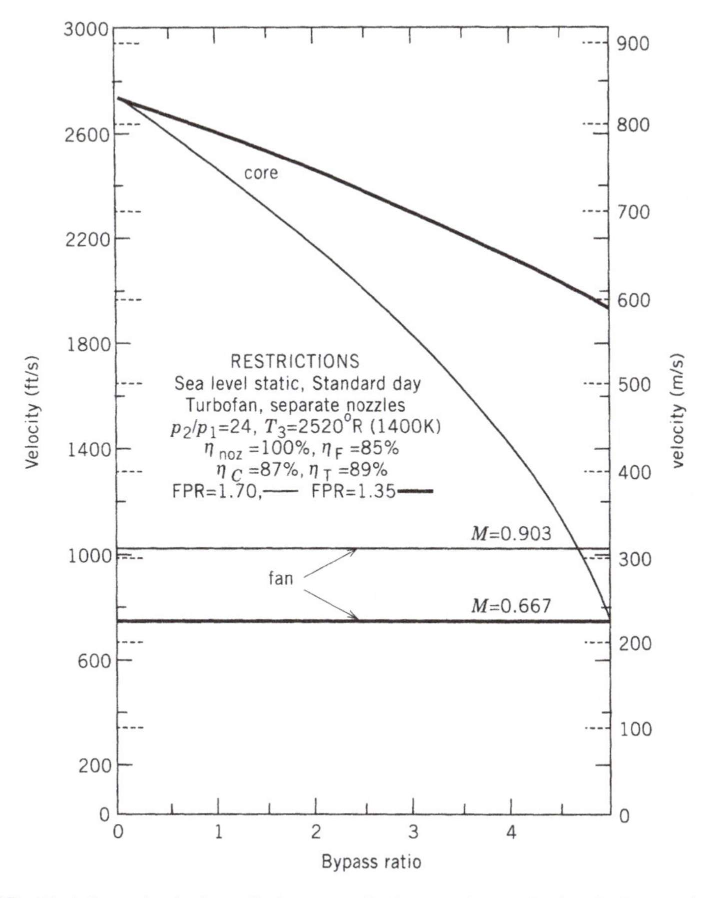

Figure 6.23. Variation of velocity with bypass ratio for an air standard turbofan engine with fixed component efficiencies for two pressure ratios.

by turboprop engines is normally expressed in equivalent shaft horsepower (ESHP). For sea level static conditions,

$$\mathrm{ESHP}_{\mathrm{static}} = \mathrm{SHP}_{\mathrm{prop}} + \frac{\mathcal{T}_{\mathrm{jet}}}{2.5} \tag{6.29}$$

The equivalent shaft horsepower in flight at a given airspeed will be the sum of the shaft horsepower and the horsepower equivalent of the *net* jet thrust. It is normally assumed for this comparison that the propeller efficiency is 80%.

Performance of a turboprop engine is expressed by the equivalent specific fuel consumption (ESFC), which is defined as

$$\mathrm{ESFC} = \frac{\dot{m}_f}{\mathrm{ESHP}} \tag{6.30}$$

TABLE 6.6. **Approximate Temperatures and Pressures for Pratt & Whitney Aircraft JT9D-7R4 Series Turbofan Engines (5)[a]**

Engine Model	Take-off Thrust (lb)	(1) p_1 (psia)	(1) T (°F)	(1.5) p (psia)	(1.5) T (°F)	(2) p (psia)	(2) T (°F)	(3) p (psia)	(3) T (°F)	(3.5) p (psia)	(3.5) T (°F)	(4) p (psia)	(4) T (°F)	(2d) p (psia)	(2d) T (°F)
−7R4A	44,300	14.7	59	33.8	218	319	889	303	2110	76.2	1320	20.9	876	21.8	128
−7R4B	46,100	14.7	59	34.3	222	330	903	314	2150	78.7	1350	21.3	894	22.1	131
−7R4C	46,900	14.7	59	35.0	226	337	912	320	2170	80.4	1370	21.3	897	22.3	132
−7R4D	48,000	14.7	59	35.4	228	340	914	322	2200	81.6	1390	21.6	915	22.5	133

[a] Values are for sea level static, standard day.

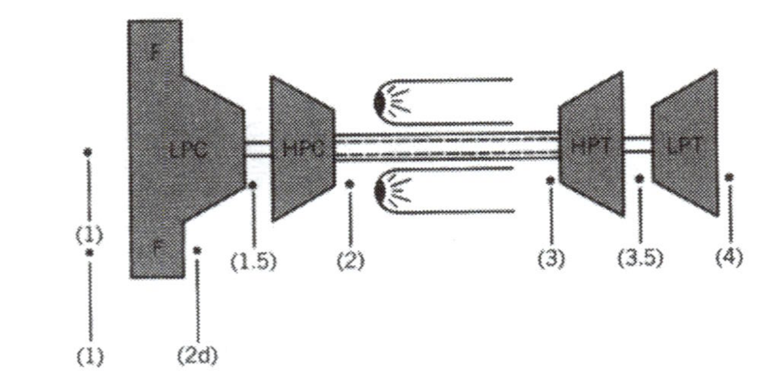

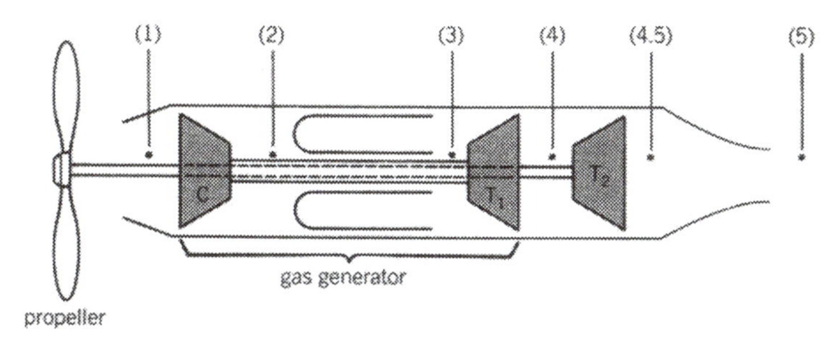

Figure 6.24. Schematic diagram of a turboprop gas turbine engine.

Example Problem 6.6

A turboprop engine is operating under the following conditions:

Flight speed at sea level, standard day	0
Air flow entering the compressor	1 lb/s (1 kg/s)
Compressor pressure ratio (total to total)	12.0
Efficiencies	
Diffuser	100%
Compressor	87%
Turbine to drive compressor	89%
Turbine to drive the propeller	89%
Nozzle	100%
Turbine inlet temperature (stagnation)	2520°R (1400 K)
Stagnation pressure leaving second turbine	25.0 psia (172.4 kPa)

Assume two turbines as illustrated in Figure 6.24. Assume an air standard cycle, but take into account the mass of fuel added. All efficiencies are based on the stagnation conditions, and the stagnation pressure remains constant from compressor outlet to turbine inlet. Calculate:

(a) The horsepower delivered by this engine to the propeller
(b) The thrust developed by the gases passing through the engine
(c) The equivalent shaft horsepower
(d) The equivalent specific fuel consumption

Compare the ESFC for this engine with the TSFC calculated in Example Problems 6.3 and 6.5.

Solution

The solution of the gas generator (through state 4) portion of this problem is the same as for Example Problem 5.5. Values for the gas generator are tabulated below.

State	T,°R(K)	p, psia (kPa)	h, Btu/lb (kJ/kg)
1	519 (288)	14.70 (101.3)	−6.0 (−14.3)
2_{oa}	1122 (623)	176.4 (1215.6)	141.7 (328.7)
3_o	2520 (1400)	176.4 (1215.6)	521.7 (1212.7)
4_{oa}	2008 (1115)	60.7 (418.5)	377.1 (876.9)

$$w_{C,a} = 147.7 \text{ Btu/lb } (343.0 \text{ kJ/kg})$$

$$f' = 0.0215$$

$$q_{in} = 410.7 \text{ Btu/lb}$$

Turbine to Drive Propeller

$$Pr_{5i} = 177.15 \left[\frac{25.0}{60.7}\right]$$

$$= 72.96$$

$$h_{o5i} = 267.5$$

$${}_4w_{5a} = 0.89(377.1 - 267.5)$$

$$= 97.5 \text{ Btu/lb thru power turbine}$$

$$\dot{w} = \frac{(1.0215)(3600)(97.5)}{2545}$$

$$= 140.9 \text{ hp}$$

Nozzle

$$\text{(a)}\ h_{o5a} = 377.1 - 97.5$$

$$= 279.6$$

$$Pr_{o5a} = 81.28$$

$$Pr_{6i} = 81.28 \left(\frac{14.70}{25.00}\right)$$

$$= 47.793$$

$$h_{6i} = 223.4$$

$$\overline{V}_{6i} = \sqrt{(279.6 - 223.4)(2)(32.174)(778)}$$

$$= 1677 \text{ ft/s}$$

(b) $\tau = \dfrac{(1.0215)(1677 - 0)}{32.174}$

$= 53.2 \text{ lb}_f$

(c) $\text{ESHP} = 140.9 + \dfrac{53.2}{2.5}$

$= 162.2 \text{ ESHP}$

(d) $\text{ESFC} = \dfrac{(0.0215)(3{,}600)}{162.2}$

$= 0.477 \text{ lb / ESHP} - \text{h}$

To compare the turbojet and turbofan engines with the turboprop engine, all values will be converted to ESHP.

$$\text{ESHP}\Big|_{\text{turbojet}} = \frac{91.1}{2.5} = 36.4 \text{ ESHP}$$

$$\text{ESFC}\Big|_{\text{turbojet}} = \frac{(0.0215)(3{,}600)}{36.4} = 2.13$$

$$\text{ESHP}\Big|_{\text{turbofan, BPR}=2} = \frac{129.5}{2.5} = 51.8 \text{ ESHP}$$

$$\text{ESFC}\Big|_{\text{turbofan, BPR}=2} = \frac{(0.0215)(3{,}600)}{51.8} = 1.49$$

$$\text{ESHP}\Big|_{\text{turbofan, BPR}=5} = \frac{185.1}{2.5} = 74.0 \text{ ESHP}$$

$$\text{ESFC}\Big|_{\text{turbofan, BPR}=5} = \frac{(0.0215)(3{,}600)}{74.0} = 1.05$$

The above calculations illustrate the advantage a turboprop engine has in fuel consumption at sea level static.

In principle, the turbofan engine is the same as the turboprop, the geared propeller being replaced by a duct-enclosed fan driven at engine speed, or, in the case of the TFE 731, at one-half the turbine speed. One fundamental operational difference between the turboprop and the turbofan is that the airflow through the fan of the turbofan engine is controlled by design so that the air velocity relative to the fan blades is unaffected by the airspeed of the aircraft. Also, the total airflow through the fan is much less than that through the propeller of a turboprop engine.

Figures 6.25 through 6.27 illustrate typical turboprop engines currently manufactured in the United States. One should observe the problem a designer faces in designing the inlet duct for a turboprop engine. The reduction gear normally is located on the same end of the engine as the propeller. The reduction gear, on the order of 9 to 1, is needed to reduce the high operating rpm of the turbine to a speed acceptable for driving the propeller. It must be capable of handling the heavy loads imposed on

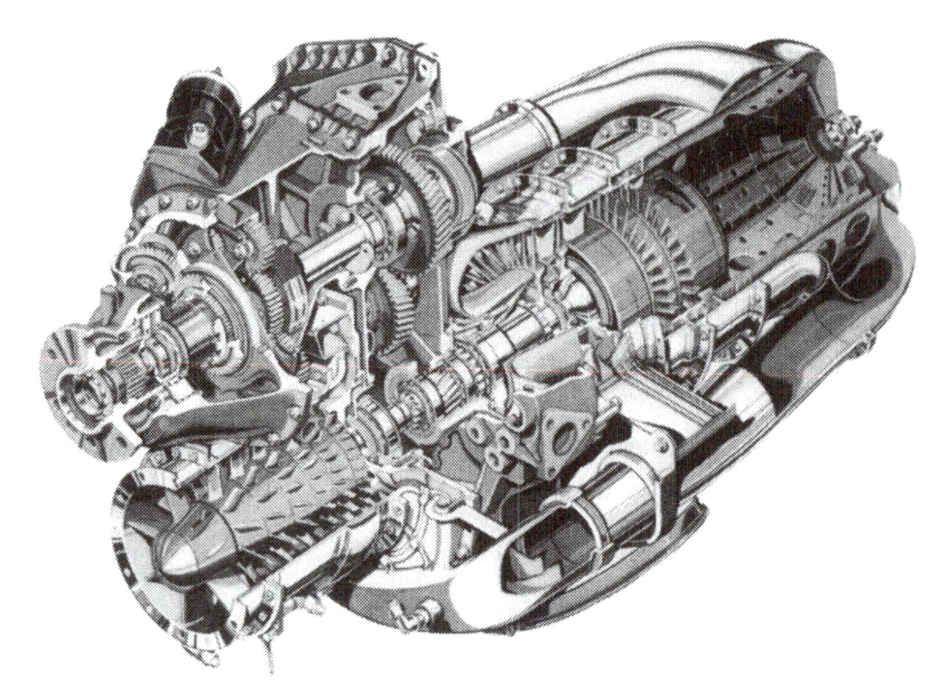

Figure 6.25. Cutaway view of the Allison Engine Company's Model 250 turboprop engine. (Courtesy of Allison Engine Company.)

it yet be light in weight and small in frontal area. The several designs used for the inlets to turboprop engines are illustrated in Figures 6.25 thorugh 6.27.

Table 6.7 lists known data for several operational U.S. turboprop engines. Note that centrifugal, axial-centrifugal, and axial-flow compressors are used, and that most of the pressure ratios are in the 6 to 9.5 range.

6.8 WATER INJECTION

It is necessary, for certain applications, to increase the thrust of an engine for short periods of time. The reason for needing this additional thrust can be to improve the take-off thrust, climb, or combat performance of an aircraft. The increased thrust could be obtained by the use of a larger engine, but this would be obtained at the expense of increased weight and frontal area.

Two ways to increase the specific power for a gas turbine were discussed in Chapter 5. The two ways were intercooling and reheat.

Both of these techniques are used on gas turbines used as propulsion devices. Intercooling is achieved by water injection and is discussed in this section. Reheat is achieved by adding an afterburner or duct heater to a turbojet or turbofan engine and is discussed in the next section.

Gas turbine engines, which are sensitive to the compressor inlet temperature, experience an appreciable power loss on hot days. It is frequently necessary, therefore, to provide some means of thrust augmentation for nonafterburning turbojet and turbofan engines during take-off on hot days.

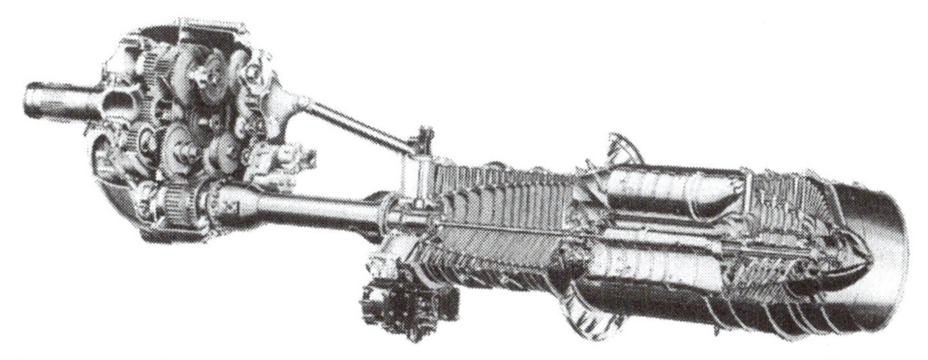

Figure 6.26. Cutaway view of the Allison Engine Company's T56 turboprop engine. (Courtesy of Allison Engine Company.)

Water may be injected into either the compressor air inlet or into the compressor diffuser case, that is, between the exit from the compressor and the inlet to the combustion chamber.

When water is injected into the compressor air inlet, added thrust is obtained principally by cooling the air entering the engine by means of vaporization of the water injected into the airstream. The reduction in compressor inlet increases the mass rate of flow into the engine, decreases the compressor work, and can change the engine "match." Engine matching is discussed in Chapter 8. This effect will be better understood once engine matching has been discussed.

Water injection into the compressor diffuser section increases the mass rate of flow through the turbine for a given compressor flow rate and lowers the combustion inlet temperature so that additional fuel may be burned without exceeding the maximum

Figure 6.27. Cutaway view of the AlliedSignal's TPE 331 turboprop gas turbine. (Courtesy of AlliedSignal.)

TABLE 6.7. **Typical U.S. Turboprop and Turboshaft Specifications (3)**

Manufacturer Model No. or Designation	Number of Compressor Stages	Maximum Power at Sea Level	Specific Fuel Consumption at Maximum Power	Pressure Ratio at Maximum rpm
Allied Signal Garrett Engine Div.				
TPE331-1[c]	2	665 shp	0.571	8.3
TPF351-20[c]	2	1300/2103 shp	0.496	13.2
General Electric Co., GE Aircraft Engines				
T58-GE-8F[b]	10	1350 shp	0.60	8.2
T64-GE-7A[b]	14	3936 shp	0.47	14.1
General Motors Allison Gas Turbine Division				
T56-A-14[b]	14	4591 shp	0.54	9.6
T56-A-427[b]	14	5250 shp	0.47	12.0
T703-A700[c]	1	650 shp	0.59	8.6
250-B17C[a]	6,1	420 shp	0.66	7.2
250-C28B,C[c]	1	550 shp	0.59	8.4

[a] Axial-centrifugal-flow compressor.
[b] Axial-flow compressor.
[c] Centrifugal-flow compressor.

turbine inlet temperature. This technique also shifts the engine match point, resulting in additional thrust.

6.9 AFTERBURNING AND DUCT HEATER ENGINES

Many times, large increases in thrust are required for short periods of time, such as during take-off or climb.

The afterburner is designed as an extension to the engine. The gases leaving the turbine are reheated in order to increase the momentum of the gas stream at the engine exit. Afterburners have been used on both turbojet and turbofan engines. When an afterburner is added to a turbofan engine, the afterburning can be done in the primary stream (gas generator stream), in the ducted stream, or in both streams. Most afterburning turbofan engines are of the mixed type, the one exception to this being the duct-burning turbofan engine. The reason for using the mixed (static pressure balanced) turbofan engine is that it eliminates a wall between two very hot streams which, if not eliminated, presents a complicated cooling problem. Schematic diagrams of the turbojet with an afterburner, turbofan with afterburner, and turbofan with a duct heater are shown in Figures 6.28, 6.29, and 6.30.

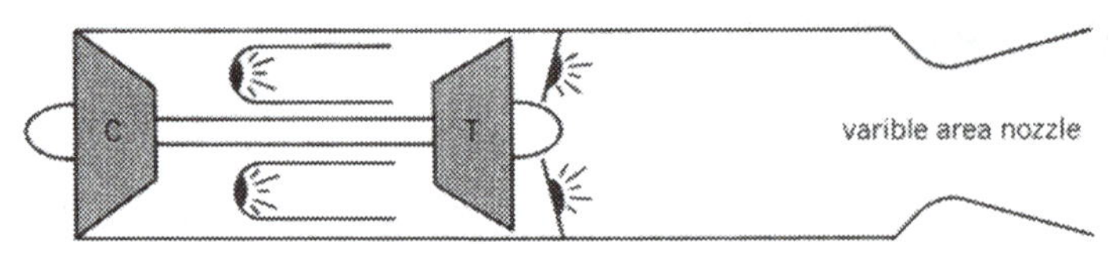

Figure 6.28. Turbojet engine with afterburner.

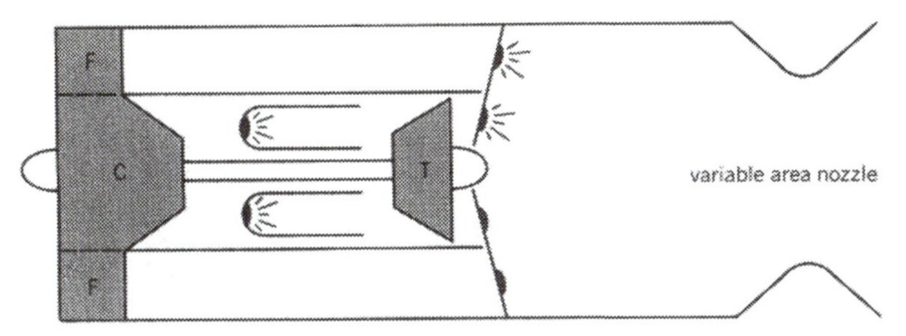

Figure 6.29. Mixed turbofan engine with afterburner.

Afterburning provides a means whereby the engine thrust can be increased well over 40% at sea level take-off to higher percentage increases at higher flight Mach numbers. The exact increase in thrust is dependent on the maximum afterburner temperature, bypass ratio, and what stream or streams are used for afterburning.

This increase in thrust is obtained at the expense of a large increase in fuel consumption and with considerable length being added to the engine. Therefore, afterburning is used only when this maximum thrust is needed for a short period of time.

The reason the turbine exhaust gases can be reheated is that only a small percentage of the oxygen available in the compressor discharge air is used in the primary burner. Also, any air passing through the fan is ducted around the primary burner and mixed with the primary gas stream behind the turbine. Therefore, a large percentage of the oxygen in the original air is available in the stream behind the turbine.

The addition of an afterburner to an engine requires some limitations on the other components, especially the turbine. First, the turbine exit diffuser must be designed to reduce the velocity to an acceptable level at the afterburner combustion chamber inlet. Turbine exit stators are usually provided to reduce the turbine exit swirl to within limits required for good afterburner performance.

The reason that afterburning increases the thrust of a jet engine can best be visualized by referring to Figure 6.31, which is an enthalpy–entropy diagram showing both an afterburning and nonafterburning turbojet engine. This figure, which is not drawn to scale, shows that for the same turbine discharge conditions, the enthalpy change across the nozzle for the afterburning engine is considerably greater than that for the nonafterburning engine. This is easily explained by the fact that constant

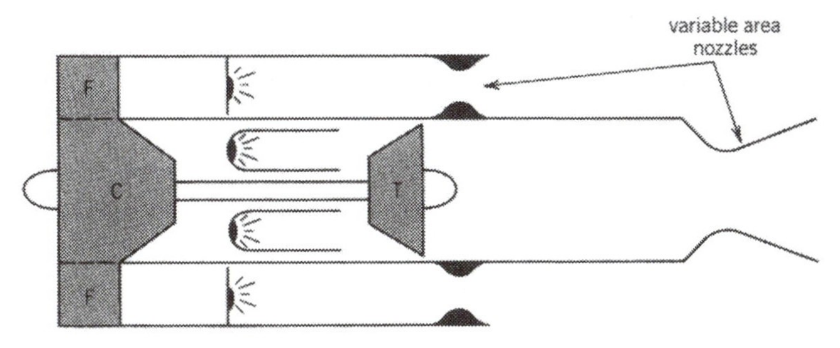

Figure 6.30. Nonmixed turbofan engine with duct heater.

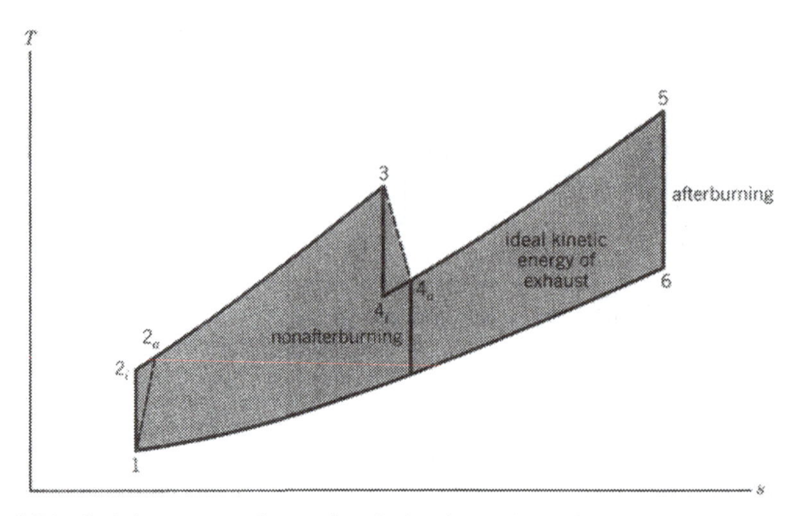

Figure 6.31. Enthalpy–entropy diagram for afterburning and nonafterburning turbojet engines.

pressure lines diverge on an enthalpy–entropy diagram as the entropy (and therefore temperature) increases as shown in Figure 6.31.

It should also be noted that the maximum temperature leaving the afterburner can be several hundred degrees above the maximum temperature permissible at the primary burner exit. This is explained by the fact that the turbine stators limit the primary burner temperature, whereas the liner cooling air limits the average afterburner temperature.

Table 6.8 lists typical specifications for U.S. turbojets with afterburners. It is of interest to compare the TSFC values given in Table 6.2 for nonafterburning turbojet engines with those in Table 6.8 for turbojet engines with the afterburner in operation. Of particular interest are the thrust and TSFC values for the JT85 with and without the afterburner in operation. Note the percent increase in thrust and TSFC.

Table 6.9 lists typical specifications for U.S. turbofan engines with afterburners. Note that two different models of the TF30 engines have thrust-specific fuel consumption values that differ by 20 percent. It is of interest to compare the values for the TF30 in Table 6.9 with those in Table 6.4

Figure 6.32 is a cutaway view of the General Electric J79 turbojet engine with afterburner. Note that this engine has a variable area nozzle.

Table 6.10 lists specifications for several J79 models.

Figure 6.33 is a cutaway view of the Pratt & Witney Aircraft F100 engine. Note that this is a mixed-flow turbofan engine. One can observe the afterburner fuel nozzle ring located after the turbine.

Example Problem 6.7

Solve Example Problem 6.3 assuming that an afterburner has been added to the engine, which increases the temperature to 1800 K (3240°R). Assume no pressure drop in the afterburner.

TABLE 6.8. **Typical Specifications for U.S. Turbojet Engines with Afterburners (3)**

Manufacturer Model No. or Designation	Number of Compressor Stages	Maximum Power at Sea Level, lb thrust	Specific Fuel Consumption at Maximum Power	Pressure Ratio at Maximum rpm
General Electric Co.				
GE Aircraft Engines				
J79-GE-8[a]	17	17,000	1.93	12.9
J79-GE-10,17[a]	17	17,820	1.93	13.4
J85-GE-5J[a]	8	3,850	2.20	6.7
United Technologies				
Pratt & Whitney				
J75-P-17[a]	15	24,500	2.15	11.9

[a] Axial-flow compressor.

Solution

The conditions entering the afterburner are from Example Problem 6.3

$$p_{o4} = p_{o5} = 418.5 \text{ kPa } (60.71 \text{ psia})$$

$$T_{o4a} = 1115 \text{ K } (2008°\text{R})$$

$$h_{o4a} = 876.9 \text{ kJ/kg } (377.1 \text{ Btu/lb})$$

$$T_{o5} = 1800 \text{ K } (3240°\text{R})$$

$$h_{o5} = 1700.5\ (731.6)$$

$$\text{Pr}_{o5} = 1310.2\ (1310.2)$$

TABLE 6.9. **Typical Specifications for U.S. Turbofan Engines with Afterburners (3)**

Manufacturer Model No. or Designation	Number of Compressor Stages	Maximum Power at Sea Level, lb thrust	Specific Fuel Consumption at Maximum Power	Pressure Ratio at Maximum rpm
General Electric Co.				
GE Aircraft Engines				
F404-GE-402[a]	3F,7C	17,700	1.79	26.0
United Technologies				
Pratt & Whitney				
TF30-P-3/P-103[a]	16	18,500	2.50	17.1
TF30-P-7/P-107[a]	16	20,350	3.01	17.5
TF30-P-100/P-111[a]	16	25,100	2.45	22.0
F100-PW-100[a]	13	23,830	2.17	24.8
F100-PW-220P[a]	13	27,000	2.06	31.4
F100-PW-229[a]	13	29,100	2.05	33.6

[a] Axial-flow compressor.

TABLE 6.10. **Specifications for the J79 Turbojet Engine**[a]

Model	J79-5C	J79-7A	J79-15	J79-17
Weight (lb)	3,685	3,575	3,685	3,835
Length (in.) (cold)	202.2	208.0	208.45	208.69
Max. diam. (in.) (cold)	38.0	38.3	38.3	39.06
Max. radius	19.0	19.2	19.2	19.5
Comp/turbine stages	17/3	17/3	17/3	17/3
Thrust/weight	4.23	4.42	4.62	4.67
Pressure ratio (MIL)	12.2	12.2	12.9	13.5
Air flow (lb/s)	162.5	162.0	169.0	170.0
RPM	7,460	7,460	7,685	7,685
T4 (T.O./cruise) (°F)	1,690	1,690	1,775	1,810
EGT limit (°F)	1,105	1,090	1,160	1,255
Max. thrust (SLS)	15,600	15,800	17,000	17,900
SFC	2.20	1.97	1.945	1.965
MIL. thrust (SLS)	10,300	10,000	10,900	11,870
SFC	0.84	0.84	0.86	0.84
Max. Mach no./alt.	2.0/35K	2.0/35K	2.4/45K	2.4/45K
Thrust (M2.0/35K)	15,700	15,600	16,700	18,700
SFC (M2.0/35K)	2.08	2.08	2.09	2.05
Cruise Mach no./alt.	0.9/35K	0.9/35K	0.9/35K	0.9/35K
Thrust	2,450	2,650	2,600	2,600
SFC	1.01	1.05	1.05	0.95

[a] Data from General Electric bulletin AEG-250-1.2, December 1968.

From Figure 5.21 (extended)

$$f' = 0.0208$$

$$Pr_{6i} = 1310.2\left[\frac{101.3}{418.5}\right]$$

$$= 317.1\ (317.2)$$

$$h_{6i} = 1077.0\ (463.4)$$

$$\overline{V}_{6i} = \sqrt{(1700.5 - 1077.0)(2)(1)(1000)}$$

$$= 1{,}117 \text{ m/s } (3{,}665 \text{ ft/s})$$

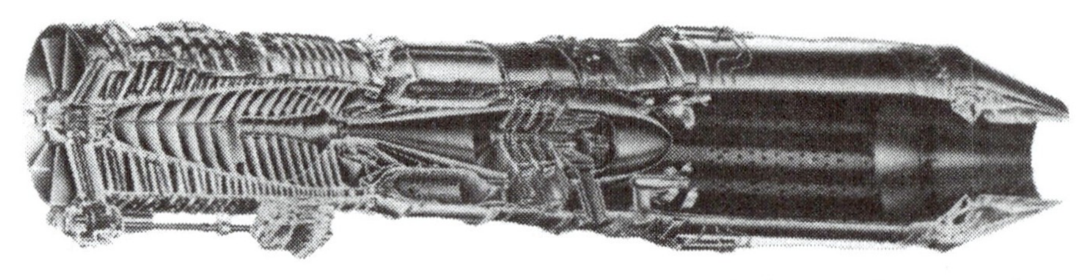

Figure 6.32. Cutaway view of the J79 turbojet engine with afterburner. (Courtesy of General Electric Co.)

Figure 6.33. Cutaway view of the Pratt & Whitney F100 engine with afterburner. (Courtesy of Pratt & Whitney Aircraft Group)

$$\tau = (1.0208)(1.0215)(1{,}117)$$
$$= 1165 \text{ N } (119 \text{ lb}_f)$$
$$\text{TSFC} = \frac{[0.0215 + (0.0208)(1.0215)]10^6}{1{,}165}$$
$$= 36.7 \text{ mg/N s } (1.295 \text{ lb / lb}_f\text{-h})$$

6.10 TURBOPROP ENGINE WITH REGENERATOR

It was noted in Chapter 5 that one way to improve the specific fuel consumption of an engine was to add a regenerator. This normally is not done with gas turbines used for propulsion because of the added weight.

One exception was the U.S. Navy's contract with Allison Division of General Motors for the development of the T78 turboprop engine. This was an attempt to build a turboprop engine with a much lower fuel consumption at partial power. The engine was to be used in antisubmarine and airborne early warning aircraft to allow them to spend more time on a search mission before returning to be refueled.

The regenerator was situated aft of the turbine to minimize the frontal area. This arrangement meant that the air, upon leaving the compressor, had to be ducted to the rear of the regenerator. It then flowed back toward the combustion chamber, absorbing energy from the exhaust gases.

The regenerator was a fixed, tubular, cross-counter-flow-type heat exchanger with the cold air inside the tubes, the hot gases outside.

REFERENCES CITED

1. *U.S. Standard Atmosphere, 1962*, U.S. Government Printing Office, Washington, D.C., 1962.
2. *Aeronautical Vest-Pocket Handbook*, 18th ed., Pratt & Whitney Aircraft Group, East Hartford, Conn., 1978.
3. *Aviation Week & Space Technology*, March 16, 1992

4. *Aviation Week and Space Technology*, October 27, 1969.
5. Pratt & Whitney Aircraft Group Commercial Products Division, August 1979.

SUGGESTED READING

Cohen, H., Rogers, G., and Saravanamuttoo, H., *Gas Turbine Theory*, John Wiley & Sons, New York, 1972.

Dusinberre, G. M., and Lester, J. C., *Gas Turbine Power*, International Textbook Co., Scranton, Pa., 1958.

Hill, P. G., and Peterson, C. R., *Mechanics and Thermodynamics of Propulsion*, Addison-Wesley Publishing Co., Reading, Mass., 1965.

Shepherd, D. G., *Aerospace Propulsion*, American Elsevier Publishing Co., New York, 1972.

Whittle, F., *Gas Turbine Aero-Thermodynamics*, Pergamon Press, Elmsford, N.Y., 1981.

NOMENCLATURE

A	area
BPR	bypass ratio
ESFC	equivalent specific fuel consumption
ESHP	equivalent shaft horsepower
f'	fuel–air ratio
$\mathbf{F}_R$	resultant force
h	specific enthalpy
I	specific thrust
k	specific heat ratio
$\dot{m}$	mass rate of flow
p	pressure
Pr	relative pressure
q	specific heat transfer
R	gas constant
SHP	shaft horsepower
T	temperature
TSFC	thrust specific fuel consumption
v	specific volume
$\overline{V}$	velocity
w	specific work
$\dot{w}$	power
η	efficiency
ρ	density
τ	thrust

Subscripts

1,2,3,4	states
a	actual, air
C	compressor
d	diffuser
e	exit
f	fuel
i	inlet
o	overall stagnation
p	propulsion
s	side
T	turbine
th	thermal
x	direction

PROBLEMS

6.1. Solve Example Problem 6.1 for a turbine inlet temperature of 2160°R (1200 K), all other values being the same.

6.2. Solve Example Problem 6.1 assuming a 3% pressure drop in the combustion chamber, all other values being the same.

6.3. Calculate, for Example Problems 6.1 and 6.2:

(a) The nozzle exit areas for the engines with converging-diverging nozzles. Compare with the exit areas of the converging nozzle.

(b) The Mach number at the exit of the converging-diverging nozzles.

6.4. Solve Example Problem 6.2 for a turbine inlet temperature of 1200 K (2160°R), all other values being the same.

6.5. Solve Example Problem 6.1b, assuming a 3% pressure drop in the combustion chamber and that the converging-diverging nozzle has an efficiency of 97%. Compare with the answer to Problem 6.2.

6.6. A turbojet engine is operating under the following conditions:

Altitude (standard day)	25,000 ft
Mach number	0.85
Air flow	1 lb/s
Compressor stagnation pressure ratio	17
Efficiencies	
Diffuser	100%
Compressor	87%
Turbine	89%
Jet nozzle	97%
Maximum stagnation temperature	2520°R

Assume an air standard cycle but take into account the mass of fuel added. All efficiencies are based on stagnation conditions, the stagnation pressure remains constant from compressor outlet to turbine inlet, and the nozzle is converging-diverging. Calculate:

(a) The thrust developed by this engine

(b) The thrust-specific fuel consumption

6.7. Solve Problem 6.6 for a turbine inlet temperature of 2160°R, all other values being the same.

6.8. A turbojet engine is operating under the following conditions:

Altitude (standard day)	12,000 m
Mach number	0.90
Air flow	1 kg/s
Compressor stagnation pressure ratio	15
Efficiencies	
Diffuser	100%
Compressor	87%
Turbine	89%
Jet nozzle	100%
Turbine inlet temperature (stagnation)	1400 K

Take into account the mass of fuel added but assume an air standard cycle. All efficiencies are based on stagnation conditions, the stagnation pressure remains constant from compressor outlet to turbine inlet, and the nozzle is converging-diverging. Calculate:

(a) The thrust developed by this engine

(b) The thrust-specific fuel consumption

6.9. Solve Problem 6.8 for a turbine inlet temperature of 1200 K, all other values being the same.

6.10. Solve Problem 6.6 assuming a converging nozzle, all other values the same.

6.11. Solve Problem 6.7 assuming a converging nozzle, all other values the same.

6.12. Solve Problem 6.8 assuming a converging nozzle, all other values the same.

6.13. Solve Problem 6.9 assuming a converging nozzle, all other values the same.

6.14. Verify the values for an ideal turbojet engine as given in Table 6.1.

6.15. A nonmixed turbofan engine is operating under the following conditions:

Flight speed at sea level (standard day)	0
Total air flow into engine	450 lb/s
Bypass ratio	1.4
Fan pressure ratio (total to total)	1.6
Compressor pressure ratio (total to total)	13
Efficiencies (based on stagnation conditions)	
Diffuser	100%
Compressor	87%
Fan	85%
Turbine	89%
Gas generator nozzle	100%
Fan nozzle	100%
Turbine inlet temperature (stagnation)	2160°R

Take into account the mass of fuel added but assume an air standard cycle. The stagnation pressure remains constant from the compressor outlet to the turbine inlet and expansion in the nozzles is back to ambient conditions. Calculate:

(a) The thrust developed by this engine

(b) The thrust-specific fuel consumption

6.16. Solve Problem 6.15 assuming that the engine has a converging nozzle, with all other values the same.

6.17. Solve Example Problem 6.5a (or b) assuming this to be a static pressure-balanced engine. This means (in this case) that the fan pressure ratio must be solved for (that is, it is unknown) and that the total pressure at states 2_d and 4 must be the same. Assume that the BPR does not change.

6.18. What are the advantages and disadvantages of the nonmixed type of turbofan engine over the mixed type of turbofan engine?

6.19. Solve Example Problem 6.5 assuming the flight condition to be

Mach number	0.9
Altitude	35,000 ft
Standard day	

6.20. Solve Example Problem 6.5 assuming a compressor pressure ratio of 24, all other values being the same.

6.21. Solve Example Problem 6.5 assuming a turbine inlet temperature of 2160°R (1200 K).

6.22. Solve Example Problem 6.5 assuming a compressor pressure ratio of 30, a bypass ratio of 6.5, and a turbine inlet temperature of 1500 K (2700°R).

6.23. Solve Problem 6.17 for the conditions of Problem 6.19.

6.24. Solve Example Problem 6.5 for a compressor pressure ratio of 22 assuming flight conditions of:

Mach number	0.9
Altitude	8,000 m
Standard day	

6.25. Performance and design characteristics for the CF6-6 turbofan engine are listed in Table 6.5. Assume the following additional information.

Fan efficiency	85%
Compressure efficiency	87%
Turbine efficiency	89%
Turbine inlet temperature	
Take-off	2760°R
Cruise	2560°R

Fan and overall pressure ratio are the same at take-off and cruise and are the values listed in Table 6.5. The configuration is a nonmixed turbofan engine with expansion in both nozzles back to ambient.

(a) Calculate the bypass ratio at both take-off and cruise. Does the ratio change with flight condition? Why? Is the bypass ratio as given in Table 6.5 for take-off or cruise?

(b) Calculate the take-off thrust and thrust-specific fuel consumption for an engine equipped with an ideal nozzle. Compare your calculated values with those listed in Table 6.5.

(c) Calculate the cruise thrust and thrust-specific fuel consumption for an engine with an ideal nozzle. Compare your values with those in Table 6.5.

(d) Why are your calculated values different than those given in Table 6.5?

(e) Calculate the power required to drive the compressor-fan combination.

6.26. Calculate, for each of the engine models listed in Table 6.6:

(a) The fan efficiency

(b) The low-pressure compressor efficiency

(c) The high-pressure compressor efficiency

(d) The high-pressure turbine efficiency

(e) The low-pressure turbine efficiency

6.27. Solve Example Problem 6.6 assuming that the pressure leaving the second turbine is 20 psia. What is the Mach number at the nozzle exit?

6.28. Information about the General Electric J79-17 turbojet engine is given in Table 6.10. Assume a turbojet engine as illustrated in Figure 6.4 and the following additional assumptions:

Diffuser efficiency—calculated by Eq. (6.23)	
Compressor efficiency	87%
Pressure drop between compressor outlet and turbine inlet	3%
Turbine efficiency	89%
Nozzle: converging-diverging	
Nozzle efficiency	97%

Assume that the air flow, as listed in Table 6.10, is for sea level static conditions. Calculate, assuming that T_4 is the turbine inlet temperature [for the sea level static (SLS) condition]:

(a) The thrust developed by this engine

(b) The thrust-specific fuel consumption

Compare your answers with those listed under MIL thrust (SLS) and SFC.

(c) Calculate the exit area and compare with the maximum diameter as given in the specifications

(d) Calculate the throat (minimum) area of the nozzle

6.29. The specifications for the General Electric J79-17 turbojet engine operating at an altitude of 35,000 ft at a Mach number of 0.9 are given in Table 6.10. Calculate, using the efficiencies as given in Problem 6.28:

(a) The air flow required to produce a thrust of 2600 lb

(b) The thrust-specific fuel consumption

Compare your answers with those listed in the specifications of Table 6.10 under "Cruise Mach no./alt."

(c) Calculate the exit area and compare with the maximum diameter as given in the specifications and the exit area calculated in Problem 6.28

6.30. Assume an afterburner has been added to the engine described in Problem 6.28. Using the information in Problem 6.28 and assuming the afterburner temperature is 3240°R, calculate:

(a) The thrust developed by this engine

(b) The thrust-specific fuel consumption

(c) The exit area assuming a converging-diverging nozzle with expansion back to ambient pressure

(d) The throat area of the nozzle. Compare this throat area with the area of the throat for the engine in Problem 6.28.

6.31. Assume that an afterburner has been added to the engine described in Problem 6.29. Using the information in Problem 6.29, the air flow determined in Problem 6.29, and assuming an afterburner temperature of 3240°R, calculate:

(a) The thrust developed by this engine

(b) The thrust-specific fuel consumption

(c) The exit area assuming a converging-diverging nozzle with expansion back to ambient pressure

(d) The throat area of the nozzle. Compare this throat area with the area determined in Problem 6.29

6.32. Solve Example Problem 6.5 assuming that the only change is a duct burner between 2_d and the fan nozzle entrance, which increases the duct air to a value of 3100°R at the fan nozzle entrance. Assume no pressure drop in the duct heater.

7

COMPRESSORS

Various gas turbine cycles were analyzed in Chapters 5 and 6. It was always assumed that each of the components had a given efficiency, that the designer could select the compressor pressure ratio and turbine inlet temperature, and that the gas turbine engine always operated at these values. No comment was made as to the design of the various components or what happened to the air flow, pressure ratio, or component efficiencies when the ambient pressure or temperature changed, the turbine inlet temperature was varied, or the gas turbine engine was operated at "off-design" conditions.

The next step in our analysis of gas turbine engines is to examine what happens to the component performance (and therefore the engine performance) when the engine operating conditions are changed. This chapter will examine the operating characteristics of axial- and centrifugal-flow compressors; Chapter 8 will deal with turbines; Chapter 9 with inlets, combustion chambers, and nozzles; and Chapter 10 will examine how the various components interact with one another. This interaction is called component matching and leads to the match point or equilibrium operating point. The series of match points for the various operating conditions is called the steady-state or equilibrium operating line.

7.1 COMPRESSORS

Efficient compression of large volumes of air is essential for a successful gas turbine engine. This has been achieved in two types of compressors, the axial-flow compressor and the centrifugal- or radial-flow compressor. Both types are discussed in this chapter.

Compressors designed for maximum efficiency would not be difficult if operation were restricted to a single operating condition. However, compressors must have good efficiency over a wide range of operating points. The object of a good compressor design is to obtain the most air through a given diameter compressor with a minimum

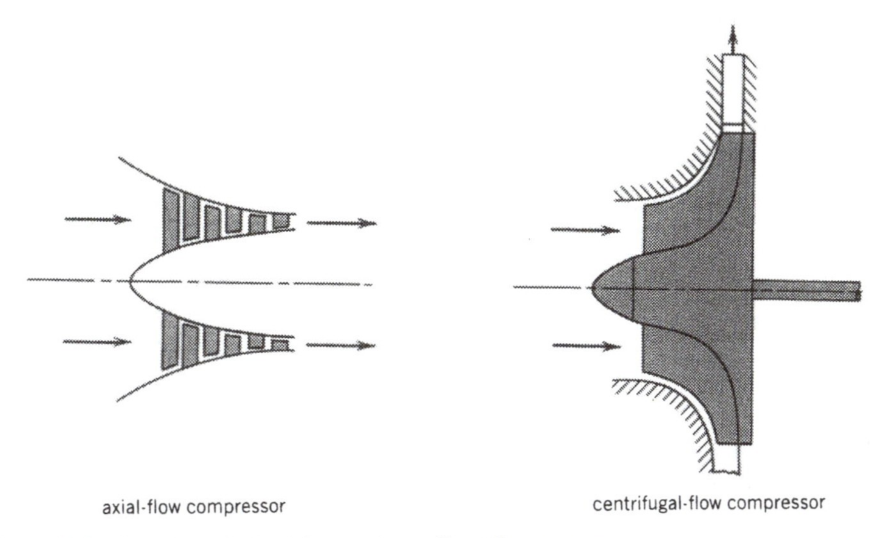

Figure 7.1. *Flow paths for axial-flow and centrifugal-flow compressors.*

number of stages while retaining relatively high efficiencies and aerodynamic stability over the operating range. The designer's freedom is usually restricted by mechanical, geometric, cost, and time constraints. The compatibility of the compressor shaft speed with that of a good turbine design must also be considered.

Axial-flow compressors were considered by the British in their early designs and were used by the Germans in the Junkers 004 and other World War II gas turbine engines. Most of the gas turbine engines designed and/or built in the United States in the 1940s and early 1950s used centrifugal-flow compressors. Today, gas turbine engines are built with axial-flow compressors, centrifugal-flow compressors, and combinations of one or more axial-flow stages followed by a centrifugal-flow compressor.

Figure 7.1 illustrates typical flow paths for axial-flow and centrifugal-flow compressors. The flow path in an axial-flow compressor is essentially parallel to the axis of rotation. Each stage includes a row of rotating blades where energy is added to the fluid. This rotor is followed by a row of fixed blades commonly referred to as a stator. Several stages are required in an axial-flow compressor to obtain the desired high pressure ratios.

In a centrifugal-flow compressor, the fluid enters at the center (eye) of the compressor and is turned radially outward. The rotating component of the centrifugal-flow compressor is followed by a diffusing passage, which may or may not incorporate stationary vanes or blades.

The advantages of the axial-flow compressor over the centrifugal-flow compressor are

1. Smaller frontal area for a given mass rate of flow.
2. Flow direction at discharge more suitable for multistaging.
3. May use cascade experiment research in developing compressor.
4. Somewhat higher efficiency at high pressure ratios.

The advantages of the centrifugal-flow compressor over the axial-flow compressor are

1. Higher stage pressure ratio.
2. Simplicity and ruggedness of construction.
3. Less drop in performance with the adherence of dust to blades.
4. Shorter length for the same overall pressure ratio.
5. Flow direction of discharge air that is convenient for the installation of an intercooler and/or heat exchanger.
6. Wider range of stable operation between surging and choking limits at a given rotational speed.

The advantages and disadvantages of each type of compressor will become more apparent when each type is discussed later in this chapter. Before examining the theory of each type of compressor, the overall compressor performance map will be discussed.

7.2 COMPRESSOR PERFORMANCE

Dimensional analysis may be applied to compressor performance to determine the significant dimensionless groups on which the performance depends. Tests have shown that the performance of both centrifugal-flow and axial-flow compressors may be described by the following quantities:

Symbol	Description	Dimensions
T_{01}	Inlet stagnation temperature	θ
p_{01}	Inlet stagnation pressure	$ML^{-1}t^{-2}$
c_p	Constant-pressure specific heat of gas	$L^2t^{-2}\theta^{-1}$
D	Characteristic dimension (usually diameter)	L
N	Rotor rotational speed	t^{-1}
$\dot{m}$	Mass flow rate of gas	Mt^{-1}
p_{02}	Exit stagnation pressure	$ML^{-1}t^{-2}$
T_{02a}	Actual exit stagnation temperature	θ
μ	Absolute viscosity of gas	$ML^{-1}t^{-1}$
ρ	Gas density at inlet	ML^{-3}

Instead of density, one of the parameters could have been the gas constant R, the molecular weight M, or the specific heat ratio k. Selecting T_{o1}, c_p, p_{o1}, and D as the repeating variables results in the following six dimensionless terms:

Variable	**Dimensionless Term**
N	$\pi_1 = \dfrac{ND}{\sqrt{c_p T_{o1}}}$
$\dot{m}$	$\pi_2 = \dfrac{\dot{m}\sqrt{c_p T_{o1}}}{p_{o1} D^2}$
p_{o2}	$\pi_3 = \dfrac{p_{o2}}{p_{o1}}$
T_{o2a}	$\pi_4 = \dfrac{T_{o2a}}{T_{o1}}$
μ	$\pi_5 = \dfrac{\mu\sqrt{c_p T_{o1}}}{p_{o1} D}$
ρ_1	$\pi_6 = \dfrac{c_p \rho_1 T_{o1}}{p_{o1}}$

Examination of each of these dimensionless terms yields the following:

- π_1 Represents the Mach number at the tip of the rotor, because D is the rotor diameter and the $\sqrt{c_p T_{o1}}$ term is proportional to the sonic velocity.
- π_2 Represents the mass flow parameter and is a function of the Mach number of the flow at the compressor inlet.
- π_3 Is the compressor pressure ratio (total to total).
- π_4 Is the actual temperature change across the compressor. When combined with π_3, it is a form of compressor efficiency.
- π_5 Represents a Reynolds number index.
- π_6 Is equal to the specific heat divided by the gas constant (c_p/R) and defines the gas being compressed.

Since air is usually the working fluid, π_6 usually is not considered or is specified with the performance data. π_5, which contains viscosity, usually is presented in the form of a curve showing the change in compressor efficiency and mass flow parameter as a function of a Reynolds number index.

In order to obtain less cumbersome and more easily understood values for the four remaining terms, the inlet total temperature T_{o1} is divided by the standard-day sea level temperature, 518.7°R (288 K), and the inlet total pressure is divided by the sea-level pressure, 14.696 psia (101.325 kPa).

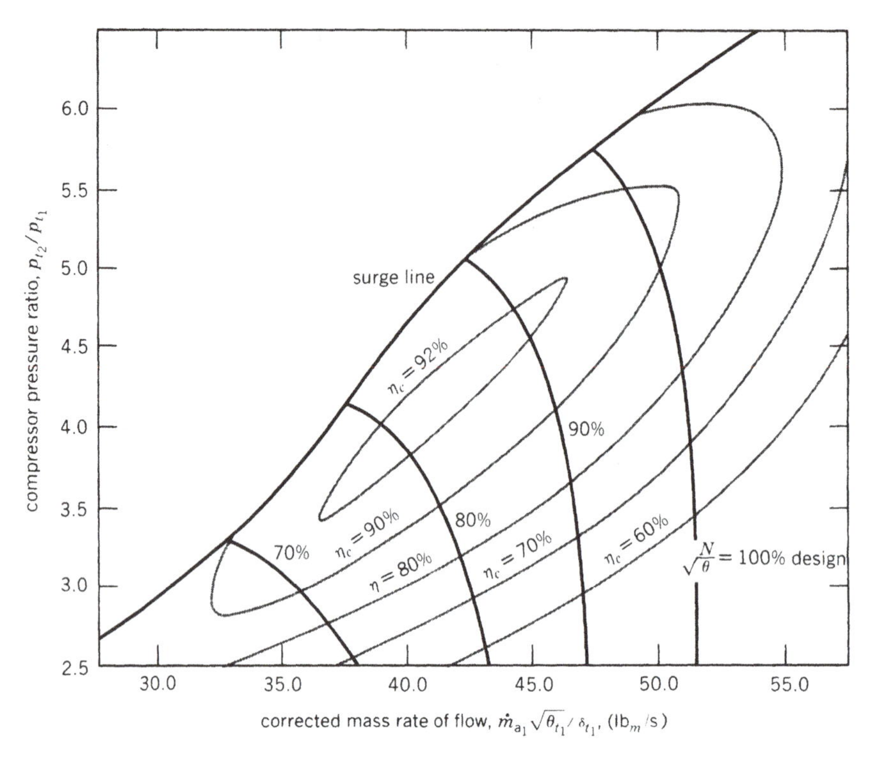

Figure 7.2. Hypothetical compressor performance map.

These terms result in θ_o, and δ_o, which are defined as follows:

$$\theta_o = \frac{\text{total temperature at compressor inlet}}{\text{sea level standard temperature}} = \frac{T_{o1}, R}{518.67} = \frac{T_{o1}, \text{K}}{288.15} \tag{7.1}$$

and

$$\delta_o = \frac{\text{total pressure at compressor inlet}}{\text{sea level standard temperature}} = \frac{P_o, \text{psia}}{14.696} = \frac{P_o, \text{kPa}}{101.325} \tag{7.2}$$

The performance of centrifugal-flow and axial-flow compressors is usually presented as a compressor map that covers the range of operation of a particular engine. The parameters usually included are

1. Compressor pressure ratio ($p_{o_{\text{outlet}}}/p_{o_{\text{inlet}}}$)
2. Inlet corrected gas flow ($\dot{m}\sqrt{\theta_{\text{o,inlet}}}/\delta_{\text{o,inlet}}$)
3. Inlet corrected rotational speed ($N/\sqrt{\theta_{\text{o,inlet}}}$)
4. Compressor adiabatic efficiency (η_c)

It should be noted that the characteristic dimension D and the specific heat c_p do not appear in the above terms. The characteristic dimension is omitted since a performance map is for a specific compressor; the specific heat is dropped since it is usually assumed to be a constant and the gas being compressed is known.

Figure 7.2 illustrates a hypothetical compressor performance map. It includes a surge line, which represents the limit of stable operation. Above and to the left of the surge line, aerodynamic instabilities become greater than can be tolerated. In the surge region, some of the compressor blades are operating stalled. One should note that the compressor map does not show the effects of Reynolds number. It should also be noted that the maximum efficiency occurs near the surge line and at a moderate speed, not at the maximum rotor speed.

There are ways, other than that illustrated in Figure 7.2, to present the performance of a compressor. This technique was selected because it is very useful in illustrating component matching.

Example Problem 7.1

Examination of Figure 6.13 shows that the specific thrust developed by the engine is 88.2 lb_f at sea level static, 83.4 lb_f at an altitude of 36,089 ft and a forward (flight) velocity of 600 ft/s. Assuming the same conditions as those used in developing these data but that the corrected mass rate of flow ($\dot{m}\sqrt{\theta_o}/\delta_o$) is the same in both cases and equal to 1 lb/s, calculate the thrust developed at the altitude point.

Solution

The only change is that, at the altitude condition, the mass rate of flow is no longer 1 lb/s. From Table C.1 at 36,089 ft,

$$T_{amb} = 389.970°R \text{ (say } 390°R)$$

$$p_{amb} = 3.283 \text{ psia}$$

From Table B.1,

$$h_{amb} = -37.0 \text{ Btu/lb}$$

$$Pr_{amb} = 0.4449$$

Therefore, at the compressor inlet,

$$h_o = -37.0 + \frac{(600)^2}{(2)(32.174)(778)} = -29.8 \text{ Btu/lb}$$

$$T_o = 420°R$$

$$Pr_o = 0.5767$$

$$p_o = 3.283 \left(\frac{0.5767}{0.4449}\right) = 4.256 \text{ psia}$$

$$\sqrt{\theta_o} = \sqrt{\frac{420}{518.7}} = 0.900$$

$$\delta_o = \frac{4.256}{14.696} = 0.290$$

Since $\dot{m}\sqrt{\theta_o}/\delta_o = 1$,

$$\dot{m} = \frac{(1)(0.290)}{0.900} = 0.322$$

or

$$\tau = (0.322)(83.4) = 26.9 \text{ lb}_f$$

One must remember that if the pressure ratio (p_{o2}/p_{o1}) and corrected mass rate of flow ($\dot{m}\sqrt{\theta_t}/\delta_t$) are the same at the two operating conditions, then the compressor is operating at the same point on the compressor map and η_c and $N/\sqrt{\theta_t}$ are the same for the two conditions. If, for Example Problem 7.1,

$$\frac{N}{\sqrt{\theta_{o_{\text{altitude}}}}} = \frac{N}{\sqrt{\theta_{o_{\text{sea level static}}}}}$$

then

$$(N)_{\text{altitude}} = (N)_{\text{SLS}}\sqrt{\frac{\theta_{o,\text{alt}}}{\theta_{o,\text{SLS}}}}$$

$$= (\text{N})_{\text{SLS}}\sqrt{\frac{0.900}{1}}$$

$$= 0.900 N_{\text{SLS}}$$

This means that the actual rotor rotational speed is different for the two operating conditions.

Figure 7.3 illustrates the variation in thrust and thrust-specific fuel consumption as a function of flight velocity and altitude for the J52-P-6A turbojet engine. One should compare this curve with the one presented in Figure 6.13.

7.3 ENERGY TRANSFER

Before discussing velocity diagrams for axial-flow and centrifugal-flow compressors, it is necessary to understand the mechanism of energy transfer between a rotor and a fluid. It is convenient, for this analysis, to choose a fixed control volume that encloses the rotor and cuts the shaft, as shown in Figure 7.4.

A gas enters the control volume with a velocity $\overline{V}_1$ at a radius r_1 and leaves with a velocity $\overline{V}_{1.5}$ at a radius r_1. The rotor angular velocity is ω in the direction shown in Figure 7.4. The velocity components are separated into an axial component, radial component, and tangential component as shown in Figure 7.4.

In a similar manner, $\overline{V}_{1.5}$, the exit velocity, may be separated into an axial component $\overline{V}_{a1.5}$, a radial component $\overline{V}_{r1.5}$, and a tangential component $\overline{V}_{u1.5}$ as shown in Figure 7.4.

It must be remembered that the following discussion assumes steady flow, that the properties at a point or circumference are uniform and invarient with time, and that

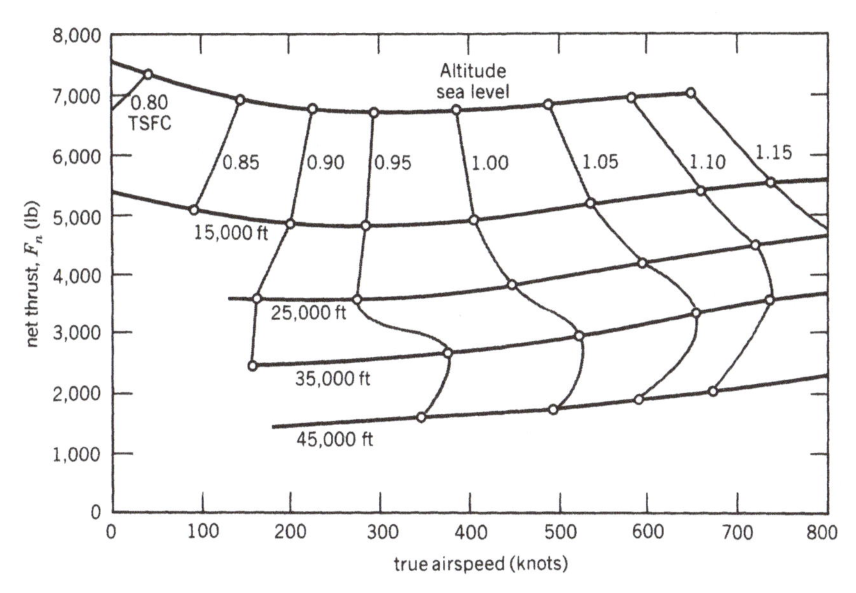

Figure 7.3. Performance data (normal rated) for the J52-P-6A turbojet engine. (Courtesy of Pratt & Whitney)

the fluid is entering and leaving with the velocities shown through a finite area evenly distributed around a circle at radius r_1 at the inlet and $r_{1.5}$ at the outlet.

The change in magnitude of the axial velocity components through the control volume (rotor) results in an axial force that must be counteracted by a thrust bearing. The change in magnitude of the radial velocity components through the rotor will result in a radial force that must be counteracted by a journal bearing unless the flow is uniformly distributed around the circumference at the inlet and exit to the rotor as assumed above.

The velocity component whose change is of importance in energy transfer is the tangential component. Equation (3.10) is the general equation for the moment of momentum for a fixed control volume. For steady-flow conditions and neglecting torques due to surface forces and the body force contribution, Eq. (3.10) becomes

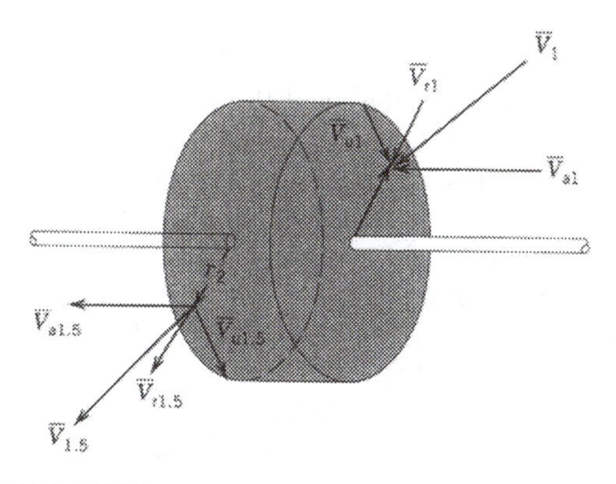

Figure 7.4. Flow through a rotor.

$$T_{shaft} = \dot{m}(r_{1.5}\overline{V}_{u1.5} - r_1\overline{V}_{u1}) \tag{7.3a}$$

The *positive direction* for $\overline{V}_u$ is in the direction of rotation. T_{shaft} in Eq. (7.3) is the torque on the shaft. If the torque is *positive*, then the machine is a *compressor*. If it is *negative*, then the machine is a *turbine*. Note that this sign convention for work is opposite in direction to that assumed for the first law.

To be consistent with the sign convention adopted earlier, Eq. (7.3a) will be changed to

$$T_{shaft} = \dot{m}(r_1\overline{V}_{u1} - r_{1.5}\overline{V}_{u1.5}) \tag{7.3b}$$

The rate of energy transfer for a rotor is the product of the torque and the angular velocity. Thus

$$\dot{P} = \omega T_{shaft} = \dot{m}(\omega r_1\overline{V}_{u1} - \omega r_{1.5}\overline{V}_{u1.5}) \tag{7.4}$$

Since the blade velocity U is

$$U = \omega r, \tag{7.5}$$

Eq. (7.4) becomes

$$\dot{P} = \dot{m}(U\overline{V}_{u1} - U\overline{V}_{u1.5}) \tag{7.6a}$$

or, per unit mass,

$$w = U(\overline{V}_{u1} - \overline{V}_{u1.5}) \tag{7.6b}$$

Equation (7.6) places no restrictions on geometry; that is, it applies to axial-flow and centrifugal-flow compressors, radial-flow and axial-flow turbines. The restrictions that apply were listed prior to Eq. (7.3).

For work to be done on the fluid, the term $U\overline{V}_{u1.5}$ must be greater than the inlet $U\overline{V}_{u1}$ term, or, if the fluid enters and leaves at the same radii, then the tangential component of the velocity at the exit of the rotor must be greater than the tangential component of the velocity at the inlet. This change in the tangential component of velocity as used in Eq. (7.6) applies across a single rotor, not across several stages of a compressor.

7.4 BLADE NOTATION FOR IDEAL AXIAL-FLOW COMPRESSORS

Attention will now be focused on the axial-flow compressor. Figure 7.5 illustrates the first three stages of an axial-flow compressor.

Many engines have inlet guide vanes prior to the first row of rotating (rotor) blades. The purpose of the inlet guide is to direct the fluid into the first row of rotor blades.

Each row of rotating blades is followed by a row of stationary (stator) blades. A stage consists of a row of rotor and a row of stator blades.

All work done on the working fluid is done by the rotating rows, the stators converting the fluid kinetic energy to pressure and directing the fluid into the next rotor.

Many recently designed gas turbine engines do not have inlet guide vanes. This eliminates one possible source of noise. This is discussed in Chapter 11, where the sources of noise are discussed.

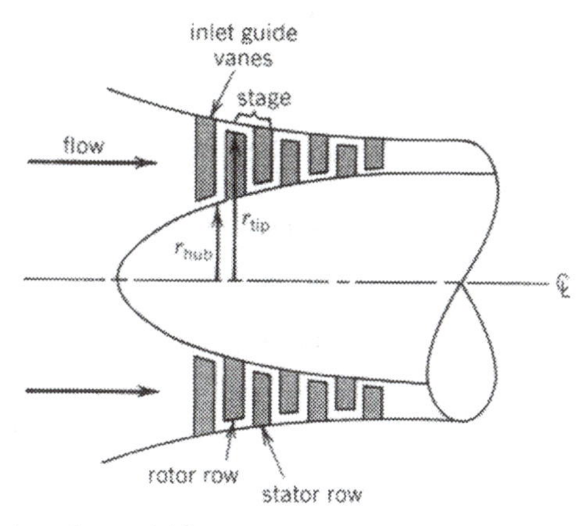

Figure 7.5. Cross section of an axial-flow compressor.

Flow through an axial-flow compressor is an extremely complicated three-dimensional problem. Solution to this general problem is beyond the scope of this book. It will be assumed that flow is one-dimensional, this being a reasonable first approximation for axial-flow compressors with high hub-to-tip ratios. The reader is referred to the Suggested Reading at the end of this chapter for a more detailed analysis of axial-flow compressors.

Two blades from a rotor row are shown in Figure 7.6. This figure represents flow

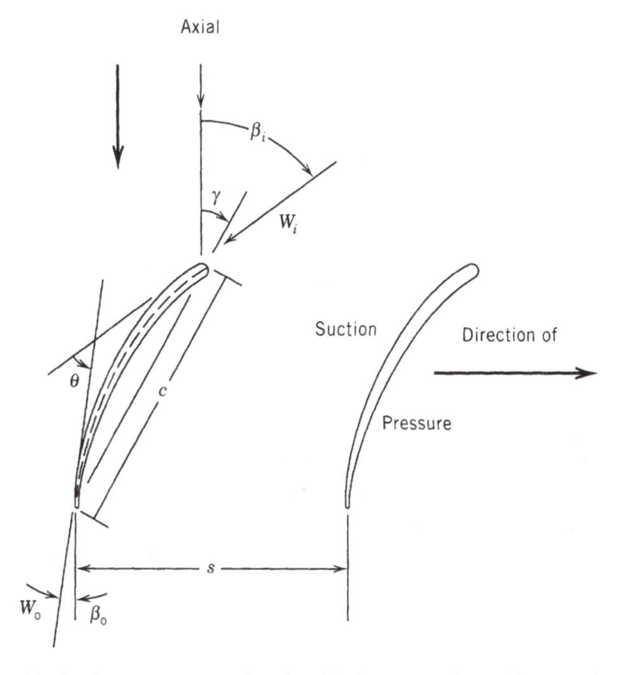

Figure 7.6. Two blades for a rotor row showing blade nomenclature for an axial-flow blade row.

at a representative radius r, halfway between the hub and the tip of the blade. The radius remains constant from inlet to outlet.

The fluid enters with a velocity W_i relative to the blade at an angle β_i and leaves with a relative velocity W_o at an angle β_o as shown in Figure 7.6. It is assumed at this time that the fluid angles β_i and β_o are the same as the blade camber line angles at inlet and outlet. The camber line is an imaginary line drawn through the center of the blade.

The fluid is, ideally, turned θ degrees as it passes through the rotor. The blades are spaced a distance s apart and have a chord length of c. The blades have a stagger angle γ, the angle between the chord line and the axial direction. The blades have a solidity σ, defined as the ratio of the chord length to the spacing. The direction of rotation and the pressure and suction surfaces are identified in Figure 7.6.

7.5 VELOCITY DIAGRAM FOR AN AXIAL-FLOW COMPRESSOR

Equation (7.6) shows that it is important to be able to calculate the change in tangential velocity across the rotor. This is best accomplished by drawing velocity diagrams that represent flow at the inlet and outlet of the rotor.

Velocity diagrams for flow through a single stage of an axial-flow compressor are shown in Figure 7.7. In this figure, the absolute flow velocity at the inlet to the rotor row and exit from the stator row is in the axial direction. It is assumed that the radius of the fluid does not change as it passes through the stage, that the rotor velocity at this radius is U, and that the relative velocity is at the same angle as the blade at both the inlet and the exit.

The relationship among the absolute velocity $\overline{V}$, relative velocity W, and blade velocity U is given by the following vector equation:

$$\mathbf{V} = \mathbf{W} + \mathbf{U} \tag{7.7}$$

This is illustrated by the velocity diagrams for the rotor in Figure 7.7.

The velocity diagrams are shown at three locations in Figure 7.7: one at the inlet to the rotor, one between the rotor and the stator, and one at the exit from the stator. It is quite common to combine these three diagrams into a single diagram as illustrated in Figure 7.8. The advantage of a single diagram is that the change in tangential velocity is clearly shown.

Example Problem 7.2

Air at 101.3 kPa (14.70 psia) and 288 K (519°R) enters an axial-flow compressor stage with a velocity of 170.0 m/s (550.0 ft/s). There are no inlet guide vanes. The rotor stage has a tip diameter of 66.0 cm (26.0 in), a hub diameter of 45.7 cm (18.0 in), and rotates at 8000 rpm. The air enters the rotor and leaves the stator in the axial direction with no change in velocity or radius. The air is turned through 15.0° as it passes through the rotor. Assuming constant specific heats with $k = 1.40$ and that the air enters and leaves the blades at the blade angles:

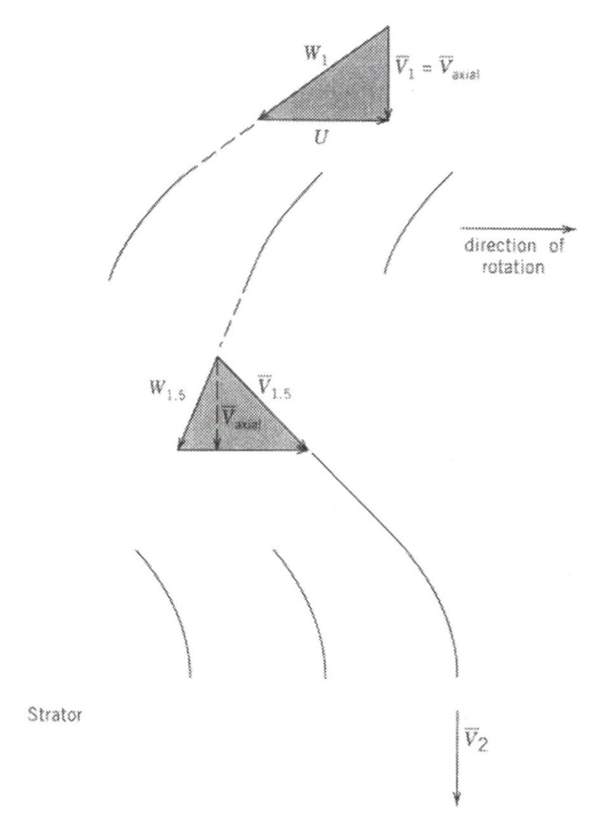

Figure 7.7. Velocity diagram for flow through a stage of an axial-flow compressor; no change in radius between inlet to rotor and outlet of stator.

(a) Construct the velocity diagrams at the mean blade height for this stage
(b) Determine the shape of the rotor and stator
(c) Calculate the mass rate of flow
(d) Calculate the power required
(e) Calculate the ideal total to total pressure ratio for this stage

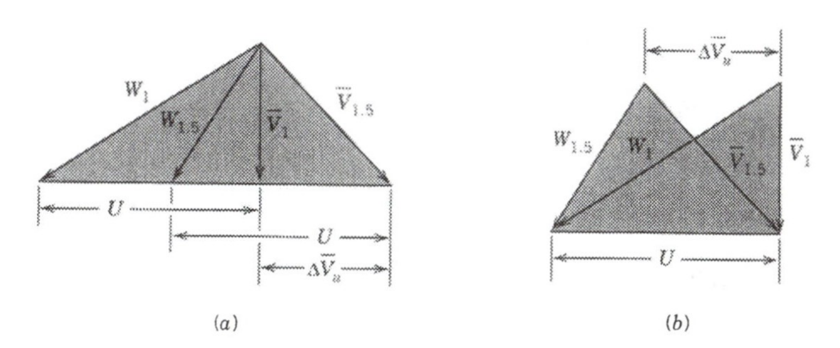

Figure 7.8. Stage velocity diagram. (a) Common apex. (b) Common blade velocity.

Solution

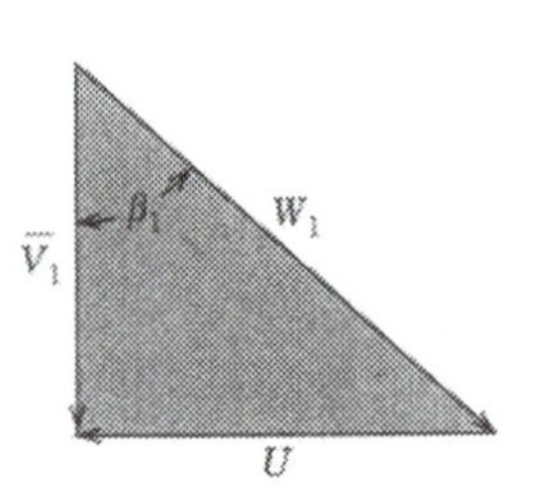

(a) $U = r\omega = \frac{1}{2}\left(\frac{45.7 + 66.0}{2}\right)\left(\frac{2\pi}{100}\right)\left(\frac{8000}{60}\right)$

$= 233.9$ m/s (767.9 ft/s)

$W_{u1} = -U = -233.9$ m/s (-767.9 ft/s)

$\beta_1 = \tan^{-1}\left(\frac{-233.9}{170.0}\right)$

$= -54.0°$ ($-54.4°$)

$W_1 = \sqrt{(170.0)^2 + (-233.9)^2}$

$= 289.2$ m/s (944.4 ft/s)

Since the rotor turns the air through 15.0°, the velocity diagram and numerical values at the outlet from the rotor are

$\beta_{1.5} = -54.0 + 15.0$

$= -39.0°$ ($-39.4°$)

$\overline{V}_{a1.5} = 170.0$ m/s (550.0 ft/s)

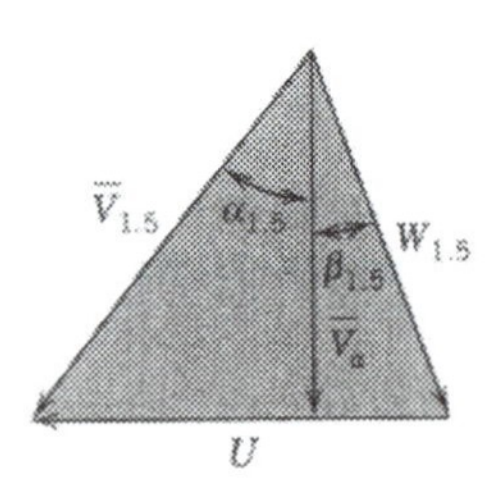

$W_{1.5} = \frac{170.0}{\cos(-39.0°)}$

$= 218.7$ m/s (711.8 ft/s)

$W_{u1.5} = V_a \tan(-39.0°)$

$= -137.7$ m/s (-451.8 ft/s)

$\overline{V}_{u1.5} = U + W_{u1.5}$

$= 233.9 + (-137.7)$

$= 96.2$ m/s (316.1 ft/s)

$\alpha_{1.5} = \tan^{-1}\left(\frac{96.2}{170.0}\right)$

$= 29.5°$ (29.9°)

$V_{1.5} = \frac{170.0}{\cos(29.5°)}$

$= 195.3$ m/s (634.4 ft/s)

The stage velocity diagram becomes

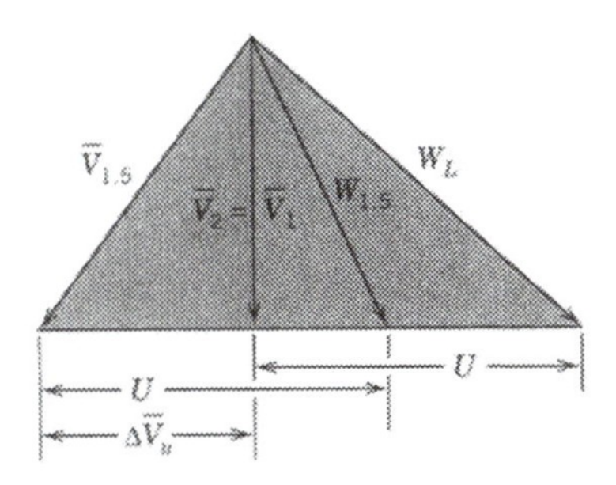

(b) The blade shapes are

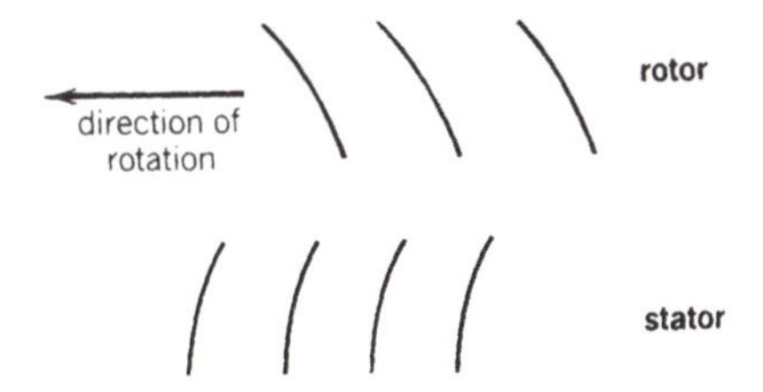

(c) $\dot{m} = \rho_1 A_1 \overline{V}_1 = \dfrac{p}{RT}\overline{V}_1\left(\dfrac{\pi D_t^2 - \pi D_h^2}{4}\right)$

$$= \frac{(101{,}300)(170.0)(28.965)\pi}{(8{,}314)(288)(4)}\,[(0.660)^2 - (0.457)^2]$$

$$= 37.10 \text{ kg/s } (80.73 \text{ lb/s})$$

(d) $T = 37.10\left(\dfrac{45.7 + 66.0}{(2)(2)(100)}\right)(0 - 96.2)$

$$= -996.6 \text{ Nm } (-727.1 \text{ ft} - \text{lb}_f)$$

$$w = T\omega = \frac{(8000)(2\pi)(-996.6)}{(60)(1000)}$$

$$= -834.9 \text{ kW } (-1108 \text{ hp})$$

$${}_1w_2 = \frac{-834.9}{37.10}$$

$$= -22.5 \text{ kJ/kg } (-9.70 \text{ Btu/lb})$$

The negative sign tells us that it is a compressor.

$$\text{(e)}\quad T_{o1} = T_1 + \frac{\overline{V}_1^2}{2c_p}$$

$$= 288 + \frac{(170.0)^2}{(2)(1000)(1.004)}$$

$$= 302.4\ \text{K}\ (544.2°\text{R})$$

$$T_{o1.5} = T_{o2} = T_{o1} - \frac{{}_1W_2}{c_p}$$

$$= 302.4 - \left(\frac{-22.5}{1.004}\right)$$

$$= 324.8\ \text{K}\ (584.6°\text{R})$$

The ideal pressure rise will occur for an adiabatic, reversible process or

$$\frac{p_{o1.5}}{p_{o1}} = \left(\frac{T_{o1.5}}{T_{o1}}\right)^{k/(k-1)} = \left(\frac{324.8}{302.4}\right)^{3.5}$$

$$= 1.284\ (1.285)$$

In this example, it should be noted that the relative velocity decreased as the air passed through the rotor and that the absolute velocity decreased as the air passed through the stator. This is the diffusing action discussed earlier. One should also note that:

1. Doubling the inlet velocity and doubling the rotor speed does not change the shape of the blades but doubles the mass rate of flow and torque, producing a fourfold increase in the power required.
2. Increasing the turning angle of the rotor increases ΔV_u, therefore increasing the torque and work per unit mass and therefore producing a higher stage pressure ratio.

This is illustrated by the values given in Table 7.1.

It is obvious from the values in Table 7.1 that increasing the turning angle increases the stage pressure ratio. Therefore, in designing an axial-flow compressor, one wants the rotor turning angle to be as large as possible. One also wants the inlet velocity as high as possible to minimize the compressor size and the hub-to-tip pressure ratio as small as possible to minimize the diameter of the compressor. It is obvious that one is not at liberty to arbitrarily select the inlet velocity, turning angle, and hub-to-tip ratio, but what are reasonable values and what other factors must be considered?

Questions that have not been answered but that must be considered are:

1. What is a typical maximum rotor tip velocity, and what limits the velocity? The maximum tip velocity is most often limited by some structual limitation. For axial flow machines, maximum *tip* velocity is usually 1300 − 1500 ft/s (400-460 m/s) with most limited to 1000 ft/s (300 m/s) or less. Tip speeds for several General Electric gas turbine engines are listed in Table 7.2.

TABLE 7.1. **Effect of Blade Turning Angle on Blade Angles, Velocities, Torque, Work, Temperature and Stage Pressure Ratio***

θ, degrees	10.0		15.0		20.0		25.0	
$\overline{V}_a$, m/s (ft/s)	170.0	(550.0)	170.0	(550.0)	170.0	(550.0)	170.0	(550.0)
U, m/s (ft/s)	233.9	(767.9)	233.9	(767.9)	233.9	(767.9)	233.9	(767.9)
$\overline{V}_1$, m/s (ft/s)	170.0	(550.0)	170.0	(550.0)	170.0	(550.0)	170.0	(550.0)
$\overline{V}_{u1}$, m/s (ft/s)	0.0	(0.0)	0.0	(0.0)	0.0	(0.0)	0.0	(0.0)
α_1, degrees	0.0	(0.0)	0.0	(0.0)	0.0	(0.0)	0.0	(0.0)
W_1, m/s (ft/s)	289.2	(944.6)	289.2	(944.6)	289.2	(944.6)	289.2	(944.6)
W_{u1}, m/s (ft/s)	−233.9	(−767.9)	−233.9	(−767.9)	−233.9	(−767.9)	−233.9	(−767.9)
β_1, degrees	−54.0	(−54.4)	−54.0	(−54.4)	−54.0	(−54.4)	−54.0	(−54.4)
$\overline{V}_{1.5}$, m/s (ft/s)	183.8	(596.0)	195.4	(634.5)	207.7	(675.1)	220.0	(715.8)
$V_{u1.5}$, m/s (ft/s)	69.8	(229.5)	96.3	(316.3)	119.3	(391.5)	139.7	(458.2)
$\alpha_{1.5}$, m/s (ft/s)	22.3	(22.7)	29.5	(29.9)	35.1	(35.4)	39.4	(39.8)
$W_{1.5}$, m/s (ft/s)	236.3	(769.7)	218.7	(711.7)	205.0	(666.5)	194.4	(631.2)
$W_{u1.5}$, m/s (ft/s)	−164.1	(−538.4)	−137.6	(−451.6)	−114.6	(−376.4)	−94.2	(−309.8)
$\beta_{1.5}$, m/s (ft/s)	−44.0	(−44.4)	−39.0	(−39.4)	−34.0	(−34.4)	−29.0	(−29.4)
T, Nm (ft · lbf)	−723.3	(−528.0)	−997.9	(−727.7)	−123.6	(−900.6)	−1448.	(−1054.)
${}_1w_2$, kJ/kg (Btu/lb)	−16.33	(−7.04)	−22.53	(−9.70)	−27.91	(−12.01)	−32.70	(−14.06)
T_{o1}, K(°R)	302.4	(544.2)	302.4	(544.2)	302.4	(544.2)	302.4	(544.2)
T_{o2}, K(°R)	318.6	(573.5)	324.8	(584.6)	330.2	(594.2)	334.9	(602.7)
P_{o2}/P_{o1}	1.201	(1.202)	1.285	(1.285)	1.360	(1.361)	1.430	(1.430)

*All values are for blade height, rotor speed, and stage inlet conditions as given in Example Problem 7.2.

TABLE 7.2. **Compressor Blade Tip Speeds for Several General Electric Compressors.***

Unit	Tip Speed (ft/s)	Compressor Tip Diameter (in.)
MS5001P	1092	49.1
MS6001B	1114	50.1
MS7001EA	1120	71.3
MS7001FA	1282	81.6
MS9001B	1092	83.5
MS9001FA	1282	97.9

*Source: Brandt (2).

2. What limits the amount of turning that can occur in a rotor? Is this dependent on rotor speed? Dependent on inlet velocity? To answer these questions, the following factors must be considered:
- Flow coefficient
- Work coefficient (blade loading coefficient)
- Mach number
- Hub-to-tip ratio
- de Haller number (diffusion factor)
- Degree of reaction

7.6 FLOW COEFFICIENT

The flow coefficient is defined as the inlet velocity (axial velocity) divided by the blade velocity or, in equation form

$$\phi = \frac{\overline{V}_a}{U} \tag{7.8}$$

This is a dimensionless parameter expressing the throughflow.

Referring to Figure 7.7, one observes that the value of the blade angle at the inlet to the rotor, β_1, remains constant as long as the flow coefficient remains constant.

7.7 WORK COEFFICIENT (BLADE LOADING COEFFICIENT)

The *work coefficient* Ψ, also called the *blade loading coefficient*, is defined as the stage (rotor) work divided by the blade kinetic energy, or, in equation form

$$\Psi = \frac{-c_p \Delta T_o}{U^2/2} \tag{7.9a}$$

For turbines, Ψ is positive. For compressors, Ψ is negative.

Some designers defined the work coefficient without the 2 factor or

$$\Psi = \frac{c_p \Delta T_o}{U^2} \tag{7.9b}$$

In this text, Eq. (7.9a) will be used.
For compressors

highly loaded stage > 1.0

lightly loaded stage < 0.6

For an axial-flow compressor (or turbine), the work coefficient may be written as

$$\begin{aligned}
\Psi &= \frac{2\overline{U}(\overline{V}_{u1} - \overline{V}_{u1.5})}{U^2} \\
&= \frac{2(\overline{V}_{u1} - \overline{V}_{u1.5})}{U} \\
&= \frac{2(\overline{V}_a \tan \alpha_1 - \overline{V}_a \tan \alpha_{1.5})}{U} \\
&= \frac{2\overline{V}_a(\tan \alpha_1 - \tan \alpha_{1.5})}{U}
\end{aligned} \tag{7.10}$$

One must remember that Eq. (7.10) is limited to axial flow compressors and turbines.
From Eq. (7.10), one observes that for the same Ψ and U value, increasing the inlet axial velocity decreases the blade turning angle and vice versa.
One can also write Eq. (7.10) in terms of the flow coefficient ϕ or

$$\Psi = 2\phi(\tan \alpha_1 - \tan \alpha_{1.5}) \tag{7.11}$$

This form shows the interdependence between the blade loading coefficient, flow coefficient, and blade turning angle.

7.8 MACH NUMBER

The Mach number was defined in Section 3.5 as

$$M = \frac{\overline{V}}{c} \tag{3.17}$$

or, for an ideal gas

$$M = \frac{\overline{V}}{\sqrt{kRT}} \tag{7.12}$$

where the temperature is the local static temperature. One must also remember that the static temperature has the same value in both the absolute and relative reference frames.

A good design will have a Mach number less than or equal to 0.9. For some transonic compressors, the tip Mach number may be 1.2.

It can be shown (see Example Problem 7.4) that for a rotor stage, the maximum Mach number occurs at the tip of the blade at the inlet to the rotor for the Mach number based on the relative velocity and at the hub (root) of the blade at the exit from the rotor for the Mach number based on the absolute velocity. One should also recognize that both the relative and absolute Mach numbers will decrease after the first stage of a multistage axial-flow compressor since the static temperature increases from inlet to exit of the compressor.

7.9 HUB-TO-TIP RATIO

The hub-to-tip ratio is the ratio of the radius at the hub (root) of the blade divided by the radius at the tip of the blade. In equation form

$$\text{RATIO} = \frac{r_h}{r_t} \tag{7.13}$$

For axial flow compressors the ratio is usually greater than or equal to 0.70 or

$$\frac{r_h}{r_t} \geq 0.70 \tag{7.14}$$

7.10 DE HALLER NUMBER

It was noted earlier that both the rotor and stator of compressors act as diffusers. The efficient diffusion, or conversion, of kinetic energy into pressure energy is an essential feature of a compressor. If the rate of diffusion is too rapid, then the boundary layer is subject to separation, resulting in a poor (inefficient) compressor design. If the rate of diffusion is very slow, the blade length becomes excessive and results in large friction losses.

Actual blade design will not be considered in this text. The reader is referred to the many books written on this subject, some of which are referenced at the end of this chapter.

An early and simple measure of the amount of diffusion was the *de Haller number*, which is defined as the exit relative velocity divided by the inlet relative velocity or, in equation form

$$\text{de Haller No.} = \frac{W_{1.5}}{W_1} \tag{7.15}$$

This method of measuring the amount of diffusion is useful in preliminary designs and will be used in this text. The de Haller number should be above 0.70 with many designers using a value of 0.72 or higher.

Example Problem 7.3

Determine for Example Problem 7.2

(a) The flow coefficient at the inlet to the compressor
(b) The hub-to-tip ratio at the inlet to the compressor
(c) The work coefficient
(d) The absolute and relative Mach numbers at the inlet to the compressor
(e) The absolute and relative Mach numbers at the exit from the rotor
(f) The de Haller number

Solution

(a) $$\phi = \frac{\overline{V}_a}{U} = \frac{550.0}{767.9}$$

$$= 0.716\ (0.727)$$

(b) $$\frac{r_h}{r_t} = \frac{18.0}{26.0} = 0.692\ (0.692)$$

(c) $$\Psi = \frac{-(0.24)(584.6 - 544.2)(2)(32.174)(778)}{(767.9)^2}$$

$$= -0.823\ (-0.822)$$

The negative sign confirms that we are dealing with a compressor.

(d) $$M_1 = \frac{550.0}{\sqrt{(1.40)(32.174)\left(\dfrac{1545}{28.965}\right)(519)}}$$

$$= \frac{550.0}{1117} = 0.493\ (0.500)$$

$$M_{r1} = \frac{944.4}{1117} = 0.846\ (0.850)$$

(e) The static temperature at the exit from the rotor is needed to determine the sonic velocity at the rotor exit. This can be determined as follows:

$$T_{1.5} = T_{01.5} - \frac{\overline{V}_{1.5}^2}{2}$$

$$T_{1.5} = 584.6 - \frac{(634.4)^2}{(0.24)(2)(32.174)(778)}$$

$$= 551.1^\circ\text{R}\ (305.8\ \text{K})$$

Therefore

$$M_{1.5} = \frac{634.4}{\sqrt{(1.40)(32.174)\left(\frac{1545}{28.965}\right)(551.1)}}$$

$$\frac{634.4}{1151} = 0.551\ (0.557)$$

$$M_{r1.5} = \frac{711.8}{1151} = 0.619\ (0.624)$$

(f) de Haller No. $= \frac{W_{1.5}}{W_1}$

$$= \frac{711.8}{944.4} = 0.754\ (0.756)$$

Earlier, good design values for an axial-flow compressor were given. It is suggested that the reader compare the values calculated in Example Problems 7.2 and 7.3 with the desired design values to judge whether this is a good design.

7.11 HUB AND TIP EFFECTS

The discussion this far has been at the mean blade height. It is important to examine what happens at the hub (root) and tip of the blade.

A frequently used design condition is to assume a constant specific work at all radii and to assume that the axial velocity remains constant. This means that the pressure ratio at the hub, mean blade diameter, and tip will be the same and that the quantity

$$r\overline{V}_u \tag{7.16}$$

will be a constant from hub to tip. This is referred to as the *free vortex condition.*

For the same specific work at all radii, the change in stagnation temperature at all radii will be the same since

$$_1w_2 = {}_1w_{1.5} = c_p(T_{o1} - T_{o1.5})$$

This means that the pressure ratio for the ideal case (isentropic flow) at all radii will be the same since

$$\frac{p_{o1.5}}{p_{o1}} = \left(\frac{T_{o1.5}}{T_{o1}}\right)^{k/(k-1)}$$

Example Problem 7.4

Solve Example Problems 7.2 and 7.3 at the hub and tip of the blade assuming that the specific work is the same at all radii and that the blade height does not change from the inlet to the rotor to the exit from the stator. Determine if the blade height remains constant.

Solution

Since the specific work is the same at the hub, mean blade height and tip

$$ {}_1w_{1.5} = U(\overline{V}_1 - \overline{V}_{1.5}) $$

Since $U_{\text{hub}} < U_{\text{mean}} < U_{\text{tip}}$, then

$$ \Delta\overline{V}_u)_{,\text{hub}} > \Delta\overline{V}_u)_{,\text{mean}} > \Delta\overline{V}_u)_{,\text{tip}} $$

Hub

$$ U = r\omega = \left(\frac{1}{2}\right)\left(\frac{18.0}{12}\right)(2\pi)\left(\frac{8000}{60}\right) $$

$$ = 628.3 \text{ ft/s } (191.4 \text{ m/s}) $$

$$ \overline{V}_1 = 550.0 \text{ ft/s } (170.0 \text{ m/s}) $$

$$ \alpha_1 = 0.0^\circ \ (0.0^\circ) $$

$$ W_{u1} = -U = -628.3 \text{ ft/s } (-191.4 \text{ m/s}) $$

$$ \beta = \tan^{-1}\left(\frac{-628.3}{550.0}\right) $$

$$ = -48.8^\circ \ (-48.4^\circ) $$

$$ W_1 = \sqrt{(550.0)^2 + (-628.3)^2} $$

$$ = 835.0 \text{ ft/s } (256.0 \text{ m/s}) $$

From Example Problem 7.2

$$ {}_1w_{1.5} = -9.70 \text{ Btu/lb } (-22.53 \text{ kJ/kg}) $$

$$ {}_1w_{1.5} = U(\overline{V}_{u1} - \overline{V}_{u1.5}) $$

Since $\overline{V}_{u1} = 0$ and $U = 628.3$ ft/s

$$ \overline{V}_{u1.5} = 386.4 \text{ ft/s } (117.7 \text{ m/s}) $$

$$ \overline{V}_{1.5} = \sqrt{(550.0)^2 + (386.4)^2} $$

$$ = 672.2 \text{ ft/s } (206.8 \text{ m/s}) $$

$$ \alpha_{1.5} = \tan^{-1}\left(\frac{386.4}{550.0}\right) $$

$$ = 35.1^\circ \ (34.7^\circ) $$

$$ W_{u1.5} = \overline{V}_{u1.5} - U $$

$$= 386.4 - 628.3$$

$$= -241.9 \text{ ft/s } (-73.7 \text{ m/s})$$

$$\beta_{1.5} = \tan^{-1}\left(\frac{-241.9}{550.0}\right)$$

$$= -23.7^\circ \ (-23.4^\circ)$$

$$\theta = \beta_{1.5} - \beta_1 = -23.7 - (-48.8^\circ) = 25.1^\circ \ (25.0^\circ)$$

$$W_{1.5} = \sqrt{(-241.9)^2 + (550.0)^2}$$

$$= 600.8 \text{ ft/s } (185.3 \text{ m/s})$$

From earlier calculations

$$T_{o1} = 544.2^\circ\text{R } (302.4 \text{ K})$$

$$T_{o1.5} = 584.6^\circ\text{R } (324.8 \text{ K})$$

$$\frac{p_{o1.5}}{p_{o1}} = 1.285 \ (1.285)$$

The method of calculating the flow coefficient, work coefficient, Mach numbers, and de Haller number are the same as illustrated in Example Problem 7.3. Therefore, only the calculated values are shown in Table 7.3.

Tip

$$U = \left(\frac{1}{2}\right)\left(\frac{26.0}{12}\right)(2\ \pi)\left(\frac{8000}{60}\right)$$

$$= 907.6 \text{ ft/s } (276.5 \text{ m/s})$$

The method for solving the balance of this problem is the same as for the hub. Therefore, only the calculated values are shown in Table 7.3.

Reviewing the values in Table 7.3, one observes that:

1. The turning angle, θ, varies from hub to tip with the maximum turning angle occurring at the hub (root) of the blade.
2. The maximum value for the absolute velocity occurs at the *root* of the blade at the *rotor outlet.*
3. The maximum value for the relative velocity occurs at the *tip* of the blade at the *rotor inlet.*
4. The flow coefficient varies from blade root to blade tip.
5. The rotor work coefficient varies from blade root to tip with the maximum value at the root of the blade.
6. The maximum absolute Mach number has its maximum value at the root of the blade at the rotor outlet.
7. The maximum relative Mach number has its maximum value at the tip of the blade at the rotor inlet.

TABLE 7.3. **Summary of Calculated Values at the Hub, Mean Blade Height Diameter and Tip for the Inlet Conditions Given in Example Problem 7.2**

Value	Hub		Mean		Tip	
$\overline{V}_1$, ft/s (m/s)	550.0	(170.0)	550.0	(170.0)	550.0	(170.0)
$\overline{V}_a$, ft/s (m/s)	550.0	(170.0)	550.0	(170.0)	550.0	(170.0)
$\overline{V}_{u1}$, ft/s (m/s)	0.0	(0.0)	0.0	(0.0)	0.0	(0.0)
U, ft/s (m/s)	628.3	(191.4)	767.9	(233.9)	907.6	(276.5)
α_1, degrees	0.0	(0.0)	0.0	(0.0)	0.0	(0.0)
W_1, ft/s (m/s)	835.0	(256.0)	944.6	(289.2)	1061.0	(324.5)
W_{u1}, ft/s (m/s)	−628.3	(−191.4)	−767.9	(−233.9)	−907.6	(−276.5)
β_1, degrees	−48.8	(−48.4)	−54.4	(−54.0)	−58.8	(−58.4)
$\overline{V}_{1.5}$, ft/s (m/s)	672.2	(206.8)	634.5	(195.4)	611.6	(188.5)
$\overline{V}_{u1.5}$, ft/s (m/s)	386.4	(117.7)	316.3	(96.3)	267.6	(81.5)
$\alpha_{1.5}$, degrees	35.1	(34.7)	29.9	(29.5)	25.9	(25.6)
$W_{1.5}$, ft/s (m/s)	600.8	(256.0)	711.7	(218.7)	843.8	(258.7)
$W_{u1.5}$, ft/s (m/s)	−241.9	(−73.7)	−451.6	(−137.6)	−640.0	(−195.0)
$\beta_{1.5}$, degrees	−23.7	(−23.4)	−39.4	(−39.0)	−49.3	(−48.9)
θ, degrees	25.1	(25.0)	15.0	(15.0)	9.5	(9.5)
${}_1W_2$, Btu/lb (kJ/kg)	−9.70	(−22.53)	−9.70	(−22.53)	−9.70	(−22.53)
T_{o1}, °R (K)	544.2	(302.4)	544.2	(302.4)	544.2	(302.4)
$T_{o1.5}$, °R (K)	584.6	(324.8)	584.6	(324.8)	584.6	(324.8)
$p_{o1.5}/p_{o1}$	1.285	(1.285)	1.285	(1.285)	1.285	(1.285)
ϕ	0.875	(0.888)	0.716	(0.727)	0.606	(0.615)
ψ	−1.23	(−1.23)	−0.823	(−0.822)	−0.59	(−0.59)
M_1	0.493	(0.500)	0.493	(0.500)	0.493	(0.500)
M_{r1}	0.748	(0.753)	0.846	(0.850)	0.950	(0.954)
$T_{1.5}$	547.0	(303.5)	551.5	(305.8)	553.5	(307.1)
$M_{1.5}$	0.587	(0.592)	0.551	(0.557)	0.530	(0.537)
$M_{r1.5}$	0.524	(0.531)	0.619	(0.624)	0.732	(0.736)
de Haller No.	0.719	(0.724)	0.754	(0.756)	0.795	(0.797)

8. The de Haller number varies from blade root to blade tip with the maximum value at the blade tip.

It was assumed that the blade height remained constant from blade root to tip. It is now important to determine if this does occur.

The blade height at the exit from the rotor can be determined from the fact that the mass rate of flow through the rotor is a constant. Therefore

$$\dot{m}_1 = \dot{m}_{1.5} = \rho_{1.5} A_{1.5} \overline{V}_{1.5} = \rho_1 A_1 \overline{V}_1$$

or, when using the ideal gas equation of state to eliminate the density, becomes

$$\frac{p_{1.5} \overline{V}_{1.5} A_{1.5}}{RT_{1.5}} = \frac{p_1 \overline{V}_1 A_1}{RT_1} \tag{7.18}$$

where the pressures and temperatures are the static values, the velocity, $\overline{V}$, is the velocity perpendicular to the direction of flow (axial velocity) and A is the annulus area.

Since the blade diameters at the hub and tip are known (see Example Problem 7.2), one can calculate the inlet area or express Eq. (7.18) in terms of the mean blade

TABLE 7.4. **Variation of Blade Height from Rotor Inlet to Stator Outlet for Example Problem 7.2.**

State	Blade Height in.	Blade Height (cm)
1.0	4.00	(10.15)
1.5	3.44	(8.74)
2.0	3.32	(8.41)

diameter and the blade height. Since the required velocities and static pressures and temperatures are known, one can calculate the annulus height (and therefore the blade height) at the rotor exit and stator inlet. In a similar manner, the blade height at the exit from the stator can be determined. Results of calculating the blade height from rotor inlet to stator outlet have been calculated and are tabulated in Table 7.4.

A computer program entitled ACOMPADU is included with this text. It will calculate the values discussed in Example Problems 7.2 and 7.3. Output for the solution for both English and SI units are shown in Tables 7.5 and 7.6, respectively. The reader is encouraged to become familiar with the ACOMPADU computer program.

7.12 DEGREE OF REACTION

A useful term for turbomachine designers is the degree of reaction. The degree of reaction for a compressor is defined as the ratio of the static enthalpy increase across the rotor to the increase in static enthalpy for the stage or, using the notation for the axial-flow stage illustrated in Figure 7.7,

$$R' = \frac{h_{1.5} - h_1}{h_2 - h_1} \tag{7.19}$$

For incompressible flow, Eq. (7.19) may be written as the ratio of static pressure rise in the rotor to static pressure increase for the stage, or

$$R' = \frac{p_{1.5} - p_1}{p_2 - p_1} \tag{7.20}$$

For the case where the axial velocity remains constant,

$$h_{1.5} - h_1 = \frac{W_1^2 - W_{1.5}^2}{2} \tag{7.21}$$

and

$$h_2 - h_1 = \frac{\overline{V}_1^2 - \overline{V}_2^2}{2} - {}_1w_2$$

TABLE 7.5. **ACOMPADU Computer Output for Example Problems 7.2 and 7.3 in English units**

Inlet temperature	=	519.0000	R
Inlet pressure	=	14.70000	psia
Inlet velocity	=	550.0000	ft/s
Blade turning angle	=	15.00000	deg
Mean blade diameter	=	22.00000	in.
Tip blade diameter at inlet	=	26.00000	in.
Hub blade diameter at inlet	=	18.00000	in.
Rotor speed	=	8000.000	rpm
No. of compressor stages	=	1.000000	
Mass flow rate	=	80.74029	lb/s
Overall compressor pressure ratio	=	1.285148	
Total power required	=	−1108.400	hp
Polytropic efficiency	=	1.000000	

State	U ft/s	V ft/s	Va ft/s	Vu ft/s	B deg	W ft/s	Wu ft/s	Alpha deg
1.0	767.9	550.0	550.0	0.0	−54.4	944.6	−767.9	0.0
1.5	0.0	634.5	550.0	316.3	−39.4	711.7	−451.6	29.9
2.0	0.0	550.0	550.0	0.0	0.0	0.0	0.0	0.0

State	Total P psia	Static P psia	Total Temp R	Static Temp R	Blade Height Inches
1.0	17.35	14.70	544.17	519.00	4.00
1.5	22.30	18.14	584.61	551.11	3.44
2.0	22.30	19.11	584.61	559.44	3.32

State	Mach No. Abs.	Mach No. Rel	Hub–Tip Ratio
1.0	0.493	0.846	0.692
1.5	0.551	0.618	0.729
2.0	0.474	0.000	0.738

Stage	Power hp	Torque ft-lbf	Work Btu/lb	Rotor Work coeff	Pressure ratio
1.0	−1108.4	−727.7	−9.70	−0.824	1.285

Stage	deHaller No.	Flow Coeff
1.0	0.753	0.716

But $\overline{V}_1 = \overline{V}_2$. Therefore

$$h_2 - h_1 = -\,{}_1w_2 = -\,{}_1w_{1.5} \tag{7.22a}$$

For the case where the flow radius does not change

$$h_2 - h_1 = -[U(\overline{V}_{u1} - \overline{V}_{u1.5})]$$

Thus, Eq. (7.19) becomes

$$R' = \frac{W_1^2 - W_{1.5}^2}{2U(V_{u1.5} - V_{u1})} \tag{7.23}$$

TABLE 7.6. **ACOMPADU Computer Output for Example Problems 7.2 and 7.3 in SI Units**

Inlet temperature	=	288.0000	K
Inlet pressure	=	101.3000	kPa
Inlet velocity	=	170.0000	m/s
Blade turning angle	=	15.00000	deg
Mean blade diameter	=	55.84988	cm.
Tip blade diameter at inlet	=	66.00000	cm.
Hub blade diameter at inlet	=	45.70000	cm.
Rotor speed	=	8000.000	rpm
No. of compressor stages	=	1.000000	
Mass flow rate	=	37.10527	kg/s
Overall compressor pressure ratio	=	1.284547	
Total power required	=	−835.9711	kW
Polytropic efficiency	=	1.000000	

State	U m/s	V m/s	Va m/s	Vu m/s	B deg	W m/s	Wu m/s	Alpha deg
1.0	233.9	170.0	170.0	0.0	−54.4	289.2	−233.9	0.0
1.5	0.0	195.4	170.0	96.3	−39.4	218.7	−137.6	29.5
2.0	0.0	170.0	170.0	0.0	0.0	0.0	0.0	0.0

State	Total P kPa	Static P kPa	Total Temp K	Static Temp K	Blade Height cm
1.0	120.14	101.30	302.38	288.00	10.15
1.5	154.33	124.98	324.81	305.81	8.74
2.0	154.33	131.71	324.81	310.43	8.41

State	Mach No. Abs.	Mach No. Rel	Hub–Tip Ratio
1.0	0.500	0.850	0.692
1.5	0.557	0.624	0.729
2.0	0.481	0.000	0.738

Stage	Power kW	Torque Newton-m	Work kJ/kg	Rotor Work coeff	Pressure ratio
1.0	−836.0	−997.9	−22.53	−0.823	1.285

Stage	deHaller No.	Flow Coeff
1.0	0.756	0.727

Since, referring to the diagram at the right

$$W_1^2 = \overline{V}_{a1}^2 + W_{u1}^2$$

$$W_{1.5}^2 = \overline{V}_{a1.5}^2 + W_{u1.5}^2$$

and

$$\overline{V}_{a1} = \overline{V}_{a1.5}$$

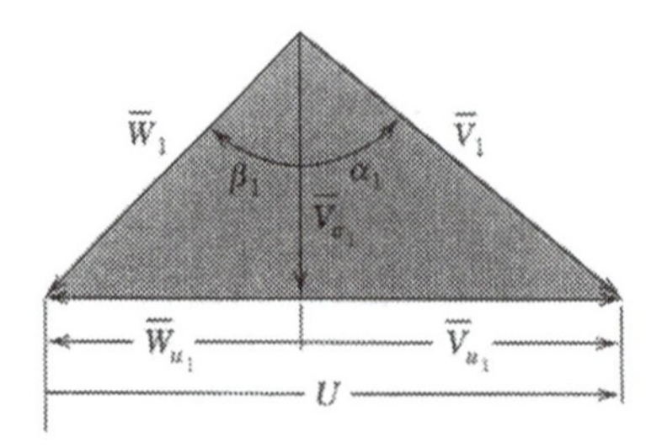

Eq. (7.22) becomes

$$R' = \frac{W_{u1}^2 - W_{u1.5}^2}{2U(\overline{V}_{u1.5} - \overline{V}_{u1})} \tag{7.24}$$

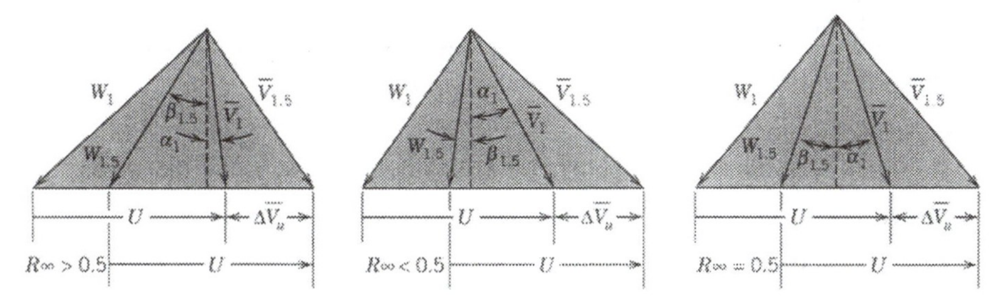

Figure 7.9. Velocity diagrams for three different degrees of reaction.

Since

$$\overline{V}_{u1} = U + W_{u1}$$

$$\overline{V}_{u1.5} = U + W_{u1.5}$$

Then

$$W_{u1} - W_{u1.5} = \overline{V}_{u1} - \overline{V}_{u1.5}$$

and Eq. (7.23) becomes

$$R' = \frac{W_{u1}^2 - W_{u1.5}^2}{2U(W_{u1.5} - W_{u1})} \tag{7.25a}$$

$$= \frac{-(W_{u1} + W_{u1.5})}{2U} \tag{7.25b}$$

Since $W_{u1} = V_{u1} - U$, Eq. (7.24b) becomes

$$R' = \frac{-(V_{u1} - U + W_{u1.5})}{2U}$$

$$= \frac{1}{2} - \left[\frac{V_{u1} + W_{u1.5}}{2U}\right] \tag{7.26a}$$

Since

$$W_{u1.5} = \overline{V}_a \tan \beta_{1.5}$$

$$\overline{V}_{u1} = V_a \tan \alpha_1$$

Eq. (7.25a) becomes

$$R' = \frac{1}{2} - \frac{\overline{V}_a}{2U}[\tan \beta_{1.5} + \tan \alpha_1] \tag{7.27}$$

Examination of Eq. (7.27) shows that the degree of reaction will be equal to 0.5 if $\tan \beta_{1.5} = -\tan \alpha_1$. When the degree of reaction is 0.5, the increase in static enthalpy (and therefore static pressure for incompressible flow) is evenly divided between the rotor and the stator. This will minimize the tendency of the blade boundary layer to separate.

Illustrated in Figure 7.9 are velocity diagrams for three different degrees of reaction. Note that when the degree of reaction is 0.5, the velocity diagram is symmetrical.

Example Problem 7.5

Calculate, for Example Problem 7.2, the degree of reaction:

(a) By using Eq. (7.27)
(b) By calculating the static enthalpy rise across both the rotor and the stator

Solution

(a) From Example Problem 7.2

$$\alpha_1 = 0°, \beta_{1.5} = -39.0°$$

Using Eq. (7.27)

$$R' = 0.500 - \frac{170.0}{(2)(233.9)}[\tan(-39.0°) + \tan(0°)]$$

$$= 0.794\ (0.794)$$

(b) $$h_{1.5} - h_1 = \frac{W_1^2 - W_{1.5}^2}{2} = \frac{(289.2)^2 - (218.7)^2}{(2)(1000)}$$

$$= 17.90 \text{ kJ/kg } (7.70 \text{ Btu/lb})$$

$$h_2 - h_{1.5} = \frac{\overline{V}_{1.5}^2 - \overline{V}_2^2}{2} = \frac{(195.3)^2 - (170.0)^2}{(2)(1000)}$$

$$= 4.62 \text{ kJ/kg } (2.00 \text{ Btu/lb})$$

$$R' = \frac{17.90}{17.90 + 4.62}$$

$$= 0.795\ (0.794)$$

Example Problem 7.6

Calculate, for Example Problem 7.4, the degree of reaction at the hub and tip of the blade.

Solution

From Example Problem 7.4

$$\text{Hub } \alpha_1 = 0°, \beta_{1.5} = -23.7°$$

$$\text{Tip } \alpha_1 = 0°, \beta_{1.5} = -49.3°$$

$$R'_{\text{hub}} = 0.500 - \frac{550.0}{(2)(628.3)}[\tan(-23.7°) + \tan(0°)]$$

$$= 0.692$$

$$R'_{\text{tip}} = 0.500 - \frac{550.0}{(2)(907.6)}[\tan(-49.3°) + \tan(0°)]$$

$$= 0.852$$

Example Problem 7.6 illustrates the fact that if the specific work is a constant from the root to the tip of the blade, the degree of reaction will vary greatly from the root to the tip of the blade.

Zero Reaction Designs
Compressors with a zero degree of reaction are normally not used in designing compressors. In fact, they should be avoided. They will be discussed in the next chapter when turbines are analyzed.

50% Reaction Designs
When the degree of reaction is 50%, the resulting velocity diagram is symmetrical, that is, $\alpha_1 = -\beta_{1.5}$ and $\alpha_{1.5} = -\beta_1$. This means that the rotor and stator have equal diffusion coefficients and that the static enthalpy change is evenly divided between the rotor and stator.

7.13 MULTISTAGE AXIAL-FLOW COMPRESSORS

The analysis so far has dealt with one stage of an axial-flow compressor. It should be noted that several stages are necessary to achieve the high pressure ratios, which exist in current axial-flow compressors.

Table 7.7 lists a few of the many gas turbines that use axial flow compressors. This table shows that the pressure ratios for the engines listed vary from 6.8 to 39.0.

It is quite common, in multistage compressors, for the blade shapes to be the same. This means that if all stages are on a common shaft, the specific work for each stage will be the same if the mean blade diameter remains constant and all stages have the same shape. It is further common to design so that the axial velocity remains constant.

One must remember that the work equation, Eq. (7.6b), applies only across one rotor, not multistages.

TABLE 7.7. **Selected Gas Turbine Engines with Axial-Flow Compressors (1)**

Manufacturer	Model No. or Designation	Number of Compressor Stages	Compressor Pressure Ratio at Maximum rpm
General Electric Co.	J79-GE-8	17	12.9
	J85-GE-4B	8	6.8
	T64-GE-7A	14	14.1
	CF6-50C	18	29.3
	GE-90B1	14	39.0
United Technologies, Pratt & Whitney Aircraft	TF30-P-3/P-103	16	17.1
	TF33-P-3	15	13.0
	F100-PW-100	13	24.8
	JT8D-11	13	16.2
	JT9D-7	15	22.3
	PW2040	17	27.6
Rolls Royce	RB211-524B2	14	28
	Spey MK.512-14	17	21
	Tyne R Ty-20 MK 801	15	13.95

Example Problem 7.7

Air at 14.70 psia (101.3 kPa) and 519°R (288 K) enters a three-stage axial-flow compressor with a velocity of 550 ft/s (170.0 m/s). There are no inlet guide vanes. Known information is as follows:

(a) Axial velocity remains constant through compressor
(b) Rotor turning angle 15°, each stage
(c) Tip blade diameter at inlet = 26.0 in. (66.0 cm)
(d) Root blade diameter at inlet = 18.0 in. (45.7 cm)
(e) Rotor speed, 8000 rpm
(f) Air enters each rotor in axial direction
(g) Air leaves each stator in axial direction

Calculate, based on calculations at the mean blade height and isentropic flow through the compressor:

(a) The stage and overall pressure ratio
(b) The power required to drive the compressor
(c) The blade height at the exit from each stator

Solution

One will note that the first stage of this three-stage compressor was solved in Example Problem 7.2. The conditions at the inlet to the second stage are the conditions at the exit from the stator of the first stage.

The ACOMPADU computer program was used to solve this problem for both English and SI units. The output from the program is shown in Tables 7.8 (English units) and 7.9 (SI units).

Reviewing the values in Table 7.8 (or 7.9), one observes that:

1. The stage pressure ratio has its maximum value in the first stage.
2. The relative and absolute velocities have their maximum values in the first stage.
3. The blade height decreases from inlet to exit.
4. The hub-to-tip ratio increases from inlet to exit.
5. The work, de Haller number and flow coefficient are the same for all three stages.

Example Problem 7.8

The JT8D-11 engine, which is listed in Table 7.7, has a 13-stage axial-flow compressor with a pressure ratio of 16.2. Calculate the *average* stage pressure ratio.

Solution

$$\text{avg. stage pressure ratio} = (16.2)^{1/13} = 1.239$$

Remember that for a multistage compressor, the stage pressure ratio will decrease from the first stage to the last stage if the turning angle is the same for all stages and the mean blade diameter remains constant. This is illustrated by the results for the three-stage axial-flow compressor, which are listed in Tables 7.8 and 7.9. For an *average* stage pressure ratio of 1.239, the first-stage pressure ratio will be much higher.

TABLE 7.8. **Results from ACOMPADU Computer Program for Example Problem 7.8 (English units)**

Inlet temperature	=	519.0000	R
Inlet pressure	=	14.70000	psia
Inlet velocity	=	550.0000	ft/s
Blade turning angle	=	15.00000	deg
Mean blade diameter	=	22.00000	in.
Tip blade diameter at inlet	=	26.00000	in.
Hub blade diameter at inlet	=	18.00000	in.
Rotor speed	=	8000.000	rpm
No. of compressor stages	=	3.000000	
Mass flow rate	=	80.74029	lb/s
Overall compressor pressure ratio	=	2.022572	
Total power required	=	−3325.201	hp
Polytropic efficiency	=	1.000000	

State	U ft/s	V ft/s	Va ft/s	Vu ft/s	B deg	W ft/s	Wu ft/s	Alpha deg
1.0	767.9	550.0	550.0	0.0	−54.4	944.6	−767.9	0.0
1.5	0.0	634.5	550.0	316.3	−39.4	711.7	−451.6	29.9
2.0	767.9	550.0	550.0	0.0	−54.4	944.6	−767.9	0.0
2.5	0.0	634.5	550.0	316.3	−39.4	711.7	−451.6	29.9
3.0	767.9	550.0	550.0	0.0	−54.4	944.6	−767.9	0.0
3.5	0.0	634.5	550.0	316.3	−39.4	711.7	−451.6	29.9
4.0	0.0	550.0	550.0	0.0	0.0	0.0	0.0	0.0

State	Total P psia	Static P psia	Total Temp R	Static Temp R	Blade Height Inches
1.0	17.35	14.70	544.17	519.00	4.00
1.5	22.30	18.14	584.61	551.11	3.44
2.0	22.30	19.11	584.61	559.44	3.32
2.5	28.18	23.24	625.05	591.55	2.88
3.0	28.18	24.40	625.05	599.87	2.79
3.5	35.09	29.29	665.48	631.98	2.44
4.0	35.09	30.66	665.48	640.31	2.37

State	Mach No. Abs.	Mach No. Rel	Hub–Tip Ratio
1.0	0.493	0.846	0.692
1.5	0.551	0.618	0.729
2.0	0.474	0.815	0.738
2.5	0.532	0.597	0.768
3.0	0.458	0.787	0.775
3.5	0.515	0.578	0.800
4.0	0.443	0.000	0.806

Stage	Power hp	Torque ft-lbf	Work Btu/lb	Rotor Work coeff	Pressure ratio
1.0	−1108.4	−727.7	−9.70	0.82	1.285
2.0	−1108.4	−727.7	−9.70	0.82	1.264
3.0	−1108.4	−727.7	−9.70	0.82	1.245

Stage	deHaller No.	Flow Coeff
1.0	0.753	0.716
2.0	0.753	0.716
3.0	0.753	0.716

TABLE 7.9. **Results from ACOMPADU Computer Program for Example Problem 7.8 (SI units)**

Inlet temperature	=	288.0000	K
Inlet pressure	=	101.3000	kPa
Inlet velocity	=	170.0000	m/s
Blade turning angle	=	15.00000	deg
Mean blade diameter	=	55.84988	cm.
Tip blade diameter at inlet	=	66.00000	cm.
Hub blade diameter at inlet	=	45.70000	cm.
Rotor speed	=	8000.000	rpm
No. of compressor stages	=	3.000000	
Mass flow rate	=	37.10527	kg/s
Overall compressor pressure ratio	=	2.020082	
Total power required	=	−2507.913	kW
Polytropic efficiency	=	1.000000	

State	U m/s	V m/s	Va m/s	Vu m/s	B deg	W m/s	Wu m/s	Alpha deg
1.0	233.9	170.0	170.0	0.0	−54.0	289.2	−233.9	0.0
1.5	0.0	195.4	170.0	96.3	−39.0	218.7	−137.6	29.5
2.0	233.9	170.0	170.0	0.0	−54.0	289.2	−233.9	0.0
2.5	0.0	195.4	170.0	96.3	−39.0	218.7	−137.6	29.5
3.0	233.9	170.0	170.0	0.0	−54.0	289.2	−233.9	0.0
3.5	0.0	195.4	170.0	96.3	−39.0	218.7	−137.6	29.5
4.0	0.0	170.0	170.0	0.0	0.0	0.0	0.0	0.0

State	Total P kPa	Static P kPa	Total Temp K	Static Temp K	Blade Height cm
1.0	120.14	101.30	302.38	288.00	10.15
1.5	154.33	124.98	324.81	305.81	8.74
2.0	154.33	131.71	324.81	310.43	8.41
2.5	194.95	160.11	347.23	328.24	7.32
3.0	195.95	168.13	347.23	332.85	7.07
3.5	242.69	201.78	369.66	350.66	6.20
4.0	242.69	211.23	369.66	355.28	6.01

State	Mach No. Abs.	Mach No. Rel	Hub–Tip Ratio
1.0	0.500	0.850	0.692
1.5	0.557	0.624	0.729
2.0	0.481	0.819	0.738
2.5	0.538	0.602	0.768
3.0	0.465	0.791	0.775
3.5	0.521	0.583	0.800
4.0	0.450	0.000	0.806

Stage	Power kW	Torque Newton-m	Work kJ/kg	Rotor Work coeff	Pressure ratio
1.0	−836.0	−997.9	−22.53	0.82	1.285
2.0	−836.0	−997.9	−22.53	0.82	1.263
3.0	−836.0	−997.9	−22.53	0.82	1.245

Stage	deHaller No.	Flow Coeff
1.0	0.756	0.727
2.0	0.756	0.727
3.0	0.756	0.727

A paper by Monhardt, et al. (3) discusses the design and testing of the FT8 low-pressure compressor. It has some excellent figures illustrating stage pressure ratio, rotor work coefficient, flow coefficient, reaction, and Mach number. The reader is encouraged to read this reference.

7.14 ACTUAL AXIAL-FLOW COMPRESSOR STAGE

The preceding sections on calculating the pressure rise for an axial-flow compressor assumed ideal flow, that is, no losses due to friction or boundary layer buildup and that the relative fluid velocity was at the same angle as the blade angles at the inlet to and exit from the blade.

The fluid velocity angle in an actual engine will not be identical to the blade angle. The angle of incidence i is the angle between the actual fluid direction at the inlet and the blade angle at the inlet. The deviation δ is the angle between the actual fluid direction at the exit and the blade angle at the exit from the blade. These two angles are shown in Figure 7.10.

Blade incidence in an actual compressor is rarely zero, and deviation is never zero. This occurs for many reasons, the most obvious being the fact that the rotor speed and mass rate of flow vary due to changing engine operating conditions. The blade camber and compressor solidity also influence the deviation angles. Each of these variations will change the velocity diagram of an axial-flow compressor stage.

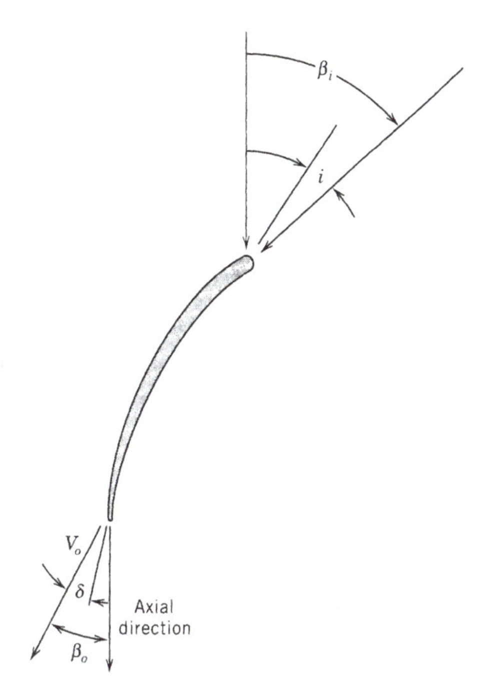

Figure 7.10. Flow diagram showing angle of incidence and deviation.

Overall compressor efficiency was defined in Section 5.2. Another compressor efficiency commonly defined is the infinitesimal or polytropic efficiency, sometimes referred to as the stage efficiency in an axial-flow compressor because of the small stage pressure ratio.

For an infinitesimal change for an adiabatic and reversible process, one may write

$$T\,ds = dh_i - v\,dp = 0 \tag{7.28}$$

The compressor polytropic efficiency, η_{p_c}, is defined by

$$\eta_{p_c} = \frac{dT_i}{dT_{\text{act}}} \tag{7.29}$$

Combining Eqs. (7.28), (7.29), (2.17), and (2.12) yields

$$\eta_{p_c} c_p \left(\frac{dT_{\text{act}}}{T}\right) = R\left(\frac{dp}{p}\right) \tag{7.30}$$

Integrating between states a and b for constant η_{p_c} and c_p yields

$$\eta_{p_c} c_p \ln\left(\frac{T_b}{T_a}\right) = R \ln\left(\frac{P_b}{P_a}\right) \tag{7.31a}$$

or

$$\frac{p_b}{p_a} = \left(\frac{T_b}{T_a}\right)^{\eta_{p_c} k/(k-1)} \tag{7.31b}$$

The overall adiabatic compressor efficiency is

$$\eta_C = \frac{w_{\text{ideal}}}{w_{\text{actual}}} = \frac{h_{o2i} - h_{o1}}{h_{o2a} - h_{o1}}$$

For constant specific heats,

$$\eta_C = \frac{T_{o2i} - T_{o1}}{T_{o2a} - T_{o1}} = \frac{(p_2/p_1)^{(k-1)/k} - 1}{(p_2/p_1)^{(k-1)/k\eta_p} - 1} \tag{7.32}$$

Example Problem 7.9

An axial-flow compressor with a pressure ratio of 10 has a polytropic compressor efficiency of 0.90. Calculate the adiabatic compressor efficiency assuming constant specific heats with $k = 1.40$.

Solution

From Eq. (7.32),

$$\eta_C = \frac{(10)^{0.4/1.4} - 1}{(10)^{0.4/(1.4)(0.9)} - 1} = 0.864$$

Figure 7.11 shows the variation in adiabatic compressor efficiency with increasing pressure ratio for several values of polytropic compressor efficiency. It illustrates that for a compressor with a pressure ratio of 25, the overall adiabatic compressor efficiency

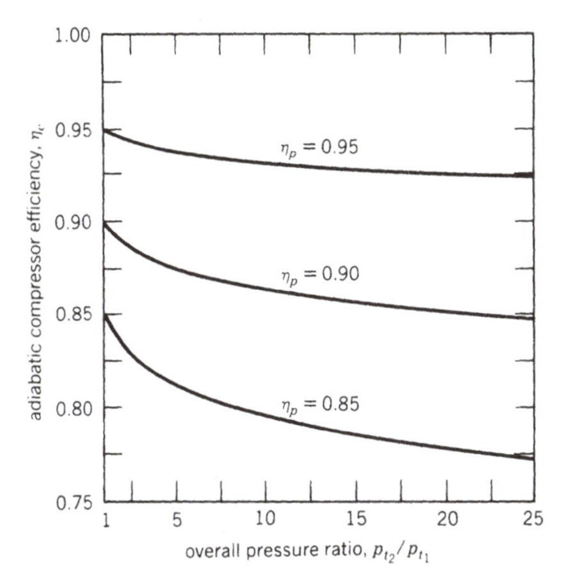

Figure 7.11. Variation of adiabatic compressor efficiency as a function of overall pressure ratio for several values of constant polytropic efficiency.

is 92.4% for a polytropic compressor efficiency of 95%, or 84.8% for a polytropic efficiency of 90%.

7.15. OFF-DESIGN PERFORMANCE OF MULTISTAGE AXIAL-FLOW COMPRESSORS

The discussion in the preceding sections has been for an axial-flow compressor stage. Axial-flow compressors, because of their relatively low stage pressure ratio, often require a significant number of stages. This was illustrated in Example Problem 7.7. Table 7.7 lists a few of the many gas turbine engines that use axial-flow compressors. This table shows that from 8 to 18 stages are required for compressor pressure ratios between 6.8 and 39. Calculations show that the average stage pressure ratios for the gas turbine engines listed in Table 7.7 range between 1.19 and 1.30.

As stated earlier, it would not be hard to design and develop an efficient axial-flow compressor if the compressor always operated at the same condition. This, of course, is not the situation, since a gas turbine engine must be started, ambient conditions vary, the desired output (and therefore operating point) will vary, and, for gas turbine engines used in aircraft, the forward velocity of the aircraft will vary and the pressure and temperature distribution at the inlet to the compressor may lead to inlet distortion. All of these variable conditions, along with others, may cause a compressor to operate in an inefficient manner, in a choked condition, or with one or more of the blades (or stages) stalled.

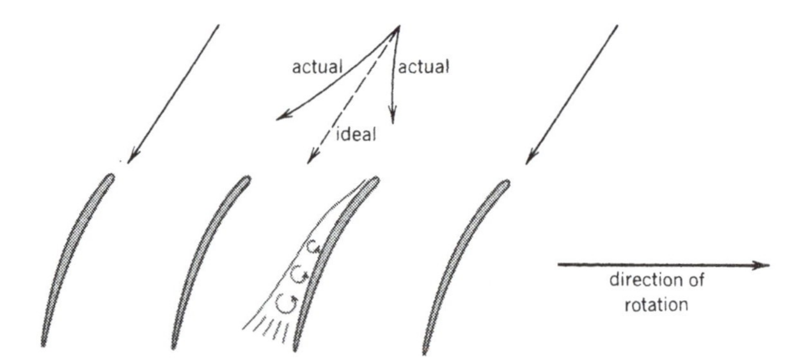

Figure 7.12. Stall.

Examination of Figure 7.2 shows that along a constant speed line, the pressure ratio decreases as the corrected mass rate of flow increases. Along most constant speed lines, the constant speed line becomes vertical. A vertical line implies that choking has occurred and the compressor has reached its maximum flow for this constant speed line. This condition can be explained by remembering that the compressor map represents the overall performance of a multistage compressor. A modern compressor is designed for high subsonic axial velocities through a compressor with fixed areas. When the overall pressure ratio decreases, the stage pressure ratio also decreases. With decreased stage pressure ratio, the density ratio across the stage will decrease. For an ideal stage this is illustrated by the relationship

$$\frac{p_{\text{out}}}{p_{\text{in}}} = \left(\frac{\rho_{\text{out}}}{\rho_{\text{in}}}\right)^k \tag{7.33}$$

This means that the axial velocity must increase, since

$$\dot{m} = \rho A\overline{V} = \text{constant}, \quad A = \text{fixed} = \text{constant} \tag{7.34}$$

thereby leading to choking in the latter stages of the compressor.

Distortion is the uneven distribution of either pressure, temperature, or velocity at the inlet to a compressor. Distortion often occurs because of the installation. This can occur when the inlet duct to the engine has a major bend a short distance before the inlet to the compressor or the inlet to the engine is in the root of the wing.

Stall, possibly caused by distortion or operating at an off-design condition, results in reduced flow between blades as shown in Figure 7.12. This reduced flow (blockage) causes the fluid to be diverted as shown in Figure 7.12. This means that the incidence angle of the blade in the direction opposite to the direction of rotation will be increased. If it is increased to a sufficient value, flow separation will occur and the stall will rotate, moving from the pressure side of the blade to the suction side. This means that the stall rotates in a direction opposite to the direction of rotation of the rotor. This situation can produce compressor vibration, which may be detrimental to the compressor if allowed to continue.

Several techniques are commonly used to improve off-design performance. In one case, the first few stator rows may be reset to change the stator inlet angle. This will reduce the tendency to stall since, when the blade angle is reset, flow will be entering the blade with a relative angle more nearly the same as the blade "metal" angle.

Example Problem 7.10

Assume, for Example Problem 7.2, that the velocity at the inlet to the stage is 180.0 m/s (590 ft/s) instead of 170.0 m/s (550 ft/s), all other conditions being the same. Calculate how many degrees the blade would have to be reset so that the direction of the relative velocity is the same as the blade angle at the mean diameter.

Solution

$$\beta_1 = \tan^{-1}\left(\frac{-233.9}{180.0}\right) = -\ 52.4^\circ\ (-52.4^\circ)$$

This shows the blade "metal" angle would have to be changed by 1.6° (2.0°).

A second method used to reduce the tendency of blades to stall is to use two-spool operation. In two-spool engines, the compression (and expansion) process is accomplished in two separate compressors, as shown in Figure 7.13.

Each spool is able to operate at its best rotor speed. Remember, in a single-spool engine, all turbine and compressor stages must operate at the same rotor speed, whereas in a two-spool engine, the two spools normally will operate at different rotor speeds. The RB211 gas turbine engine is a three-spool engine, the fan operating at one speed, the low-pressure compressor (LPC) and high-pressure compressor (HPC) each operating at its own rotor speed.

A third technique employed to reduce the tendency to stall is to use interstage bleed. In this situation, bleed ports are located between blade rows or at the exit from the compressor. As an example, a bleed port might be located between the LPC and HPC in Figure 7.13. When the bleed port is open, some of the air that has been compressed up to this point is removed from the compressor. Referring to Eq. (7.34), when the mass rate of flow is reduced, the velocity into the next stage, and therefore

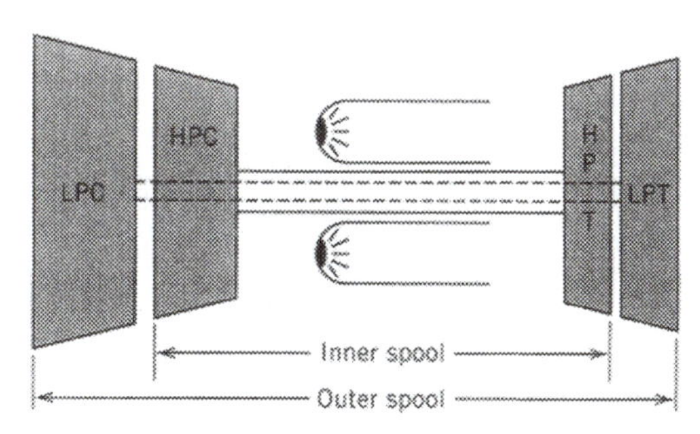

Figure 7.13. Two-spool gas turbine.

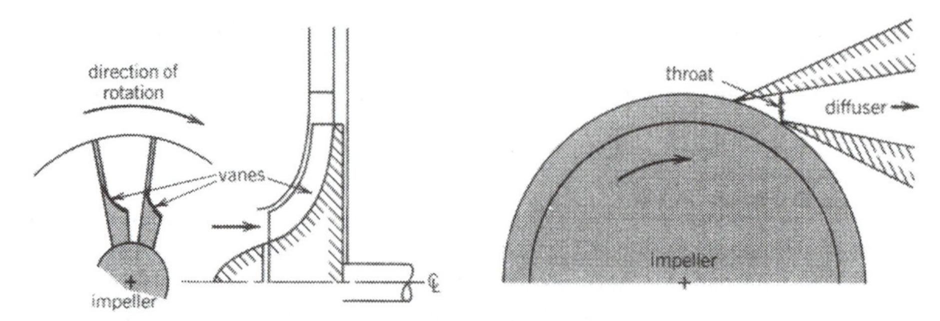

Figure 7.14. Schematic diagram of a centrifugal-flow compressor.

the relative velocity and angle, are changed, bringing the angle of the relative velocity closer to that of the blade, thereby reducing the tendency to choke or stall.

7.16 CENTRIFUGAL-FLOW COMPRESSORS

The preceding sections have been concerned mainly with axial-flow compressors. As stated earlier, many compressors have flow entering the compressor in an axial direction but leaving in the radial direction. These are called centrifugal-flow compressors.

Side and front views of a typical centrifugal-flow compressor are shown in Figure 7.14.

The two basic components of a centrifugal-flow compressor are the impeller (rotor), where work is done on the working fluid, and the diffuser (stator), where the velocity is decreased and the static pressure increases.

The inlet to the impeller is normally referred to as the eye or inducer as illustrated in Figure 7.15.

If there are no prewhirl vanes in front of the impeller, the fluid flow into the impeller will be in the axial direction. This means that the impeller blades must be curved in the direction of rotation for smooth entry, the amount of curvature (twist) varying from hub to tip at the inlet and depending on the axial velocity and the impeller rotational speed.

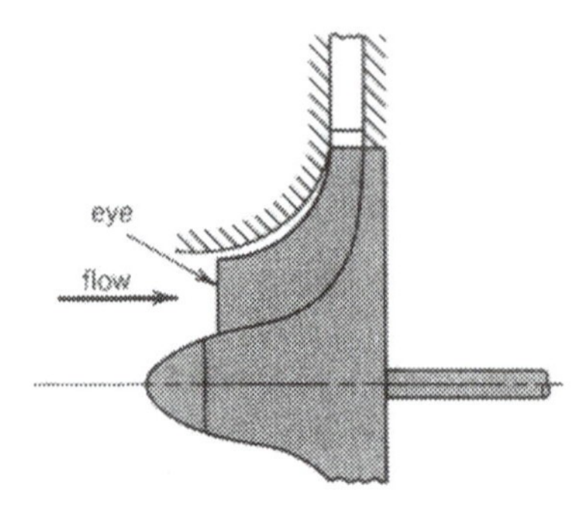

Figure 7.15. Schematic diagram of a centrifugal-flow compressor.

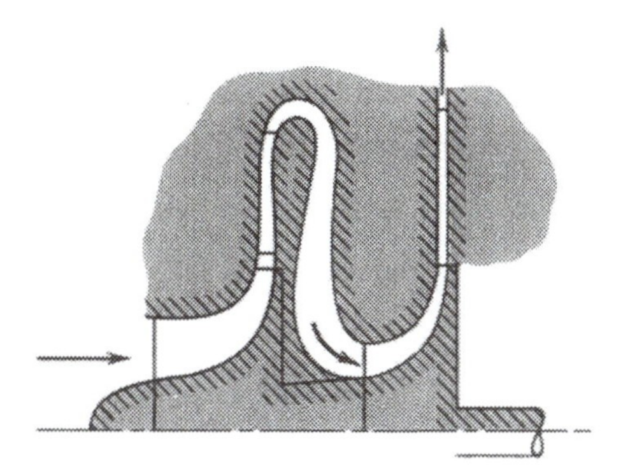

Figure 7.16. Two-stage centrifugal-flow compressor.

Sometimes it is desirable to have a higher pressure ratio than can be achieved in a single-stage centrifugal-flow compressor. One way to achieve a higher overall pressure ratio is to use two centrifugal-flow compressors. This is shown schematically in Figure 7.16.

One should observe the amount of turning (return or crossover) that must be done to direct the flow from the outlet of the diffuser of the first compressor into the impeller of the second compressor.

Example Problem 7.11

The mass rate of flow, at 519°R (288 K) and 14.70 psia (101.3 kPa), at the inlet to the impeller of a centrifugal-flow compressor is 4.0 lb/s (1.814 kg/s). The inlet flow is in the axial direction, the impeller eye has a minimum diameter of 1.5 in. (3.81 cm) and a maximum diameter of 5 in. (12.7 cm), and rotates at 35,000 rpm. Assuming no blockage due to the blades, calculate the ideal blade angle at the hub and the tip at the inlet to the impeller.

Solution

$$A_{\text{inlet}} = \frac{\pi}{4}\left[\left(\frac{5}{12}\right)^2 - \left(\frac{1.5}{12}\right)^2\right] = 0.124 \text{ ft}^2 \ (0.0115 \text{ m}^2)$$

$$\overline{V}_{\text{avg,inlet}} = \frac{\dot{m}}{\rho A} = \frac{(4.0)(53.3)(519)}{(0.124)(14.7)(144)}$$

$$= 421.3 \text{ ft/s } (128.4 \text{ m/s})$$

$$U_{\text{hub}} = r\omega = \frac{(1.5)(35{,}000)(2\pi)}{(2)(12)(60)}$$

$$= 229.1 \text{ ft/s } (69.8 \text{ m/s})$$

$$U_{\text{tip}} = \frac{(5)(35{,}000)(2\pi)}{(2)(12)(60)} = 763.6 \text{ ft/s } (232.7$$

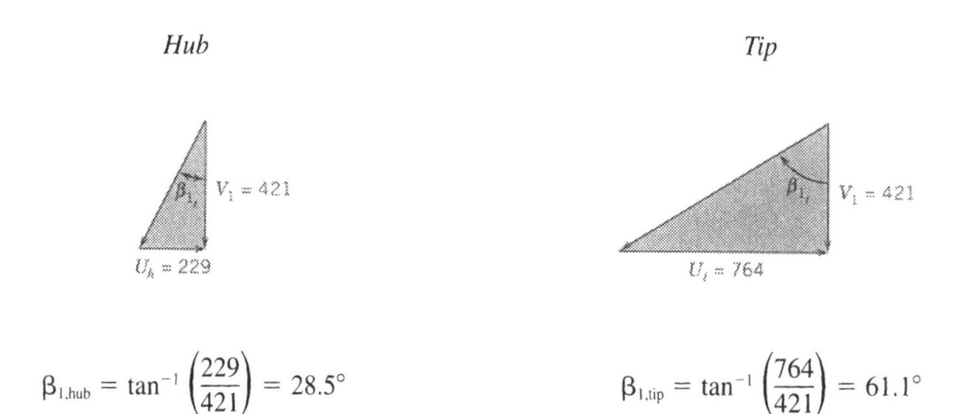

$$\beta_{1,\text{hub}} = \tan^{-1}\left(\frac{229}{421}\right) = 28.5°$$

$$\beta_{1,\text{tip}} = \tan^{-1}\left(\frac{764}{421}\right) = 61.1°$$

The above example illustrates the amount of twist between the hub and tip at the inlet to the impeller. The hub-to-tip ratio of a centrifugal-flow compressor is much lower than that of an axial-flow compressor. Factors that influence the hub-to-tip ratio are

1. Desired mass rate of flow
2. Maximum absolute inlet velocity
3. Maximum inlet diameter (keep as small as possible to keep overall compressor diameter to a minimum)

The work done on the air, and therefore the increase in pressure across the rotor, depends on the change in whirl (tangential) velocity. Impellers may be designed with backward-curved, radial-flow, or forward-curved exit blade angles. The forward-facing blades yield the highest increases in pressure but also lead to the highest absolute velocities at the outlet from the impeller and are rarely used. Backward, radial, and forward-facing blade situations are illustrated in Figure 7.17.

It must be remembered that the diffuser entrance angle must be at approximately the same angle as the absolute flow direction of the fluid as it leaves the rotor. Also, impeller tip velocities normally are limited to 2000 to 2200 ft/s tip velocities due to material stress.

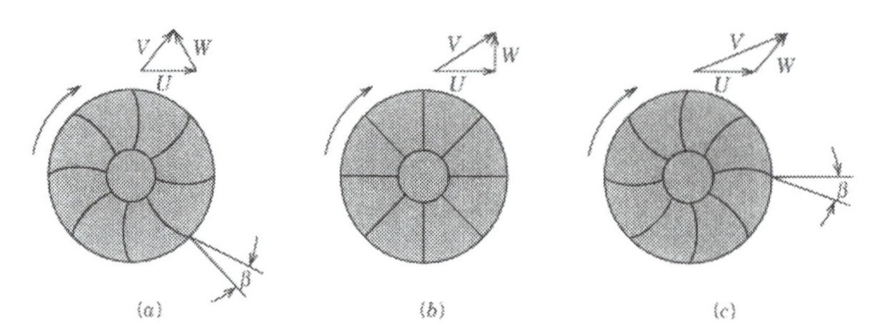

Figure 7.17. Schematic drawing for three impeller exit angles. (a) Backward-facing. (b) Radial-flow. (c) Forward-facing.

Example Problem 7.12

The impeller described in Example Problem 7.11 has the following exit conditions:

max. diam. = 8.0 in. (20.3 cm)

radial velocity = 400 ft/s (121.9 m/s)

Exit blade angle is in the radial direction ($\beta_{1.5} = 0$). Calculate:

(a) Exit velocity diagram
(b) Work done on air, Btu/lb
(c) Exit Mach number
(d) Exit static and stagnation pressures
(e) Space width at impeller exit assuming no blade blockage

Solution

$$U_{1.5} = \left(\frac{8}{2}\right)\left(\frac{35{,}000}{12}\right)\left(\frac{2\pi}{60}\right) = 1122 \text{ ft/s (372.4 m/s)}$$

$$\overline{V}_{\text{radical}} = W_{1.5} = 400 \text{ ft/s (121.9 m/s)}$$

(a) Exit velocity diagram:

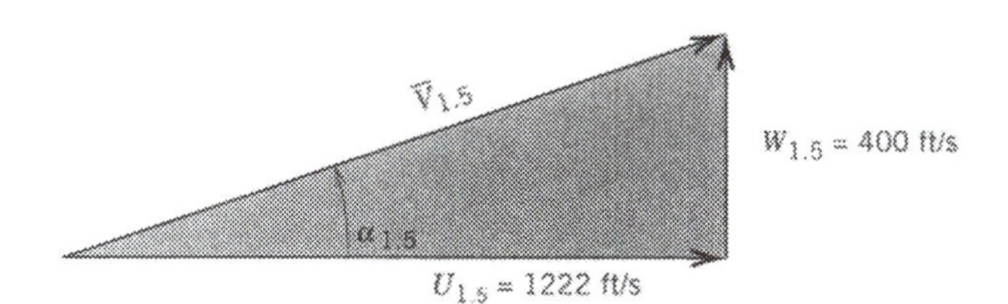

$$\overline{V}_{1.5} = \sqrt{(1222)^2 + (400)^2}$$

$$= 1286 \text{ ft/s (391.8 m/s)}$$

$$\alpha_{1.5} = \tan^{-1}\left(\frac{400}{1222}\right) = 18.1°$$

$$\overline{V}_{u1.5} = U_{1.5} = 1222 \text{ ft/s (372.4 m/s)}$$

(b) $${}_1w_{1.5} = \frac{U_1\overline{V}_{u1} - U_{1.5}\overline{V}_{u1.5}}{g_c}$$

$$= \frac{0 - (1222)(1222)}{(32.174)(778)} = -59.63 \text{ Btu/lb (138.7 kJ/kg)}$$

(c) From the first law, assuming constant specific heats,

$$T_{1.5} = T_1 + \frac{\overline{V}_1^2 - \overline{V}_{1.5}^2}{2g_c c_p} - \frac{{}_1w_{1.5}}{c_p}$$

$$= 519 + \frac{(421.3)^2 - (1286)^2}{(2)(32.174)(778)(0.24)} - \frac{-59.63}{0.24}$$

$$= 644.4°\text{R } (358 \text{ K})$$

$$c_{1.5} = \sqrt{(1.4)(32.174)(53.35)(644.6)}$$

$$= 1244 \text{ ft/s } (379 \text{ m/s})$$

$$M_{1.5} = \frac{1286}{1244} = 1.03$$

$$M_{r1.5} = \frac{400}{1244} = 0.32$$

(d) Assuming no losses in the impeller,

$$\frac{p_{1.5}}{p_1} = \left(\frac{T_{1.5}}{T_1}\right)^{k/(k-1)} = \left(\frac{644}{519}\right)^{1.4/0.4} = 2.14$$

$$p_{1.5} = (2.14)(14.70) = 31.46 \text{ psia } (216.8 \text{ kPa})$$

$$T_{o1.5} = T_{1.5} + \frac{\overline{V}_{1.5}^2}{2g_c c_p}$$

$$= 644 + \frac{(1286)^2}{(0.24)(2)(32.174)(778)}$$

$$= 782°\text{R } (434 \text{ K})$$

$$\frac{p_{o1.5}}{p_1} = \left(\frac{782}{519}\right)^{3.5} = 4.21$$

$$p_{o1.5} = (4.21)(14.70)$$

$$= 61.9 \text{ psia } (426.5 \text{ kPa})$$

$$\frac{p_{o1.5}}{p_{1.5}} = \frac{4.21}{2.14} = 1.97$$

(e) The space width x at the exit from the impeller can be determined from the continuity equation

$$\dot{m} = \frac{A\overline{V}p}{RT}$$

where A is the exit area, $\overline{V}$ is the absolute velocity in the radial direction, p is the static pressure, and T the static temperature at the impeller exit.

From above

$$p_{1.5} = 31.46 \text{ psia } (216.8 \text{ kPa})$$

$$T_{1.5} = 664.4°\text{R } (358 \text{ K})$$

$$\overline{V}_{r1.5} = 400 \text{ ft/s } (121.9 \text{ m/s})$$

$$D_{1.5} = 8.0 \text{ in } (20.3 \text{ cm})$$

$$\dot{m} = 4.0 \text{ lb/s } (1.814 \text{ kg/s})$$

$$A_{1.5} = \frac{(4.0)(53.35)(644.4)}{(400)(31.46)} = 10.93 \text{ in}^2$$

$$x = \frac{10.93}{8.0\pi} = 0.44 \text{ in } (1.11 \text{ cm})$$

The previous discussion has dealt with the impeller only. One should note that in Example Problem 7.12, the Mach number of the exit absolute velocity was greater than 1.0 and was at an angle of 71.9° with the radial direction.

The fluid next enters the diffuser, where the velocity is decreased and the static pressure increases. Most diffusers, to avoid separation, have a half-divergence angle of 4°–6° with a constant thickness. This is shown schematically in Figure 7.18.

There is normally a small space between the impeller outlet and the diffuser inlet. In this region, since the area is increasing, the velocity will decrease. If friction is neglected, the angular momentum will be constant, or

$$r_{1.5}\overline{V}_{u1.5} = r_{1.7}\overline{V}_{u1.7}$$

where $r_{1.5}$ and $r_{1.7}$ are the radii at the impeller outlet and diffuser inlet, respectively. Remember, the radial velocity will decrease since the area is increasing.

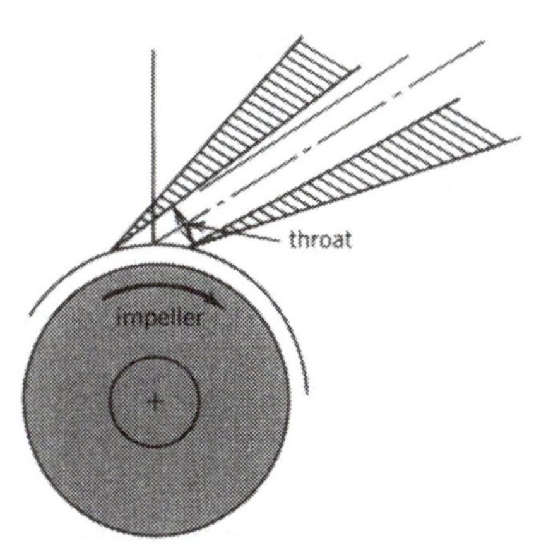

Figure 7.18. Impeller-diffuser flow path for a centrifugal-flow compressor.

TABLE 7.10. **Several Gas Turbine Engines That Use Centrifugal-Flow Compressors (1)**

Manufacturer Model	No. of Stages	Pressure Ratio
Allied-Signal Garrett Engine Div.		
T76G10, 12, 420, 421	2	8.5
TPE331-8/9	2	10.3
TPF351-20	2	13.2
General Motors Allison Gas Turbine Div.		
T703-A700	1	8.6
250-C30M	1	8.6
Teledyne CAE		
J69-T-25A	1	3.9
304	1	5.5
312-1	1	5.7
Williams International Corp.		
WR2-6	1	4.2
Pratt & Whitney Canada, Inc.		
PW119, 119A	2	11.8
PW123	2	14.4

7.17 AXIAL-CENTRIFUGAL COMPRESSORS

Table 7.10 illustrates the fact that most gas turbine engines that have been built with centrifugal compressors use only one stage and the pressure ratio for most single-stage centrifugal-flow compressors is approximately 4.0 or lower.

In many engines, it is desirable to have a higher maximum pressure ratio. This was illustrated in Chapter 5. One way to accomplish a higher overall pressure ratio is to use one or more axial-flow stages in front of a centrifugal-flow compressor.

Example Problem 7.13

A centrifugal-flow compressor has a pressure ratio of 4.0. How many axial-flow stages, each with a pressure ratio of 1.21, would be required to raise the overall pressure ratio to 7.0?

Solution

Let x = number of stages. Then

$$(1.21)^x(4.0) = 7.0$$

$$x = 3$$

Table 7.11 lists several gas turbine engines that have one or more axial-flow stages followed by a centrifugal-flow compressor. One should compare the pressure ratios listed in Table 7.10, which are for engines with only a centrifugal-flow compressor, with those for engines listed in Table 7.11. Note those in Table 7.11 have one to six axial-flow stages followed by a centrifugal-flow compressor.

It is usually necessary to include some form of variable geometry in an axial-

TABLE 7.11. **Several Gas Turbine Engines That Use Axial-Centrifugal Flow Compressors (1)**

Manufacturer Model	No. of Stages		Pressure Ratio
	Axial	Cent.	
General Motors Allison Gas Turbine Div.			
250-B17C	6	1	7.2
250-B17F, F1, F2	4	1	7.9
Teledyne CAE			
J69-T-29	1	1	5.3
J402-CA-702	2	1	8.5
312-2	1	1	8.1
Textron Lycoming Stratford Div.			
LTP101-700A-1A	1	1	8.6
GLC 38	5	1	22.
Pratt & Whitney Canada, Inc.			
PT 6A-15AG	3	1	6.5
PT 6A-42	3	1	8.0
PT 6A-65AG	4	1	9.7

centrifugal flow compressor to provide optimum matching between the axial-flow stages and the centrifugal-flow stage. This variable geometry may include:

1. Variable-angle stators
2. Interstage bleed ports

Including variable geometry does increase the complexity, cost, and reliability of the compressor.

REFERENCES CITED

1. *Aviation Week & Space Technology*, March 16, 1992.
2. Brandt, D. E., *Gas Turbine Design Philosophy*, GER-3434A, 35th GE Turbine State-of-the-Art Technology Seminar.
3. Monhardt, R. J., Richardson, J. H. and Boettcher, J. M., Design and Testing of the FT8 Gas Turbine Low-Pressure Compressor, *Journal of Turbomachinery*, ASME, New York, NY, April 1990.

SUGGESTED READING

Dixon, S. L., *Fluid Mechanics, Thermodynamics of Turbomachinery*, 3rd ed., Pergamon Press, Elmsford, N.Y., 1978.

Horlock, J.H., *Axial Compressors, Fluid Mechanics and Thermodynamics*, R. E. Krieger Publishing Co., Huntington, N.Y., 1973.

Shepherd, D. G., *Principles of Turbomachinery*, Macmillan Co., New York, 1956.

NOMENCLATURE

A	area
c	sonic velocity, chord
c_p	constant pressure specific heat
D	diameter, characteristic dimension
h	enthalpy
i	angle of incidence
k	specific heat ratio
$\dot{m}$	mass flow rate
M	Mach number, molecular weight
N	rotor rotational speed
p	pressure
$\dot{P}$	power
Pr	relative pressure
r	radius
R	gas constant
R'	degree of reaction
T	temperature, torque
U	blade velocity
$\overline{V}$	absolute velocity
w	work
W	relative velocity
α	angle (absolute velocity)
β	angle (relative velocity)
δ	pressure ratio, deviation angle
η	efficiency
γ	blade stagger angle
μ	absolute viscosity
ω	angular velocity
ϕ	flow coefficient
Ψ	work coefficient
σ	solidity
ρ	density
θ	temperature ratio, turning stagger angle

Subscripts

1, 1.5, 2	states
a	axial
amb	ambient
C	compressor
h	hub (root)
o	stagnation
p	polytropic
r	relative
t	tip
u	tangential component

PROBLEMS

7.1. The data used in constructing Figure 6.13 are tabulated below.

Flight Velocity ft/s	Sea Level	18,000 ft	36,089 ft
0	88.2	—	—
100	—	90.5	—
200	82.3	—	92.8
300	—	85.0	—
400	77.0	—	87.8
600	72.0	77.8	83.4
800	67.4	73.4	79.4
950	—	—	76.4
1000	62.8	69.1	—

These values were calculated assuming a mass rate of flow of 1 lb/s entering the compressor for all conditions.
Assuming that the corrected mass rate of flow $\dot{m}\sqrt{\theta_o}/\delta_o$ is the same for all

cases, calculate the thrust developed for the above flight conditions. Then plot thrust versus flight speed for the three altitudes and compare with Figures 6.13 and 7.2.

7.2. The corrected mass rate of flow $\dot{m}\sqrt{\theta_o}/\delta_o$ at the inlet to a compressor of a gas turbine engine is 100 lb/s (45 kg/s). Determine the actual mass rate of flow at the compressor inlet:

(a) For a hot day, sea level static

(b) At an altitude of 30,000 ft (9000 km), Mach number of 0.85

7.3. Solve Example Problem 7.2 assuming that the air enters the compressor with a velocity of 650 ft/s (200 m/s), all other values the same. Verify results using ACOMPADU computer program.

7.4. Solve Example Problem 7.2 assuming that the air enters the compressor with a velocity of 150 m/s (500 ft/s), all other values the same. Verify results using ACOMPADU computer program.

7.5. Solve Example Problem 7.2 assuming that the air is turned through 20.0°, all other values the same. Compare with results listed in Table 7.1.

7.6. Solve Example Problem 7.2 assuming that the air is turned through 10.0°, all other values being the same. Compare with results listed in Table 7.1.

7.7. Air at 14.70 psia (101.3 kPa) and 519°R (288 K) enters a seven-stage axial-flow compressor with a velocity of 500 ft/s (150 m/s). There are no inlet guide vanes. The first rotor stage has a tip diameter of 22.0 in. (56.0 cm), a hub (root) diameter of 14.0 in. (35.6 cm), and rotates at 8000 rpm. The air enters each rotor and leaves each stator in the axial direction with no change in axial velocity. All stages have the same shape, and the mean blade diameter is constant. The air is turned through 12.0° as it passes through each rotor stage. Assuming constant specific heats with $k = 1.40$ and that the air enters and leaves the blades at the blade angles:

(a) Construct the velocity diagrams at the mean blade height for the first stage

(b) Determine the shape of the rotor and stator

(c) Calculate the mass rate of flow into the compressor

(d) Calculate the ideal total-to-total pressure ratio for each stage and the overall pressure ratio

(e) Calculate the total power required to drive this compressor

7.8. Solve Problem 7.7 assuming the rotor turning angle is 15.0°, all other values the same.

7.9. Air at 101.3 kPa (14.70 psia) and 288 K (519°R) enters a three-stage axial-flow compressor with a velocity of 170 m/s (550 ft/s). There are no inlet guide vanes. Known information is:

(a) Axial velocity remains constant through compressor

(b) Rotor turning angle 15.0°, each stage

(c) Blade height, inlet, first rotor stage, 8.9 cm (3.5 in)

(d) Mean blade diameter, 50.8 cm (20.0 in.)

(e) Rotor speed, 9900 rpm

Calculate, based on calculations at the mean blade diameter and assuming that the air enters each rotor in the axial direction, constant specific heats, and that the air enters and leaves the blades at the blade angles and a compressor polytropic efficiency of 1.00:

(a) The stage and overall compressor pressure ratio

(b) The power required to drive the compressor

(c) The blade height at the exit from the third stator stage

7.10. Solve Problem 7.9 assuming the inlet velocity is 600 ft/s (183 m/s), all other values the same.

7.11. Determine, for Problem 7.7

(a) The flow coefficient at the inlet to the compressor

(b) The hub-to-tip ratio at the inlet to the compressor

(c) The absolute and relative Mach numbers at the inlet to the first-stage rotor

(d) The work coefficient

(e) The absolute and relative Mach numbers at the exit from the first-stage rotor

7.12. Determine, for Problem 7.8

(a) The flow coefficient at the inlet to the compressor

(b) The hub-to-tip ratio at the inlet to the compressor

(c) The absolute and relative Mach numbers at the inlet to the first-stage rotor

(d) The work coefficient

(e) The absolute and relative Mach numbers at the exit from the first-stage rotor

7.13. Determine, for Problem 7.9

(a) The flow coefficient at the inlet to the compressor

(b) The hub-to-tip ratio at the inlet to the compressor

(c) The absolute and relative Mach numbers at the inlet to the first-stage rotor

(d) The work coefficient

(e) The absolute and relative Mach numbers at the exit from the first-stage rotor

7.14. Solve Problem 7.7 and 7.11 at the hub (root) and tip of the blade assuming that the specific work is the same at all radii and that the blade height does not change from the inlet to the rotor to the exit from the stator.

7.15. Solve Problem 7.9 and 7.13 at the hub (root) and tip of the blade assuming that the specific work is the same at all radii and that the blade height does not change from the inlet to the rotor to the exit from the stator.

7.16. Calculate, for Problem 7.7, the degree of reaction.

7.17. Calculate, for Problem 7.9, the degree of reaction.

7.18. Determine, for Problem 7.7 (7.14), the degree of reaction at the hub and tip of the first stage.

7.19. Determine, for Problem 7.9 (7.15), the degree of reaction at the hub and tip of the first stage.

7.20. Verify, by calculations, the blade heights as listed in Table 7.4.

7.21. Determine, using ACOMPADU computer program, the blade heights at the exit from each rotor and stator for Problem 7.7.

7.22. Determine, using the ACOMPADU computer program, the blade heights at the exit from each rotor and stator for Problem 7.9.

7.23. Air at 14.70 psia (101.3 kPa), 519°R (288 K) enters a three-stage axial-flow compressor with a velocity of 550 ft/s (170 m/s). There are no inlet guide vanes. Known information is:

(a) Axial velocity remains constant through compressor

(b) Rotor turning angle is 14.0°, each stage

(c) Blade height, inlet, first rotor stage, 3.5 in. (8.9 cm)

(d) Mean blade diameter, 20.0 in. (50.8 cm)

(e) Rotor speed, 9900 rpm

Calculate, based on calculations at the mean blade diameter and assuming that the air enters each rotor in the axial direction, constant specific heats, and that the air enters and leaves the blades at the blade angles and a compressor polytropic efficiency of 0.90:

(a) The stage and overall compressor pressure ratio

(b) The power required to drive the compressor

(c) The blade height at the exit from the third stator stage

7.24. Solve Problem 7.7 assuming a compressor polytropic efficiency of 0.90.

7.25. Solve Problem 7.9 assuming a compressor polytropic efficiency of 0.90.

7.26. Calculate, for Example Problem 7.2, the blade turning angle if the degree of reaction for the stage is 0.50.

7.27. Is it possible to have a degree of reaction greater than 1.0?

7.28. Solve Example Problem 7.11 assuming that the impeller blade exit angle is facing 15.0° in the backward direction.

7.29. Solve Example Problem 7.11 assuming that the impeller blade exit angle is facing 15.0° in the forward direction.

7.30. A single-stage centrifugal-flow compressor operates under the following conditions:

Inlet temperature	288 K (519°R)
Inlet pressure	101.3 kPa (14.70 psia)
Inducer tip diameter	14.0 cm (5.50 in.)
Inducer hub diameter	4.0 cm (1.57 in.)
Impeller tip diameter	21.0 cm (8.27 in.)
Impeller tip velocity	450 m/s (1475 ft/s)
Diffuser leading edge diameter	25.0 cm (9.84 in.)
Air flow rate into compressor	2.0 kg/s (4.4 lb/s)
Impeller blade exit angle	0°
Impeller exit radial velocity	125 m/s (410 ft/s)

Calculate, assuming constant specific heats and that the air enters the inducer in the axial direction:

(a) The blade angle at the hub and tip of the inducer

(b) The Mach number at the inlet and exit to the impeller for both the absolute and relative velocities

(c) The static pressure leaving the impeller

(d) The total pressure leaving the impeller

(e) The power required to drive the compressor

(f) The Mach number and optimum angle for the flow at the inlet to the diffuser

7.31. A single-stage centrifugal-flow compressor operates under the following conditions:

Inlet temperature	519°R (288 K)
Inlet pressure	14.70 psia (101.3 kPa)
Inducer tip diameter	3.40 in. (8.64 cm)
Inducer hub diameter	1.45 in. (3.68 cm)
Impeller tip diameter	5.10 in. (13.0 cm)
Impeller tip velocity	1800 ft/s (550 m/s)
Diffuser leading edge diameter	5.70 in. (14.5 cm)
Air flow rate into compressor	2.0 lb/s (0.90 kg/s)
Impeller blade exit angle (backward)	40°
Impeller exit radial velocity	507 ft/s (155 m/s)

Calculate, assuming constant specific heats and that the air enters the inducer in the axial direction:

(a) The blade angle at the hub and tip of the inducer

(b) The Mach number at the inlet and exit to the impeller for both the absolute and relative velocities

(c) The static pressure leaving the impeller

(d) The total pressure leaving the impeller

(e) The power required to drive the compressor

(f) The Mach number and optimum angle for the flow at the inlet to the diffuser

7.32. A single-stage centrifugal-flow compressor operates under the following conditions:

Impeller			
Inlet	Hub	Mean	Tip
Radius, inches	2.06	2.50	2.94
Blade velocity, ft/s	602	731	859
Axial velocity, ft/s	460	460	460
Total pressure, psia	29.36	29.36	29.36
Total temperature, °R	632	632	632
Static pressure, psia	26.60	26.60	26.60
Static temperature, °R	615	615	615
Exit			
Radius, inches	5.10		
Blade velocity, ft/s	1491		
Radial velocity, ft/s	460		
Blade angle	0°		

Assuming constant specific heats, determine, at the exit from the impeller:

(a) The velocity diagram

(b) The work done on the air, Btu/lb

(c) The static and stagnation temperatures, °R

(d) The static and stagnation pressures, psia

(e) The absolute and relative Mach numbers

(f) The blade width at the impeller exit assuming no blade blockage

7.33. Write a general computer program that will calculate the values asked for in Example Problems 7.11 and 7.12. Assume constant specific heats and ideal conditions, that is, adiabatic and reversible flow.

7.34. Air at 519°R (288 K), 14.70 psia (101.3 kPa) enters a multistage axial-flow compressor, which is followed by a single-stage centrifugal-flow compressor. Known information is as follows:

(a) Axial velocity remains constant from inlet to first axial-flow stage to inlet to the impeller

(b) Overall stagnation pressure ratio is 7.1

(c) Centrifugal compressor ratio is 4.0

(d) Hub-to-tip ratio at *inlet* to first axial-flow stage is 0.58

(e) All axial-flow rotors and the centrifugal impeller are on the same shaft, which rotates at 33,500 rpm

(f) The mean blade radius remains constant from inlet to the first axial-flow rotor to inlet to the impeller and has a value of 2.50 in. (6.35 cm)

(g) Centrifugal-flow impeller tip radius is 5.10 in. (13.0 cm)

(h) Diffuser leading edge radius is 5.30 in. (13.5 cm)

(i) Impeller exit radial velocity is same as the axial velocity at the inlet to the axial-flow compressor.

(j) All axial-flow stages have the same shape.

(k) Flow coefficient at inlet to the first axial-flow rotor is 0.63.

(l) Other axial-flow stage constraints:
 (i) Work coefficient ≤ 0.80.
 (ii) Ideal case (no losses).
 (iii) Absolute and relative Mach numbers ≤ 0.95.
 (iv) de Haller number ≥ 0.70.

Assuming constant specific heats with $c_p = 0.24$ Btu/(lb°R), $k = 1.40$, that the air enters and leaves each stage at the blade angle, and no losses, calculate:

(a) The mass rate of flow into the compressor

(b) The mean diameter velocity diagrams, Mach numbers, work and flow coefficients, and de Haller numbers for the axial-flow stages

(c) Number of axial-flow stages required

(d) The hub, mean, and tip velocity diagrams at the *inlet* to the impeller

(e) The impeller exit angle along with the absolute and relative velocities and Mach numbers

(f) The *static* pressure and temperature at the exit from the impeller

(g) The power required to drive the complete compressor

7.35. Solve Problem 7.34 for the overall pressure ratio assuming that the pressure ratio for the three-stage compressor is the maximum possible for the axial-flow compressor restrictions listed in item (*l*), all other values the same.

7.36. Modify the computer program of Problem 7.33 so that an *n*-stage axial-flow compressor is possible prior to the centrifugal-flow compressor.

TURBINES

8.1 GENERAL

The preceding chapter discussed the fundamentals of axial-flow and centrifugal-flow compressors and one of several ways in which compressor performance is presented. It is next important to examine turbines in a similar manner. The thermodynamics and fluid mechanics are basically the same as that used in understanding compressors.

Equation (7.6b),

$$w = U_1\overline{V}_{u1} - U_{1.5}\overline{V}_{u1.5} \tag{7.6b}$$

was a general equation for the work done on or by a compressor or turbine. It placed no restrictions on the geometry of the rotor. If the $U_{1.5}\overline{V}_{u1.5}$ term is greater than the $U_1\overline{V}_{u1}$ term, then work will be done on the fluid and it is a compressor. If $U_{1.5}\overline{V}_{1.5}$ is less then $U_1\overline{V}_{u1}$, then work will be done by the fluid and it is a turbine. This means that, if the fluid enters and leaves the rotor at the same radius, it is a compressor if $\Delta\overline{V}_u$ is positive (or increases) and a turbine if $\Delta\overline{V}_u$ is negative (or decreases).

In a compressor, a stage consists of a rotor and a stator or diffuser. In a turbine, a stage also consists of a stationary and a rotating member, the stationary row, commonly called a nozzle, and which precedes the rotating (blade) row.

There are two general types of turbines, the radial-flow and axial-flow turbines. The radial-flow turbine is similar to a centrifugal-flow compressor except that flow is inward instead of outward. Figure 8.1 illustrates side and front views of a radial-flow turbine.

Radial-flow turbines are used only for extremely low powers or where compactness is more important than performance. The reader is referred to the illustrations in earlier chapters to observe when, if ever, a radial-flow turbine was used even when a centrifugal-flow compressor was used.

Axial-flow turbines are almost always used in gas turbine engines. An axial-flow turbine may consist of one or more stages, each stage consisting of a nozzle row and a rotor row. The relative velocities in an axial-flow turbine are, in general, substantially

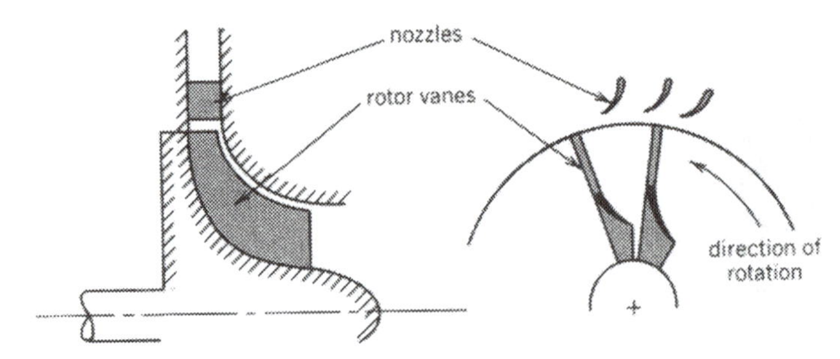

Figure 8.1. Radial-flow turbine.

higher than occur in axial-flow compressors, with a greater change in enthalpy per stage. In the nozzle row, the tangential velocity is increased in the direction of rotation with a corresponding drop in the static pressure. In the rotor row, the tangential velocity is decreased. Considerably fewer stages are needed in an axial-flow turbine than in an axial-flow compressor because in an axial-flow compressor the flow is decelerating (diffusing) in the passageways with a corresponding rise in pressure, whereas the gas is accelerating in a turbine. The diffusing action of a compressor allows only moderate changes in the compressor passageways to avoid separation.

8.2 TURBINE PERFORMANCE

Once again, the principles of dimensional analysis will be applied to flow through a turbine to determine the significant dimensionless groups used to describe the performance of a turbine. It will be assumed that the turbine inlet state is 3 and the exit state is 4.

Symbol	Description	Dimensions
T_{o3}	Inlet stagnation temperature	θ
p_{o3}	Inlet stagnation pressure	$ML^{-1}t^{-2}$
c_p	Constant-pressure specific heat of gas	$L^2t^{-2}\theta^{-1}$
D	Characteristic dimension (usually diameter)	L
N	Rotor rotational speed	t^{-1}
$\dot{m}_g$	Mass flow rate of gas	Mt^{-1}
p_{o4}	Exit stagnation pressure	$ML^{-1}t^{-2}$
T_{o4a}	Actual exit stagnation temperature	θ
μ	Absolute gas viscosity	$ML^{-1}t^{-1}$
ρ	Gas density at inlet	ML^{-3}

The gas constant R, molecular weight M, or gas specific heat ratio k could have been used instead of the density. It is also necessary to include the effect of turbine cooling, since most gas turbine engine turbines use some form of turbine cooling.

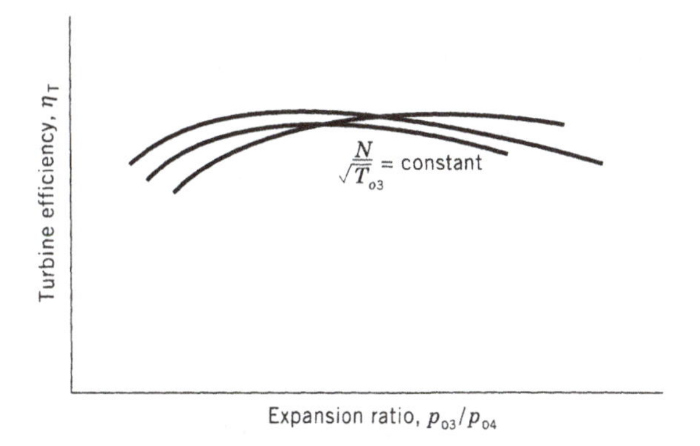

Figure 8.2. Plot of turbine efficiency versus expansion ratio for several values of corrected rotor speed.

The parameters used may take on several forms. The ones that will be used in this book are

1. Expansion ratio (p_{o3}/p_{o4})
2. Inlet flow parameter ($\dot{m}_g\sqrt{T_{o3}}/p_{o3}A$)
3. Rotational speed ($N/\sqrt{T_{o3}}$)
4. Turbine adiabatic efficiency (η_T)
5. Turbine cooling airflow

The performance of turbines usually is presented using two plots instead of the single "map" used for compressors. One set of plots is illustrated in Figures 8.2 and 8.3.

Observe that in Figure 8.3, the curve closely approximates that for flow through a converging nozzle (Figure 3.6), the main difference being that the pressure ratio in Figure 8.3 is a total-to-total ratio, in Figure 3.6 a total-to-static ratio. This similarity

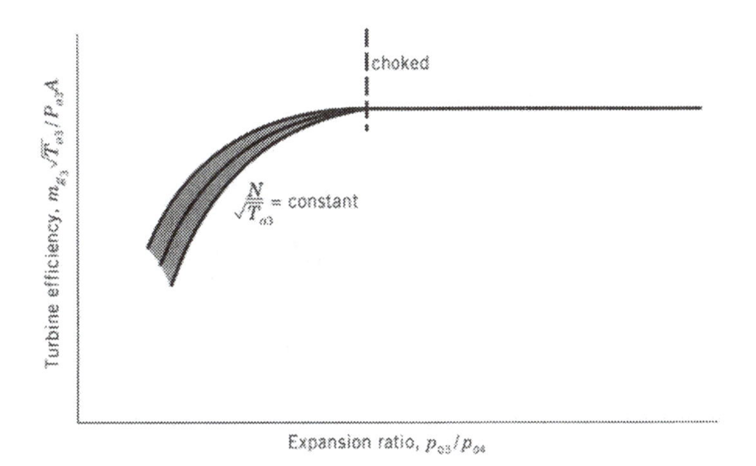

Figure 8.3. Plot of turbine inlet flow parameter versus expansion ratio for several values of corrected rotor speed.

occurs because choking can occur in either the nozzle or rotor of the turbine, thereby fixing the flow parameter for the turbine. One should remember that when a turbine is choked, the flow parameter remains constant. This does not mean that the expansion ratio (and therefore work) is fixed, since the expansion ratio may continue to change even though the flow parameter is constant. Also observe that the flow parameter is only slightly dependent on rotor speed. Both of these facts will be important when component matching is discussed in Chapter 10.

The area used in the flow parameter is the minimum flow area in the first turbine nozzle row. This will occur at the throat of the turbine nozzles.

8.3 BLADE NOTATION FOR IDEAL AXIAL-FLOW TURBINES

Several blades from a turbine stage are shown in Figure 8.4. This figure represents flow at a representative radius r.

The blade nomenclature for an axial-flow turbine is quite similar to that of an axial-flow compressor. Once again, the positive direction for $\overline{V}_u$ is in the direction of rotation. Angles measured in the direction of rotation are positive, those in the direction opposite to rotation negative.

One should note that for both the nozzle and rotor row, the area decreases from inlet to the blade to the throat.

Ideally, the fluid enters at an angle tangent to the camber line at the leading edge and leaves at an angle tangent to the camber line at the trailing edge.

8.4 STAGE VELOCITY DIAGRAM FOR AN AXIAL-FLOW TURBINE

It is important to be able to calculate the change in the tangential velocity across the rotor. This is best accomplished by drawing velocity diagrams that represent flow at the inlet and outlet of the rotor.

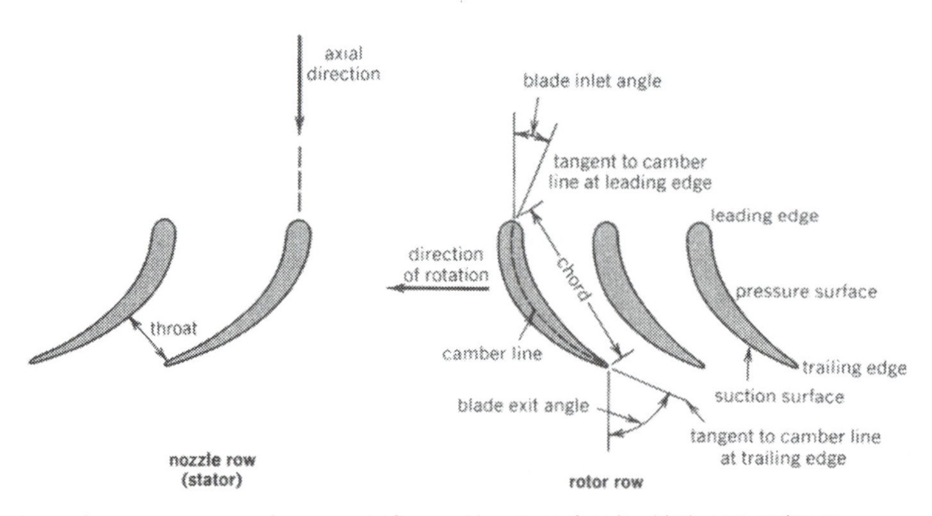

Figure 8.4. Several blades from an axial-flow turbine stage showing blade nomenclature.

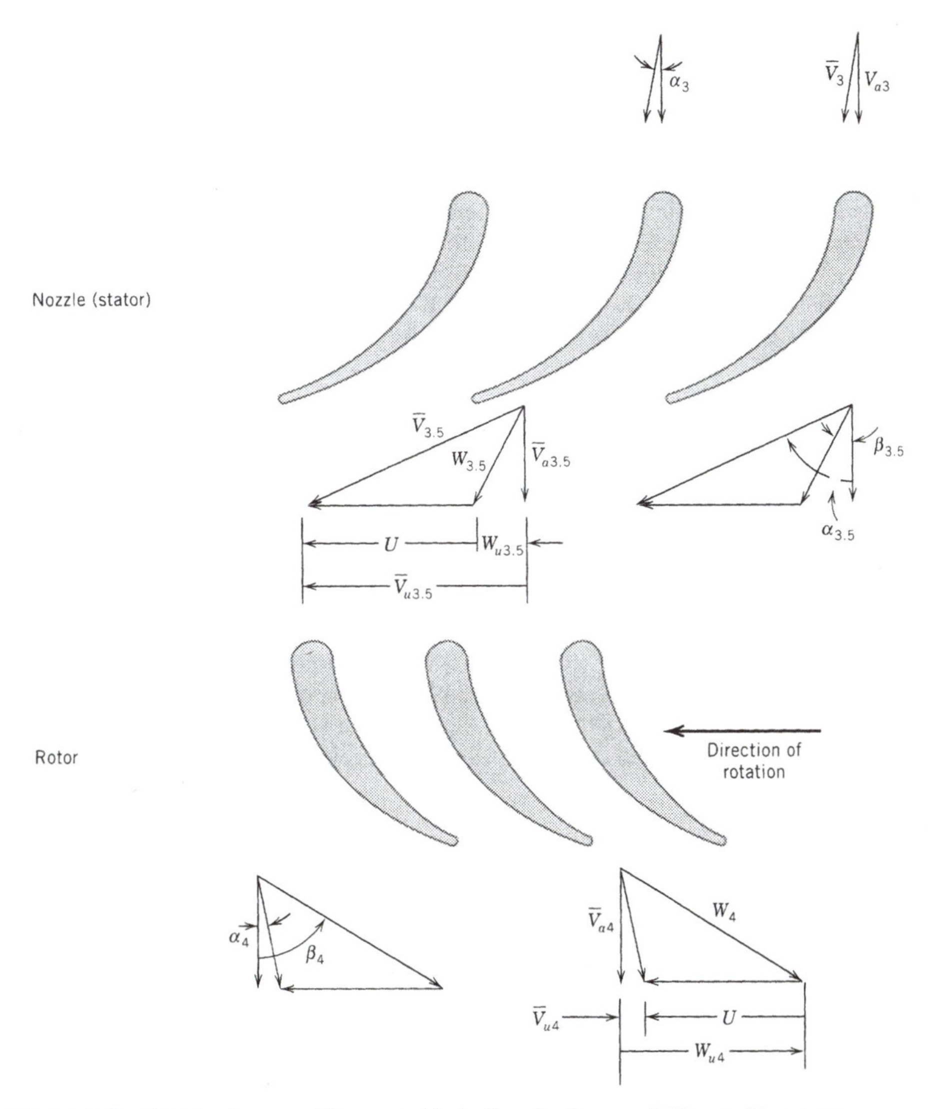

Figure 8.5. Velocity diagram at the mean blade diameter for an axial-flow turbine stage.

Velocity diagrams for flow through a single stage of an axial-flow turbine are shown in Figure 8.5.

The gas enters the nozzle row with a pressure p_3, temperature T_3, and a velocity $\overline{V}_3$ at an angle α_3. It expands in the nozzle to a pressure $p_{3.5}$, temperature $T_{3.5}$, and leaves with an absolute velocity $\overline{V}_{3.5}$ at an angle $\alpha_{3.5}$.

The gas next enters the rotor with a relative velocity $W_{3.5}$ at an angle $\beta_{3.5}$. The rotor blade inlet angle is selected so that $\beta_{3.5}$ is tangent to (or nearly so) the tangent to the camber line at the leading edge.

The gas is turned, in the rotor, and leaves at a pressure p_4, a temperature T_4, and a relative velocity W_4 at an angle β_4. The absolute velocity at the rotor exit is $\overline{V}_4$ and is at an angle α_4.

In many turbines, $\overline{V}_3$ is in the axial direction ($\alpha_3 = 0$). Therefore, $\overline{V}_3 = \overline{V}_{a3}$. In a multistage turbine, the angle and velocity with which the gas leaves the first stage ($\overline{V}_4$ at α_4 in the above example) define the conditions for the next row of nozzles. It is common for $\overline{V}_3$ and α_3 to be equal to $\overline{V}_4$ and α_4 so that the blade shapes are the same for successive stages.

The blade speed U increases with increasing radius. This means that the shape of the velocity diagram varies from the hub to the tip of the blade. In many axial-flow turbines, the hub-to-tip ratio is near 1.0. In this case, a good first approximation is to assume that the velocity diagrams drawn at the mean blade diameter closely approximate the conditions at the hub and tip. This assumption will be used in this chapter.

8.5 ENERGY TRANSFER

Equation (7.6), as was noted in Chapter 7, applies to a turbine rotor. The discussion that follows will be restricted to an axial-flow turbine where the mean radius does not change for flow through the rotor row. This means that the blade velocity is the same at the inlet and outlet of the rotor. Therefore, the greater the change in the tangential component of the absolute (or relative) velocity, the larger the torque produced; therefore, the larger the work output from a stage.

Figure 8.6 represents the case where the axial velocity remains constant. One should observe that

$$\frac{U}{\overline{V}_a} = \tan \alpha_{3.5} - \tan \beta_{3.5} \tag{8.1a}$$

$$= \tan \beta_4 - \tan \alpha_4 \tag{8.1b}$$

From Eq. (7.6b) (applied to the turbine states of 3.5 at the inlet to the rotor and

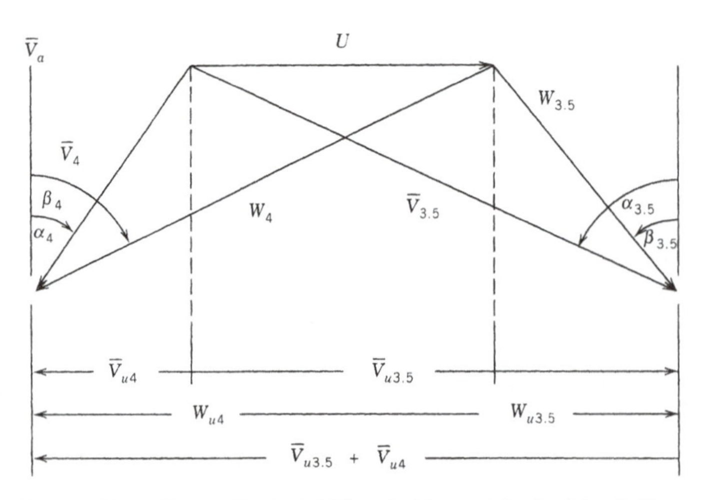

Figure 8.6. Stage velocity diagram for an axial-flow turbine constant axial velocity.

4 at the exit from the rotor), remembering that $\overline{V}_{u3.5}$ is positive in the direction of rotation and that positive work is for a turbine, the work for the stage becomes

$$w_{stage} = U(\overline{V}_{u3.5} - \overline{V}_{u4}) \tag{8.2a}$$

$$= U\overline{V}_a(\tan \alpha_{3.5} - \tan \alpha_4) \tag{8.2b}$$

From the steady-flow equation,

$$w_{stage} = h_{o3} - h_{o4} \tag{8.3a}$$

or, for constant specific heats,

$$w_{stage} = c_p\,(T_{o3} - T_{o4}) \tag{8.3b}$$

The ideal stage expansion ratio is, for variable specific heat,

$$\frac{p_{o3}}{p_{o4}} = \frac{Pr_{o3}}{Pr_{o4}} \tag{8.4a}$$

or, for constant specific heats

$$\frac{p_{o3}}{p_{o4}} = \left(\frac{T_{o3}}{T_{o4}}\right)^{k/(k-1)} \tag{8.4b}$$

Example Problem 8.1

Air, at a total temperature of 2520°R (1400 K) and a total pressure of 323.3 psia (2230 kPa), enters the nozzle row of a turbine stage in the axial direction at the rate of 80.0 lb/s (36.4 kg/s). Other known data are (see Figure 8.5 for nomenclature):

Turbine rotor speed	14,000 rpm
Mean blade diameter	19.0 in. (48.3 cm)
$\overline{V}_{a3.5}$	550.0 ft/s (170.0 m/s)
$\alpha_{3.5}$	+72.0°
α_4	0.0°

Assuming variable specific heats, that the air enters and leaves the blades at the blade angle and that the axial velocity remains constant throughout the stage:

(a) Construct the velocity diagrams at the mean blade diameter for this stage
(b) Calculate the Mach number of the absolute velocity at the exit from the nozzle (state 3.5)
(c) Calculate the work and power developed by this stage based on mean diameter conditions
(d) Calculate the total-to-total expansion ratio, p_{o3}/p_{o4} for this stage
(e) Calculate the numerical value of $\dfrac{\dot{m}\sqrt{T_{o3}}}{p_{o3}A}$ where A is the minimum (throat) area of the nozzle row

The velocity diagram is

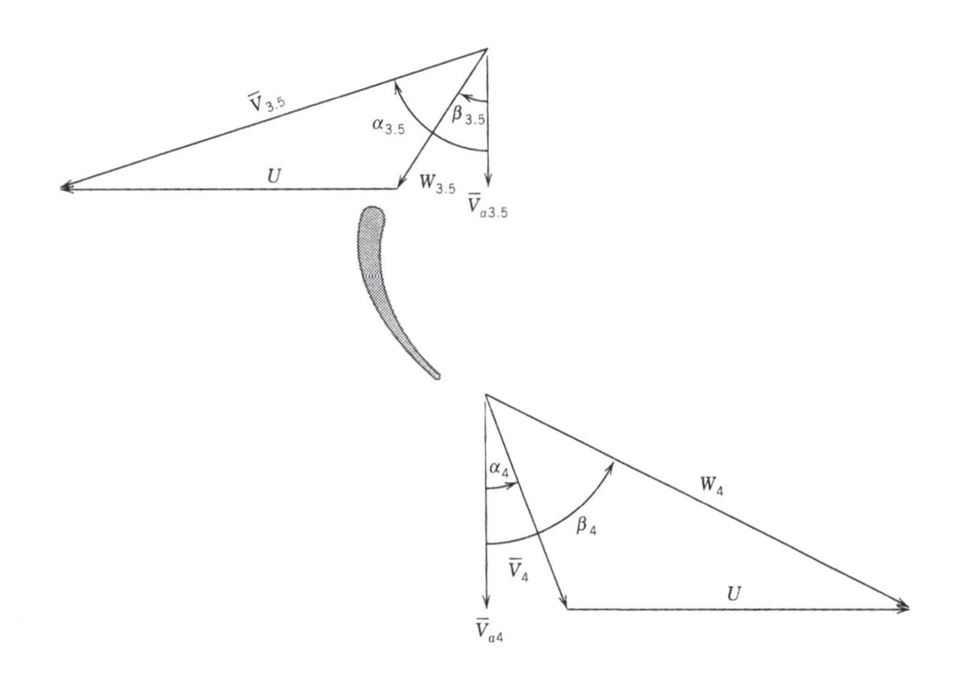

(b) $h_{o3.5} = h_{o3} = h_{3.5} + \dfrac{\overline{V}^2_{3.5}}{2\,g_c}$

$$T_{o3} = 2520°\text{R (1400 K)}$$

$$h_{o3} = 521.7\ (1212.7)$$

$$Pr_{o3} = 450.9\ (450.9)$$

Therefore

$$h_{3.5} = 521.7 - \frac{1780^2}{2g_c\,J}$$

$$= 458.4 \text{ Btu/lb (1061.5 kJ/kg)}$$

$$T_{3.5} = 2298°\text{R (1273 K)}$$

$$Pr_{3.5} = 307.3\ (303.9)$$

$$k_{3.5} = 1.320\ (1.320)$$

$$c_{3.5} = \sqrt{(1.320)(g_c)(53.35)(2298)}$$

$$= 2282 \text{ ft/s (694 m/s)}$$

$$M_{3.5} = \frac{1780}{2282} = 0.780\ (0.754)$$

(c) $_3w_4 = {}_{3.5}w_4 = \dfrac{1161(1693 - 0)}{(32.174)(778)}$

$$= 78.5 \text{ Btu/lb } (185 \text{ kJ/kg})$$

The positive sign tells us that it is a turbine.

$$\dot{w} = \frac{(80)(3600)(78.5)}{2545}$$

$$= 8886 \text{ hp } (6740 \text{ kw})$$

(d) To determine the expansion ratio, one needs to know the stage exit total pressure and temperature.

$$h_{o4} = 521.7 - 78.5$$

$$= 443.2\ (1027.7)$$

$$Pr_{o4} = 278.8\ (276.8)$$

$$T_{o4} = 2244°\text{R } (1244 \text{ K})$$

For an ideal stage (adiabatic and reversible)

$$\frac{p_{o3}}{p_{o4}} = \frac{Pr_{o3}}{Pr_{o4}} = \frac{450.9}{278.8} = 1.62\ (1.63)$$

Note that the axial-flow turbine stage expansion ratio is much higher than the pressure ratio for an axial-flow compressor stage.

(e) To calculate $\dot{m}\dfrac{\sqrt{T_{o3}}}{p_{o3}A}$, one needs to know the throat area of the first nozzle row.

$$p_{3.5} = p_{o3}\left(\frac{Pr_{3.5}}{Pr_{o3}}\right)$$

$$= 323.3\left(\frac{307.3}{450.9}\right)$$

$$= 220.3 \text{ psia } (1503 \text{ kPa})$$

$$A_{3.5} = A_{min} = \frac{\dot{m}\,R\,T_{3.5}}{p_{3.5}\,\overline{V}_{3.5}}$$

$$= 25.0 \text{ in}^2\ (169 \text{ cm}^2)$$

$$\dot{m}\frac{\sqrt{T_{o3}}}{p_{o3}\,A_{min}} = (80)\frac{\sqrt{2520}}{(323.3)(25.0)}$$

$$= 0.497 \text{ lb°R/(lb}_f\text{s)}$$

Reviewing the results of Example Problem 8.1, one observes that

1. The maximum velocity that occurs in the stage is the *absolute* velocity at the nozzle exit (state 3.5).
2. Increasing $\overline{V}_a$ will increase $\overline{V}_{3.5}$ resulting in a high Mach number.

3. Having α_4 in the negative direction will result in more work being done by the stage but results in the absolute velocity having a swirl component ($\overline{V}_u$ component), which is not beneficial after the turbine rotor and will result in frictional losses.
4. Adding a second stage, same $\overline{V}_a$, results in same $\overline{V}_{3.5}$ but higher Mach number since the static temperature will be lower.

Several quantities which turbine designers find useful are the flow coefficient ϕ, the blade loading coefficient Ψ, the speed-work parameter λ, and the blade-jet speed ratio ν.

The *flow coefficient* is the ratio of the inlet axial velocity divided by the blade velocity or

$$\phi = \frac{\overline{V}_a}{U} \tag{8.5}$$

The blade loading coefficient was defined in Section 7.7. It is defined as the stage (rotor) work divided by the blade kinetic energy, or, in equation form

$$\Psi = -\left(\frac{2\, c_p \Delta T_0}{U^2}\right) \tag{8.6}$$

For turbines, Ψ is positive. As was pointed out earlier, some designers define the blade loading coefficient without the 2 factor. In this text, Eq. (7.9a) will be used.

For turbines,

highly loaded stage	$\Psi > 3.0$
lightly loaded stage	$\Psi < 2.0$

The reciprocal of the blade loading coefficient, called the *speed-work* parameter, is

$$\lambda = \frac{1}{\Psi} = \frac{U^2}{2\, w_{\text{stage}}} \tag{8.7}$$

Another parameter often used is the *blade-jet speed ratio* ν, defined as the ratio of blade velocity to the jet or spouting velocity, $\overline{V}_j$. The jet or spouting velocity is defined as the velocity that would result from ideal expansion for the inlet total pressure to the stage exit static pressure. Therefore,

$$\nu = \frac{U}{\overline{V}_j} \tag{8.8}$$

Example Problem 8.2

Determine, for Example Problem 8.1,

(a) The blade loading coefficient.
(b) The speed-work parameter.

Solution

$$\text{(a)}\ \Psi = \left(\frac{2(521.7 - 443.2)(32.174)(778)}{1161^2}\right)$$

$$= 2.92$$

$$\text{(b)}\ \lambda = \frac{1}{2.92} = 0.343$$

8.6 DEGREE OF REACTION

The degree of reaction for a stage is defined as the change in static enthalpy across the rotor as a fraction of the change in the absolute total enthalpy across the stage. Therefore, referring to Figure 8.5 for state locations,

$$R' = \frac{h_{3.5} - h_4}{h_{o3} - h_{o4}} = \frac{h_{3.5} - h_4}{h_{o3.5} - h_{o4}} \tag{8.9}$$

For the case where the axial velocity remains constant, the change in static enthalpy across the rotor becomes

$$h_{3.5} - h_4 = \frac{W_4^2 - W_{3.5}^2}{2} = \frac{W_{u4}^2 - W_{u3.5}^2}{2} \tag{8.10}$$

But

$$\tan \beta_4 = \frac{W_{u4}}{V_a} \tag{8.11a}$$

and

$$\tan \beta_{3.5} = \frac{W_{u3}}{\overline{V}_a} \tag{8.11b}$$

Therefore

$$h_{3.5} - h_4 = \frac{\overline{V}_a^2(\tan \beta_4 - \tan \beta_{3.5})(\tan \beta_4 + \tan \beta_{3.5})}{2} \tag{8.12}$$

The change in total enthalpy becomes

$$h_{o3} - h_{o4} = h_{o3.5} - h_{o4} = {}_3w_4 = U(\overline{V}_{u3} - \overline{V}_{u4}) \tag{8.13}$$

$$= U\overline{V}_a(\tan \alpha_{3.5} - \tan \alpha_4) \tag{8.14}$$

But

$$\tan \alpha_{3.5} = \frac{U}{\overline{V}_a} + \tan \beta_{3.5} \tag{8.15a}$$

$$\tan \alpha_4 = \frac{U}{\overline{V}_a} + \tan \beta_4 \tag{8.15b}$$

Therefore

$$h_{o3.5} - h_{o4} = U\overline{V}_a(\tan \beta_{3.5} - \tan \beta_4) \tag{8.16}$$

Combining Eqs. (8.9), (8.12) and (8.16) yields

$$R' = -\frac{\overline{V}_a}{2U}(\tan \beta_4 + \tan \beta_{3.5}) \tag{8.17}$$

Combining Eqs. (8.17) and (8.15b) yields

$$R' = \frac{1}{2} - \frac{\overline{V}_a}{2U}(\tan \beta_4 + \tan \alpha_{3.5}) \tag{8.18a}$$

$$= \frac{1}{2} - \frac{\overline{V}_a}{2U}(\tan \beta_{3.5} + \tan \alpha_4) \tag{8.18b}$$

Reaction may be positive, negative, or zero.

8.7 IMPULSE TURBINE

A turbine with a zero degree of reaction is called an *impulse turbine*. In an impulse turbine, there is no change in the static enthalpy across the stage. All work done by the stage results from a change in the absolute kinetic energy across the rotor.

Examination of Eq. (8.17) shows that if the degree of reaction is zero,

$$\tan \beta_4 + \tan \beta_{3.5} = 0$$

or $\beta_{3.5}$ and β_4 are equal in magnitude but opposite in sign. Figure 8.7 illustrates the blade shapes and velocity diagrams for an impulse turbine.

A turbine stage with equal static enthalpy change across the rotor and the stator (nozzle) will have a degree of reaction of 50%. In this case.

$$\tan \beta_{3.5} + \tan \alpha_4 = 0$$

or the velocity diagrams at the inlet and exit to the rotor will be symmetrical. This is illustrated in Figure 8.8.

One must remember that Eqs. (8.17) and (8.18) are limited to the case of constant axial velocity. It must also be emphasized that designing a turbine stage is a three dimensional problem where the degree of reaction may vary continuously.

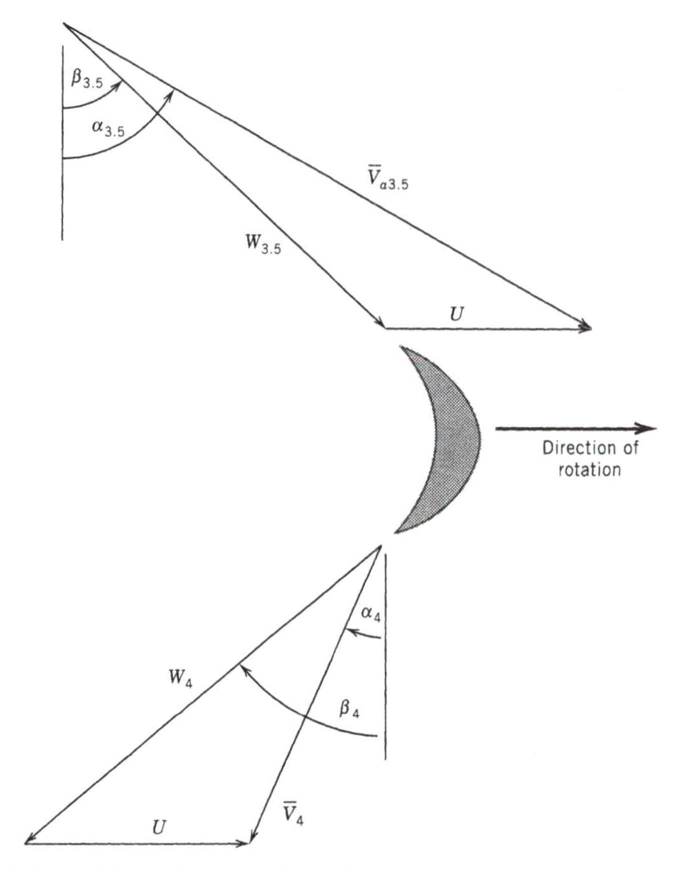

Figure 8.7. Velocity diagram for an impulse turbine.

Example Problem 8.3

Calculate, for Example Problem 8.1, the degree of reaction.

Solution

From Example Problem 8.1

$$_3w_4 = 78.5 \text{ Btu/lb } (185 \text{ kJ/kg})$$

$$h_{o3} = 521.7\ (1212.7)$$

$$h_{3.5} = 458.4\ (1061.5)$$

$$h_{o4} = 443.2\ (1027.7)$$

$$h_4 = 443.2 - \frac{(550)^2}{(2)(32.174)(778)}$$

$$= 437.2\ (1013.3)$$

$$R' = \frac{458.4 - 437.2}{78.5}$$

$$= 0.270\ (0.261)$$

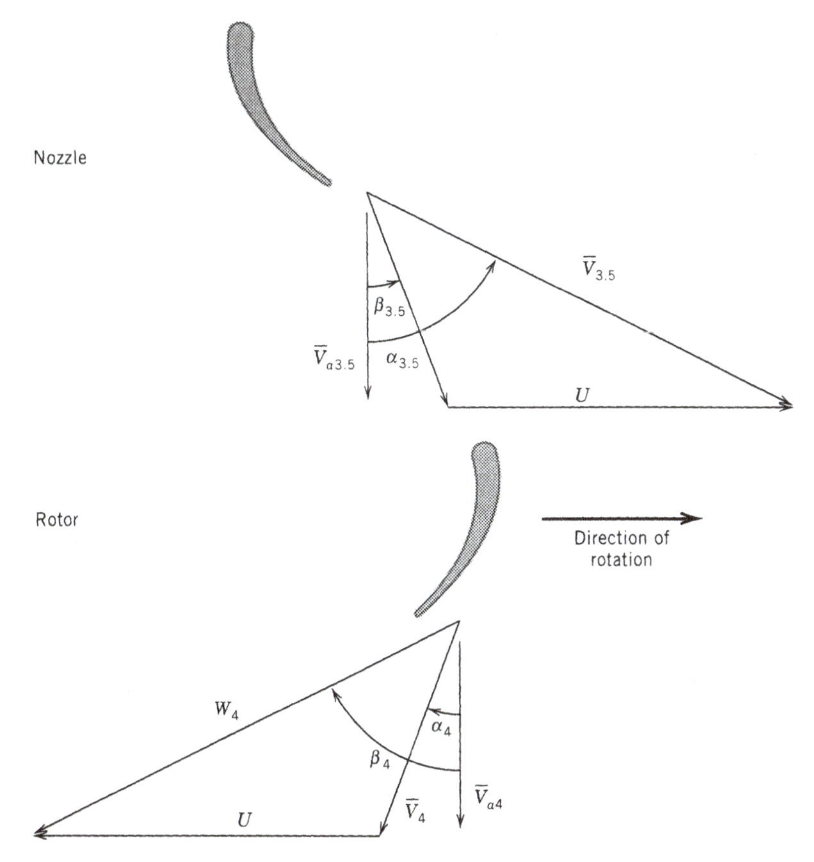

Figure 8.8. Turbine stage with 50% degree of reaction.

Since the axial velocity remains constant, one of several forms developed in sections for calculation the degree of reaction can be used.

From Eq. (8.17),

$$R' = \frac{-550}{(2)(1161)} [\tan(-64.7°) + \tan(44.1°)]$$

$$= 0.272\ (0.260)$$

From Eq. (8.18a),

$$R' = \frac{1}{2} - \frac{550}{(2)(1161)} [\tan(-64.7°) + \tan(72.0°)]$$

$$= 0.272\ (0.260)$$

8.8 VELOCITY DIAGRAM TYPES FOR AXIAL-FLOW TURBINES

Velocity diagrams will have different sizes and shapes depending on

1. The diagram type
2. The value of the speed-work parameter

Common types of diagrams are

1. zero-exit-swirl diagram
2. impulse diagram
3. symmetrical diagram

One common type of velocity diagram is one where the velocity at the exit from the rotor is in the axial direction; that is, $\overline{V}_{u4} = 0$. Stage velocity diagrams for speed-work parameter values of 0.5, 0.25, and 0.125 are shown in Figure 8.9. These are all drawn for the same axial velocity and show the effect on the blade and air angles and velocities.

According to Glassman (1), zero-exit-swirl diagrams are seldom used when the speed-work parameter value is less than 0.25.

The relationship between stage reaction and speed-work parameter is shown in Figure 8.10. Note that the stage reaction is zero for a speed-work parameter value of 0.25 and negative below this value decreasing to a negative 1.000 for a speed-work parameter value of 0.125.

It can be shown that the relationship between stage reaction R' and speed-work parameter λ is

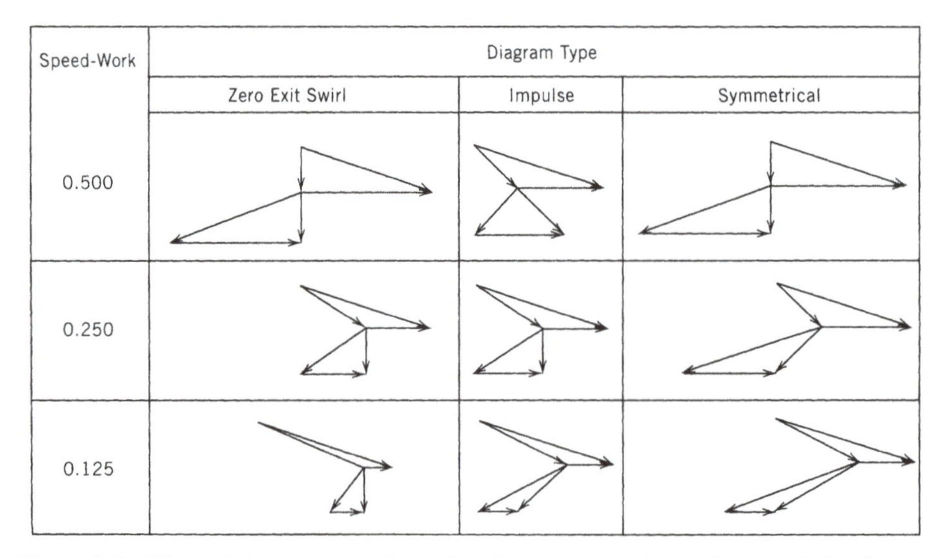

Figure 8.9. Effects of diagram type and speed-work parameter on shape of stage velocity diagram. All are for the same axial velocity.

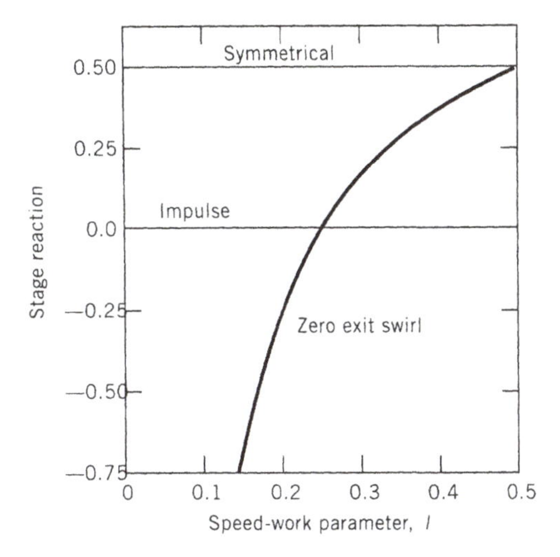

Figure 8.10. Effect of speed-work parameter and diagram type on stage reaction.

$$R' = 1 - \frac{1}{4\lambda} \tag{8.19}$$

Figure 8.11 shows the relation between exit swirl and speed-work parameter. For an impulse turbine the stage reaction is zero, and the relative velocity remains constant ($W_{3.5} = W_4$). This means that for constant axial velocity, $\beta_{3.5} = -\beta_4$.

Velocity diagrams for speed-work parameter values of 0.50, 0.25, and 0.125 are shown in Figure 8.9. Note that for a λ value of 0.25, the velocity at the exit is in the axial direction ($\overline{V}_u = 0$).

For a symmetrical velocity diagram, $\alpha_{3.5} = -\beta_4$ and $\alpha_4 = -\beta_{3.5}$. Velocity diagrams for speed-work parameter values of 0.50, 0.25, and 0.125 are shown in Figure 8.9. The relation between stage reaction and speed-work parameter are shown in Figure

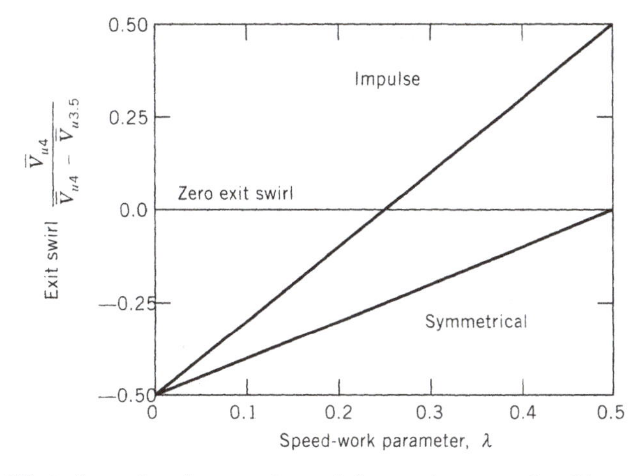

Figure 8.11. Effect of speed-work parameter and diagram type on exit swirl.

8.10 with the relation between exit swirl velocity and speed-work parameter in Figure 8.11.

8.9 HUB AND TIP EFFECTS

The discussion so far has been at the mean blade diameter. It is now important to investigate the change in the blade height as the air passes through the rotor. This is best illustrated by an example problem.

Example Problem 8.4

Determine, for Example Problem 8.1, the blade height and hub-to-tip ratio at the inlet and exit from the rotor.

Solution

Known values from Example Problem 8.1 at state 3.5, the inlet to the rotor, are:

$$T_{3.5} = 1273 \text{ K } (2298°\text{R})$$

$$p_{3.5} = 1503 \text{ kPa } (220.3 \text{ psia})$$

$$\overline{V}_{3.5} = 170 \text{ m/s } (550 \text{ ft/s})$$

$$\dot{m} = 36.4 \text{ kg/s } (80.0 \text{ lb/s})$$

$$\text{r}_\text{m} = 24.15 \text{ cm } (9.50 \text{ in.})$$

From the one-dimensional continuity equation

$$A_{a3.5} = \frac{\dot{m}\, R\, T_{3.5}}{\overline{V}_{3.5}\, p_{3.5}}$$

Assuming a blade height of 2x

$$A_{a3.5} = ((\text{r}_\text{m} + x)^2 - (\text{r}_\text{m} - x)^2)\, \pi$$

or

$$x = \frac{\dot{m}\, R\, T_{3.5}}{4\, \pi\, \text{r}_\text{m}\, \overline{V}_{a3.5}\, p_{3.5}}$$

This results in

$$x_{3.5} = 1.72 \text{ cm } (0.68 \text{ in.})$$

or a blade height at state 3.5 of 3.44 cm (1.36 in.).
The hub-to-tip ratio at state 3.5 is

$$\frac{r_h}{r_t} = \frac{24.15 - 1.72}{24.15 + 1.72} = 0.867\ (0.866)$$

At state 4, the exit from the rotor

$$T_4 = 1232 \text{ K } (2222°\text{R})$$
$$p_4 = 1314.7 \text{ kPa } (192.2 \text{ psia})$$
$$\overline{V}_{a4} = 170 \text{ m/s } (500 \text{ ft/s})$$
$$\dot{m} = 36.4 \text{ kg/s } (80.0 \text{ lb/s})$$
$$\text{r}_\text{m} = 241.5 \text{ cm } (9.50 \text{ in.})$$

This results in

$$x_4 = 1.90 \text{ cm } (0.75 \text{ in.})$$
$$\text{blade height} = 3.80 \text{ cm } (1.50 \text{ in.})$$
$$\frac{r_h}{r_t} = 0.854 \ (0.854)$$

One should observe the increase in the blade height through this rotor.

8.10 ACTUAL AXIAL-FLOW TURBINE STAGES

It was assumed in the preceding sections that the gas entered and left the blade at the blade camber angle and, in calculating the expansion ratio, that the expansion process was an adiabatic and reversible process. This, of course, does not occur in an actual engine.

The gas velocity angle at the inlet to a blade will not be identical to the blade angle. It will be at an incidence angle i, as illustrated in Figure 8.12. The angle of incidence is defined as the angle between the actual fluid direction at the inlet to the blade and the blade angle at the inlet.

The gas velocity angle at the exit from a blade will also be different from the blade exit angle as illustrated in Figure 8.12. The deviation, δ, is defined as the angle between the actual fluid direction at the exit from the blade and the blade angle at the exit from the blade.

In an actual stage, flow will be approximately adiabatic but will not be reversible. There are several efficiencies that are defined for a turbine row, stage, and stages (overall).

Since the stator row is a series of nozzles, the stator efficiency is defined as

$$\eta_{st} = \frac{\overline{V}_{3.5}^2}{\overline{V}_{3.5,\text{ideal}}^2} \tag{8.20}$$

where $\overline{V}_{3.5}$ is the actual absolute velocity leaving the stator (nozzle) and $\overline{V}_{3.5,\text{ideal}}$ is the velocity that would have occurred if expansion had been adiabatic and reversible to the same pressure. This is illustrated in Figure 8.13. It should be remembered that the total enthalpy will remain constant across a nozzle row since the process is adiabatic. The total pressure will decrease across a nozzle row unless the process is reversible.

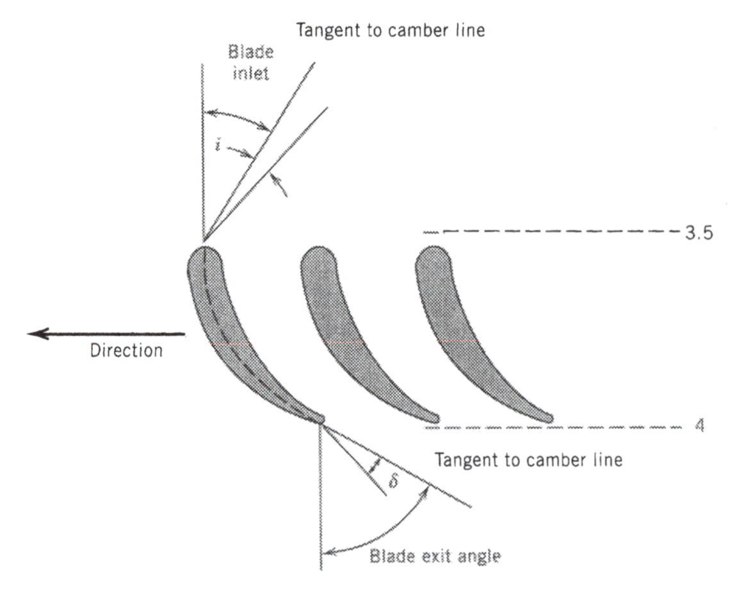

Figure 8.12. Flow diagram showing angle of incidence and deviation.

The rotor efficiency may be defined in a similar manner using *relative* velocities, or

$$\eta_{r_0} = \frac{\overline{w}_4^2}{\overline{w}_{4,\text{ideal}}^2} \tag{8.21}$$

One should remember that the total enthalpy, based on relative velocities, will remain constant, since for an observer riding with the rotor, no work is done and it behaves

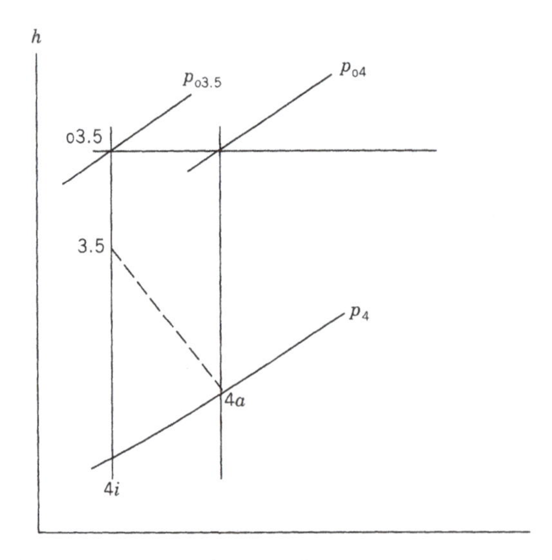

Figure 8.13. Enthalpy–entropy diagram for a nozzle row.

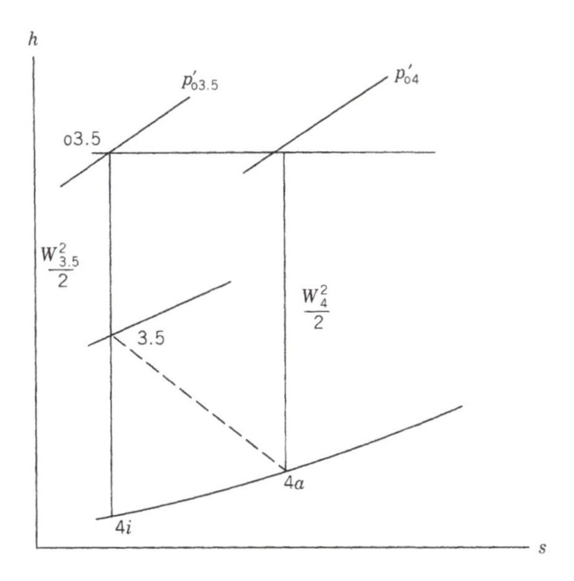

Figure 8.14. Enthalpy–entropy diagram showing the relative states across a rotor.

like a nozzle. $\overline{W}_{4,\text{ideal}}$, as used in Eq. (8.21), is defined as the velocity that would result if expansion were adiabatic and reversible. The relative ideal and actual states for the rotor are shown in Figure 8.14.

Figure 8.15 is an enthalpy–entropy diagram for a turbine stage where inlet to the stator (nozzle) is state 3, inlet to the rotor state 3.5 and exit from the rotor state 4.

The overall efficiency, based on total conditions, is defined as the ratio of the actual work to the ideal work, or

$$\eta_T = \frac{w_{\text{act}}}{w_{\text{ideal}}} = \frac{\Delta h_{\text{o,act}}}{\Delta h_{\text{o,ideal}}} \tag{8.22}$$

This is commonly called the overall adiabatic or isentropic turbine efficiency.

Figure 8.16 is an enthalpy–entropy diagram for a multistage turbine. The overall turbine efficiency for a multistage turbine is defined as

$$\eta_{\text{overall}} = \frac{\Sigma \Delta h_{\text{act}}}{\Delta h_{\text{ideal,overall}}} \tag{8.23}$$

This becomes, for the two-stage turbine illustrated in Figure 8.16,

$$\eta_{\text{overall}} = \frac{\Delta h_{\text{act,1}} + \Delta h_{\text{act,2}}}{\Delta h_{\text{ideal,overall}}} \tag{8.24}$$

The overall turbine efficiency for a multistage turbine will be greater than that of the stage efficiency, since $\Sigma \Delta h_{\text{ideal,stages}}$ is greater than $\Delta h_{\text{ideal,overall}}$. This is commonly called the reheat effect.

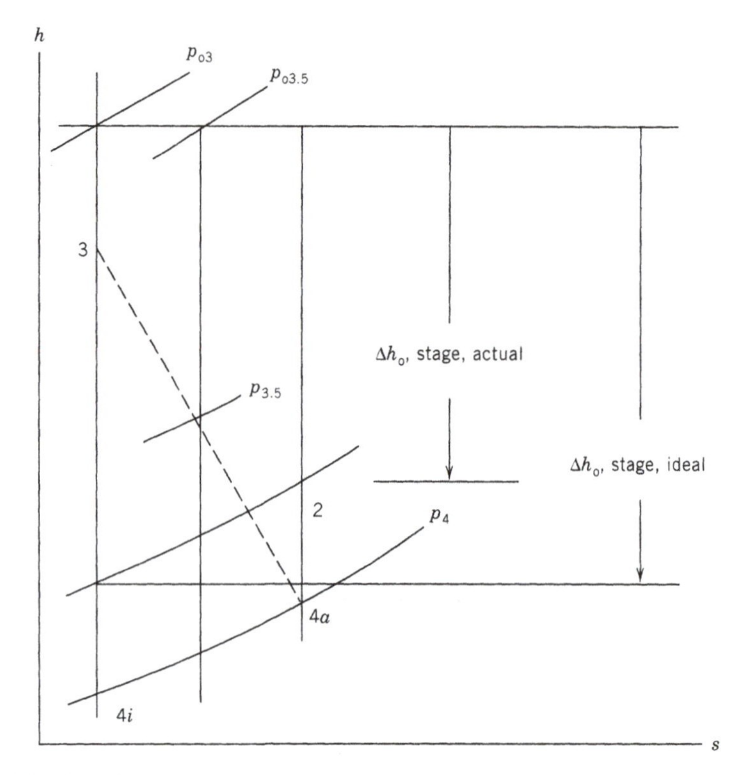

Figure 8.15. Enthalpy–entropy diagram for a turbine stage.

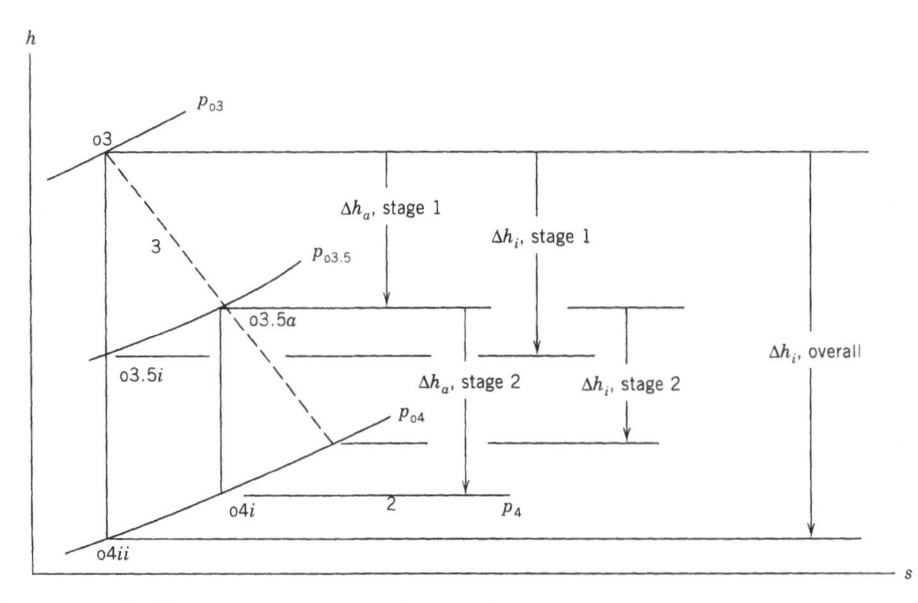

Figure 8.16. Enthalpy–entropy diagram for a two-stage turbine.

Example Problem 8.5

Air enters a two-stage axial-flow turbine at a total temperature of 1400 K (2520°R) and a total pressure of 2230 kPa (323.3 psia). The *actual work* developed by each stage is 185.0 kJ/kg (78.5 Btu/lb), and each stage has an adiabatic efficiency of 87.0%. Calculate:

(a) The total pressure at the exit from each stage
(b) The overall adiabatic efficiency

Solution

The important states are identified on the following temperature-entropy diagram. All states shown are total (stagnation) states. The inlet to the first turbine stage is state 3, the exit from the first turbine stage and inlet to the second turbine stage is state 3.5 and the exit from the second turbine stage is state 4.

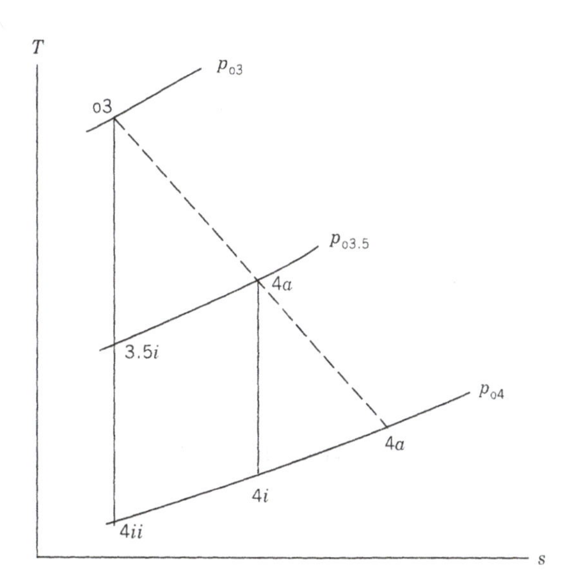

$$T_{o3} = 1400 \text{ K } (2520°\text{R})$$

$$p_{o3} = 2230 \text{ kPa } (323.3 \text{ psia})$$

$$h_{o3} = 1212.7\ (521.7)$$

$$Pr_{o3} = 450.9\ (450.9)$$

First Stage

$$_3w_{3.5i} = \frac{w_a}{0.87} = \frac{185.0}{0.87}$$

$$= 212.6 \text{ kJ/kg } (90.2 \text{ Btu/lb})$$

$$h_{o3.5i} = h_{o3} - {}_3w_{3.5i} = 1000.1\ (431.5)$$

$$Pr_{o3.5i} = 256.1\ (260.5)$$

$$p_{o3.5i} = \frac{(2230)(256.1)}{450.9}$$

$$= 1267\ \text{kPa}\ (186.8\ \text{psia})$$

Second Stage

The properties are the inlet to the second stage are the *actual* properties at the exit from the first stage.

$$p_{o3.5} = 1267\ \text{kPa}\ (186.8\ \text{psia})$$

$$h_{o3.5a} = 1212.7 - 185.0 = 1027.7\ (443.2)$$

$$Pr_{o3.5a} = 276.8\ (278.8)$$

$$_{3.5}w_{4i} = \frac{185.0}{0.87} = 212.6\ \text{kJ/kg}\ (90.2\ \text{Btu/lb})$$

$$h_{o4i} = 1027.7 - 212.6 = 815.1\ (353.0)$$

$$Pr_{o4} = 146.06\ (148.04)$$

$$p_{o4} = \frac{(1267)(146.1)}{276.8} = 669\ \text{kPa}\ (99.2\ \text{psia})$$

$$Pr_{o4ii} = \frac{(450.9)(669)}{2230} = 135.3\ (138.3)$$

$$h_{o4ii} = 793.2\ (344.1)$$

$$\Delta h_{oii,\text{overall}} = 1212.7 - 793.2 = 419.5\ \text{kJ/kg}\ (177.6\ \text{Btu/lb})$$

$$\eta_{\text{overall}} = \frac{(2)(185.0)}{419.5} = 0.882\ (0.884)$$

Note that the overal turbine efficiency is higher than the stage efficiency.

Another turbine efficiency commonly defined is the infinitesimal or polytropic efficiency. The turbine polytropic efficiency, η_{p_t}, is defined by

$$\eta_{p_t} = \frac{dT_{\text{act}}}{dT_i} \tag{8.25}$$

Combining Eqs. (7.17), (8.25), (2.12), and (2.17) yields

$$\frac{c_p\,dT_{\text{act}}}{\eta_{p_t}T} = R\frac{dp}{p} \tag{8.26}$$

Integrating between states a and b for constant η_{p_t} and c_p yields

$$\frac{c_p}{\eta_{p_t}} \ln \frac{T_b}{T_a} = R \ln \frac{p_b}{p_a} \tag{8.27a}$$

or

$$\frac{p_b}{p_a} = \left(\frac{T_b}{T_a}\right)^{k/\eta_{p_t}(k-1)} \tag{8.27b}$$

The overall turbine efficiency was defined by Eq. (8.22). Combining Eqs. (8.22) and (8.27b) results in a relationship between overall turbine efficiency η_t and turbine polytropic efficiency η_{p_t} that is

$$\eta_T = \frac{1 - (p_{ob}/p_{oa})^{\eta_{p_t}(k-1)/k}}{1 - (p_{ob}/p_{oa})^{(k-1)/k}} \tag{8.28}$$

Example Problem 8.6

An axial-flow turbine with an expansion ratio of 10 has a polytropic turbine efficiency of 0.90. Calculate the adiabatic (overall) turbine efficiency assuming constant specific heats with $k = 1.40$.

Solution

From Eq. (8.28),

$$\eta_T = \frac{1 - (0.1)^{0.257}}{1 - (0.1)^{0.286}} = 0.927$$

Figure 8.17 illustrates the variation in adiabatic turbine efficiency with increasing

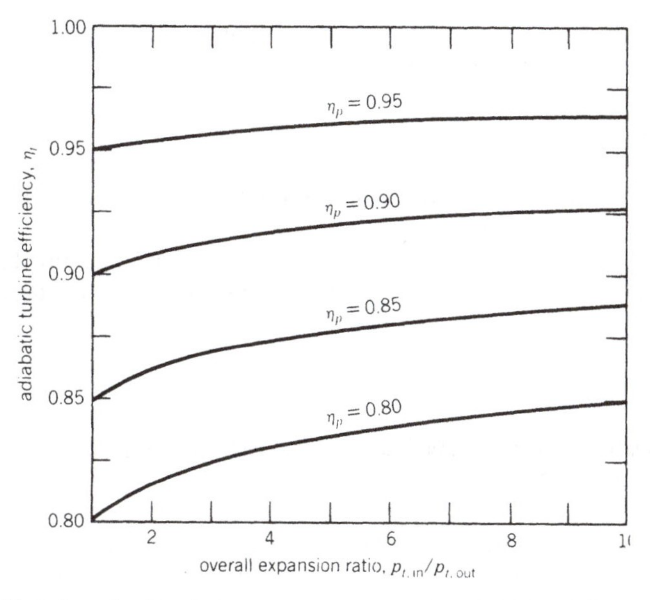

Figure 8.17. Variation of adiabatic turbine efficiency as a function of overall expansion ratio for several values of constant polytropic efficiency.

a turbine with an expansion ratio of 16, the overall adiabatic turbine efficiency is 96.7% for a polytropic efficiency of 95%, or 93.2% for a polytropic efficiency of 90%. Note that for a constant polytropic turbine efficiency, the overall turbine efficiency increases with increasing expansion ratio as illustrated in Figure 8.17.

8.11 TURBINE COOLING

The preceding sections discussed radial-and axial-flow turbines, their performance parameters and stage velocity diagrams.

Gas turbine engine designers are constantly trying to increase the thrust (or power) output from a given engine, reduce its weight, and/or decrease the fuel consumption for a specified output.

It was illustrated in Chapters 5 and 6 that one way to achieve these improvements is to operate at as high a turbine inlet temperature as possible. The turbine inlet temperature is limited by current materials. Improvements in materials have allowed the maximum turbine inlet temperature to increase slowly with time, but engine designers have had to seek ways of increasing the turbine inlet temperature at a rate faster than materials will allow.

One solution has been the use of turbine cooling, which has allowed the turbine designer to increase the turbine inlet temperature while maintaining a constant blade (material) temperature.

Turbine cooling was considered in German designs in 1935, with concentrated efforts during World War II. Since that time, most, if not all, engine designers have considered and/or used turbine cooling in their designs. Over the years, alloys have been developed that have allowed turbine inlet temperatures to increase at the rate of approximately 18°F per year. When turbine cooling was introduced, this allowed the rate at which the turbine inlet temperatures increased with time to approximately double.

When considering turbine cooling, one must first decide what fluid will be used as a coolant. Air is the most logical choice as a coolant, since it is readily available. It can be extracted (bled) from the compressor, ducted to the turbine blade (stator or rotor), and used as a coolant.

Another choice of coolant is a liquid coolant such as water. Section 8.12 discusses turbine cooling techniques using air as the coolant. Section 8.13 discusses turbine blade cooling using water (or other liquid) as the coolant.

8.12 TURBINE COOLING TECHNIQUES USING AIR AS COOLANT

Air is the main coolant that has been used with turbine blade cooling. The four methods used are convention cooling, impingement cooling, film cooling, and transpiration cooling. Each of these methods is discussed below.

Convection cooling is the simplest and was the first turbine cooling method used.

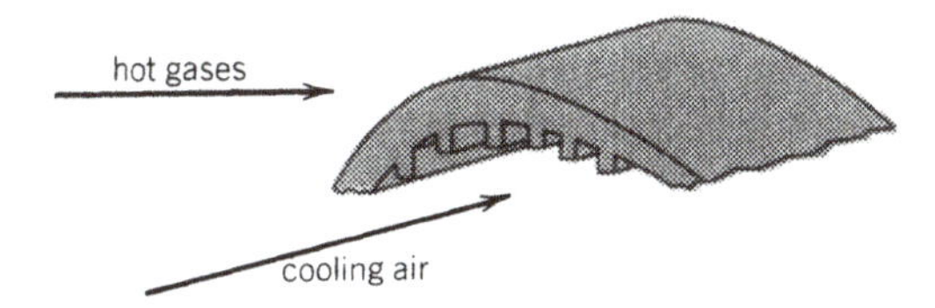

Figure 8.18. General convection cooling technique.

With convection cooling, the coolant (air) flows outward from the base of the turbine blade to the end through internal passages within the blade.

The general concept of convection cooling is illustrated in Figure 8.18. Several possible blade configurations using convection cooling are shown in Figure 8.19.

The effectiveness of convection cooling is limited by the size of the internal

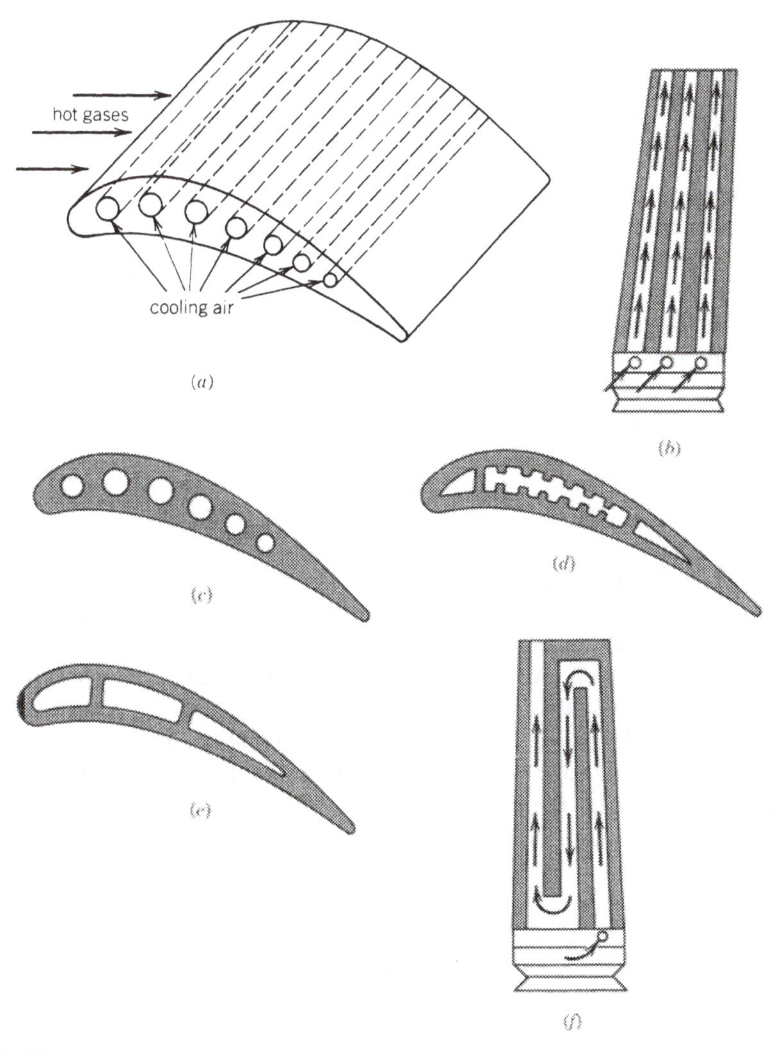

Figure 8.19. Several blade designs for convection cooling.

Figure 8.20. General impingement cooling technique.

passages within the blade and the restriction on the quantity of cooling air available. Air as a coolant requires large internal surface area and high velocity. In designing the internal passageways, one must decide between the most effective geometry from a heat transfer standpoint and the most economical from a manufacturing standpoint.

One of the major shortcomings of the early convection cooling techniques, in addition to the large amount of cooling air required, was the fact that this method failed to cool effectively the thin trailing edge of the blade—no cooling air passed through this portion of the blade since it was so thin.

Early blade designs used circular passageways of approximately the same diameter as shown in Figure 8.19a. Later designs, as illustrated in Figures 8.19d and 8.19e, divided the blades into several passageways of varying size and shape with the amount of cooling air to each passageway being controlled, the leading and trailing edge regions of the blade receiving more cooling air than the midsection because more cooling is required in these two regions. Some blades had a flow path similar to that shown in Figure 8.19f.

Impingement cooling is a form of convection cooling, the main difference being that instead of the air flowing radially through one or more sections of the blade, the cooling air is brought radially through a center core of the blade, then turned normal to the radial direction, and passed through a series of holes so that it impinges on the inside of the blade, usually just opposite from the stagnation point of the blade. This is shown schematically in Figure 8.20, with one possible design using impingement cooling illustrated in Figure 8.21.

Impingement cooling is a very effective method in local areas and is easily adapted to stator (nozzle) blades. This method may be used in rotor blades if sufficient space is available to include the required hardware inside the blade. This method is usually employed at the leading edge of the blade, but it may be used in other areas if desired.

Film cooling involves the injection of a secondary fluid (air) into the boundary

Figure 8.21. One possible impingement cooling design.

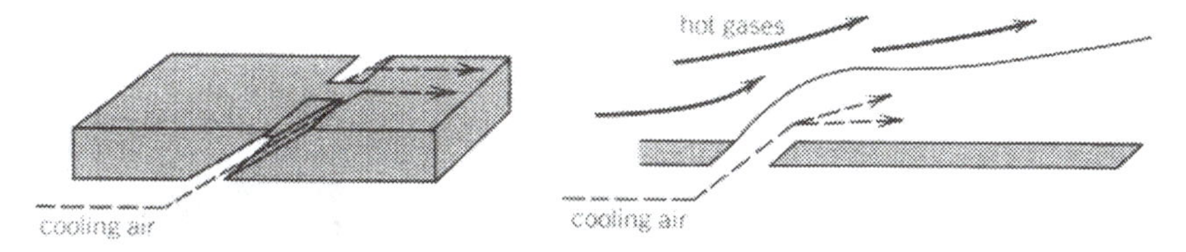

Figure 8.22. General film cooling technique.

layer of the primary fluid (hot gas). This is an effective way to protect the surface from the hot gases by directing the cooling air into the boundary layer to provide a protective, cool film along the surface. This method is shown schematically in Figure 8.22, with one possible design illustrated in Figure 8.23. The cool air forms a relatively cool insulating film, which maintains a lower blade material temperature than if no film cooling had been used.

The injection of the coolant air into the boundary layer causes turbine losses, which tend to reduce some of the advantages of using higher turbine inlet temperatures. Also, if too much air is injected into the boundary layer or the velocity is too high, the cooling air may penetrate the boundary layer, defeating the purpose of using film cooling. If the holes are placed too close together, stress concentrations, which could be detrimental to engine performance and reliability, may occur.

Film cooling is more effective than normal convection cooling or impingement cooling. The cooling air absorbs energy as it passes inside the blade and through the holes, then further reduces the metal (blade) temperature by reducing the amount of energy transferred from the hot gases to the blade. It should be noted that a large number of small holes are required in the blade, because the cooling effect of the film is quickly dissipated by downstream mixing of the film air with mainstream gases and the heating of the film air by the hot gases. One must also remember that the air used for film cooling must be at a high pressure. This is not always possible, especially for the leading (stagnation) point of the first nozzle.

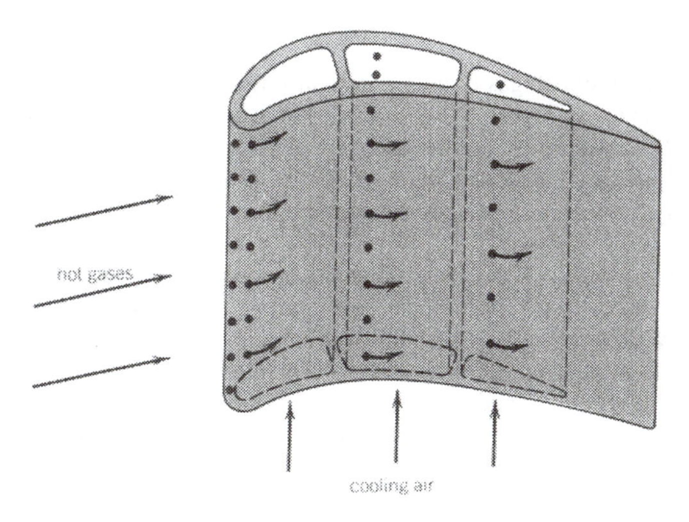

Figure 8.23. One possible film cooling design.

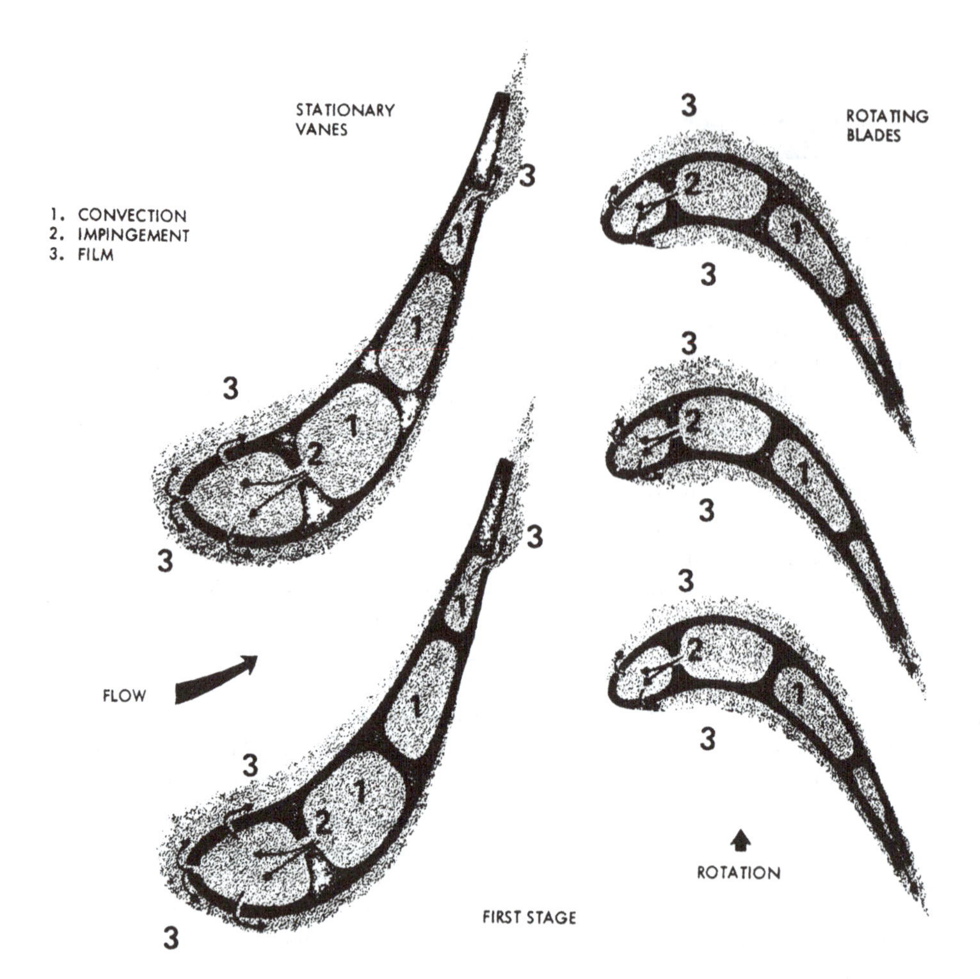

Figure 8.24. *Turbine blade and vane cooling techniques. (Courtesy of General Electric Co.)*

Most recently designed gas turbine engine turbines use several of the above methods. This is shown in Figure 8.24, which illustrates the use of all three methods. Figure 8.25 illustrates three possible rotor cooling configurations for a turbine rotor. The most complicated methods are used in the first-stage nozzle, the techniques decreasing in complexity as the gas temperature decreases with passage through the turbine stages.

Transpiration cooling is the most efficient air-cooling technique available. This is sometimes referred to as a full-blade film cooling or porous blade cooling. This method involves the use of a porous material through which the cooling air is forced into the boundary layer to form a relatively cooling, insulating film or layer. For effective cooling by the transpiration method, the pores (openings) should be small. This can lead to pore blockage due to oxidation and foreign material. A typical blade using transpiration cooling is illustrated in Figures 8.26 and 8.27.

should remember that although turbine blade cooling allows higher turbine

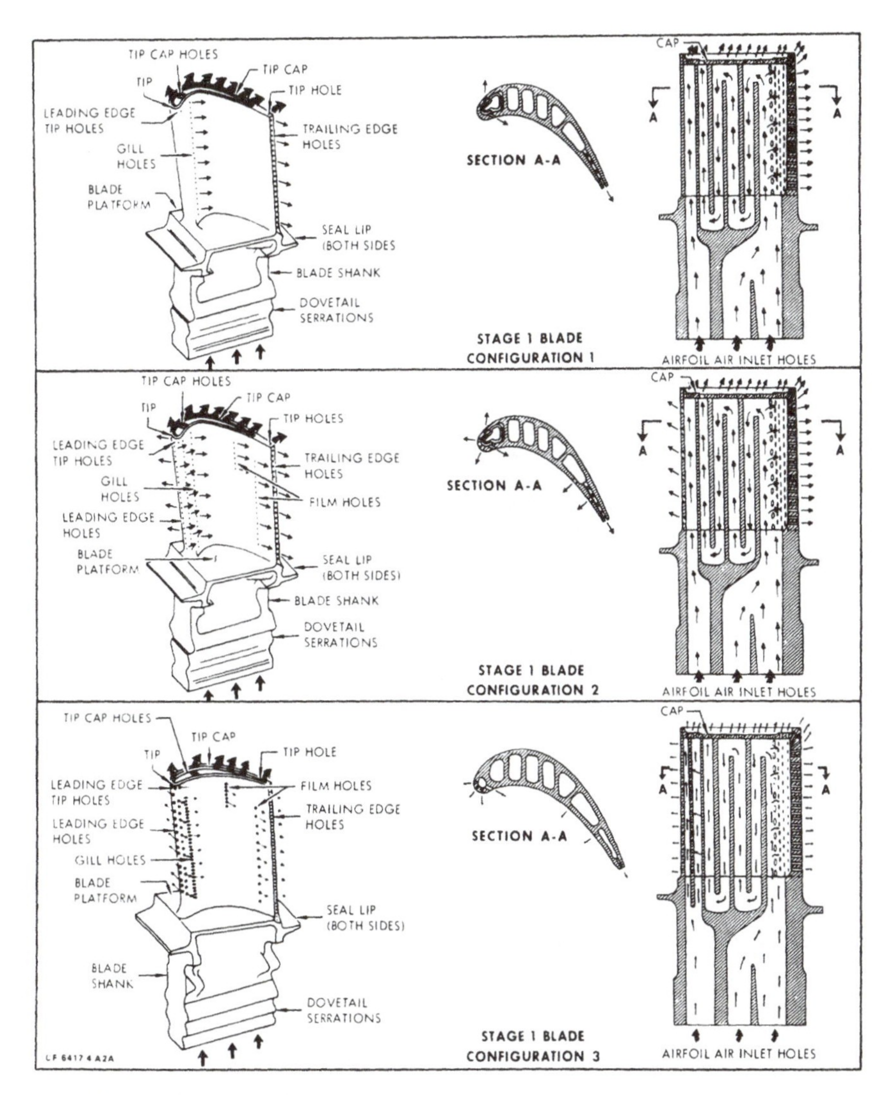

Figure 8.25. Three possible high-pressure rotor cooling configurations. (Courtesy of General Electric Co.)

Figure 8.26. Transpiration cooling technique.

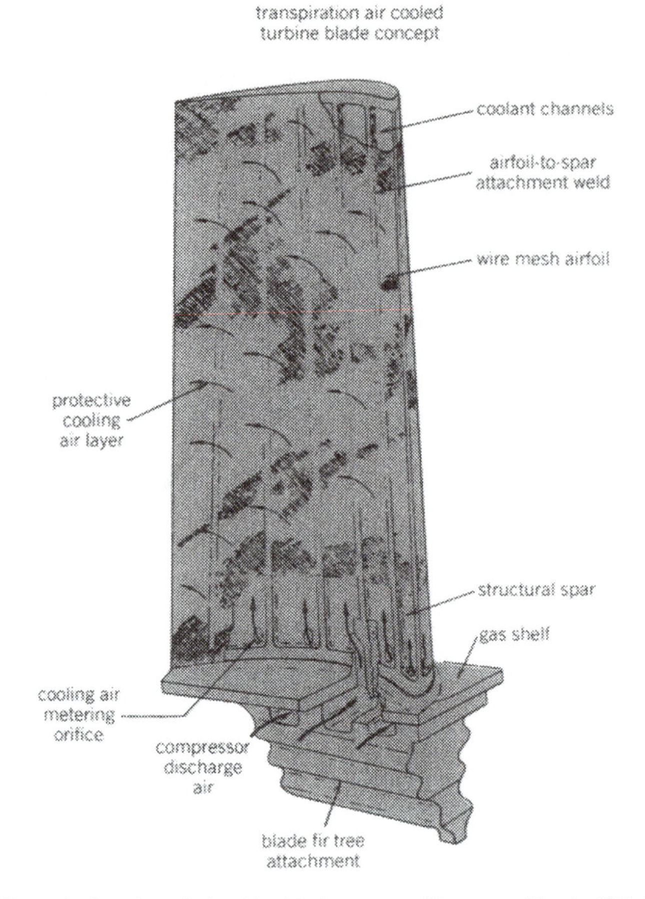

Figure 8.27. *Transpiration air-cooled turbine blade concept. (Courtesy of Curtiss-Wright Corporation)*

First, one must select a cooling technique that is efficient and has the least affect on the overall system. The detrimental effects for all cooling techniques include:

1. Added cost of producing turbine blades.
2. Turbine blade reliability
3. Loss of turbine work due to the cooling air bypassing one or more of the turbine stages
4. Loss due to the cooling air being mixed with the hot gas stream
5. Decrease in steam enthalpy when the cooling air is mixed with the hot gas stream

There is a limit to the maximum amount of cooling air that can be used for a net increase in power or thrust. The usual technique is to extract the cooling air at the outlet of the compressor, where its pressure is a maximum. Air may be extracted at a point prior to the compressor outlet if it is going to be used in a turbine blade where this supply pressure will be adequate.

8.13 LIQUID-COOLED TURBINE BLADES

A more effective way of cooling turbine blades is to use a liquid coolant. A liquid coolant has a much higher specific heat and provides the opportunity for evaporative cooling.

This method may be considered only for use in stationary power plants or other on-ground applications, since the size of the necessary heat exchanger for an aircraft application would prohibit its use there. On-ground applications must have available large quantities of clean water and can be designed with the necessary heat exchanger.

Difficulties do exist, even in stationary power plants, in transferring the liquid coolant to and from the moving rotor blades, maintaining a leakproof yet simple system, and eliminating the effects of corrosion and surface scaling.

SUGGESTED READING

Dixon, S. L., *Fluid Mechanics, Thermodynamics of Turbomachinery*, 3rd ed., Pergamon Press, Elmsford, N.Y., 1978.

Glassman, A. J. (ed.), *Turbine Design and Application*, Vol. 1, NASA SP-290, 1972.

Glassman, A. J. (ed.), *Turbine Design and Application*, Vol. 2, NASA SP-290, 1973.

Glassman, A. J. (ed.), *Turbine Design and Application*, Vol. 3, NASA SP-290, 1975.

Horlock, J. H., *Axial Flow Turbines, Fluid Mechanics and Thermodynamics*, R. E. Krieger Publishing Co., Huntington, N.Y., 1973.

Shepherd, D. G., *Principles of Turbomachinery*, Macmillan Co., New York, 1956.

NOMENCLATURE

A area
c_p constant pressure specific heat
D characteristic dimension
h enthalpy
i angle of incidence, ideal
$\dot{m}_g$ mass flow rate of gas
N rotor rotational speed
p pressure
Pr relative pressure
R gas constant
R' degree of reaction
T temperature
U blade velocity
$\overline{V}$ absolute velocity
w work
$\dot{w}$ power
W relative velocity
x length
α angle (absolute velocity)
β angle (relative velocity)
δ deviation angle
η efficiency
λ speed-work parameter
μ absolute gas viscosity
ϕ flow coefficient
Ψ work coefficient
ρ gas density

Subscripts

3, 3.5, 4, etc. states
a axial
h hub
m mean
o stagnation
p polytropic
t tip
T turbine
u tangential component

PROBLEMS

8.1. Solve Example Problem 8.1 assuming constant specific heats (c_p = 0.25 Btu/lb°R, k = 1.40).

8.2. The air in Example Problem 8.1 enters a second stage. The mean blade diameter of this second stage is 48.3 cm (19.0 in.). Other known data are as follows:

Turbine rotor speed	14,000 rpm
$\alpha_{4.5}$ (exit second-stage nozzle)	+72.0°
α_5 (exit second-stage rotor)	0°
$V_{a4.5} = V_{a5}$	170.0 m/s (550 ft/s))

Determine, assuming variable specific heats and that the pressure, temperature, and velocity at the inlet to the second stage are the exit conditions from the stage in Example Problem 8.1:

(a) The velocity diagrams at the mean blade diameter

(b) The Mach number of the absolute and relative velocities at states 4.5 and 5

(c) The work and power developed by this stage based on the mean blade diameter conditions and a mass rate of flow of 36.4 kg/s (50.0 lb/s)

(d) The ideal total-to-total expansion ratio (p_{o4}/p_{o5}) for this stage

Compare your answers with those of Example Problem 8.1.

8.3. The following information is known about a two-stage, axial-flow turbine:

Mass rate of flow into stage	15.0 kg/s (33.1 lb/s)
p_{o3} (total pressure, turbine inlet)	1013 kPa (147.0 psia)
T_{o3} (total temperature, turbine inlet)	1200 K (2160°R)
Mean blade diameter (=constant)	25.0 cm (9.84 in.)
Turbine rotor speed	26,500 rpm
$\overline{V}_a$ = constant	180.0 m/s (600 ft/s)
$\alpha_{3.5} = \alpha_{4.5}$	+70.0°
$\alpha_4 = \alpha_5$	0°

Assuming variable specific heats and that the axial velocity remains constant:

(a) Construct the velocity diagrams for both stages

(b) Calculate the work per stage and the power developed based on mean blade diameter values

(c) Calculate the absolute and relative Mach numbers at state 3.5, 4, 4.5, and 5

(d) Calculate the ideal total-to-total expansion ratios for each stage

8.4. Solve Example Problem 8.1 assuming $\alpha_{3.5}$ = +65.0°, all other values the same.

8.5. Solve Example Problem 8.1 assuming the axial velocity is 180.0 m/s (600 ft/s), all other values the same.

8.6. Solve Example Problem 8.1 assuming α_4 = −10°, all other values the same.

8.7. Solve Example Problem 8.1 assuming a turbine rotor speed of 11,000 rpm, all other values the same.

8.8. Calculate, for Problem 8.3

(a) The degree of reaction of each stage

(b) The value of the speed-work parameter for each stage

(c) The blade height at the exit from each nozzle and each rotor

8.9. Determine, for Problem 8.2, the annulus area and blade heights at states 4.5 and 5. Compare with the answer to Example Problem 8.4.

8.10. Determine, for the turbine in Problem 8.3:

(a) The stage expansion ratios if each stage has an efficiency of 87.0%

(b) The overall turbine efficiency if each stage has an efficiency of 87.0%

8.11. Solve Problem 8.3 assuming that instead of α_4 and α_5 being known, it is known that the degree of reaction of each stage is 50%.

8.12. Solve Problem 8.3 assuming that $\alpha_4 = \alpha_5 = -10°$, all other values being the same.

8.13. The following information is known at the mean blade diameter of an axial-flow turbine stage:

Blade velocity	900 ft/s (275 m/s)
Axial velocity	550 ft/s (170 m/s) = constant
Mean blade diameter	constant
Nozzle exit angle, $\alpha_{3.5}$	+72.0°

Determine, for the three diagram types listed below, the stage velocity diagrams (angles and magnitude), the speed-work parameter, and the degree of reaction.

(a) Zero exit swirl

(b) Impulse

(c) Symmetrical

8.14. A two-stage, axial-flow turbine is being designed. Known turbine design data are:

Mean blade radius	28.7 in. (73.0 cm) = constant
Rotational speed, N	5100 rpm
Turbine inlet total temperature, T_{03}	2340°R (1300 K)
Turbine inlet total pressure, p_{03}	123.2 psia (850 kPa)
Mass rate of flow into turbine	239 lb/s (108 kg/s)
Axial velocity (V_a = constant)	550 ft/s (170 m/s)
$\alpha_{3.5} = \alpha_{4.5}$	+75.4°
α_4	−27.3°
α_5	0°

The velocity leaving the turbine nozzles is at an angle of +75.4°. Calculate, assuming variable specific heats, no losses (adiabatic and reversible expansion) and that the axial velocity remains constant:

(a) The velocity diagrams for both stages

(b) The work, Btu/lb, for each state based on mean blade diameter values.

(c) The absolute and relative Mach numbers at states 3.5, 4, 4.5, and 5.

(d) The ideal total-to-total expansion ratio for each stage.

(e) The required blade height at states 3.5, 4, 4.5, and 5.

(f) The stage speed-work parameter.

(g) The value of $\dot{m}\sqrt{T_o}/p_oA$ where T_o and p_o are the stagnation values at the inlet to the first nozzle and A is the area at the exit from the first nozzle.

8.15. An axial-flow turbine is being designed to drive an axial-flow compressor. Turbine design data are:

Turbine inlet total temperature, T_{o3}	2520°R (1400 K)
Turbine inlet total pressure, p_{o3}	115 psia (790 kPa)
Rotor speed, N	9000 rpm
Mean blade diameter	18.0 in. (45.7 cm)
Total turbine work output	102 Btu/lb (237 kJ/kg)

The velocity leaving the turbine nozzle(s) is 1300 ft/s (396 m/s) at an angle of +68.0°. Calculate, assuming constant specific heat, no losses (adiabatic and reversible expansion) and that the axial velocity remains constant throughout the turbine:

(a) The velocity diagram and absolute and relative Mach numbers for a single-stage turbine

(b) The number of stages required if the velocity at the exit from each rotor is in the axial direction. Results must include velocity diagrams, total pressures, and total temperatures at the exit from each rotor. Assume the same shape for all stages.

(c) The stage speed-work parameter for (a) and (b).

(d) The maximum mass rate of flow for this turbine if the ratio r_{hub}/r_{tip} at the *exit* from the last rotor is 0.85.

(e) The value of $\dot{m}\sqrt{T_{o3}}/p_{o3}\,A$/for $\dot{m}$ = 100 lb/s (233 kg/s).

8.16. A turbojet engine operates under the following conditions:

Mach number	0.90
Altitude	11,000 m (36,089 ft)
Air flow into engine	1 kg/s (1 lb/s)
Compressor total pressure ratio	24
Compressor efficiency	87.0%
Turbine efficiency	89.0%
Nozzle efficiency	100%
Turbine inlet total temperature, T_{o3}	1200 K (2160°R)
Pressure leaving nozzle is ambient	

Assuming an air standard cycle, no pressure drop between compressor outlet and turbine inlet, and taking into account the mass of fuel added, calculate:

(a) The thrust developed by this engine

(b) The thrust-specific fuel consumption

8.17. Solve Problem 8.16 assuming that the turbine inlet temperature is 2520°R (1400 K) and that 6% of the air leaving the compressor bypasses the combustion chamber and turbine and is added back into the main stream at the turbine exit total pressure (see the following diagram); that is, $p_{o4} = p_{o5}$.

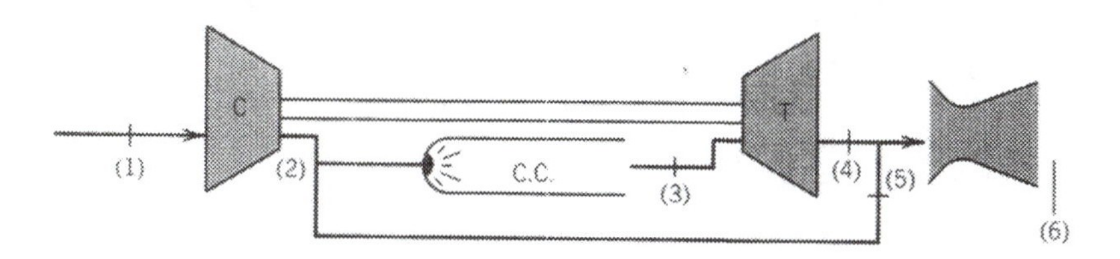

8.18. Solve Problem 8.17 assuming that the turbine inlet temperature is 1300 K (2340°R) and that 4% of the air leaving the compressor bypasses the combustion chamber and turbine.

9

INLETS, COMBUSTION CHAMBERS, AND NOZZLES

9.1. GENERAL

Compressors and turbines were analyzed in the two previous chapters. It is now important to analyze three additional components: inlets, combustion chambers, and nozzles. These will be examined in this chapter.

The purposes of any aircraft gas turbine engine inlet is to provide a sufficient air supply to the compressor with as low a loss in total pressure as possible and with as small a drag force on the airplane as possible. It also has the purpose of being a diffuser that reduces the velocity of the entering air as efficiently as possible. Current inlets on subsonic aircraft usually contain noise-absorbing materials, a topic discussed in Chapter 11.

9.2. SUBSONIC INLETS

It was observed in the photographs in the earlier chapters that there are a number of different types of air inlet ducts for aircraft gas turbine engines. With turboprop engines, the inlet design is complicated by the propeller and gear box at the inlet to the engine. Aircraft engines may be located under the wings of the aircraft, at the base of the vertical stabilizer, or in the fuselage of the aircraft with the inlet located in the root of the wing or under the fuselage. Each of these installations can cause problems associated with subsonic inlets, namely, distortion at the compressor inlet and total pressure losses. Inlets also may be classified as single-entrance, such as occurs with engines installed beneath the wings of an aircraft, or divided entrance, such as occurs in fighter aircraft when the inlets are located at the roots of the aircraft

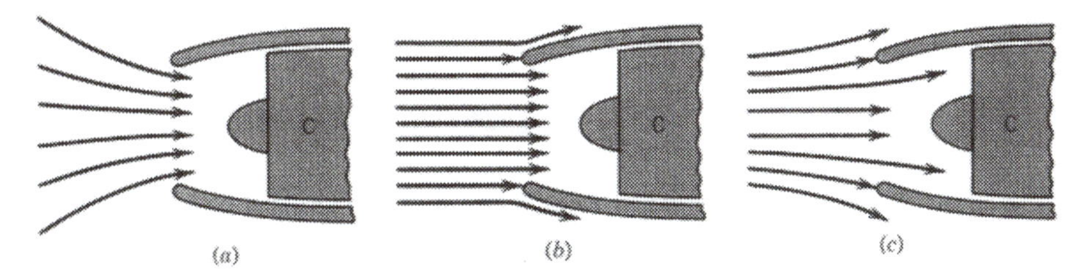

Figure 9.1. Air flow pattern for several forward speeds. (a) Static operation. (b) Low forward speed. (c) High forward speed.

wings. The divided-entrance configuration may lead to distortion (pressure and/or temperature variation) at the inlet to the engine.

Subsonic inlets are of fixed design, although inlets for some high-bypass-ratio engines have been designed with blow-in doors, which are spring-loaded parts installed in the perimeter of the inlet duct designed to deliver additional air to the face of the gas turbine engine at high power output and low aircraft forward speed. The internal surface of a subsonic inlet is a diffusing section ahead of the compressor. The air pattern at the inlet at zero aircraft forward speed (static operation) is shown in Figure 9.1a, at low forward speed is shown in Figure 9.1b, and for a high forward speed in Figure 9.1c. It is essential that the inlet be designed so that boundary-layer separation does not occur.

Figure 9.2 illustrates the effect of changing the angle of attack at the inlet to the duct.

9.3. SUPERSONIC INLETS

Supersonic inlets are much more difficult to design than are subsonic inlets. The inlet used on a supersonic aircraft is the design that optimizes performance for the mission for which the aircraft is designed. The inlet must provide adequate subsonic performance, good pressure distribution at the compressor inlet, high pressure recovery ratios, and must be able to operate efficiently at all ambient pressures and temperatures during take-off, subsonic flight, and at its supersonic design condition.

Both axisymmetric and two-dimensional inlets have been designed and used. Vari-

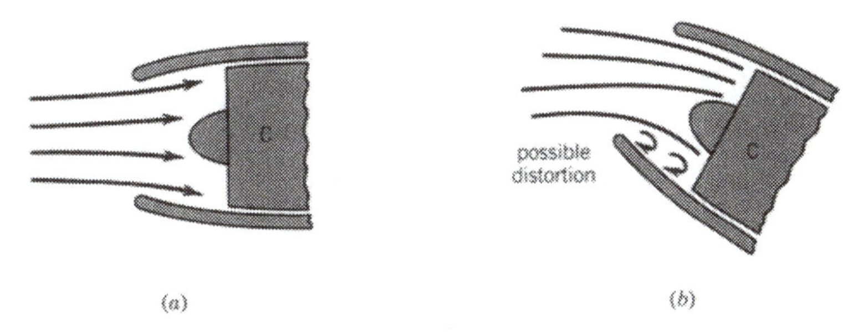

Figure 9.2. Effect of angle of attack on inlet air flow pattern. (a) Normal operation. (b) High angle of attack.

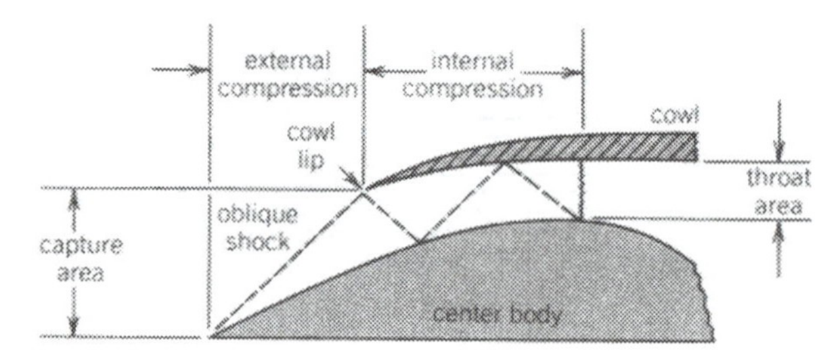

Figure 9.3. Supersonic inlet.

able geometry center bodies, which change the inlet geometry for better off-design operation, and boundary-layer bleed have been incorporated into supersonic inlets.

The inlet performance characteristics that are used to assess the performance of supersonic inlets and that have the largest influence on aircraft performance and range are

1. Total pressure recovery
2. Cowl drag
3. Boundary-layer bleed flow
4. Capture-area ratio (mass flow ratio)
5. Weight

Supersonic inlets are usually classified by their percent of internal compression. Internal compression refers to the amount of supersonic area change that takes place between the cowl lip and the throat. This is illustrated in Figure 9.3, which identifies the center body, cowl lip, capture area, and throat for a supersonic inlet.

The total supersonic area change is the difference between the capture area and the minimum (throat) area. The area change that occurs in front of the cowl lip is called external compression; the amount of *supersonic* area change that occurs between the cowl lip and the throat is called internal compression. Inlets are usually classified as external compression inlets, mixed compression inlets, and internal compression inlets.

The *external* compression inlet completes the supersonic diffusion process outside the covered portion of the inlet. The normal shock where the flow changes from supersonic to subsonic and the throat are ideally located at the cowl lip as illustrated in Figure 9.4.

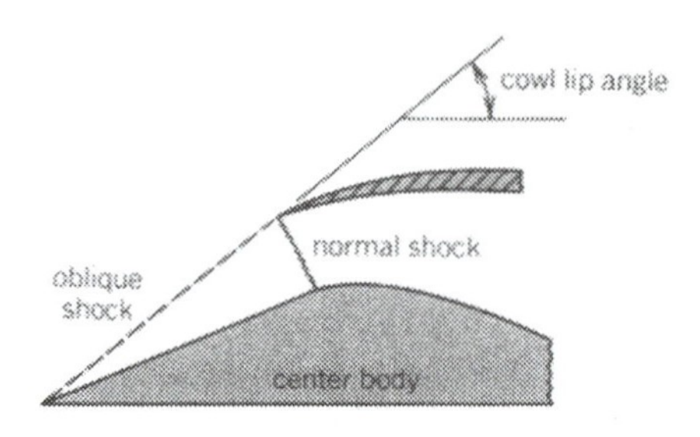

Figure 9.4. Supersonic inlet, all external compression.

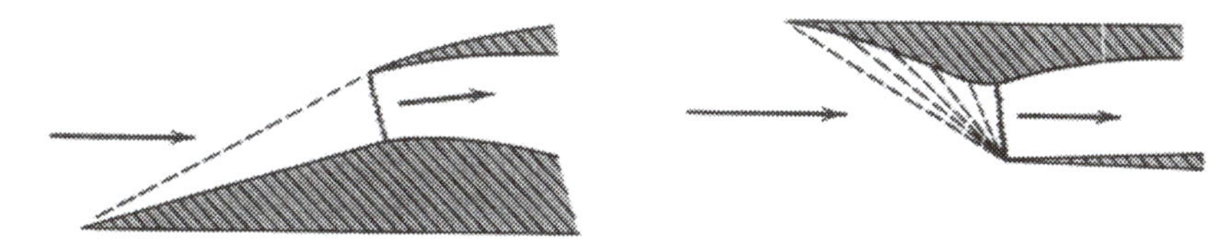

Figure 9.5. External compression inlets with critical flow.

The fixed external compression inlet is designed so that the oblique shock or shocks intersect the cowl lip as shown in Figure 9.5. The normal shock is located at the cowl lip and is referred to as critical flow.

Example Problem 9.1

Air enters a two-dimensional supersonic diffuser at a pressure of 2.039 psia (14.102 kPa), a temperature of 390°R (217 K), and with a Mach number of 3.0. The two-dimensional oblique shock diffuser has an oblique shock angle of 27.8°, which is followed by a normal shock as shown in the diagram. Determine, assuming constant specific heats:

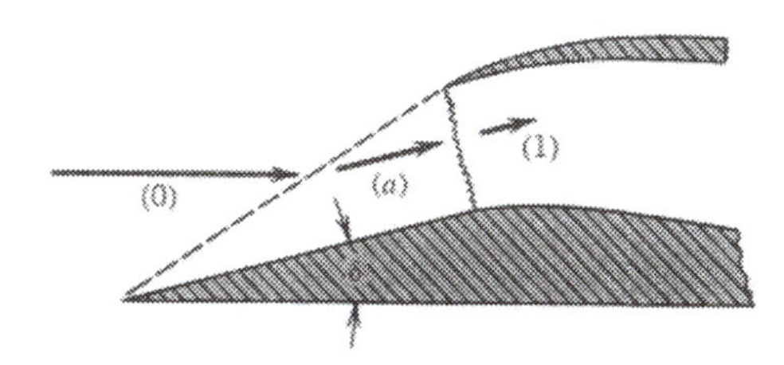

(a) The velocity of the air entering the oblique shock
(b) The total temperature and pressure of the air entering the oblique shock
(c) The Mach number after the oblique shock
(d) The total pressure after the oblique shock
(e) The flow deflection angle
(f) The static pressure and temperature after the oblique shock
(g) The Mach number after the normal shock
(h) The total pressure after the normal shock
(i) The static pressure and temperature after the normal shock

Solution

From Table B.3, $k = 1.400$.

$$\text{(a)}\quad \overline{V}_0 = 3.0\sqrt{kg_cRT_0}$$

$$= 3.0\sqrt{(1.4)(32.174)(53.35)(390)}$$

$$= 2904 \text{ ft/s } (885 \text{ m/s})$$

(b) $T_{00} = T_0 + \dfrac{\overline{V}_0^2}{c_p 2 g_c}$

$$= 390 + \frac{(2904)^2(28.965)}{(6.963)(2)(32.174)(778)}$$

$$= 390 + 701 = 1091°\text{R (606 K)}$$

$$P_{o0} = p_0\left(\frac{T_{o0}}{T_0}\right)^{k/(k-1)} = 2.039\left(\frac{1091}{390}\right)^{1.4/0.4}$$

$$= 74.66 \text{ psia (516.3 kPa)}$$

The conditions across the oblique shock were calculated in Example Problem 3.4. Therefore, from Example Problem 3.4:

(c) $M_a = 1.75$

(d) $P_{oa} = p_{o0}\,(0.95819) = 71.54$ psia (494.8 kPa)

(e) $\delta = 10.3°$

(f) $p_a = (2.12)(2.039) = 4.323$ psia (29.90 kPa)
$T_a = (390)(1.2547) = 489°$R

(g) The Mach number after the normal shock may be calculated using Eq. (3.33):

$$M_1 = \sqrt{\frac{(0.4)(1.75)^2 + 2}{(2)(1.4)(1.75)^2 - 0.4}} = 0.628$$

(h) The total pressure ratio across the normal shock may be calculated from Eq. (3.38) and is

$$\frac{P_{o1}}{p_{o0}} = \left[\frac{(2.4)(1.75)^2}{2 + (0.4)(1.75)^2}\right]^{3.5}\left[\left(\frac{2.8}{2.4}\right)(1.75)^2 - \frac{0.4}{2.4}\right]^{-2.5}$$

$$= 0.8346$$

$$p_{o1} = (0.8346)(71.54) = 59.71 \text{ psia (413.0 kPa)}$$

(i) The static temperature and pressure ratio across the normal shock may be calculated using Eqs. (3.39) and (3.40), respectively. The results are

$$\frac{T_1}{T_a} = \left[\frac{2}{(2.4)(1.75)^2} + \frac{0.4}{2.4}\right]\left[\left(\frac{2.8}{2.4}\right)(1.75)^2 - \frac{0.4}{2.4}\right]$$

$$= 1.4946$$

$$T_1 = (1.4946)(489) = 731°\text{R (406 K)}$$

$$\frac{p_1}{p_a} = 1 + \frac{2.8}{2.4}[(1.75)^2 - 1] = 3.4063$$

$$p_1 = (3.4063)(4.323) = 14.73 \text{ psia (101.8 kPa)}$$

At air velocities below the design velocity or for high internal flow resistance, the normal shock occurs in front of the cowl lip as shown in Figure 9.6a. This is referred to as subcritical flow, and flow spillage occurs. At air velocities above design, the oblique shock may change as shown in Figure 9.6b, with the shock impinging inside the cowl lip and the normal shock moving to the diverging section. This type of operation is referred to as supercritical operation. Ways to eliminate these problems are discussed later.

Example Problem 9.2

Air enters the supersonic diffuser of Example Problem 9.1 at a Mach number of 2.0. The oblique shock angle is 39.6°. Calculate:

(a) The total pressure ratio across the oblique shock.
(b) The static temperature ratio across the oblique shock.
(c) The static pressure ratio across the oblique shock.
(d) The Mach number after the oblique shock.
(e) Verify that the oblique shock angle would be 39.6° for a deflection angle of 10.3°.

Solution

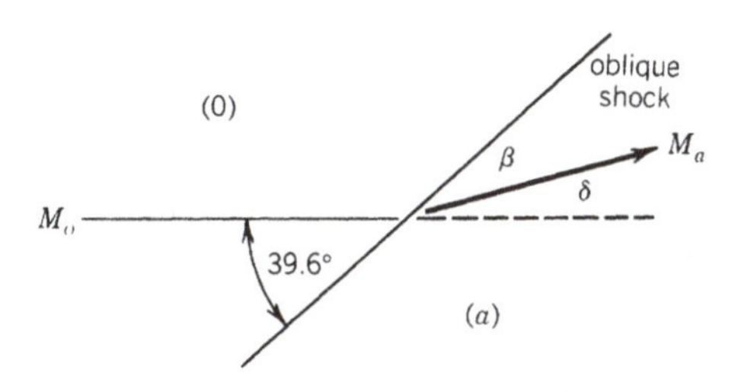

$$M_{0,n} = (2)(\sin 39.6°) = 1.275$$

$$M_{0,t} = (2)(\cos 39.6°) = 1.541$$

From Eq. (3.33),

$$M_{a,n} = \sqrt{\frac{(0.4)(1.275)^2 + 2}{(2)(1.4)(1.275)^2 - 0.4}} = 0.7990$$

From Eq. (3.38),

$$\frac{p_{oa}}{p_{o0}} = \left[\frac{(2.4)(1.275)^2}{2 + (0.4)(1.275)^2}\right]^{3.5} \left[\left(\frac{2.8}{2.4}\right)(1.275)^2 - \frac{0.4}{2.4}\right]^{-2.5}$$

$$= 0.9835$$

From Eq. (3.39),

$$\frac{T_a}{T_0} = \left[\frac{2}{(2.4)(1.275)^2} + \frac{0.4}{2.4}\right]\left[\left(\frac{2.8}{2.4}\right)(1.275)^2 - \frac{0.4}{2.4}\right]$$

$$= 1.1751$$

$$\frac{p_a}{p_0} = 1 + \left(\frac{2.8}{2.4}\right)\left[(1.275)^2 - 1\right] = 1.7299$$

$$M_{a,t} = 1.541\sqrt{\frac{1}{1.1751}} = 1.4216$$

$$M_a = \sqrt{(1.4216)^2 + (0.7990)^2} = 1.631$$

$$\beta = \tan^{-1}\left(\frac{0.7990}{1.4216}\right) = 29.3°$$

$$\delta = 39.6 - 29.3 = 10.3°$$

Example Problem 9.2 illustrates the fact that as the Mach number of the air entering the supersonic diffuser decreases, the angle of the oblique shock (for a fixed wedge angle) changes. This means that for a fixed geometry and center body location, the capture area changes. This was illustrated in Figure 9.6.

Example Problem 9.1 illustrated the fact that if an external compression diffuser uses a single oblique shock followed by a normal shock, the ratio of the total pressure after the normal shock to the total pressure before the oblique shock is quite low. For the situation illustrated by Example Problem 9.1, the ratio was

$$\frac{p_{o1}}{p_{o0}} = \frac{59.71}{74.66} = 0.80$$

a very low value.

The external compression inlet uses oblique shocks for a portion of the supersonic

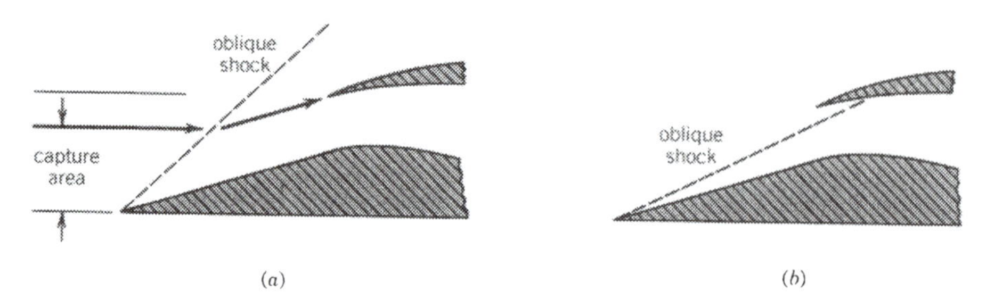

Figure 9.6. Subcritical and supercritical operation of a supersonic inlet. (a) Subcritical operation. (b) Supercritical operation.

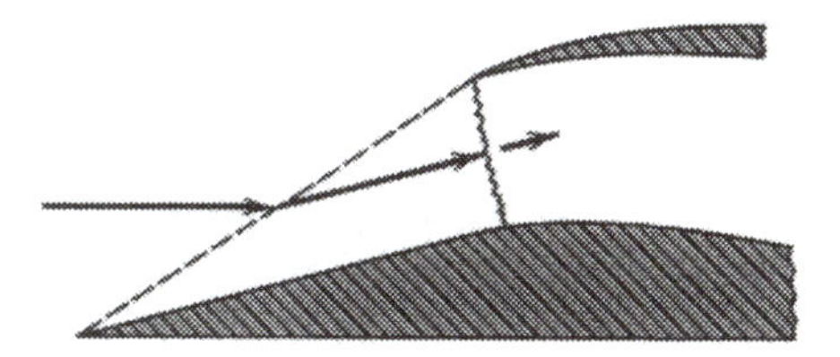

Figure 9.7. Single wedge external compression supersonic inlet.

flow diffusion. The oblique shocks are followed by a normal shock across which the flow changes from supersonic to subsonic.

The external compression inlet has a center body that extends forward of the cowl lip. The center body may be a cone or wedge that produces a single oblique shock as illustrated in Figure 9.7; a double cone or wedge that produces two oblique shocks that intersect at the cowl lip as illustrated in Figure 9.8; or a multiangle center body that produces a series of weak shocks as illustrated in Figure 9.9. A diffuser that uses an infinite number of infinitesimal oblique shocks to reduce the supersonic flow to sonic flow would be an isentropic compression process and would be the most efficient.

The internal compression inlet locates all of the shocks within the covered passageway as illustrated in Figure 9.10. The terminal (normal) shock is usually located near the throat of the inlet. The inlet is said to be operating in the critical mode when the terminal shock is located at the throat, in subcritical operation when the normal shock is located upstream of the throat, and in supercritical operation when the terminal shock is located downstream of the throat. The terminal shock, for subcritical operation, is located in the converging section of the inlet and is unstable. It will move ahead of the cowl lip, producing a condition called unstart. When this happens, the pressure recovery of the inlet drops and flow spills over the cowl, producing a high drag and aircraft control problems. For this reason, supersonic inlets are designed to operate supercritically instead of critically. This is done to maintain a margin of stability in the event of a sudden change in the inlet flow. It must be remembered that the highest total pressure recovery occurs when the terminal shock is located at the throat.

The *mixed compression* inlet uses a combination of external and internal compression. This type of inlet is illustrated in Figure 9.11.

The type of inlet selected for an aircraft designed to operate at supersonic speeds depends on the aircraft mission—that is, what fraction of the flying time will be at supersonic speeds, the cruise Mach number, and cruise altitude. Most, if not all, supersonic inlets currently in use are either external compression or mixed compression inlets.

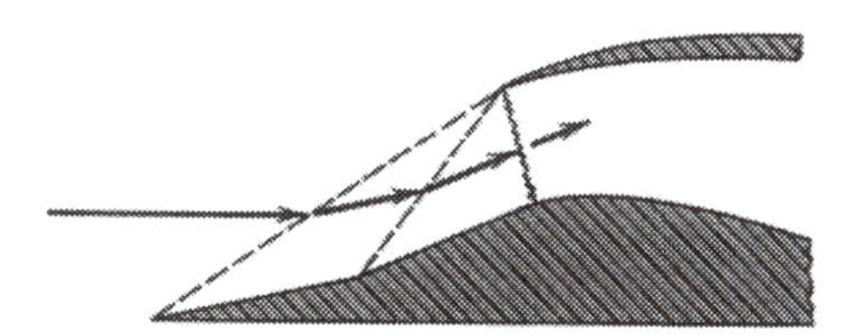

Figure 9.8. Double cone or wedge external compression supersonic inlet.

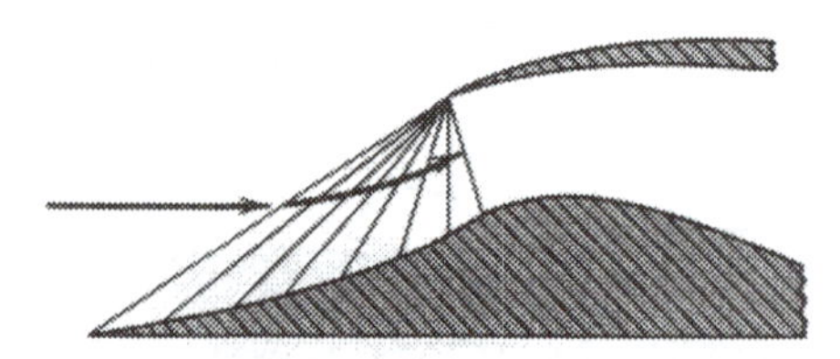

Figure 9.9. Isentropic compression supersonic inlet.

Inlets are designed for one flight condition but must provide stable operation and an adequate match with the compressor during off-design operation. Several ways of controlling the inlet flow to provide stable operation and a proper match between the inlet and engine are to use variable geometry and boundary-layer bleed and to incorporate flow bypass into the inlet design.

Variable inlet geometry may involve the use of a translating center body, use of a cowl where the lip angle may be varied, variable ramp angles, and/or variable throat areas. Varying the geometry allows the inlet to operate with the optimum shock location, thereby providing optimum off-design performance. Using variable geometry inlets requires the use of sensors, which adds complexity and weight to the inlet.

Bleeding involves the extraction of air from the low-momentum boundary layer. This reduces the possibility of flow separation in the diffuser and improves engine matching. This is normally extracted at the points where the shock wave reflects from the wall.

Bypass (dump) doors are also used. They allow excess inlet air to spill in cases of engine throttling or shut-down.

9.4 COMBUSTION CHAMBERS

Efforts to produce an efficient, compact, low-emission combustion chamber for a gas turbine engine, especially for aircraft application, are hampered by the wide range of operating conditions over which the system must operate. The operating range must include starting, idle, acceleration and deceleration, and full-power operation. Aircraft engines must operate at sea level conditions and at high altitudes.

Requirements for a gas turbine engine combustion chamber system include:

1. Release of the fuel chemical energy in the smallest space (length and diameter) possible
2. Minimum pressure drop over the operating spectrum

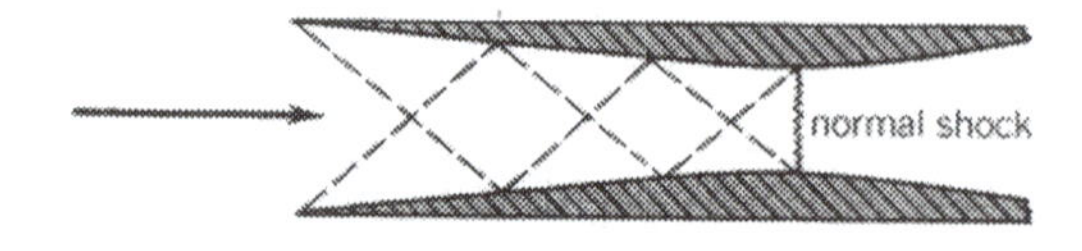

Figure 9.10. Internal compression supersonic inlet.

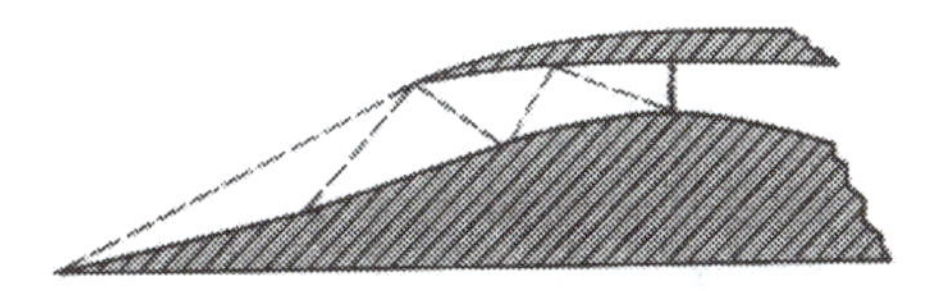

Figure 9.11. Mixed compression inlet.

3. Stable and efficient operation over a wide range of fuel-air ratios, altitudes, flight speeds, and/or power levels
4. Reliability equal to or greater than the overhaul life of the engine
5. Relight capability at altitude for aircraft engines
6. Good (near uniform) temperature distribution at the inlet to the turbine stator (nozzles)
7. Low emissions (high combustion efficiency)

A typical combustion chamber is shown in Figure 9.12. Three zones, the diffuser, primary, and secondary/dilution, are identified in this figure.

The *diffuser zone* is a transition zone between the compressor outlet and combustion chamber inlet. It is important to reduce the velocity since the pressure loss is a function of the velocity squared. Maintaining typical compressor axial velocities of 500–600 ft/s (150–170 m/s) would lead to excessive pressure losses.

The function of the *primary zone* is manyfold. First, this is the region where fuel is injected and ignition occurs. The fuel must be injected in a manner that provides approximately a stoichiometric mixture of air and fuel that is uniformly distributed. The fuel injection system must be able to do this over the entire operating range, that is, from idle to full power. The air velocity, for all operating conditions, must be below the flame velocity so that the flame is not carried out of the combustion chamber. Fuel droplet size, which is a function of the fuel supply pressure, is important. A liquid fuel must be evaporated before it can be burned. The evaporation rate is improved if the liquid fuel has a large surface area and if the fuel is injected with a high velocity. Injecting a fuel at high velocity breaks the fuel into smaller droplets, thereby enhancing evaporation. An alternative method is to heat the fuel.

If the droplets are too small, they will not penetrate well into the airstream. Alternatively, if the droplets are too large, the evaporation time is increased, leading to poor combustion.

Primary air enters either through or near the fuel nozzle. Additional air, sometimes called secondary air, is introduced as shown in Figure 9.12 to ensure complete combustion. The fuel-air ratio must be within certain limits for combustion to occur. These limits vary with pressure, air temperature, and velocity, but usually are between an equivalence ratio of 0.6 and 2.5. The equivalence ratio is defined as

$$\phi = \frac{(f/a)_{\text{actual}}}{(f/a)_{\text{stoich}}} \tag{9.1}$$

Note that if the equivalence ratio is less than 1.0, it is a lean mixture; if the equivalence ratio is greater than 1.0, it is a rich mixture.

The function of the secondary/dilution zone is to introduce the remaining air to

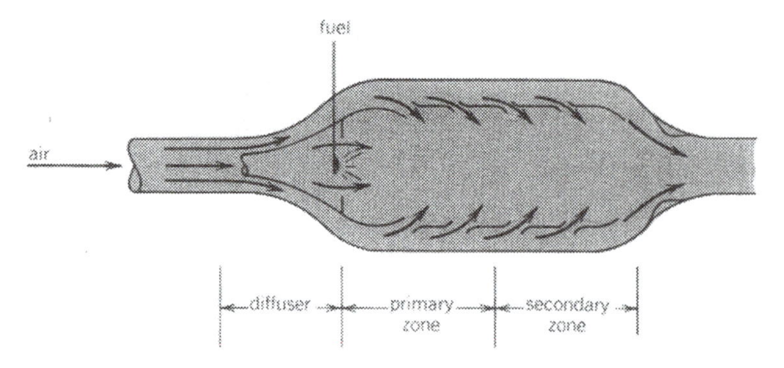

Figure 9.12. Conventional gas turbine combustion chamber.

reduce the combustion chamber gases to the desired turbine inlet temperature and to provide adequate mixing to obtain a uniform temperature distribution at the turbine nozzle inlet to avoid "hot spots."

Most gas turbines have combustion systems with a fixed number of fuel nozzles of fixed size. This means that fuel at both idle and full power is injected through the same number of fuel nozzles. Variations under development are discussed in Section 9.5. Injecting fuel through the same number of nozzles at both idle and full power leads to undesirable conditions as discussed below.

At low power levels, typically engine idle, single-stage combustion chambers as illustrated in Figure 9.12 have the following undesirable characteristics:

1. Poor atomization and distribution
2. Low combustion stability, primarily because of the low air inlet temperature and pressure
3. Possible quenching of the combustion gases before combustion is complete

At low power levels, the turbine inlet temperature is low. This means that single burning stage combustion chambers will be fuel-rich in the primary zone with poor vaporization and mixing with extensive dilution in the secondary zone leading to the formation of carbon monoxide and total hydrocarbons.

At high power levels, typically full power for on-ground application or take-off for aircraft gas turbine engines, the combustion efficiency is virtually 100% and carbon monoxide and total hydrocarbon levels are extremely low. Oxides of nitrogen become a problem because of the high maximum combustion chamber temperatures.

9.5 TYPES OF COMBUSTION SYSTEMS

There are three basic types of combustion chamber systems used in gas turbine engines. The three types are can, annular, and can-annular or cannular, a combination of the can and annular types of combustion chambers.

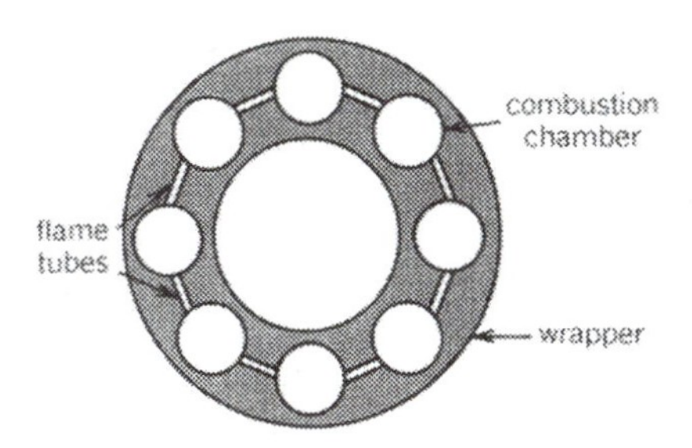

Figure 9.13. Gas turbine engine with canannular combustion chamber.

A *can-type* combustion chamber is composed of a number of small-diameter units with individual liners and wrapper (casing). Can-type combustion chambers are most often used on gas turbine engines with centrifugal compressors. As the air leaves the compressor diffuser, it is divided and ducted to the individual cans.

Can-type combustion chambers have excellent structural strength without excessive weight. Individual cans may be inspected, removed, and replaced, and it is easier to produce a number of small cans than to make a single large system.

The can-type combustion chamber, for the same air flow, is longer, since the individual can diameter must be held to a minimum. The individual cans are wasteful of the available space, since they use the individual cross-sectional area inefficiently. This type of combustion chamber also subjects the turbine nozzles to a wide variation of temperatures. Hot spots are common if the dilution air is not properly mixed with the combustion gases.

The *annular* combustion chamber consists of one or two continuous shrouds. The fuel is introduced through nozzles at the inlet to the shroud, with secondary air entering through holes. This secondary air keeps the flame away from the shroud and dilutes the combustion chamber gases to the desired turbine inlet temperature.

The annular combustion chamber has the advantage of using the available space effectively and of providing a near-uniform gas mixture at the inlet to the turbine nozzles and the lowest pressure drop. Its main disadvantages are that it is impossible to disassemble the burner liner without completely removing the engine, it has a tendency to warp, and it is structurally weak.

The *can-annular* type of combustion chamber has been used on many of the gas turbine engines currently in use. The can-annular design consists of individual cans placed inside a cylindrical combustion chamber as shown in Figure 9.13.

Each can has primary air admitted near the fuel nozzle and is perforated so that secondary cooling air may be admitted downstream of the primary zone. There usually is a flame tube that joins the cans, enabling the flame to pass from can to can during starting.

The can-annual type of combustion chamber utilizes the cross-sectional area more effectively than does the can-type combustion chamber and allows the use of a number of flame tubes, which eliminates the development problems inherent in the annular combustion chamber. It is structurally stronger than the annular design and usually has a shroud that is removable, permitting access to the cans without removing the engine to service. This design provides a fairly even temperature distribution at the turbine nozzle inlet, which reduces the danger of hot spots.

The main combustion chamber performance parameters are

1. Pressure drop
2. Combustion efficiency
3. Outlet temperature distribution
4. Stability

The ideal combustion chamber would have zero pressure drop, a combustion efficiency of 100%, a uniform exit temperature, and be stable over the entire engine operating range with relight capability under all conditions. This, of course, is impossible, and any design is a compromise. The ideal combustion chamber configuration is thought to be the full annular combustion chamber.

One of the most important factors affecting pressure loss is the obstructions placed in the gas flow stream that are used to stabilize the flame and provide proper mixing. The flame holder velocity must be low and the secondary air usually passes through small openings, both of which lead to pressure losses.

The trade-off that must occur in a combustion chamber becomes apparent when size versus performance is considered. As the combustion chamber size is reduced and velocities increase, the pressure loss increases. This, of course, has an adverse effect on performance. The mixing must be intense, but this increases the turbulence and combustion chamber loss.

Combustion efficiency is a measure of the completeness of combustion. High combustion chamber efficiencies are achieved when the fuel–air mixture is mixed adequately, the fuel is completely vaporized, and adequate time is provided for the chemical reaction to occur before the secondary (dilution) air is introduced, which slows the reaction by lowering the mixture temperature, in many cases lowering the temperature to a point where it stops the chemical reaction, freezing the chemical composition.

When operating conditions are changed from the design conditions, combustion efficiency decreases. This occurs because the fuel spray pattern is changed, atomization is poor because the droplet size increases or the velocity is decreased, leading to poor vaporization and fuel–air mixing, and the reaction time increases because of reduced inlet temperature and/or pressure.

The flame tends to grow longer as the combustion chamber inlet pressure is reduced. This means that since the combustion chamber length is fixed by a point at which the secondary (dilution) air is introduced, the flame is quenched long before combustion is complete, leading to the formation and emission of carbon monoxide and total hydrocarbons. This is very pronounced in aircraft engines at altitude because of the low secondary air temperatures.

The outlet temperature distribution affects the time between overhauls of a gas turbine engine and engine performance. A gas turbine engine is designed for a maximum turbine inlet temperature. A poor temperature distribution produces hot spots at the turbine nozzle inlet, leading to mechanical problems and thermal stresses. Effective mixing usually requires some means of ducting one stream across the other.

There is, for every combustion chamber, a maximum (rich) fuel–air ratio and a minimum (lean) fuel–air ratio. If the fuel–air ratio is outside these limits, the flame becomes unstable.

The preceding discussion has involved single-stage combustion chambers. Fuel, in a single-stage combustion chamber, is added to the airstream under all operating conditions through one group of nozzles. This means that each nozzle must be capable of injecting a varying quantity of fuel, since the amount of fuel required at idle is only a fraction of that required at full power. When the quantity of fuel injected is decreased, the fuel supply pressure must be lower, leading to poor atomization, penetration, and vaporization.

Several combustion chamber changes have been investigated to alleviate the above-mentioned problem associated with single-stage combustion chambers. Most changes involve using one group of fuel nozzles at idle, an additional group of fuel nozzles for full power. This allows the fuel to be injected into the airstream at a high pressure at idle, providing a means of supplying the additional fuel required at fuel power, both groups of nozzles operating at the best fuel pressure for good penetration, atomization, and vaporization.

9.6 EXHAUST NOZZLES

The general characteristics of converging and converging-diverging nozzles were discussed in Chapter 3. Designing an exhaust system for a subsonic commercial airplane usually involves use of a fixed-area converging nozzle, whereas designing an exhaust system for a supersonic airplane usually involves an exhaust system with variable geometry. The exhaust system selected for a supersonic aircraft is a compromise among weight, complexity, and performance.

Many types of nozzles have been used in aircraft designs. These include:

1. Fixed-area converging nozzles
2. Fixed-area converging-diverging nozzles
3. Variable-area converging-diverging nozzles
4. Plug nozzles
5. Two-dimensional nozzles

The exhaust nozzle on an aircraft gas turbine engine should:

1. Be matched to the other engine components for all engine operating conditions
2. Provide the optimum expansion ratio
3. Have minimum losses at both design and off-design conditions
4. Have low drag
5. Provide reverse thrust if necessary
6. Be able to incorporate noise-absorbing material

The simplest exhaust system is the converging nozzle with fixed area. This type of exhaust system has no moving parts, needs no control mechanism, and usually is used on subsonic commercial aircraft. Almost any smooth contour in the converging region will provide good performance because of the favorable pressure gradient in this region.

It was noted in Chapter 6 that additional thrust may be achieved if a converging-diverging nozzle is used. A fixed-area nozzle is designed for one expansion ratio and mass rate of flow. For all other expansion ratios, the nozzle, will either overexpand or underexpand. A fixed-area converging-diverging nozzle adds weight, length, and possibly drag to the exhaust system, which may offset any additional thrust that would be achieved by using a fixed-area converging-diverging nozzle.

Many aircraft gas turbine engines, including all afterburning engines, require an exhaust system where the throat area of the nozzle may be varied. The reason that a variable area nozzle is required is illustrated by the following example problem.

Example Problem 9.3

The conditions leaving the turbine of an afterburning turbojet engine are shown below.

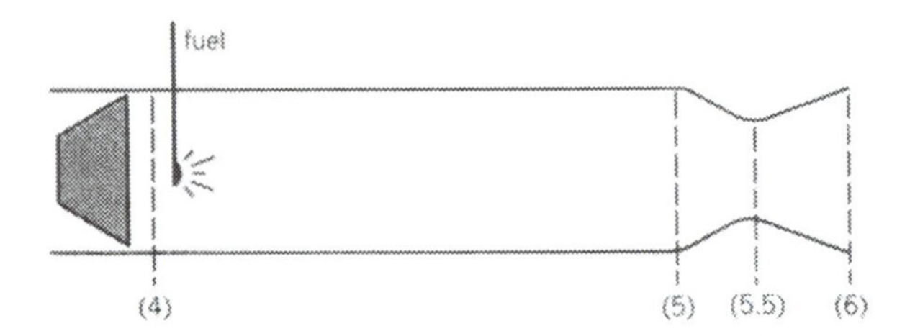

$$p_{o4} = 419.1 \text{ kPa } (60.75 \text{ psia})$$

$$T_{o4} = 1116 \text{ K } (2008^\circ\text{R})$$

Calculate the ratio of the nozzle throat area with and without the afterburner in operation. Assume that the nozzle flow rate is the same in both cases. Assume that the maximum temperature with the afterburner in operation is 1800 K (3240°R), there is no pressure drop between states 4 and 5, constant specific heats, and that the ambient pressure is 101.325 kPa (14.696 psia).

Solution

The nozzle expansion ratio is

$$\frac{p_{o4}}{p_{amb}} = \frac{419.1}{101.3} = 4.14$$

This value is well above the choking expansion ratio (see Figure 3.6). Therefore,

$$\frac{\dot{m}\sqrt{T_{o5}}}{p_{o5}A_{5.5}} = \text{constant}$$

The mass rate of flow and total pressure are the same with and without the afterburner in operation. Therefore,

$$\frac{A_{5.5}|_{\text{A.B.}}}{A_{5.5}|_{\text{N.A.B.}}} = \sqrt{\frac{1800}{1116}} = 1.27$$

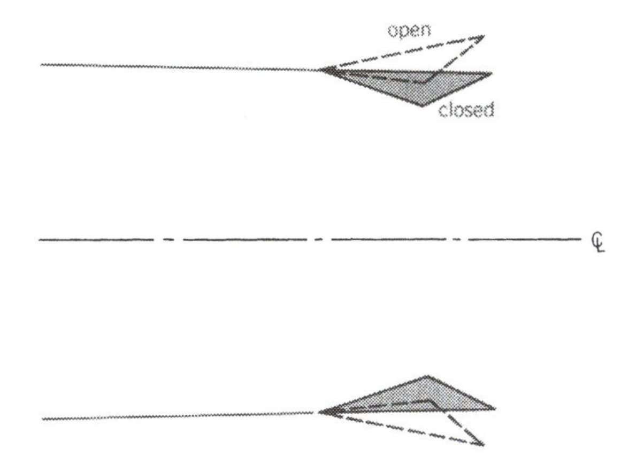

Figure 9.14. Schematic diagram of a variable-area converging-diverging nozzle.

This means that the throat area must be 27% larger when the afterburner is in operation to handle the same mass rate of flow. It also means that some kind of iris mechanism with necessary controls must be used to achieve this area variation. One way of achieving this area variation is shown schematically in Figure 9.14.

Another type of nozzle that has been investigated extensively when a variable-area nozzle is required is the plug nozzle, which is shown schematically in Figure 9.15. The throat area, in a plug nozzle, may be varied by an axial translation of the plug, the outer casing or by an iris arrangement.

The major disadvantages of a plug nozzle are that it requires cooling and it is structurally weak.

The nozzles discussed above all are axisymmetric nozzles. Another type of nozzle that has been considered is the two-dimensional nozzle. Two-dimensional nozzles have the potential for reduced cruise drag and thrust vectoring for increased maneuverability and are simpler to produce. Design problems inherent with two-dimensional nozzles are that they are heavier and have a lower nozzle efficiency at subsonic velocities.

The choice of the exhaust system to be used in a given airplane requires an extensive analysis of the aircraft mission and compromises between minimum weight and maximum performance.

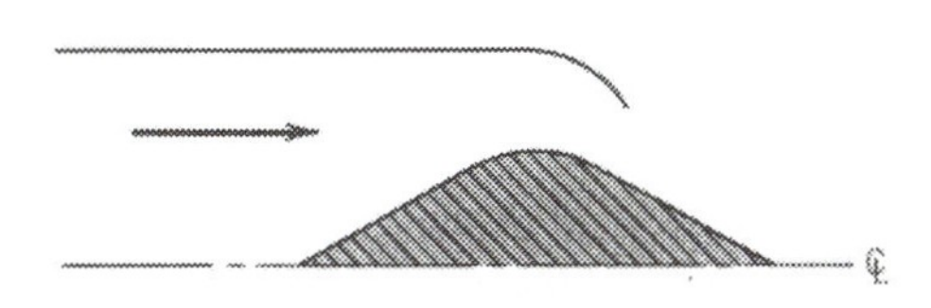

Figure 9.15. Schematic diagram of a plug nozzle.

PROBLEMS

9.1. Air enters a two-dimensional supersonic diffuser at a pressure of 1.765 psia, a temperature of 390°R, and with a Mach number of 3.0. The two-dimensional oblique shock diffuser has an oblique shock angle of 35° which is followed by a normal shock. Determine, assuming constant specific heats:

(a) The velocity of the air entering the oblique shock

(b) The total temperature and pressure of the air entering the oblique shock

(c) The Mach number after the oblique shock

(d) The total pressure after the oblique shock

(e) The flow deflection angle

(f) The static temperature and pressure after the oblique shock

(g) The Mach number after the normal shock

(h) The total pressure after the normal shock

(i) The static pressure and temperature after the normal shock

9.2. Solve Problem 9.1 assuming a Mach number of 3.6, an oblique shock angle of 22°, and all other values the same.

9.3. Solve Problem 9.1 for a pressure of 5.924 kPa, a temperature of 217 K, a Mach number of 2.6, and an oblique shock angle of 30°.

9.4. Solve Problem 9.3 for a Mach number of 3.0 and an oblique shock angle of 25°.

9.5. Air enters a two-dimensional supersonic diffuser at a pressure of 2.039 psia, a temperature of 390°R, and with a Mach number of 3.0. The two-dimensional shock diffuser is a double-wedge diffuser as shown in the diagram. It has oblique shock angles as shown in the figure. Determine:

(a) The deflection angles δ_1 and δ_2.

(b) The total pressure ratio, p_{o1}/p_{o0}

(c) The static pressure ratio, p_1/p_0.

(d) The static temperature ratio, T_1/T_0.

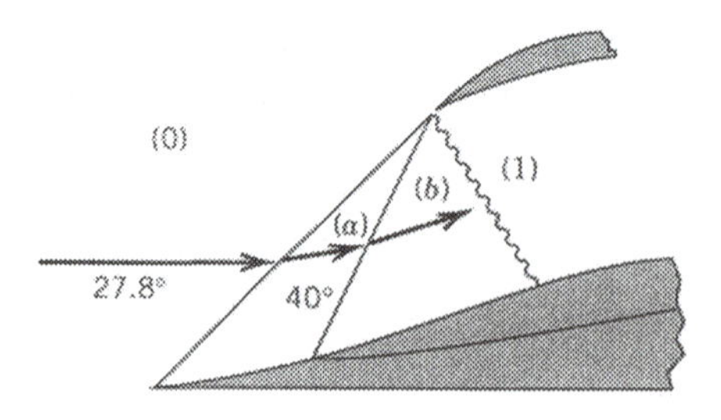

9.6. Solve Problem 9.5 assuming that the second oblique shock angle is 37° instead of 40°, all other values being the same.

9.7. Solve Problem 9.5 assuming a pressure of 6.410 kPa, a temperature of 217 K, and initial Mach number of 3.6, an initial oblique shock angle of 30°, and a second oblique shock angle of 45°.

9.8. Solve Problem 9.7 assuming that the second oblique shock is 40° instead of 45°.

9.9. Assume a two-dimensional oblique shock inlet with an inlet Mach number of 2.6 that is followed by a converging-diverging section. The oblique shock angle is at 30°.

(a) What is the necessary wedge angle?

(b) What is the Mach number after the oblique shock?

9.10. Solve Example Problem 6.1b assuming a jet nozzle efficiency of 97% instead of 100%, all other values being the same.

9.11. Solve Example Problem 6.2b assuming a jet nozzle efficiency of 97% instead of 100%, all other values being the same.

9.12. Solve Example Problem 9.3 taking into account the mass of fuel added and variable specific heats.

9.13. Determine, for Example Problem 6.2b, the effect of a change to:

(a) Diffuser efficiency of 98%, all other values the same

(b) Compressor efficiency of 85%, all other values the same

(c) Turbine efficiency of 87%, all other values the same

(d) Nozzle efficiency of 98%, all other values the same

9.14. Solve Example Problem 6.4 assuming that an afterburner has been added which increases the temperature to 1800 K (3240 R). Assuming no pressure drop in the afterburner, calculate the percent increase in the exhaust nozzle throat area for the same mass rate of flow.

10

COMPONENT MATCHING

10.1 GENERAL

Engine components and the manner in which component performance is presented were discussed in the three preceding chapters. Next it is important to investigate the matching aspects of the gas turbine engine. A matching study is an investigation of the interplay of the engine geometry and engine parameters such as pressure ratio, airflow, rotor speed, component efficiencies, pressure drops, areas, and so on. Such a study must be conducted to answer questions about the steady-state and transient operation of a gas turbine engine.

Most of the discussion in this chapter deals with the steady-state operation of a single-spool turbojet engine and attempts to answer such questions as the following:

1. For a fixed engine geometry, what happens to the component parameters and resulting component match when a gas turbine is operated at an off-design condition?
2. How does changing the turbine inlet temperature influence the resulting component match?
3. Does the component match point change when the jet nozzle area of a turbojet engine is changed? If so, how does one determine the new match point?
4. What effect does afterburner operation have on the match point for the gas generator of an afterburning turbojet engine? Is it possible for an afterburning turbojet engine gas generator to operate at the same match point with the afterburner both in operation and inoperable?
5. How does diffuser water injection or operation with a low-Btu gas influence the engine match point?
6. How does variation of the turbine nozzle area influence the engine match?
7. How does engine bleed, horsepower extraction, and/or turbine blade cooling influence the engine match?

Answers to these questions involve the interplay of a large number of variables. Therefore, the analysis that follows is a simplification of the type of analysis that must be performed on a gas turbine engine. The following discussion of engine matching is intended to give answers, in a general way, to many of the above questions. The equations developed to answer these questions contain a number of assumptions that would be questionable if used in an actual engine. They are made here to simplify the problem so that one can see the trends without becoming deeply involved in a number of fine details. Engine matching, because of its complexity, is done almost exclusively on high-speed digital computers but, since all new computer programs must be checked for accuracy and are constantly being modified, one must understand the principles of matching techniques.

The geometry and flow areas of a given power plant are established at the "design point." At all other operating conditions, the components must be "matched" to determine the pressure ratio, airflow, rotor speed, efficiency, and so on. The "match point" is defined as the steady-state operating point for a gas turbine when the compressor and turbine are balanced in rotor speed, power, and flow, the operating points at the various power settings defining the operating line for the given engine configuration.

No matter what type of engine is being considered, the conservation of mass, energy, and momentum must be satisfied. The first two lead to the "match point," the third to the thrust developed by the engine.

Satisfaction of the conservation of mass requires that:

1. The flow through the turbine must equal the flow through the compressor plus the fuel added minus any air extracted. Care must be taken to account for any air extracted from the system and exactly where it reenters the system.
2. The exhaust system (nozzle) flow characteristics must be satified.

Satisfaction of the conservation of energy requires that the power developed by each turbine equal the power required to drive each compressor plus losses and power extracted.

Since only general matching trends of a single-spool turbojet engine are being sought, several simplifying assumptions are made. These assumptions hold true unless stated otherwise and include the following:

1. No burner or exhaust system pressure losses will be included; that is, it is assumed the pressure remains constant through the combustion chamber and from the exit of the turbine to the inlet of the exhaust nozzle.
2. The mass rate of flow is assumed to be the same through the compressor, turbine, and exhaust nozzle. This means that the mass of fuel added is being neglected, no air is extracted from or after the compressor, and turbine cooling is not being used on the engine being studied.
3. No power is extracted and no losses occur between the turbine and the compressor; that is, bearing losses are neglected.
4. The turbine nozzle and jet nozzle areas are constant values as determined for the design point conditions.

5. It is assumed that air is the working fluid throughout and that the specific heat ratio has a constant value of 1.4.

10.2 CONSERVATION OF MASS

To facilitate the understanding of engine matching, the single-spool turbojet engine will be considered. The results of this study will be used as a building block to help understand the other types.

Figure 10.1 is a schematic diagram of a single-spool turbojet engine showing the various components and station numbers.

The turbine flow characteristics for the turbine that will be used in this analysis are shown in Figure 10.2. Note in Figure 10.2 that the variation of turbine flow parameter as a function of expansion ratio is shown as a single curve with no corrected turbine rotational speed lines. This type of turbine flow characteristics was assumed:

1. To simplify the analysis
2. Because of the small effect that turbine rotational speed has on the turbine flow parameter and efficiency

The single-spool turbojet engine turbine operates choked over a wide region of its operating spectrum. Thus, most of the time, the turbine operating point will be to the right of point *A* (choking expansion ratio) as shown in Figure 10.2.

When the turbine flow parameter becomes a constant (the turbine is choked), the turbine expansion ratio can still continue to increase. For a fixed turbine inlet temperature and turbine efficiency, the turbine work is dependent on the turbine expansion ratio. Whether the "match point" for the turbine is at *A, B, C, D*, or another point depends on the flow characteristics of what follows the turbine, this being the flow characteristics of the exhaust nozzle for the single-spool turbojet engine.

The relationship between the turbine flow characteristics and compressor flow characteristics is expressed algebraically by the following equality:

$$\frac{\dot{m}_{g3}\sqrt{T_{o3}}}{p_{o3}A_3} = \left[\frac{\dot{m}_{a1}\sqrt{\theta_{o1}}}{\delta_{o1}}\right]\sqrt{\frac{T_{o3}}{\theta_{o1}}}\left(\frac{\dot{m}_{g3}}{\dot{m}_{a1}}\right)\left[\frac{1}{(p_{o3}/p_{o2})(p_{o2}/p_{o1})p_{\text{std}}A_3}\right] \tag{10.1}$$

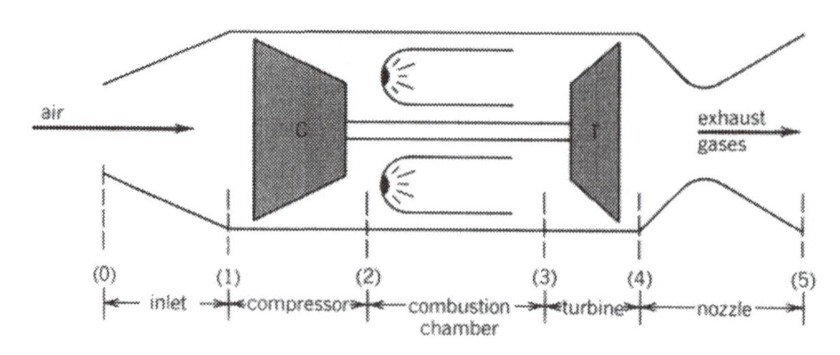

Figure 10.1. Schematic diagram of a single-spool turbojet engine.

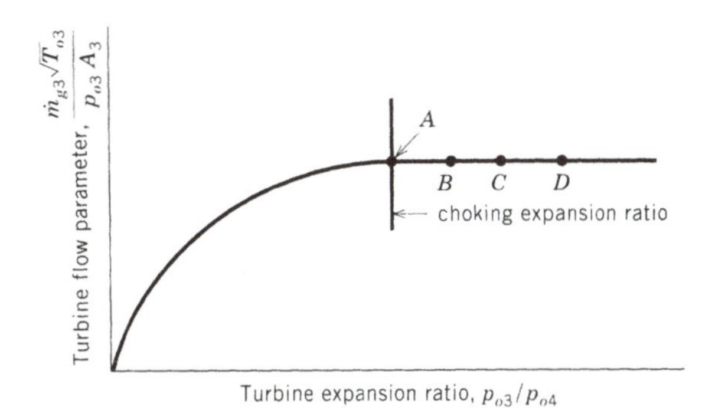

Figure 10.2. Turbine flow characteristics.

For a constant pressure loss in the combustion chamber, mass rate of flow through the compressor the same as through the turbine ($\dot{m}_{g3} = \dot{m}_{a1}$), fixed turbine nozzle area (A_3 = constant), and a choked turbine ($\dot{m}_{g3}\sqrt{T_{o3}}/p_{o3}A_3$), Eq. (10.1) becomes

$$\frac{p_{o2}}{p_{o1}} = C_1\left[\frac{\dot{m}_{a1}\sqrt{\theta_{o1}}}{\delta_{o1}}\right]\sqrt{\frac{T_{o3}}{\theta_{o1}}} \tag{10.2}$$

where C_1 is a constant.

For a fixed T_{o3}/θ_{o1}, Eq. (10.2) reduces to

$$\frac{p_{o2}}{p_{o1}} = C_2\frac{\dot{m}_{a1}\sqrt{\theta_{o1}}}{\delta_{o1}} \tag{10.3}$$

which is the equation of a straight line passing through the point (0.0, 0.0). The slope of this straight line, C_2, increases for increasing values of T_{o3}/θ_{o1}, that is, turbine inlet temperature over compressor inlet temperature.

Plotting Eq. (10.3) on a typical compressor map yields the results shown in Figure 10.3.

Equation 10.3 shows that the pressure ratio is zero when the airflow is zero. This, of course, is impossible. What is wrong? Equation (10.3) assumes a choked turbine. At low air flows, the turbine unchokes so that the constant T_{o3}/θ_{o1} lines curve into a pressure ratio of 1.0 at zero airflow. This is shown in Figure 10.3 by the dashed lines.

For a given T_{o3}/θ_{o1}, where is the operating point? Is it at A, B, C, or D as shown in Figure 10.3? This is fixed by the compressor work required, which, for an energy balance, is fixed by the turbine work, the turbine work being fixed by the turbine inlet temperature and expansion ratio, the turbine expansion ratio being fixed by the nozzle flow characteristics. This is discussed in the following sections.

10.3 COMPRESSOR WORK

The ideal work of compression for the single-spool turbojet engine illustrated in Figure 10.1 is given in Eq. (5.1), which, for constant specific heats, becomes

$$w_{C,i} = \Delta h_{oC,i} = c_{pC}(T_{o2i} - T_{o1}) \tag{10.4}$$

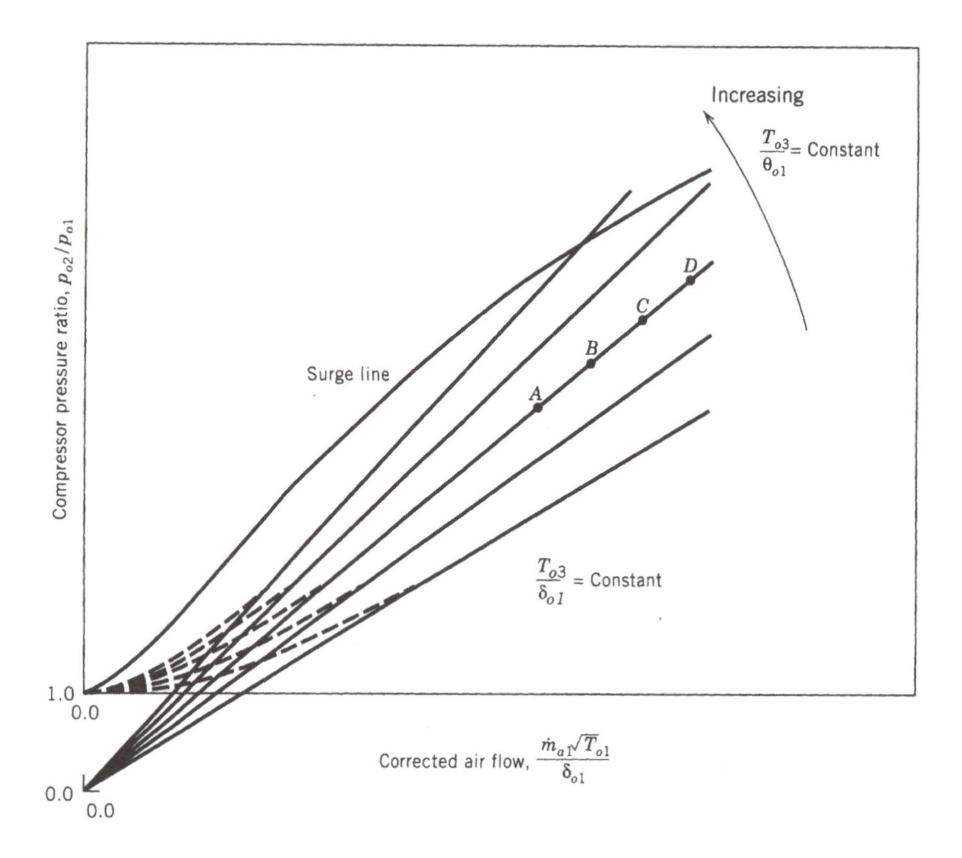

Figure 10.3. Equation (10.3) plotted on a typical compressor map.

Equation (10.4), when divided by the compressor inlet temperature over the standard temperature, becomes, for constant specific heats

$$\frac{\Delta h_{oC,i}}{\theta_{o1}} = c_{pC}\, T_{\text{std}} \left[\left(\frac{p_{o2}}{p_{o1}} \right)^{(k-1)/k} - 1 \right] \tag{10.5}$$

Equation (10.5) shows that the ideal work of compression divided by θ_{o1} is a function of pressure ratio only. When variable specific heats are considered, the ideal work of compression divided by θ_{o1} is approximately a function of pressure ratio.

Example Problem 10.1

Calculate, for pressure ratios of 2.5, 5.0, and 7.5:

(a) The ideal work of compression for compressor inlet temperatures of 450°R (250 K), 540°R (300 K), and 585°R (325 K)
(b) The values of ideal work over theta for the above pressure ratios and compressor inlet temperatures

Assume constant specific heats with $\bar{c}_p = 6.954$ Btu/lb-mol°R ($\bar{c}_p = 29.104$ kJ/kmol K), $k = 1.400$.

Solution

$$w_{C,i} = c_{pC}\,(T_{o2} - T_{o1})$$

$$= c_{pC}\,T_{o1}\left[\left(\frac{p_{o2}}{p_{o1}}\right)^{(k-1)/k} - 1\right]$$

$$\theta_{o1} = \frac{T_{o1}}{518.67}\left(\frac{T_{o1}}{288.15}\right)$$

The results are, for the specified pressure ratios and compressor inlet temperatures

	T_{o1}		$w_{c,i}$		$w_{c,i}/\theta_{o1}$	
p_{o2}/p_{o1}	(°R)	(K)	(Btu/lb)	(kJ/kg)	(Btu/lb)	(kJ/kg)
2.5	450	250	32.3	75.2	37.3	86.7
	540	300	38.8	90.2	37.3	86.7
	585	325	42.0	97.7	37.3	86.7
5.0	450	250	63.1	146.7	72.7	169.0
	540	300	75.7	176.0	72.7	169.0
	585	325	82.0	190.6	72.7	169.0
7.5	450	250	84.1	195.5	96.9	225.3
	540	300	100.9	234.6	96.9	225.3
	585	325	109.3	254.2	96.9	225.3

The actual compressor work depends on the compressor efficiency. Therefore,

$$\frac{\Delta h_{oC,a}}{\theta_{o1}} = \frac{\Delta h_{oC,i}}{\theta_{o1}}\left(\frac{1}{\eta_C}\right)$$

$$= \frac{c_{pC}\,T_{std}\,[(p_{o2}/p_{o1})^{(k-1)/k} - 1]}{\eta_C} \tag{10.6}$$

10.4 TURBINE WORK

The ideal turbine work, for the turbine of the single-spool turbojet engine illustrated in Figure 10.1, is, for constant specific heats,

$$w_{T,i} = \Delta h_{oT,i} = c_{pT}(T_{o3} - T_{o4i}) \tag{10.7}$$

$$= c_{pT}\,T_{o3}\left[1 - \left(\frac{p_{o3}}{p_{o4}}\right)^{(1-k)/k}\right] \tag{10.8}$$

or

$$\frac{\Delta h_{oT,i}}{\theta_{o3}} = c_{pT} T_{std} \left[1 - \left(\frac{p_{o3}}{p_{o4}} \right)^{(1-k)/k} \right] \tag{10.9}$$

Equation (10.8) illustrates that the ideal work developed by a turbine is a function of the expansion ratio and the turbine inlet temperature. Equation (10.9) illustrates that the ideal work developed by a turbine divided by theta (T_{o3}/T_{std}) is a function of the expansion ratio only.

The actual turbine work depends on the turbine efficiency, or

$$\frac{\Delta h_{oT,a}}{\theta_{o3}} = \frac{\Delta h_{oT,i}}{\theta_{o3}} \eta_T \tag{10.10}$$

$$= c_{pT} T_{std} \eta_T \left[1 - \left(\frac{p_{o3}}{p_{o4}} \right)^{(1-k)/k} \right] \tag{10.11}$$

10.5 TURBINE-COMPRESSOR ENERGY BALANCE

In Section 10.2 it was shown that, for a choked turbine, the compressor pressure ratio is a function of turbine inlet temperature over theta and the compressor corrected mass rate of flow, or

$$\frac{p_{o2}}{p_{o1}} = C_1 \left(\frac{\dot{m}_{a1} \sqrt{\theta_{o1}}}{\delta_{o1}} \right) \sqrt{\frac{T_{o3}}{\theta_{o1}}} \tag{10.2}$$

This results in several straight lines as illustrated in Figure 10.3, each straight line being for a different turbine inlet temperature over theta. Equation (10.2) is based on the conservation of mass and, as illustrated in Figure 10.3, does not allow a person to determine whether the engine will be operating at point A, B, C, or D.

The actual steady-state operating point occurs where the turbine power is equal to the power required to drive the compressor; that is, it is determined by an energy balance. Based on the assumptions stated earlier (no power extraction, no losses between compressor and turbine, and $\dot{m}_{a1} = \dot{m}_{g3}$),

$$\frac{\Delta h_{oC,a}}{\theta_{o1}} = \frac{\Delta h_{oT,a}}{\theta_{o3}} \left(\frac{T_{o3}}{T_{o1}} \right) \tag{10.12}$$

Combining Eqs. (10.6), (10.11), and (10.12) yields, for the case where $c_{p,C} = c_{p,T}$,

$$\left[\left(\frac{p_{o2}}{p_{o1}} \right)^{(k-1)/k} - 1 \right] = \eta_C \eta_T \left[1 - \left(\frac{p_{o3}}{p_{o4}} \right)^{(1-k)/k} \right] \left(\frac{T_{o3}}{T_{o1}} \right) \tag{10.13}$$

This equation illustrates the fact that the compressor pressure ratio is a function of compressor efficiency, turbine efficiency, turbine expansion ratio, and the ratio of turbine inlet temperature to compressor inlet temperature, or

$$\frac{p_{o2}}{p_{o1}} = f\left(\eta_C, \eta_T, \frac{p_{o3}}{p_{o4}}, \frac{T_{o3}}{T_{o1}}\right) \tag{10.14}$$

Equation (10.14) illustrates that, for a fixed T_{o3}/T_{o1} and constant values of η_C and η_T, the compressor pressure ratio is a function of the turbine expansion ratio. The turbine expansion ratio, for a specified T_{o3}/T_{o1}, is fixed by the exhaust nozzle flow characteristics.

10.6 EXHAUST NOZZLE

The flow characteristics of a nozzle were discussed in Section 3.9 and were presented in Figure 3.6. The nozzle flow characteristics are reproduced in Figure 10.4, the only difference being that the exhaust expansion ratio, p_{o4}/p_{amb}, is used instead of p_o/p_s as in Figure 3.6.

The turbine and exhaust nozzle flow parameters, when combined, yield the following expression:

$$\frac{\dot{m}_{g4}\sqrt{T_{o4a}}}{p_{o4}\,A_4} = \frac{\dot{m}_{g3}\sqrt{T_{o3}}}{p_{o3}\,A_3}\left(\frac{p_{o3}}{p_{o4}}\right)\sqrt{\frac{T_{o4a}}{T_{o3}}}\left(\frac{A_3}{A_4}\right) \tag{10.15}$$

For a given turbine polytropic efficiency, the turbine actual temperature ratio is related to the turbine expansion ratio according to the following equation:

$$\frac{T_{o4a}}{T_{o3}} = \left(\frac{p_{o4}}{p_{o3}}\right)^{\eta_p(k-1)/k} \tag{10.16}$$

Combining Eqs. (10.15) and (10.16) and assuming that $\dot{m}_{g3} = \dot{m}_{g4} = \dot{m}_g$ yields

$$\frac{\dot{m}_g\sqrt{T_{o4a}}}{p_{o4}\,A_4} = \frac{\dot{m}_g\sqrt{T_{o3}}}{p_{o3}\,A_3}\left(\frac{p_{o3}}{p_{o4}}\right)^{1-\frac{\eta_p(k-1)/k}{2}}\left(\frac{A_3}{A_4}\right) \tag{10.17}$$

Equation 10.17 shows that, for exhaust and turbine nozzles of fixed area, the expansion ratio has a constant value once the turbine and exhaust nozzles become

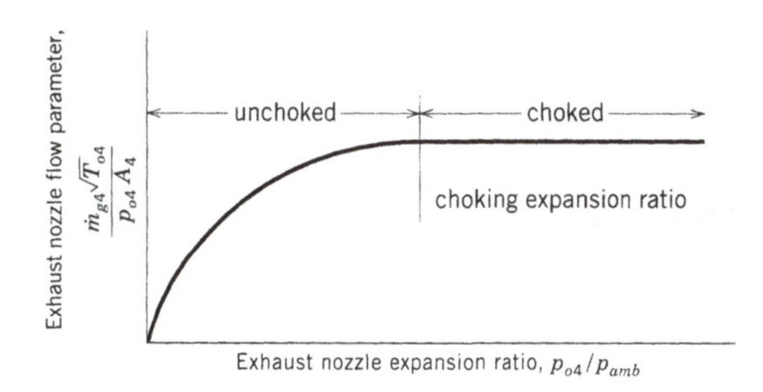

Figure 10.4. Typical exhaust nozzle flow characteristics.

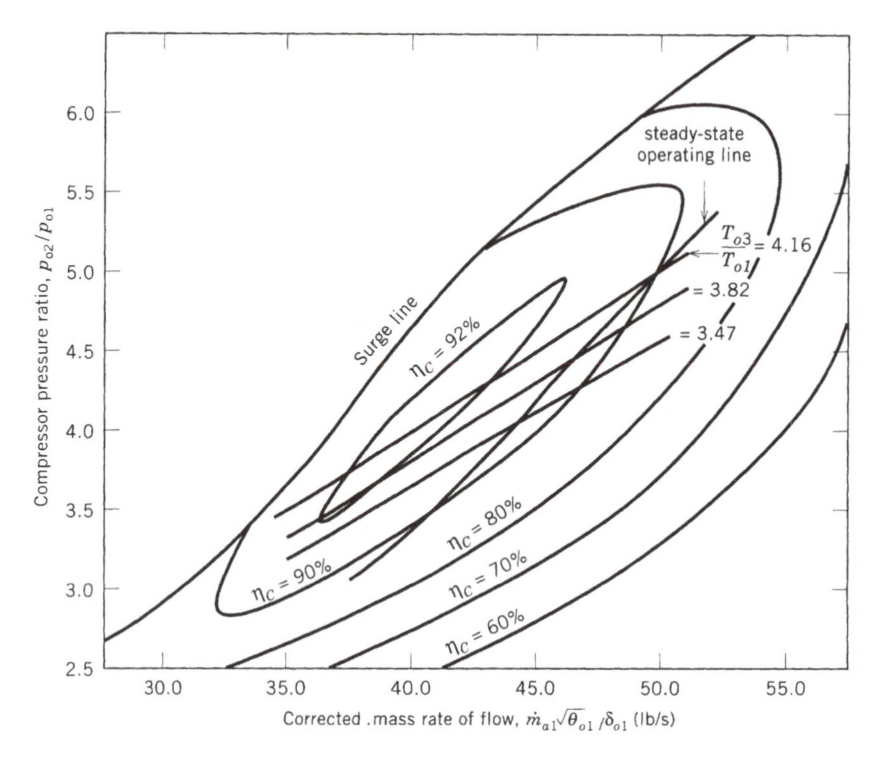

Figure 10.5. Hypothetical compressor map with steady-state operating line.

choked. This holds true only if the turbine polytropic efficiency (and therefore the overall turbine efficiency) is a constant.

10.7 OPERATING LINE

It is now important to determine what happens to the steady-state operating conditions of an engine as the turbine inlet temperature and/or flight conditions are changed. This will be done by assuming a hypothetical compressor map for a single-spool turbojet engine, fixing the areas at the design point, then determining what happens to the engine operation as the turbine inlet temperature is varied. This is illustrated by the following example problems, with the results illustrated on the hypothetical compressor map in Figure 10.5.

Example Problem 10.2

A single-spool turbojet engine, having the compressor performance illustrated in Figure 10.5, has the following design point conditions at sea level static on a standard day.

$$\frac{p_{o2}}{p_{o1}} = 5.0$$

$$\frac{\dot{m}_{a1}\sqrt{\theta_{o1}}}{\delta_{o1}} = 50.0 \text{ lb/s (22.7 kg/s)}$$

$$T_{o3} = 2160°\text{R (1200 K)}$$

$$\eta_T = 92\% = \text{constant}$$

Calculate, neglecting the mass of fuel added, the turbine expansion ratio. Assume no pressure drop in the combustion chamber and constant specific heats with $c_p = 0.240$ Btu/lb°R ($c_p = 1.004$ kJ/kg K), $k = 1.40$.

Solution

From the compressor map (Figure 10.5) at the design point conditions, it is determined that the compressor efficiency η_C is 90%. Therefore, from Eq. (10.6), since $\theta_{o1} = 1.0$,

$$w_{C,a} = \Delta h_{C,a} = \frac{(1)\,(0.240)\,(518.67)\,[(5)^{(0.4/1.4)} - 1]}{0.90}$$

$$= 80.75 \text{ Btu/lb (187.7 kJ/kg)}$$

$$\Delta h_{T,i} = \frac{\Delta h_{T,a}}{\eta_T} = \frac{\Delta h_{C,a}}{\eta_T} = \frac{80.75}{0.92}$$

$$= 87.77 \text{ Btu/lb (204.0 kJ/kg)}$$

$$\theta_{o3} = \frac{2160}{518.67} = 4.16$$

From Eq. (10.9),

$$\frac{p_{o3}}{p_{o4}} = \left[1 - \frac{87.77}{(4.16)\,(0.24)\,(518.67)}\right]^{-(1.4/0.4)}$$

$$= 1.91$$

The design point, which was illustrated in the preceding example problem, determines the exhaust nozzle area. Note that the expansion ratio across the exhaust nozzle, for the conditions assumed, is

$$\frac{P_{o4}}{p_{\text{amb}}} = \frac{5.0}{1.91} = 2.62$$

which is well above the pressure ratio required for the exhaust nozzle to be choked.

It is next important to determine the operating line for this engine. This is illustrated in Example Problem 10.3.

Example Problem 10.3

Determine, for the engine defined in Example Problem 10.2, the steady-state operating points when the turbine inlet temperature is

(a) 1980°R (1100 K)
(b) 1800°R (1000 K)

Solution

(a) For $T_{o3} = 1980°R$ (1100 K), $T_{o1} = 518.67$ (288.15 K), $\theta_{o1} = 1.0$.
From Eq. (10.11),

$$\frac{\Delta h_{T,a}}{\theta_{ol}} = \frac{(3.82)\,(0.24)\,(518.67)\,(0.92)}{1.0}\,[1 - (1.91)^{(-0.4)/1.4}]$$

$$= 73.84 \text{ Btu/lb } (171.6 \text{ kJ/kg})$$

At the design point conditions and using Eq. (10.2), one may determine C_1.

$$C_1 = \frac{5.0}{(50)\,(\sqrt{2160})} = 0.00215(0.00636)$$

Therefore, the relationship between compressor pressure ratio and corrected mass rate of flow is, in English units,

$$\frac{p_{o2}}{p_{o1}} = 0.00215\left(\frac{\dot{m}_{a1}\sqrt{\theta_{o1}}}{\delta_{o1}}\right)\sqrt{\frac{T_{o3}}{\theta_{o1}}} \qquad (10.18)$$

or, for $\theta_{o3} = 3.82$, the relationship becomes

$$\frac{p_{o2}}{p_{o1}} = 0.0957\left(\frac{\dot{m}_{a1}\sqrt{\theta_{o1}}}{\delta_{o1}}\right) \qquad (10.19)$$

Equation (10.19), which is an expression of the conservation of mass, is plotted on Figure 10.5 and is labeled $T_{o3}/T_{o1} = 3.82$.

Since both the turbine and exhaust nozzle are choked and the turbine efficiency is a constant, the turbine expansion ratio is a constant and equal to 1.91, the value determined in Example Problem 10.2.

The actual turbine work, as determined by Eq. (10.11), is

$$\frac{\Delta h_{T,a}}{\theta_{o3}} = (0.24)\,(518.67)\,(0.92)\,[1 - (1.91)^{(-0.4/1.4)}]$$

$$= 19.33 \text{ Btu/lb } (44.9 \text{ kJ/kg})$$

Note that this is a constant value for this engine as long as the turbine *and* exhaust nozzle are choked.

For $T_{o3} = 1980°R$ (1100 K),

$$\frac{\Delta h_{T,a}}{\theta_{o1}} = \frac{\Delta h_{C,a}}{\theta_{o1}} = (19.33)\,(3.82)$$

$$= 73.84 \text{ Btu/lb } (171.6 \text{ kJ/kg})$$

It is now necessary to determine the compressor pressure ratio that satisfies $\Delta h_{C,a}/\theta_{o1} = 73.84$.

The ideal compressor work over θ_{o1} is determined from Eq. (10.5). For a pressure ratio of 4.4, this result is

$$\frac{\Delta h_{C,i}}{\theta_{o1}} = (0.24)(518.67)[(4.4)^{(0.4/1.4)} - 1]$$

$$= 65.60 \text{ Btu/lb } (152.5 \text{ kJ/kg})$$

The compressor efficiency, for an ideal work of 65.60 and an actual work of 73.84 is

$$\eta_C = \frac{65.60}{73.84} = 0.888$$

The results for several pressure ratios are, for an actual work of 73.84

P_{o2}/p_{o1}	$\Delta h_{C,i}/\theta_{o1}$ Btu/lb (kJ/kg)	η_C
4.5	66.83 (155.3)	0.905
4.4	65.60 (152.5)	0.888
4.3	64.36 (149.6)	0.872
4.2	63.09 (146.6)	0.854
4.1	61.81 (143.6)	0.837

Therefore, a $\Delta h_{C,a}/\theta_{o1} = 73.84$ line may be constructed on the compressor map. This is shown in the following diagram, which is a portion of the compressor map.

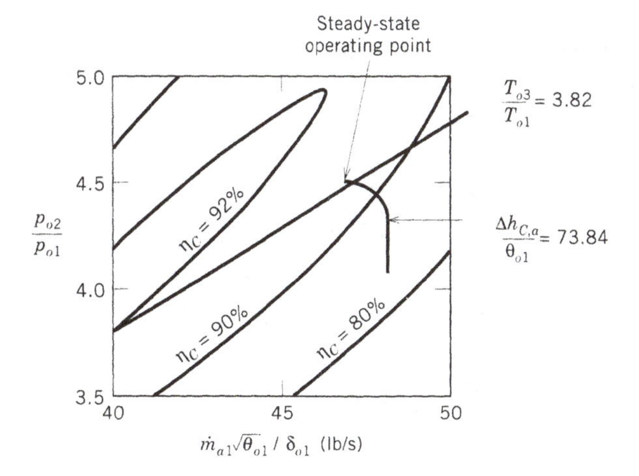

The point of intersection between the $T_{o3}/T_{o1} = 3.82$ line and $\Delta h_{C,a}/\theta_{o1} = 73.84$ line is the steady-state operating point at sea level static on a standard day for a turbine inlet temperature of 1980°R (1100 K). From the plot, this point is

$$\frac{p_{o2}}{p_{ol}} = 4.49$$

$$\eta_C = 90.4\%$$

$$\frac{\dot{m}_{al}\sqrt{\theta_{ol}}}{\delta_{ol}} = 47.1 \text{ lb/s (21.4 kg/s)}$$

(b) For $T_{o3} = 1800°\text{R}$ (1000 K),

$$\theta_{o3} = \frac{1800}{518.67} = 3.47$$

$$\frac{\Delta h_{T,a}}{\theta_{ol}} = \frac{\Delta h_{C,a}}{\theta_{o1}} = (19.33)(3.47)$$

$$= 67.08 \text{ Btu/lb (155.8 kJ/kg)}$$

P_{o2}/p_{o1}	$\Delta h_{C,i}/\theta_{o1}$ Btu/lb (kJ/kg)	η_C
4.1	61.81 (143.6)	0.921
4.0	60.50 (140.6)	0.902
3.9	59.16 (137.5)	0.882
3.8	57.81 (134.3)	0.862

Once again, plotting on a portion of the compressor map, one determines that the steady-state operating point is

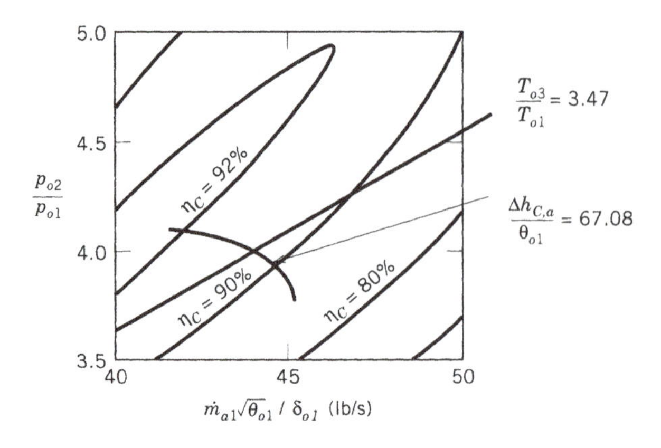

$$\frac{p_{o2}}{p_{o1}} = 4.05$$

$$\eta_C = 90.3\%$$

$$\frac{\dot{m}_{a1}\sqrt{\theta_{o1}}}{\delta_{o1}} = 44.3 \text{ lb/s } (20.1 \text{ kg/s})$$

Three points of the steady-state operating line have now been determined. The steady-state operating line is shown on the compressor map (Figure 10.5).

The exhaust nozzle is choked above a compressor pressure ratio of approximately 3.6. Below this compressor pressure ratio, the exhaust nozzle will not be choked. This means that even though the turbine may be operating choked, the turbine expansion ratio will change as the turbine inlet temperature varies if the steady-state operating point is below a compressor pressure ratio of 3.6 for the engine illustrated in Figure 10.5.

It is important to keep in mind the many simplifying assumptions in the preceding analysis. These included:

1. The mass rate of flow through each component is the same.
2. There is no pressure drop in the combustion chamber or between the turbine exit and exhaust nozzle inlet.
3. No power is extracted, and bearing losses are neglected.
4. The turbine and exhaust nozzle areas have fixed values.
5. There are constant specific heats with $k = 1.4$.
6. There is constant turbine efficiency.
7. The turbine and exhaust nozzle are choked.

One should next consider what happens when one or more of these simplifying assumptions is changed.

1. How would the match point for a specified T_{o3}/T_{o1} change if the mass of fuel added had been taken into account?

In Eq. (10.1), $\dot{m}_{g3}/\dot{m}_{a1} > 1.0$. Therefore, the constant C_1 of Eq. (10.2) will increase, since all other quantities in Eq. (10.1) remain constant. This means that the slope of each of the T_{o3}/T_{o1} lines will increase.

Since the power developed by the turbine must be equal to the power required to drive the compressor,

$$\dot{m}_{a1}\,\Delta h_{C.a} = \dot{m}_{g3}\,\Delta h_{T.a} \tag{10.20}$$

or Eq. (10.12) becomes

$$\frac{\Delta h_{C.a}}{\theta_{o1}} = \left(\frac{\dot{m}_{g3}}{\dot{m}_{a1}}\right)\left(\frac{\Delta h_{T.a}}{\theta_{o3}}\right)\left(\frac{T_{o3}}{T_{o1}}\right) \tag{10.21}$$

This means that the match point for specified T_{o3}/T_{o1} will shift as illustrated below, where the original conservation of mass and energy lines are shown as solid lines, the new ones as dashed lines.

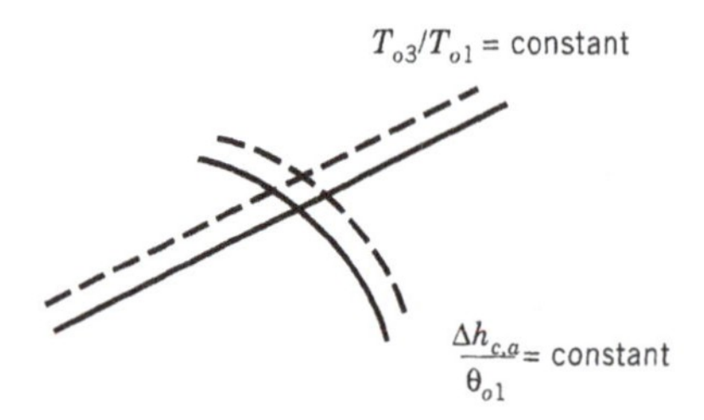

2. How would the match point for a specified T_{o3}/T_{o1} change if there is a 5% pressure drop in the combustion chamber?

In Eq. (10.1), $p_{o3}/p_{o2} < 1.0$. Therefore, since all other quantities in Eq. (10.1) remain constant, the constant C_1 in Eq. (10.2) will increase. Equation (10.12) will not change. Therefore, the match point will shift as illustrated below.

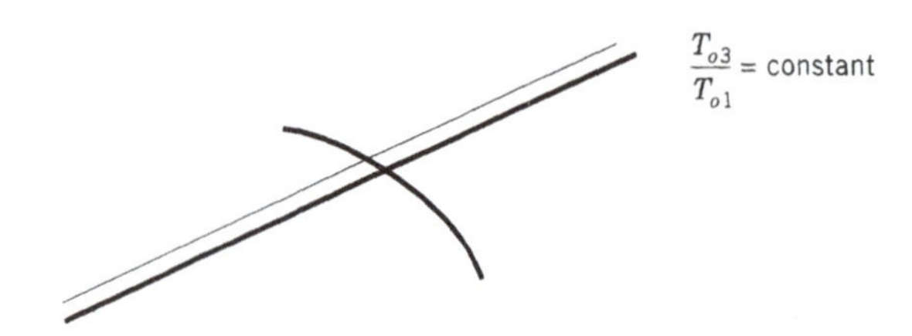

The preceding two examples illustrate how the conservation of mass and conservation of energy "lines" change when one of the assumptions is changed. The reader now should be able to determine how the match points, and therefore the operating line, will shift if the exhaust nozzle area is increased, what must be changed and how for the gas generator to operate at the same "match point" when an afterburner is in operation, and/or how diffuser water injection or operation with a low-Btu gas will influence the engine match point.

10.8 GENERAL MATCHING PROCEDURE

The preceding section discussed component matching of a single-spool turbojet with many simplifying assumptions. The general matching procedure must take into account variable specific heats, the actual products of combustion, the fact that air may be extracted at an intermediate stage or at the exit from the compressor, turbine cooling may be used, the mass of fuel added, the pressure drop in the combustion chamber, turbine flow characteristics as a function of rotor speed, the fact that turbine efficiency

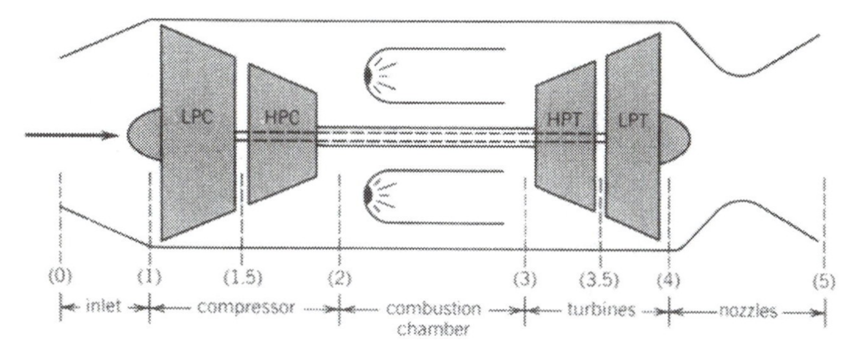

Figure 10.6. Schematic diagram of a two-spool turbojet engine.

is not a constant, and so on. It is obvious that the general matching problem is very complicated, yet it must be done to select the best engine design for a given application.

A general matching procedure is outlined below for a two-spool turbojet engine. It is assumed that the two-spool turbojet engine has the station numbers shown in Figure 10.6 and that the compressor, turbine, and nozzle performance characteristics as shown in Figure 10.7 are known and stored in a high-speed digital computer where the matching study is being performed.

The following discussion describes a general matching procedure. It is assumed that the flight conditions (altitude and velocity) and turbine inlet temperature are known.

1. Knowing the flight condition fixes T_{o1} and p_{o1}.
2. Assume an LPC operating point (assume $p_{o1.5}/p_{o1}$ and $\dot{m}_{a1}\sqrt{\theta_{o1}}/\delta_{o1}$. Since η_{LPC} and $N_{LPC}/\sqrt{\theta_{o1}}$ are known from the LPC map (Figure 10.7*a*), N_{LPC}, $\dot{m}_{a1}$, $\Delta h_{LPC,a}$, $\dot{m}_{a1.5}$, $T_{o1.5,a}$ and $p_{o1.5}$ may be calculated.
3. Assume an LPC pressure ratio, $p_{o2}/p_{o1.5}$. Calculate $\dot{m}_{a1.5}\sqrt{\theta_{o1.5}}/\delta_{o1.5}$, then read from the HPC map (Figure 10.7*b*) η_{HPC}, $N_{HPC}/\sqrt{\theta_{o1.5}}$. Calculate N_{HPC}, $\Delta h_{LPC,a}$, $\dot{m}_{a2}$, T_{o2a}, and p_{o2}.
4. Since T_{o3} is known, determine the fuel–air ratio and combustion chamber pressure drop. Calculate $\dot{m}_{g3}$ and p_{o3}.
5. Assume a HPT expansion ratio. Calculate $N_{HPT}/\sqrt{\theta_{o3}}$. Determine from the turbine performance characteristics η_{HPT} (Figure 10.7*d*) and $\dot{m}_{g3}\sqrt{T_{o3}}/p_{o3}A_3$ (Figure 10.7*c*). Calculate $\Delta h_{HPT,a}$ and $\dot{m}_{g3}$.
6. Check to determine if $\Delta h_{HPT,a}$ and $\dot{m}_{g3}$ are within a preset tolerance. If not, repeat steps 3 through 5 until a match does exist.
7. Once the high-pressure spool has been matched, $\dot{m}_{g3.5}$, $p_{o3.5}$, $T_{o3.5a}$, and N_{LPC} are known. Assume an LPT expansion ratio, $p_{o3.5}/p_{o4}$. Determine η_{LPC} (Figure 10.7*f*) and $\dot{m}_{g3.5}\sqrt{T_{o3.5}}/p_{o3.5}A_{3.5}$ (Figure 10.7*e*). Calculate $\Delta h_{LPT,a}$ and $\dot{m}_{g3.5}$.
8. Check to determine if $\Delta h_{LPT,a}$ and $\dot{m}_{g3}$ are within a preset tolerance. If not, repeat steps 2 through 7 until a match does exist.
9. Once the low-pressure spool has been matched, $\dot{m}_{g4}$, p_{o4}, and T_{o4a} are known. Determine, from the exhaust nozzle flow characteristics (Figure 10.7*g*), $\dot{m}_{g4}\sqrt{T_{o4}}/p_{o4}A_4$. Calculate A_4 and compare with the known exhaust nozzle A_4. An engine match exists and the thrust, thrust-specific fuel consumption, and other

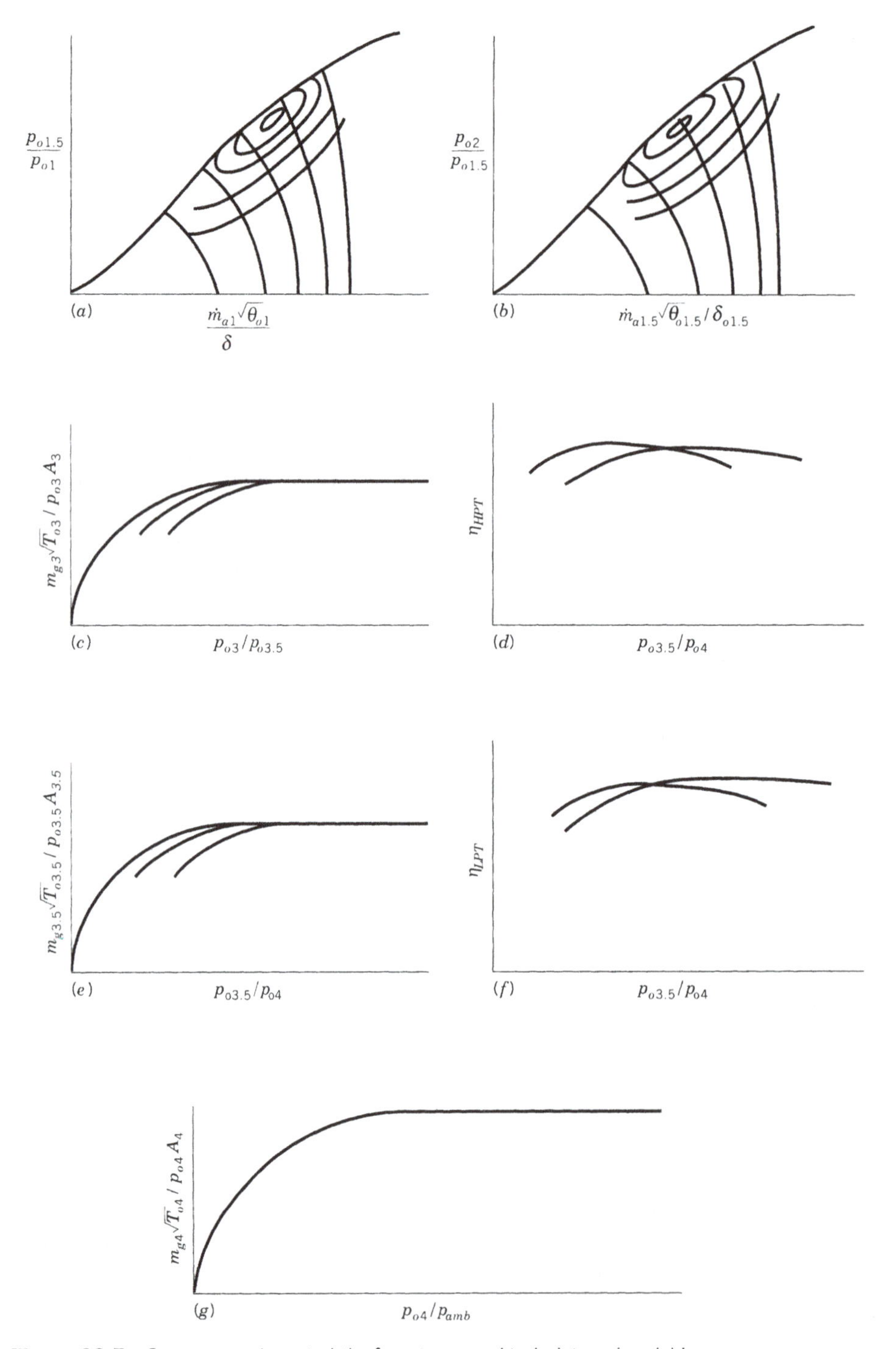

Figure 10.7. Component characteristics for a two-spool turbojet engine. (a) Low-pressure compressor. (b) High-pressure compressor. (c) High-pressure turbine. (d) High-pressure turbine. (e) Low-pressure turbine. (f) Low-pressure turbine. (g) Exhaust nozzle.

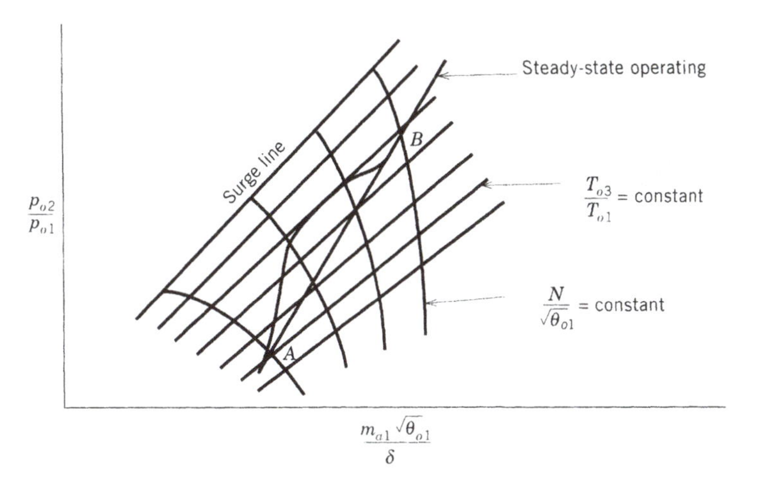

Figure 10.8. Transient operation.

desired values may be calculated with a preset tolerance. If a match does not exist, repeat 2 through 9 until a match has been achieved.

It should be obvious, from the matching procedure described above, that the component matching study for a gas turbine is a complex problem. The procedure described is for a two-spool turbojet engine. It should be obvious that the problem complexity increases as one studies a turbofan engine. Keep in mind that three-spool static pressure-balanced engines are currently being built. A matching procedure for this type of engine is left to the reader.

10.9 TRANSIENT OPERATION

The discussion so far in this chapter has dealt with the steady-state operation of a gas turbine. It is next important to determine what happens when the engine undergoes a power change.

Consider the case where a single-spool gas turbine engine, with the steady-state operating line shown in Figure 10.8, is to be accelerated from point A (low power setting) to point B (high power setting).

To change the steady-state operating point from A to B, the mass rate of flow of fuel must be increased. Increasing the mass rate of flow of fuel will initially increase the engine to a higher T_{o3}/T_{o1} line than the initial equilibrium value without appreciably changing the corrected mass rate of airflow ($\dot{m}_{a1}\sqrt{\theta_{o1}}/\delta_{o1}$) or the corrected rotor speed ($N/\sqrt{\theta_{o1}}$).

Next, the engine will start to increase in rotor speed (N), which will increase the corrected mass rate of flow. The exact path that the engine will follow in accelerating from A to B depends on the design characteristics of the engine components and the manner in which the fuel flow to the combustion chamber is changed. One possible acceleration path between steady-state operating points A and B is shown in Figure 10.8. Acceleration and deceleration paths will take the compressor to opposite sides of the steady-state operating line.

One must be careful during transient operations to schedule the fuel flow change so that the compressor does not surge during the change.

NOMENCLATURE

A area
C constant
c_p constant pressure specific heat
h specific enthalpy
k specific heat ratio
$\dot{m}$ mass rate of flow
N rotor speed
p pressure
T temperature
w specific work
δ pressure divided by ambient pressure
η efficiency
θ temperature over ambient temperature

Subscripts

1,2,3,4 states
a air
C compressor
g gas
o stagnation
p polytropic
T turbine

PROBLEMS

10.1. Solve Example Problem 10.1 assuming variable specific heats.

10.2. Calculate, for turbine expansion ratios of 2.0, 3.0, and 4.0:

(a) The ideal work of expansion for turbine inlet temperatures of 1000 K (1800°R), 1100 K (1980°R), and 1200 K (2160°R)

(b) The values of ideal work over theta (T_{03}/T_{std})

First assume constant specific heats, then variable specific heats.

10.3. Solve Example Problem 10.2:

(a) Assuming a 5% pressure drop in the combustion chamber

(b) Taking into account the mass of fuel added

(c) Assuming a 5% pressure drop and taking into account the mass of fuel added.

10.4. Solve Example Problem 10.3 assuming a 5% pressure drop in the combustion chamber, all other values being the same.

10.5. Determine, for Example Problem 10.3, the thrust developed by the engine and thrust-specific fuel consumption for sea level static conditions on a standard day, turbine inlet temperatures of 2160°R (1200 K), 1980°R (1100 K), and 1800°R (1000 K).

10.6. Assume, for Example Problem 10.2, that an afterburner is added to the turbojet engine. Calculate, assuming no change in the gas generator match point, the percent increase in the throat (minimum) area of the exhaust nozzle.

10.7. Assume that the sea level static, standard day operating line, as determined in Example Problems 10.2 and 10.3, does not shift at other altitudes or flight conditions. Determine, for an altitude of 11,000 m (36,089 ft) and a Mach number of 0.85 with the turbojet engine operating with a pressure ratio of 5.0:

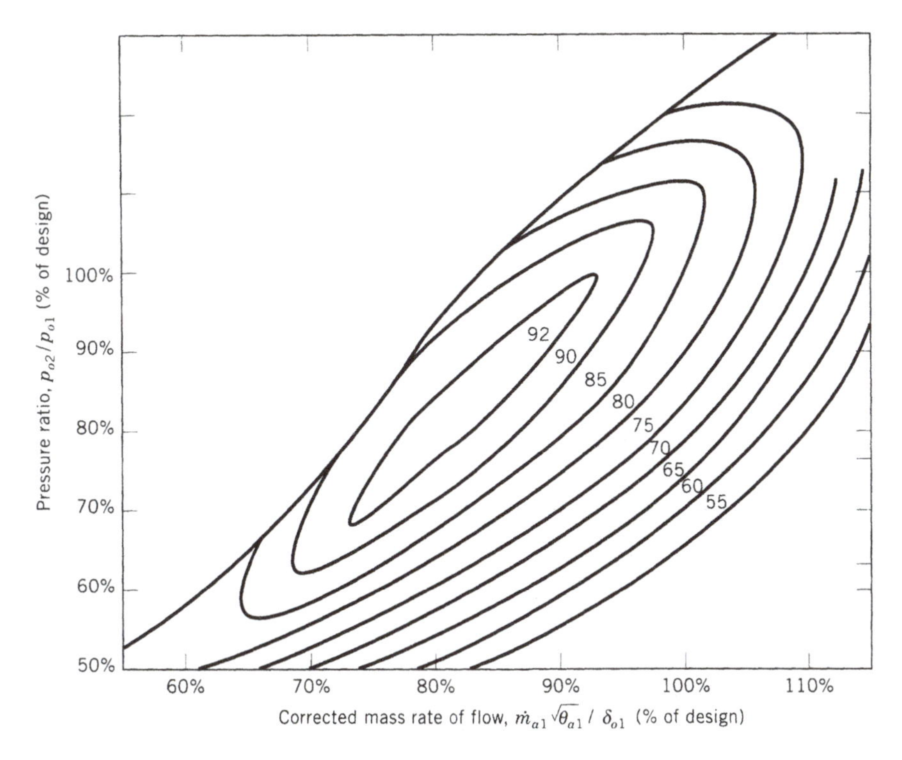

Figure 10.9. Compressor map for Problem 10.10.

(a) The percent change in rotor speed

(b) The mass rate of flow of air into the compressor

(c) The thrust developed by the engine

10.8. Solve Problem 10.7 assuming that the turbojet engine is operating at the altitude point with the same rotor speed (N).

10.9. Assume that the sea level static, standard day operating line as determined in Example Problem 10.3, does not shift at other altitudes or flight conditions. Determine, at an altitude of 18,000 ft (5,500 m) and a flight Mach number of 0.85, the thrust developed by the engine if the turbine inlet temperature is 1980°R (1100 K).

10.10. Determine, for the hypothetical compressor map shown in Figure 10.9, the sea level static, standard day, steady-state operating line for the single-spool turbojet engine if the compressor design point values are $p_{o2}/p_{o1} = 6.5$, $\dot{m}_{a1}\sqrt{\theta_{o1}}/\delta_{o1} = 35.0$ kg/s (77.16 lb/s), and the design point turbine inlet temperature is 1300 K (2340°R). Assume a constant turbine efficiency of 90%, neglect the mass of fuel added, assume no pressure drop in the combustion chamber, and constant specific heats.

10.11. Solve Problem 10.10 assuming a 5% pressure drop in the combustion chamber, all other values being the same.

10.12. A company wants to operate the engine described in Problem 10.10 on a refuse-derived, low-Btu gas. Known data about the refuse-derived fuel is as follows:

Analysis	(% volume)
H_2	35%
CO_2	10%
CH_4	5%
CO	50%
	100%
HHV	250 Btu/ft^3

Calculate, for this fuel, the thrust developed by this engine. *Do not* neglect the mass of fuel added.

10.13. Determine, for Example Problem 10.2, the change if the turbine efficiency is 90%, all other values being the same.

10.14. Solve Example Problems 10.2 and 10.3 assuming that the exhaust nozzle area is increased 10%.

10.15. Outline a matching procedure for a three-spool static pressure balance turbofan engine. Specify known component characteristics and necessary assumptions.

10.16. Solve Problem 10.15 assuming a nonmixed turbofan engine.

10.17. Determine, for Example Problem 10.2, the turbine inlet temperature, thrust and thrust-specific fuel consumption if the compressor pressure ratio is 5.0 and the ambient temperature and pressure are 540°R (360K) and 14.7 psia (101.3 kPa), respectively.

ENVIRONMENTAL CONSIDERATIONS

11.1 GENERAL

The ten preceding chapters have covered the history of the gas turbine, examined possible gas turbine engine cycles, considered the various components and factors that influence their performance, and then examined the manner in which these components influence one another to determine the engine operating point.

One last area will be considered, this being a brief examination of gas turbine air and noise emissions, current gas turbine air and noise regulations, and engine modifications that can be made to reduce the quantity of air pollutants and noise emitted by a gas turbine engine.

11.2 AIR POLLUTION

Air pollution episodes and problems existed long before any ambient levels or gas turbine exhaust stream measurements were conducted.

Most of the early air pollution episodes, such as occurred in London in 1873, 1952, and 1956, in the Meuse Valley of Belgium in 1930, or in Donora, Pennsylvania, in 1948, were associated with particulate and sulfur dioxide emissions and occurred in valleys or during atmospheric inversions in highly industrialized areas.

The first federal legislation enacted was the Air Pollution Control Act of 1955 (Public Law 84-159, July 14, 1955). It was narrow in scope and considered prevention and control of air pollution to be primarily the responsibility of state and local governments.

Congress, because of worsening conditions in urban areas due to air pollution by mobile sources, directed the surgeon general to conduct a thorough study of motor vehicle exhaust effects on human health. As a result, the act of 1955 was amended.

The two acts were the Air Pollution Control Act Amendments of 1960 (Public Law 86-493, June 6, 1960) and Amendments of 1962 (Public Law 87-761, October 9, 1962).

The Clean Air Act of 1963 (Public Law 88-206, December 1963) provided for the development of air quality criteria; encouraged state, regional, and local programs for the control and abatement of air pollution; and, for the first time, provided for federal financial aid for research and technical assistance.

The Air Quality Act of 1967 (Public Law 90-148, November 21, 1967) initiated a two-year study on the concept of national emission standards from stationary sources which served as the basis for the 1970 legislative act. The 1967 act also required

> . . . development and issue to the States such criteria of air quality as in his judgment may be requisite for the protection of the public health and welfare.

This provision led to a series of documents for common air pollutants such as carbon monoxide, sulfur oxides, nitrogen oxides, hydrocarbons, and so on, one document on each pollutant for air quality criteria which summarized what science at that time was able to measure of the effects of air pollution on humans and the environment, the second document summarizing information on techniques for controlling certain emissions.

The 1967 act also required that:

> The Secretary shall conduct a full and complete investigation and study on the feasibility and practicability of controlling emissions from jet and piston aircraft engines and of establishing national emission standards with respect thereto, and report to Congress the results of such study and investigation within one year from the date of enactment of the Air Quality Act of 1967, together with his recommendations.

The major provisions of the Clean Air Amendments of 1970 (Public Law 91-604, December 31, 1970) included:

1. That each state had the primary responsibility for assuring air quality within the entire geographic area comprising such state.
2. A requirement that national ambient air quality standards (both primary and secondary) were to be established by the Environmental Protection Agency (EPA).
3. A requirement that standards of performance for new stationary sources were to be established and implemented.
4. A requirement that industry was to monitor and maintain emission records and make them available to EPA officials.
5. A requirement establishing aircraft emission standards. These included:

> (1) Within 90 days after the date of enactment of the Clean Air Amendments of 1970, the Administrator shall commence a study and investigation of emissions of air pollutants from aircraft in order to determine—
>
> (a) The extent to which such emissions affect air quality in air quality control regions throughout the United States, and
>
> (b) the technological feasibility of controlling such emissions.
>
> (2) Within 180 days after commencing such study and investigation, the Administrator shall publish a report of such study and investigation and shall issue proposed emission standards applicable to emissions of any air pollutant from

any class or classes of aircraft or aircraft engines which in his judgment cause or contribute to or are likely to cause or contribute to air pollution which endangers the public health or welfare.
(3) The Administrator shall hold public hearings with respect to such proposed standards. Such hearing shall, to the extent practicable, be held in air quality control regions which are most seriously affected by aircraft emissions.
(a) Within 90 days after the issuance of such proposed regulations, he shall issue such regulations with such motivations as he deems appropriate. Such regulations may be revised from time to time.
(b) Any regulation prescribed under this section (and any revision thereof) shall take effect after such period as the Administrator finds necessary (after consultation with the Secretary of Transportation) to permit the development and application of the requisite technology, giving appropriate consideration to the cost of compliance within such period.
(c) Any regulations under this section, or amendments thereto, with respect to aircraft, shall be prescribed only after consultation with the Secretary of Transportation in order to assure appropriate consideration for aircraft safety.

The Clean Air Act Amendments of 1977 included many amendments. The one affecting gas turbines was Sec. 225, which stated:

(c) Any regulations in effect under this section on date of enactment of the Clean Air Act Amendments of 1977 or proposed or promulgated thereafter, or amendments thereto, with respect to aircraft shall not apply if disapproved by the President, after notice and opportunity for public hearing, on the basis of a finding by the Secretary of Transportation that any such regulation would create a hazard to aircraft safety. Any such finding shall include a reasonably specific statement of the basis upon which the finding was made.

11.3 AIRCRAFT EMISSION STANDARDS

The Clean Air Amendments of 1970 required that national and primary ambient air quality standards were to be established by EPA. These standards were published in the *Federal Register* in 1971 (1) and set maximum concentration limits for carbon monoxide, hydrocarbons, nitrogen dioxide, sulfur dioxide, particulates, and oxidant.

The Clean Air Amendments of 1970 also required the establishment of aircraft emission standards. Proposed emission standards were published on December 12, 1972 (2) with the final emission standards and test procedures being published on July 17, 1973 (3). The standards published in 1973 have undergone minor modifications over the last twenty years. It is recommended that any reader needing the latest standards consult the *Code of Federal Regulations* (5). The original (4) and current (5) standards are discussed below.

Of the pollutants generated in any combustion process, only carbon monoxide (CO), hydrocarbons (HC), nitrogen oxides (NO_x), and smoke have created the most concern in aircraft gas turbine engines and for which emission standards have been issued.

The first step in developing aircraft and aircraft engine standards was deciding how to classify the engines. They had to be classified in a manner consistent with

TABLE 11.1. **Original Engine Classification System for EPA Standards (4)**

EPA Class	Description
T1	All aircraft turbofan and turbojet engines of rated power less than 8000 lb thrust, except engines in Class T5
T2	All aircraft turbofan and turbojet engines of rated power of 8000 lb thrust or greater, except engines in Class T3, T4, or T5
T3	All aircraft gas turbine engines of the JT3D model family
T4	All aircraft gas turbine engines of the JT8D model family
T5	All aircraft gas turbine engines for propulsion of aircraft designed to operate at supersonic flight speeds
P1	All aircraft piston engines except radial engines
P2	All aircraft turboprop engines
APU	Any engine installed in or on an aircraft exclusive of the propulsion engines

their design, performance, and construction, giving consideration to their potential for reducing their emissions and the need to do so.

Eight classes were originally defined. These are listed in Table 11.1.

A separate class was selected for turboprop gas turbine engines because the proposed equivalency between shaft horsepower and jet thrust was not considered acceptable over the landing-take-off cycle.

The Pratt and Whitney JT3D and JT8D engines were given separate categories in order to be able to require separate schedules of smoke retrofits.

Supersonic gas turbine engines were included in a separate category because:

1. Some employ afterburners during take-off and acceleration which, because of the combustion pressure and mixing methods, result in higher hydrocarbon and carbon monoxide emissions.
2. The cycle pressure ratios for engines used in supersonic aircraft usually operate at lower pressure ratios than do engines designed for subsonic flight. At low flight speed and low altitude, these engines do not benefit from the ram effect they experience at high flight speed.
3. They usually do not employ high-bypass-ratio turbofan engines.

The current engine classes are slightly different than those listed in Table 11.1. The current classifications are listed in Table 11.2 (5).

The next step before aircraft gas turbine engine emission standards could be set

TABLE 11.2. **Current Engine Classification System for EPA Standards (5)**

EPA Class	Description
TP	All aircraft turboprop engines
TF	All turbofan or turbojet aircraft engines except engines of Class T3, T8, and TSS
T3	All aircraft gas turbine engines of JT3D model family
T8	All aircraft gas turbine engines of the JT8D model family
TSS	All aircraft gas turbine engines employed for propulsion of aircraft designed to operate at supersonic flight speeds

TABLE 11.3. **Gas Turbine Engine Power Settings for Emission Measurements (5)**

Aircraft Operating Mode	Engine Class		
	TP	TF, T3, T8	TSS
Taxi/idle	[1]	[1]	[1]
Takeoff	100%	100%	100%
Climbout	90%	85%	65%
Descent	NA	NA	15%
Approach	30%	30%	34%

[1] 7% rated thrust.

was to determine the engine operating conditions that would be used for determining the air pollutant emissions. The current exhaust emission test selected is designed to measure hydrocarbons, carbon monoxide, and carbon dioxide emissions for a simulated aircraft landing–takeoff (LTO) based on the time in each mode during a high-activity period at major airports.

The standards specify that the gas turbine engine be tested in each of the operating modes listed in Table 11.3. The values given in Table 11.3 are the percentage of rated power settings on a standard day, which is defined as a temperature of 15°C, a pressure of 101325 Pa, and a specific humidity of 0.00 kg H_2O/kg dry air. Emission testing is conducted on a warmed-up engine, which has achieved a steady operating temperature.

The time in each mode shall be that listed in Table 11.4.

The 1979 standards (4) set gaseous emission standards for hydrocarbons, carbon monoxide, nitric oxide, and smoke. The 1994 standards (5) set gaseous emission standards for hydrocarbons, smoke exhaust emissions, and fuel venting standards.

The 1994 engine fuel venting emission standards apply to all new aircraft gas turbine engines of classes T3, T8, TF, and TSS with rated output equal to or greater than 36 kilonewtons (8090 lb) manufactured after January 1, 1974, and to all in-use aircraft gas turbine engines of classes T3, T8, TF, and TSS with rated output equal to or greater than 36 kilonewtons (8090 lb) manufactured after February 1, 1974. These engine fuel venting emission standards also apply to all new aircraft gas turbine

TABLE 11.4. **Landing–Take-off Cycle for Aircraft Engine Emission Standards (5)**

Aircraft Operating Mode	Engine Class		
	TP min	TF, T3, T8 min	TSS min
Taxi/idle	26.0	26.0	26.0
Takeoff	0.5	0.7	1.2
Climbout	2.5	2.2	2.0
Descent	NA	NA	1.2
Approach	4.5	4.0	2.3

TABLE 11.5. **Smoke Exhaust Emission Standard—New Aircraft Gas Turbine Engines (5)**

Engine Class	Rated Output	Manufactured On or After	Maximum Smoke Number
T8	all	February 1, 1974	30
TF	≥ 129 kilonewtons (29,000 pounds)	January 1, 1976	83.6 $(rO^*)^{-0.274}$
T3	all	January 1, 1978	25
T3, T8, TSS, TF	≥ 26.7 kilonewtons (6000 pounds)	January 1, 1984	83.6 $(rO^*)^{-0.274}$ with max of 50
TF	≥ 26.7 kilonewtons (6000 pounds)	August 9, 1985	83.6 $(rO^*)^{-0.274}$ with max of 50
TP	≥ 1000 kilowatts	January 1, 1984	187 $(rO^{**})^{-0.168}$

* rO is rated output in kilonewtons.

** rO is rated output in kilowatts.

engines of class TF with rated outputs less than 36 kilonewtons (8090 lb), to all in-use turboprop engines of class TP manufactured on or after January 1, 1975, and to all in-use aircraft gas turbine engines of class TF with rated outputs less than 36 kilonewtons (8090 lb) and class TP turboprop engines manufactured after January 1, 1975.

The engine fuel venting emission standards require that no fuel be discharged into the atmosphere from any new or in-use gas turbine engine after dates listed above. The purpose of this standard is to eliminate the direct discharge into the atmosphere of fuel drained from the engine fuel nozzle manifolds after the engine has been shut down.

The current smoke exhaust emission standards are listed in Table 11.5. The effective date, maximum smoke number, engine class, and rated output where the standards apply are listed in this table.

The 1979 standards are listed in Tables 11.6 and 11.7. Note at that time there were different standards for newly manufactured and newly certified engines and that there were HC, CO, and NO_x standards.

The 1994 gaseous exhaust emission hydrocarbon limits are listed in Table 11.8. It should be noted that the hydrocarbon emission standards are based on mass of pollutant

TABLE 11.6. **Gaseous Emission Standards Applicable to Newly Manufactured Aircraft Turbine Engines (4)**

Engine Class	HC[a]	CO[a]	NO_x[a]
T1	1.6	9.4	3.7
T2, T3, T4	0.8	4.3	3.0
P2	4.9	26.8	12.9
T5	3.9	30.1	9.0

[a] "T" standards are pounds pollutant/1000 lb-thrust hours/cycle. "P" standards are pounds pollutant/1000 hp-hours/cycle.

TABLE 11.7. **Gaseous Emission Standards Applicable to Newly Certified Aircraft Gas Turbine Engines (4)**

Engine Class	HC[a]	CO[a]	NO_x[a]	Effective Date
T2, T3, T4	1.0	7.8	5.0	January 1, 1983
T5	1.0	7.8	5.0	January 1, 1984

[a] Pounds pollutant/1000 lb-thrust hours/cycle.

emitted to thrust-hours over the landing–take-off cycle that is specified in Table 11.4 and at the power settings that are listed in Table 11.3.

Other ways of expressing gaseous emissions include:

1. Pollutant concentration. This has the advantage of being easy to use but provides no guide as to the mass of pollutant being emitted by the gas turbine engine.
2. Ratio of mass of pollutants emitted to mass of fuel consumed. This provides a guide as to how "clean" the combustion system is in the gas turbine engine but does not reveal the air pollution effects of the complete engine since different engines have different fuel consumption characteristics.
3. Total mass of pollutants emitted over the LTO cycle. This would be the most useful one if one was interested in estimating total airport emissions.

The one selected by EPA normalizes emissions over the LTO cycle.

11.4 STATIONARY GAS TURBINE ENGINE EMISSION STANDARDS

The primary federal environmental laws applicable to power generating as turbine engines are the Clean Air Act Amendments of 1970, 1977, and 1990. The national and primary ambient air quality standards established by the Clean Air Amendments of 1970, which were discussed in Section 11.3, are very important when considering standards that apply to power generating gas turbine engines.

The current standards are published in the *Code of Federal Regulations* (6) and will be discussed in this section. Subpart GG-Standards of Performance for Stationary

TABLE 11.8. **Hydrocarbon Gaseous Exhaust Emission Standards—New Commercial Aircraft Gas Turbine Engines Manufactured on or after January 1, 1984 (5)**

Engine Class	Rated Output	Maximum Emissions Grams/Kilonewton Rated Output
T3, T8, TF	≥ 26.7 kilonewtons	19.6
TSS	all	$140\ (0.92)^{rPR*}$

* rPR is the engine rated pressure ratio.

TABLE 11.9. **Gas Turbine Engine New Source Performance Standards (6)**

Input Energy (Based on lower heating value of fuel)	Use	Emission Limit*	
		NO_x, ppmv	SO_2, ppmv
Greater than 107.2 gigajoules/h (100 million Btu/h)	Electric Utility	75**	150
Between 10.7 and 107.2 gigajoules/h (Between 10 and 100 million Btu/h)	All uses	150***	150
Base load equal to or less than 30 megawatts	Nonutility	150***	150
Less than 10.7 gigajoules/h (Less than 10 million Btu/h)	All uses	None	150
Greater than 30 megawatts	Nonutility	None	150
Regenerative cycle, input less than 107.2 gigajoules/h (100 million Btu/h)	All uses	None	150
All input levels	Emergency Fire Fighting Military	None	150

* All values corrected to 15% oxygen, dry basis.

** Use Equation 11.1.

*** Use Equation 11.2.

Gas Turbines of Part 60-Standards of Performance of New Stationary Sources (6) apply to all gas turbine engines with a heat input at peak load equal to or greater than 10.7 gigajoules per hour (GJ/h) (10 million Btu/h) for gas turbine engine installations, which commenced construction, modification, or reconstruction after October 3, 1977. Maximum emission limits for the oxides of nitrogen, NO_x, and sulfur dioxide, SO_2, are specified in the standard based on input energy to the gas turbine unit and intended use; that is, utility or industrial application.

A unit is considered by EPA an industrial gas turbine engine if less than one-third of the potential electric output capacity is sold and a utility gas turbine engine if more than one-third of the potential electric output capacity is sold through a utility system.

Current gas turbine engine New Source Performance Standards (NSPS) are listed in Table 11.9. The emission limits as listed in Table 11.9 are the NSPS limits when on a dry basis and converted to 15% oxygen.

The limits are calculated using the following formulas

(a) For electric utility applications when the heat input is greater than 107.2 GJ/h (100 million Btu/h)

$$NO_{x,\max} = 0.0075\left(\frac{14.4}{Y}\right) + F \tag{11.1}$$

(b) For all uses where the heat input is between 10.7 and 107.2 GJ (10 million and 100 million Btu/h) or for base loaded nonutility applications where the output is less than 30 mW

$$NO_{x,\max} = 0.0150\left(\frac{14.4}{Y}\right) + F \tag{11.2}$$

In Equations (11.1) and (11.2), Y is the manufacturer's rated heat rate at rated load, kilojoules per watt hour, and F is an allowance for fuel-bound nitrogen.

The value of Y cannot exceed 14.4 kilojoules per watt hour (kJ/wh). The value of 14.4 assumes that the gas turbine engine has a thermal efficiency of 25%. This means that gas turbine engines with thermal efficiencies higher than 25% will be allowed to emit more NO_x than is listed in Table 11.9.

No fuel-bound allowance is made if the percent by weight nitrogen in the fuel is 0.015% or less. It is recommended that the reader consult the latest standards if they are interested in the fuel-bound nitrogen allowance for fuels with higher nitrogen contents.

The sulfur dioxide standards are the same for all gas turbine engines. All are limited to 150 ppmv at 15% oxygen and on a dry basis. All operators shall never burn fuel that contains sulfur in excess of 0.8% by weight.

There are several exceptions to the emission limits listed in Table 11.9. Two of these are:

1. Stationary gas turbine engines that use water or steam injection for control of NO_x emissions are exempt when ice fog is deemed a traffic hazard.
2. Stationary gas turbine engines with a peak load heat input equal to or greater than 10.7 GJ/h (10 million Btu/h) and less than 107.2 GJ/h (100 million Btu/h) that commenced construction prior to October 3, 1982.

The emission limits as listed in Table 11.9 are the maximum levels for new sources. Levels can be set at lower levels to prevent significant deterioration in areas that meet the National Ambient Air Quality Standards (NAAQS) or in areas that do not meet the NAAQS air quality levels.

A Prevention of Significant Deterioration (PDS) permit is required for a new gas turbine engine facility in an attainment area if there is a potential for the facility to emit at least 250 tons per year of any regulated pollutant. Best Available Control Technology may be required. For new gas turbine engine facilities in NAAQS nonattainment areas, facilities are required to use the Lowest Achievable Rate controls available.

11.5 NO_x FORMATION

It was noted in the previous section that the oxides of nitrogen, NO_x, are the predominant emission from stationary gas turbine engines and the one that is controlled by standards. The most prevalent NO_x emissions are nitric oxide, NO, and nitrogen dioxide, NO_2.

Nitric oxide is the one mainly formed in the combustion chamber. Factors that influence the amount of NO formed are:

1. Maximum temperature
2. Percent excess air
3. Pressure
4. Time at the maximum temperature
5. Fuel-bound nitrogen.

The maximum temperature occurs when the fuel is burned with the stoichiometric (theoretical) amount of air. The higher the temperature of the air at the inlet to the combustion chamber, the higher the resulting equilibrium adiabatic flame temperature. This was illustrated by the values tabulated in Table 4.7.

Burning the fuel with excess air lowers the maximum temperature but increases the availability of oxygen and nitrogen in the products of combustion. The values in Table 4.7 show that for a fixed air supply temperature and combustion chamber pressure, the amount of NO formed for equilibrium conditions increases from 0% excess air to 30% excess air, then starts to decrease even though the adiabatic equilibrium flame temperature decreases continuously from a temperature of 4390°R for 0% excess air to 3898°R for 30% excess air.

The results tabulated in Table 4.7 show that increasing the combustion chamber pressure increases the equilibrium adiabatic flame temperature but decreases the amount of NO formed.

The preceding analysis assumes that equilibrium has been reached. It is next important to determine the rate at which the products will reach equilibrium. The basic mechanism presently used to predict the formation of NO had its origin in the work of Zeldovich and coworkers around 1946. The reader is referred to Wark and Warner (7) for a development of one mechanism that can be used to determine the rate at which NO is formed. The equation developed by Wark and Warner (7) is

$$(1 - Y)^{C+1}(1 + Y)^{C-1} = e^{-Mt} \tag{11.3}$$

where

$$Y = [NO]/[NO]_e \tag{11.4}$$

$$C = \frac{(2.1 \times 10^4)[x_{N_2}]^{0.5}\, e^{-7750/T}}{T[x_{O_2}]^{0.5}} \tag{11.5}$$

$$M = \frac{(5.4 \times 10^{15})(p)^{0.5}(x_{N_2})^{0.5}\, e^{-58330/T}}{T} \tag{11.6}$$

In the above equations

$[NO]$ = concentration of NO

$[NO]_e$ = concentration of NO at equilibrium

x_{N_2} = mole fraction of N_2 in the products

x_{O_2} = mole fraction of O_2 in the products

T = temperature, K

p = pressure, atmospheres

It is of interest to determine the length of time it takes to form a given amount of NO for various temperatures, pressures, and percent excess air values. This is illustrated by Example Problem 11.1.

Example Problem 11.1

Liquid n-octane, C_8H_{18}, is burned with the stoichiometric amount of dry air in a steady-flow process at a pressure of 1 atmosphere. The *n*-octane and dry air are supplied at 537°R (298 K). The resulting equilibrium temperature is 4076°R (2264 K) and the equilibrium products have the following mole fractions.

$$x_{N_2} = 0.7179$$

$$x_{02} = 0.005961$$

$$x_{NO} = 0.002456 \text{ or } 2456 \text{ ppm}$$

Determine the length of time required to form 150 ppm.

Solution

Using Eq. (11.6), (11.5), (11.4) and (11.3).

$$M = \frac{(5.4 \times 10^{15})(1)^{0.5}(0.7179)^{0.5}e^{(-58330/2264)}}{2264}$$

$$= 13.07$$

$$C = \frac{(2.1 \times 10^{4})(0.7179)^{0.5}e^{(-7750/2264)}}{2264\ (0.005961)^{0.5}}$$

$$= 3.319$$

$$Y = \frac{150}{2456} = 0.06107$$

$$(1.0 - 0.06107)^{4.319}(1.0 + 0.06107)^{2.319} = e^{(-13.07t)}$$

$$t = 0.009357 \text{ s}$$

$$= 9.36 \text{ milliseconds (ms)}$$

If the same equilibrium composition were at a temperature 100 K lower (3896°R, 2164 K), the time required to form 150 ppm of NO increases to 29.4 ms. This shows the strong effect of temperature on the length of time it takes to form nitric oxide, NO.

Table 11.10 shows the effect of excess air on the equilibrium amount of nitric oxide and the time to form 150 ppm of NO. All values tabulated in Table 11.10 are for liquid *n*-octane, C_8H_{18}, as the fuel and a pressure of ten atmospheres.

Note that as the percent excess air increases from 0% to 30%, the equilibrium adiabatic flame temperature decreases from 4390°R (2439 K) to 3898°R (2166 K), the equilibrium amount of NO *increases* from 3236 ppm to 5348 ppm (it peaks at 20% excess air), and the time required to form 150 ppm increases from 0.38 ms to

TABLE 11.10. **Effect of Percent Excess Dry Air on the Length of Time to Form 150 ppm of NO_x When Liquid *n*-Octane is Burned Adiabatically and at Constant Pressure with Dry Air**

Pressure, atm	10	10	10	10	10
Air Supply Temp., °R(K)	1000(556)	1000(556)	1000(556)	1000(556)	1000(556)
Percent Excess Air Supplied	0	10	20	30	40
Adiabatic Flame Temp.,					
Equilibrium Composition, °R(K)	4390(2439)	4252(2362)	4072(2262)	3898(2166)	3738(2077)
Equilibrium Product Composition					
Total moles/mole fuel					
Mole fraction N_2	0.7179	0.7255	0.7306	0.7347	0.7381
Mole fraction O_2	0.005213	0.01700	0.03037	0.04286	0.05423
Mole fraction NO	0.003236	0.005083	0.005566	0.005348	0.004865
Time to Form 150 ppmv, ms	0.38	0.51	1.32	4.12	23.7

4.12 ms. Once again it shows the strong temperature dependence on the length of time it takes to form nitric oxide.

11.6 EMISSION REDUCTION, AIRCRAFT GAS TURBINE ENGINES

The National Aeronautics and Space Administration (NASA) in mid-1971 decided to begin a major program in aircraft gas turbine emissions reduction technology. Their program consisted of in-house research on low-emission concepts as well as contracted research with the major aircraft gas turbine engine manufacturers. The results of the contract efforts are reported in NASA Conference Publication 2021 (9).

The objectives of the NASA program were

1. To investigate new combustor concepts that had the potential for significantly lower emission levels.
2. To measure, once the new concepts had been developed to their full potential, the combustor emission reduction in a test engine

NASA awarded several multiphase contracts to engine manufacturers. The phases consisted of combustion research as follows:

Phase 1. Screening combustor concepts to determine which, if any, had the potential for lower emissions.

Phase 2. The best concepts would be further developed with emphasis on combustion performance as well as emission reduction.

Phase 3. Install the best or most engine-ready combustor in an engine and test the engine for performance and emission levels.

Representative engines in each of the EPA classes, as identified in Table 11.11, were selected and contracts awarded.

TABLE 11.11. **Representative Aircraft Gas Turbine Engines Selected for Emission Reduction Program**

EPA Engine Class	Gas Turbine	Manufacturer
T1	TFE-731-2	Garrett AiResearch
T2	CF6-50	General Electric
	JT9D-7	Pratt & Whitney
T4	JT8D-17	Pratt & Whitney
P2	501-D22A	Detroit Diesel Allison Div.

It was necessary to determine, for each engine, its emission levels when compared with the 1979 EPA standards which, for the engines being discussed, were the same as the 1983 emission standards listed in Table 11.4. The results of tests are summarized in Table 11.12.

Other program performance goals were as follows:

1. During all engine operatings modes, the combustor would have a combustion efficiency of 99%+.
2. That the pressure loss at cruise be no greater than 6%.
3. The new combustor design durability be adequate at all engine conditions.
4. That all combustor concepts must fit in the present engine combustor casing envelope.

General combustor characteristic showed that virtually all of the THC and CO emissions are generated at low power, primarily at engine idle, and that NO_x emissions are lowest at engine idle and increase as the gas turbine power increases. To reduce NO_x emissions, the maximum flame temperature must be reduced and/or the residence time of the combustion gases at the maximum temperature must be decreased.

Results for the JT9D and CF6 gas turbine engines, two of the five engines selected by NASA for its emission reduction program, are summarized below.

TABLE 11.12. **Emission Levels of the Gas Turbine Engines Selected by NASA for Their Emission-Reduction Program**

		THC		CO		NO_x		Smoke	
Engine	Engine	Std	Prod[a]	Std	Prod[a]	Std	Prod[a]	Std	Prod[a]
P2	501-D22A	4.9	306	26.8	118	12.9	48	29	189
T1	TFE-731	1.6	331	9.4	180	3.7	162	40	118
T4	JT8D-17	0.8	500	4.3	356	3.0	260	25	120
T2	JT9D-7	0.8	488	4.3	198	3.0	197	20	50
T2	CF6-50	0.8	538	4.3	251	3.0	257	19	68

[a] Production values as a percent of EPA standards.

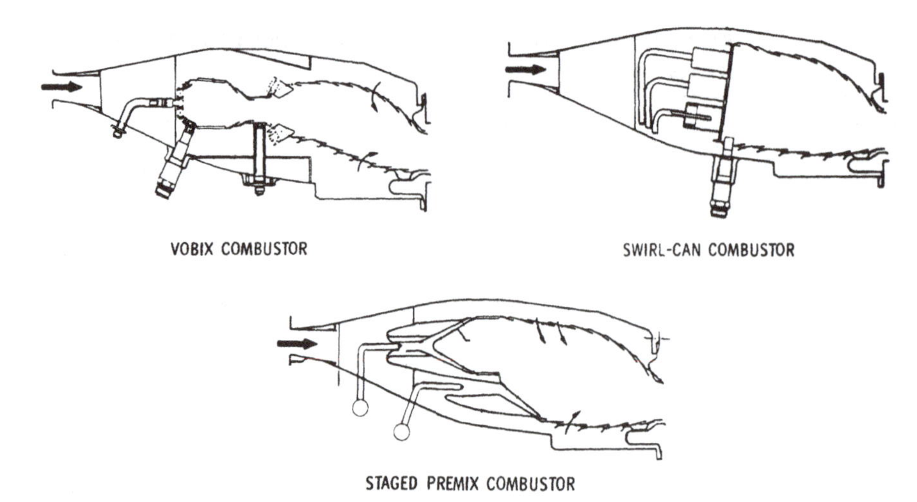

Figure 11.1. Phase I combustor concepts for the JT9D engine. Reproduced by permission from Ref. (9).

Figure 11.1 shows the various combustors tested by Pratt & Whitney, Figure 11.2 the various combustors tested by General Electric. Pratt & Whitney selected the Vorbix combustor for phase II testing, General Electric the double-annular concept selected for phase II testing.

Figures 11.3 and 11.4 show the production and phase II combustors for the JT9D and CF6 engines, respectively.

The results of the phase II testing for these two engines are shown in Table 11.13.

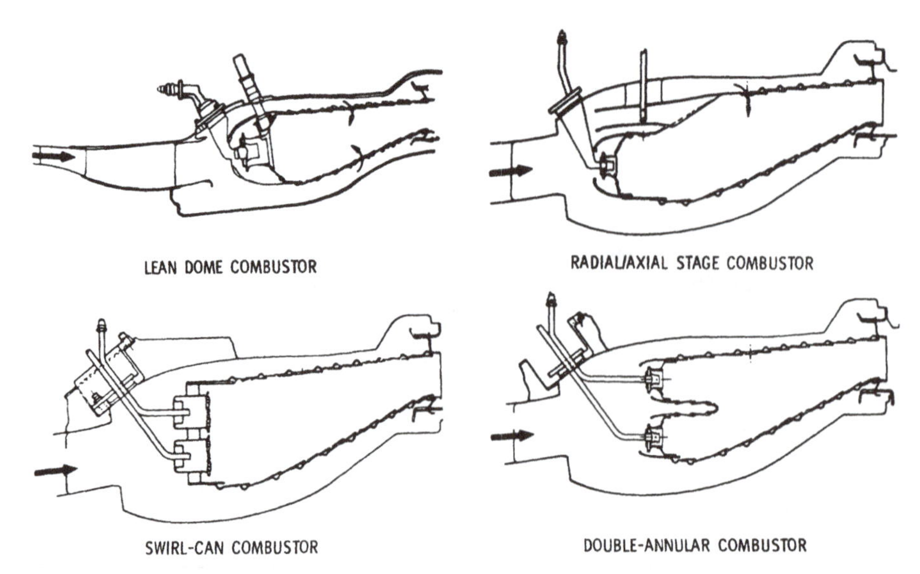

Figure 11.2. Phase I combustor concepts for the CF6 engine. Reproduced by permission from Ref. (9).

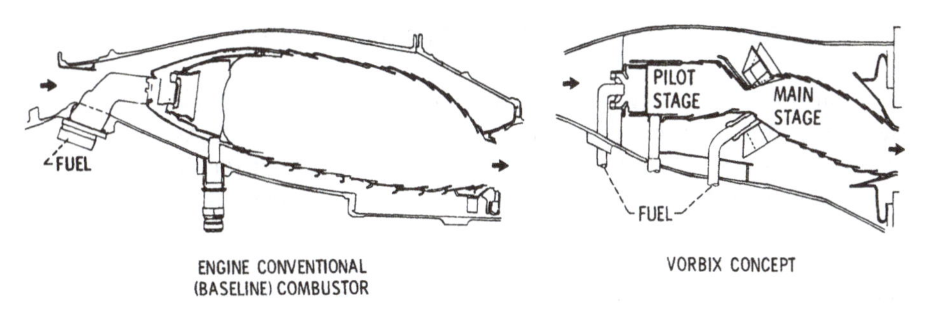

Figure 11.3. Phase II combustor concept for the JT9D-7 engine. Reproduced by permission from Ref. (9).

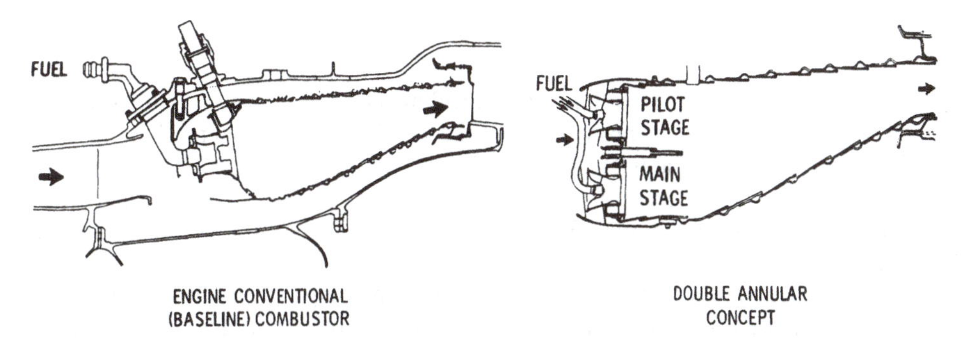

Figure 11.4. Phase II combustor for the CF6-50 engine. Reproduced by permission from Ref. (9).

TABLE 11.13. **Summary of Phase II Combustors for the JT9D-9 and CF6-50 Gas Turbine Engines**

	THC[a]	CO[a]	NO_x[a]
Pratt & Whitney JT9D-7			
Conventional combustor	488	198	197
Vorbix combustor	38	151	73
General Electric CF6-50			
Conventional combustor	538	251	257
Double-annular combustor	38	70	142

[a] Values listed are emission levels as a percent of 1979 EPA standards (10).

11.7 NO_x REDUCTION, STATIONARY GAS TURBINE ENGINES

It was illustrated in Section 11.5 that the higher the temperature and the longer the gases are at that temperature, the more nitric oxide is formed. As shown by the emission limits in Table 11.9, NO_x is the main pollutant from stationary gas turbine engines. The amount of SO_2 emitted is limited by the amount of sulfur in the fuel since this is the only source of sulfur.

Prior to NO_x emission controls, gas turbine engine combustion chambers were designed so that the fuel–air ratio in the primary zone was approximately the stoichiometric value; that is, the percent excess air in the primary zone was 0%. This resulted in maximum temperature.

The maximum temperature can be reduced by designing the combustion chamber so that the primary zone either operates fuel rich (insufficient air for complete combustion) or fuel lean (excess air). Both of these conditions can result in increased smoke (fuel rich) or increased carbon monoxide and total hydrocarbon emissions (fuel lean).

Several methods can be used to reduce NO_x emissions. These include

1. Water or steam injection
2. Staged combustion
3. Selective catalytic reduction

The most commonly used method of controlling NO_x emissions is with *water* or *steam injection* into the primary zone of the combustion chamber. The water (or steam) injected acts as a heat sink, resulting in a lower maximum temperature, thereby reducing the amount of NO_x formed. The rate at which water is injected is approximately 50% of the fuel flow. Steam rates are usually 100%–200% of the fuel flow.

An advantage of this method is that it increases the output from the power turbine due to the increased flow through the gas generator and power turbines and the higher specific heat of water. The reader is referred to Example Problem 5.8 for calculations illustrating the increase expected.

Disadvantages of using water or steam injection are

1. Need a constant supply of pure water
2. Can increase CO emissions
3. Combustion chamber pressure oscillations, especially with water injection
4. Heat rate penalty.

Schorr (8) reported the following results when using water and steam injection

1. An NO_x level of 75 ppmvd for an oil-fired simple cycle with water flow of 0.5 the fuel flow. Output was increased 3% when compared with no water injection, and the unit had a heat rate penalty of 1.8%.
2. An NO_x level of 42 ppmvd for a natural gas fired simple cycle with water flow equal to the fuel flow. Output increased 5% and the heat rate penalty was 3%.
3. An NO_x level of 42 ppmvd for a natural gas fired combined cycle with steam flow of 1.4 times the fuel flow. Output increased 5%, and the heat rate penalty was 2%.
4. An NO_x level of 25 ppmvd for a natural gas fired simple cycle with GE's Quiet

Combustor. Water injection was used with water flow 1.2 times fuel flow. The output increased 6%, and the heat rate penalty was 4%.

5. An NO_x level of 25 ppmvd for a natural gas fired combined cycle with steam flow 1.3 times fuel flow. Output increased 5.5% with a heat rate penalty of 3%.

Staged combustion is currently being tested by a number of manufacturers. It provides a way of achieving NO_x emission levels of 25 ppmvd or less at 15% oxygen without using water or steam injection. Most of the systems being tested use a two-stage premixed combustor for use with natural gas. The resulting mixture is lean so the amount of NO_x is low.

The reader is referred to an article by Schorr (8) for a description of the General Electric dry low NO_x combustor.

Selective catalytic reduction involves injecting ammonia into the gas turbine engine exhaust stream. The exhaust gases then pass over a catalyst where the NO_x reacts with the ammonia, NH_3, oxygen, O_2, and nitrogen, N_2, to form water, H_2O, and nitrogen, N_2. When combined with water or steam injection, Schorr (8) reports that NO_x levels of 10 ppm or less can be achieved.

One major disadvantage is that the reaction is very temperature dependent. Schorr (8) states that for a vanadium pentoxide type catalyst, the exhaust gas temperature range for best operation is 600°F–750°F. For this reason, the selective catalytic reduction method for reducing NO_x emissions is limited to combined cycles.

11.8 NOISE

Aircraft noise, since the introduction of jet-powered commercial airplanes, has been of concern. A continuous effort has been made to develop the technology to design a quiet gas turbine engine.

One indication of the public's concern about aircraft noise is the large number of major airports around the world that have noise restrictions in the form of curfews, night-time limitations, flight restrictions, and/or preferred runaways or routings that take aircraft over water or sparsely populated areas. The major problem with aircraft noise, in terms of number of people exposed and the frequency with which they are exposed, occurs in the vicinity of airports.

Noise, unlike emissions, whose sole source is the combustion chamber of the gas turbine engine, has many sources. The two main sources of noise are

1. Propulsion system noise
2. Aircraft noise other than propulsion system noise, including sonic boom, flap noise, etc.

Only propulsion system noise will be considered.

Propulsion system noise may be separated into two categories:

1. Externally generated noise associated with the exhaust gases from the propulsion system, the propeller of a turboprop-powered aircraft
2. Internally generated noise associated with the rotating machinery and the combustion process

An important distinction between internally generated and externally generated noise is that internally generated noise can be suppressed, whereas externally generated noise cannot be suppressed.

When gas turbine engine-powered commercial aircraft were introduced, there was virtually no concession in the engine design to noise reduction. The first commercial aircraft were powered by turboprop engines; then, in the later 1950s, commercial aircraft powered by turbojet engines, in which the noise was predominantly from the mixing of the high-velocity exhaust stream with the ambient air, entered service. With the introduction of the high-bypass-ratio turbofan engines in the late 1960s, the predominate source of noise changed from the exhaust stream (an external source) to the turbomachinery and combustion chamber (an internal source.)

The sources of noise will be divided into three main groups:

1. Jet or exhaust noise
2. Fan noise
3. Core noise

Each of these groups is discussed below.

Jet noise results from the mixing of the high-velocity exhaust stream with the ambient air. A considerable amount of turbulence is generated when these two streams at different velocities mix, with the intensity of the turbulence, and hence the noise, increasing as the velocity difference increases. Researchers have found that the magnitude of the jet noise increases as the *eighth power* of the velocity.

The dominating noise of the early turbojet engines, the jet roar, was generated behind the jet engine exhaust nozzles where the high exhaust stream mixed with the ambient air.

With the introduction of the nonmixed turbofan engine, there were two exhaust streams, therefore two sources of external noise. One source was the turbulent mixing of the fan exhaust steam with the ambient air. The other source was the turbulent mixing of the core exhaust stream with the fan exhaust stream and the ambient air.

It was noted in Chapter 6 that when a turbojet engine is converted to a turbofan engine with the same core pressure ratio and turbine inlet temperature, the core velocity decreases, the amount of decrease and difference in velocity between the core and fan exhaust streams depending on the fan pressure ratio and bypass ratio.

Fan noise is one of the major, if not the predominent, sources of noise in a high-bypass-ratio turbofan engine. Fan noise has different characteristics depending on whether the tip speed of the fan rotor blades is subsonic or supersonic. Fan noise usually is separated into three categories:

1. Broad-band noise, which is essentially the noise generated from the turbulence in the air and by the airload fluctuations as it passes across the blade (rotor and stator) surfaces.
2. Discrete tone noise, which is noise generated by the fluctuating pressures generated by the interaction of the rotor blades and the stationary blades. The frequency of this noise may be predicted by the rotor rotational speed.
3. Multiple-tone noise, which is associated with the shock waves on the rotor blades caused by supersonic relative flow over the blades.

As higher-bypass-ratio engines are built, the exhaust velocity and therefore the jet noise are reduced. Under these conditions, the turbofan *core noise* becomes more important. Core noise consists of compressor noise, combustion noise, and turbine noise.

Compressor and turbine noise are similar to fan noise, resulting mainly from the interaction of the rotating and stationary blades. Combustion noise results from the turbulence generated by the burning of the fuel.

11.9 NOISE STANDARDS

Increased use of jet-powered commercial aircraft, along with a growing concern for the quality of the environment, has resulted in considerable emphasis on the reduction of noise from gas turbine engines since the mid-1960s.

The initial national noise regulations were those in Public Law 90-411 (10) in 1969, which are commonly known as FAR 36. These were modified in 1973 (11). It is suggested that the reader consult the *Code of Federal Regulations* under Title 14, Part 36 (12), for the latest regulations. Below is a condensed version of these regulations.

The noise regulations depend on the number of engines, when application is made, and other engine design conditions.

For subsonic aircraft with turbofan engines with a bypass ratio of 2 or more, the current noise limits are as follows:

1. If application was made before January 1, 1967, it had to meet stage 2 noise limits or be reduced to the lowest levels economically and technologically possible.
2. If application was made on or after January 1, 1967, and before November 5, 1975, it had to meet stage 2 noise limits.
3. If application was made on or after November 5, 1975, it must meet stage 3 noise levels.

For aircraft with turbofan engines with a bypass ratio less than 2, the current noise limits are as follows:

1. If application was made before December 1, 1969, the noise level had to be the lowest level reasonably obtainable through use of procedures and information developed for the flight crew.
2. If application was made on or after December 1, 1969, and before November 5, 1975, the noise levels has to be no greater than stage 2 noise limits.
3. If application is made after November 5, 1975, it must meet stage 3 noise limits.

For the Concorde airplane, the noise levels must be reduced to the lowest levels that are economically reasonable, technologically practicable, and appropriate for the Concorde design.

Section A36 of the code prescribes in detail the conditions under which the aircraft noise certification tests must be conducted, the measurement procedures that must be used to measure the noise levels, and the corrections that must be applied for variations in atmospheric conditions or flight path.

The measurement points are as follows:

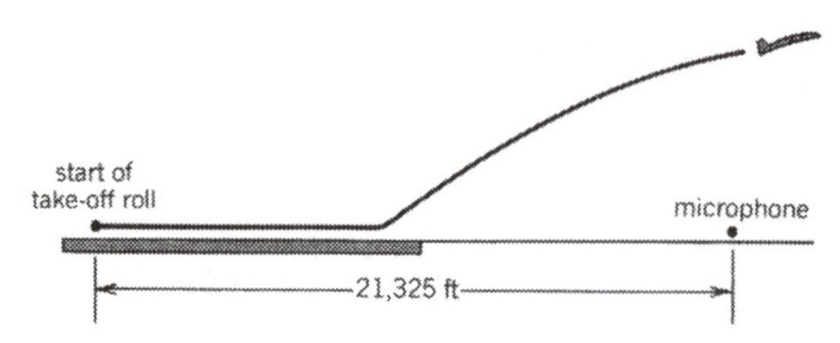

Figure 11.5. Typical take-off profile for noise measurements.

1. For take-off, at a point 21,325 ft (6500 m) from the start of the take-off roll on the extended centerline of the runaway. A typical profile is shown in Figure 11.5. The take-off profile is defined by five parameters, including length of take-off roll, climb angle, and thrust change points.
2. For approach, measurements are taken at a point 6562 ft (2000 m) from the touch-down point on the extended centerline of the runaway. A typical profile is shown in Figure 11.6.
3. For sideline, measurements are made at a point parallel to and 1476 feet (450 meters) from the extended centerline of the runaway where the noise level after liftoff is the greatest. The exception is that for Stage 1 and 2 compliance for aircraft powered by more than three turbojet engines, measurements are made at 0.35 nautical miles from the centerline.

Stage 2 take-off, sideline, and approach noise limits are shown in Figure 11.7. The Stage 2 noise limits are independent of the number of engines.

Stage 3 take-off noise limits are shown in Figure 11.8. Note that for maximum weights of 850,000 lb or more, the limits are constant but depend on the number of engines. In all cases the allowable noise limits are reduced by 4 EPNdB for each halving of the 850,000 lb maximum weight until a maximum noise limit of 89 EPNdB is reached. This occurs for maximum weights of 44,673 lb for aircraft with more than three engines, 63,177 lb for aircraft with three engines, and 106,250 lb for aircraft with fewer than three engines.

Stage 3 sideline noise limits are shown in Figure 11.9. The allowable sideline noise limits are independent of the number of engines.

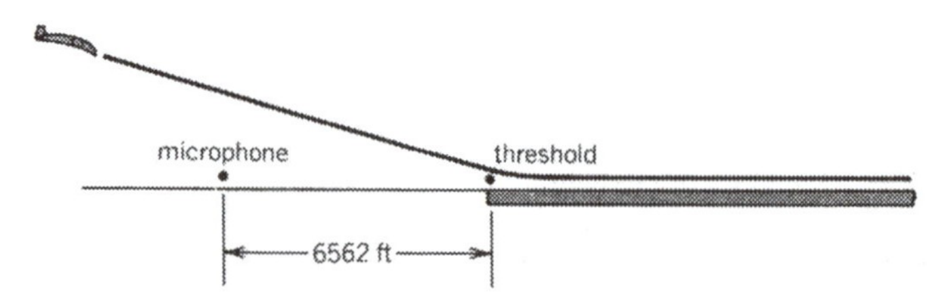

Figure 11.6. Typical approach profile for noise measurements.

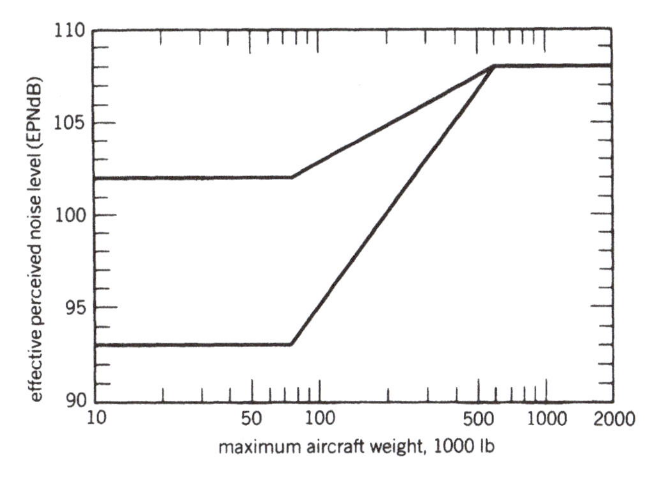

Figure 11.7. *Stage 2 take-off, sideline, and approach noise limits.*

Stage 3 approach noise limits are shown in Figure 11.10. These noise limits, like the sideline limits, are independent of the number of engines.

11.10 NOISE REDUCTION

Noise sources and noise standards were discussed in the two preceding sections. It is now important to examine what has or can be done to reduce the noise emitted from a gas turbine engine.

One must keep in mind that engine changes that reduce noise emissions usually add weight, length, and cost to the gas turbine engine. Quite often, features desirable in reducing noise are in conflict with the best aerodynamic design.

External noise is caused by the mixing of the exhaust stream with the ambient air,

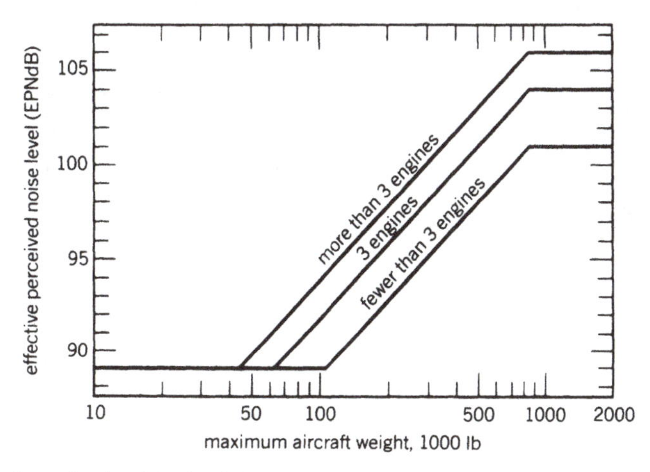

Figure 11.8. *Stage 3 take-off noise limits.*

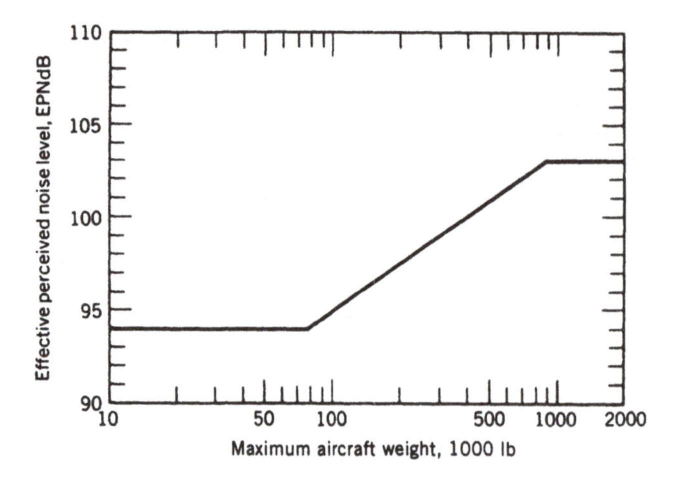

Figure 11.9. Stage 3 sideline noise limits.

with the noise level increasing by approximately the eighth power of velocity. Early commercial airplanes powered by turbojet engines used noise suppressors. Shortly thereafter, low-bypass-ratio turbofan (BPR $\equiv$ 1) were installed on commercial aircraft, which slightly lowered the exhaust stream velocity. Most of these aircraft were certified and entered service long before the current noise regulations. For these aircraft, exhaust (external) noise dominates. The only effective way to reduce the noise level of these aircraft is to decrease the exhaust stream velocity. This, of course, can be done by replacing the engine on the aircraft with a turbofan engine with a higher bypass ratio.

An effective way to reduce the maximum exhaust velocity, and therefore the exhaust jet noise, on mixed (pressure-balanced) turbofan engines is to use an exhaust-gas mixer behind the turbine. This device is shown on the JT8D-209 engine in Figure

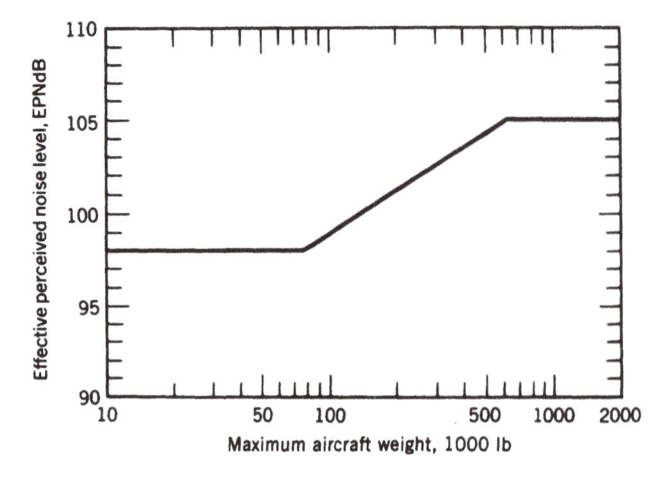

Figure 11.10. Stage 3 approach noise limits.

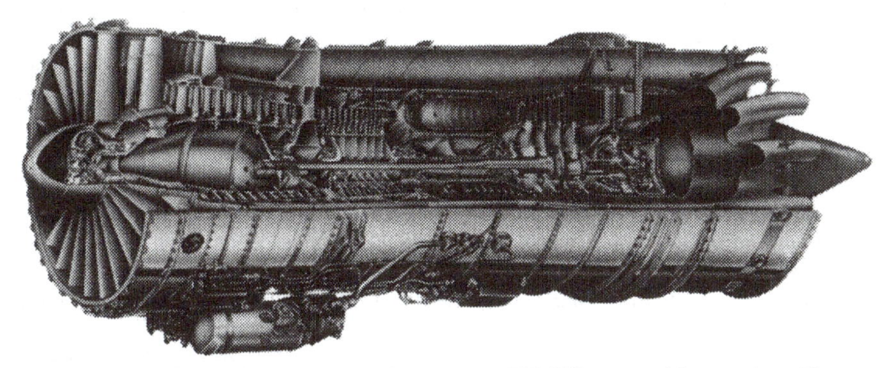

Figure 11.11. Cutaway view of Pratt & Whitney JT8D-209 gas turbine engine. (Courtesy of Pratt & Whitney)

11.11. An exhaust-gas mixer mixes the high-velocity core exhaust gas stream with the lower-velocity cold stream that has passed through the fan but bypassed the combustion chamber and turbine. It reduces the maximum exhaust stream velocity and eliminates the external turbulent mixing of these two streams. It does add weight to the engine.

Turbofan engines with bypass ratios of 5 or higher have been installed on many of the commercial aircraft in the last twenty years. With high-bypass-ratio turbofan engines, the exhaust velocity, and therefore the exhaust noise, is reduced considerably. With the reduction in the exhaust noise, additional internal noise is heard and becomes a problem.

Internal noise, unlike external noise, may either be eliminated or confined.

One way to confine the noise is to install an acoustical liner at the inlet to the engine and/or in the exhaust duct. Conventional acoustical treatment is quite effective and is currently used but does add weight to the airplane.

Other ways to reduce internal noise and the disadvantages are to:

1. Decrease the fan tip speed. The lowest-noise fans generally use subsonic fans. Low-speed fans usually require more compressor stages, which leads to a heavier and more costly engine.
2. Increase the spacing between the rotor and stator. Noise is reduced as the spacing increases, but large spacings tend to increase the length and weight of the engine.
3. Eliminate the inlet guide vanes, which eliminates one of the rotor-stator interactions.
4. Change the number of rotor and/or stator blades. Changing the ratio changes the frequency of the noise. If the number of blades is decreased, it may affect the turbomachinery performance.

REFERENCES CITED

1. Environmental Protection Agency, National Primary and Secondary Ambient Air Quality Standards, *Federal Register*, Vol. 36, No. 84, April 30, 1971.
2. Title 40—Protection of the Environment; Part 871—Ground Operation of Aircraft to Control

Emissions, Advanced Notice of Proposed Rule Making, *Federal Register*, Vol. 37, No. 239, December 12, 1972.

3. Environmental Protection Agency, Control of Air Pollution for Aircraft Engines—Emission Standards and Test Procedures for Aircraft, *Federal Register*, Vol. 38, No. 136, July 17, 1973.
4. *Code of Federal Regulations*, Title 14, U.S. Government Printing Office, Washington, D.C., 1979.
5. *Code of Federal Regulations*, Title 14, Part 34, U.S. Government Printing Office, Washington, D.C., January 1, 1994.
6. *Code of Federal Regulations*, Title 40, Part 60, U.S. Government Printing Office, Washington, D.C. July 1, 1993.
7. Wark, K. and Warner, C., *Air Pollution, Its Origin and Control*, IEP A Dun-Donnelley Publisher, New York, NY, 1976.
8. Schorr, M., NO_x Emission Control for Gas Turbines: A 1991 Update on Regulations and Technology (Part II), *Turbomachinery International*, Nov/Dec 1991.
9. *Aircraft Engine Emissions*, NASA Conference Publication 2021, May 1977.
10. Environmental Protection Agency, Noise Standards: Aircraft Type Certification, Federal Aviation Regulations Part 36, *Federal Register*, 34 FR 18364, November 18, 1969.
11. Environmental Protection Agency, Noise Standards for Newly Produced Airplanes of Older Type Designs, Federal Aviation Regulations Part 21.183 (e) and Part 36 Amended, *Federal Register*, 38 FR 29569, October 26, 1973.
12. *Code of Federal Regulations*, Title 14, Part 36, U.S. Government Printing Office, Washington, D.C.

TABLE A.1. **Properties of Argon (Ar) at 1 atm (English)**

Temp. °R	$\overline{C}_p^\circ$ Btu/lb-mol°R	$\overline{h}^\circ$ Btu/lb-mol	$\overline{s}^\circ$ Btu/lb-mol°R	$\overline{g}^\circ$ Btu/lb-mol	Pr
536.67	4.968	0.0	36.982	−19846.9	1.2085
600.	4.968	314.6	37.536	−22206.8	1.5973
700.	4.968	811.4	38.302	−25999.7	2.348
800.	4.968	1308.2	38.965	−29863.8	3.279
900.	4.968	1805.0	39.550	−33790.1	4.402
1000.	4.968	2301.8	40.074	−37771.7	5.728
1100.	4.968	2798.6	40.547	−41803.1	7.269
1200.	4.968	3295.4	40.979	−45879.8	9.035
1300.	4.968	3792.2	41.377	−49997.8	11.037
1400.	4.968	4289.0	41.745	−54154.2	13.284
1500.	4.968	4785.8	42.088	−58346.0	15.784
1600.	4.968	5282.6	42.408	−62571.0	18.548
1700.	4.968	5779.4	42.710	−66827.1	21.583
1800.	4.968	6276.2	42.994	−71112.4	24.899
1900.	4.968	6773.0	43.262	−75425.3	28.502
2000.	4.968	7269.8	43.517	−79764.3	32.402
2100.	4.968	7766.5	43.759	−84128.3	36.605
2200.	4.968	8263.3	43.991	−88515.9	41.120
2300.	4.968	8760.1	44.211	−92926.0	45.953
2400.	4.968	9256.9	44.423	−97357.8	51.11
2500.	4.968	9753.7	44.626	−101810.3	56.60
2600.	4.968	10250.5	44.820	−106282.7	62.44
2700.	4.968	10747.3	45.008	−110774.2	68.61
2800.	4.968	11244.1	45.189	−115284.0	75.14
2900.	4.968	11740.9	45.363	−119811.7	82.03
3000.	4.968	12237.7	45.531	−124356.5	89.29
3100.	4.968	12734.5	45.694	−128917.8	96.92
3200.	4.968	13231.3	45.852	−133495.1	104.92
3300.	4.968	13728.1	46.005	−138088.0	113.31
3400.	4.968	14224.9	46.153	−142695.9	122.09
3500.	4.968	14721.7	46.297	−147318.5	131.27
3600.	4.968	15218.5	46.437	−151955.3	140.85
3700.	4.968	15715.3	46.573	−156605.8	150.83
3800.	4.968	16212.1	46.706	−161269.8	161.23
3900.	4.968	16708.9	46.835	−165946.8	172.05
4000.	4.968	17205.7	46.961	−170636.6	183.29
4100.	4.968	17702.4	47.083	−175338.9	194.96
4200.	4.968	18199.2	47.203	−180053.2	207.1
4300.	4.968	18696.0	47.320	−184779.3	219.6
4400.	4.968	19192.8	47.434	−189517.1	232.6
4500.	4.968	19689.6	47.546	−194266.1	246.1

TABLE A.1. **Properties of Argon (Ar), at 1 atm (SI)**

Temp. K	$\overline{C}_p^\circ$ kJ/kmol K	$\overline{h}^\circ$ kJ/kmol	$\overline{s}^\circ$ kJ/kmol K	$\overline{g}^\circ$ kJ/kmol	Pr
298.15	20.786	0.0	154.731	−46133.0	1.2085
300.	20.786	38.5	154.859	−46419.3	1.2274
350.	20.786	1077.7	158.063	−54244.4	1.8045
400.	20.786	2117.0	160.839	−62218.6	2.520
450.	20.786	3156.3	163.287	−70322.9	3.382
500.	20.786	4195.6	165.477	−78543.0	4.402
550.	20.786	5234.9	167.458	−86867.2	5.586
600.	20.786	6274.2	169.267	−95285.9	6.943
650.	20.786	7313.5	170.931	−103791.4	8.481
700.	20.786	8352.8	172.471	−112377.0	10.208
750.	20.786	9392.1	173.905	−121036.8	12.129
800.	20.786	10431.4	175.247	−129765.9	14.253
850.	20.786	11470.7	176.507	−138560.1	16.585
900.	20.786	12510.0	177.695	−147415.4	19.133
950.	20.786	13549.3	178.819	−156328.5	21.902
1000.	20.786	14588.5	179.885	−165296.3	24.899
1050.	20.786	15627.8	180.899	−174316.1	28.129
1100.	20.786	16667.1	181.866	−183385.4	31.598
1150.	20.786	17706.4	182.790	−192502.0	35.312
1200.	20.786	18745.7	183.675	−201663.8	39.276
1250.	20.786	19785.0	184.523	−210868.9	43.496
1300.	20.786	20824.3	185.338	−220115.5	47.977
1350.	20.786	21863.6	186.123	−229402.2	52.72
1400.	20.786	22902.9	186.879	−238727.3	57.74
1450.	20.786	23942.2	187.608	−248089.6	63.04
1500.	20.786	24981.5	188.313	−257487.7	68.61
1550.	20.786	26020.8	188.994	−266920.5	74.47
1600.	20.786	27060.1	189.654	−276386.8	80.63
1650.	20.786	28099.3	190.294	−285885.6	87.07
1700.	20.786	29138.6	190.914	−295415.9	93.82
1750.	20.786	30177.9	191.517	−304976.8	100.87
1800.	20.786	31217.2	192.103	−314567.3	108.23
1850.	20.786	32256.5	192.672	−324186.7	115.91
1900.	20.786	33295.8	193.226	−333834.3	123.90
1950.	20.786	34335.1	193.766	−343509.1	132.21
2000.	20.786	35374.4	194.293	−353210.7	140.85
2050.	20.786	36413.7	194.806	−362938.2	149.82
2100.	20.786	37453.0	195.307	−372691.0	159.12
2150.	20.786	38492.3	195.796	−382468.6	168.76
2200.	20.786	39531.6	196.274	−392270.4	178.74
2250.	20.786	40570.9	196.741	−402095.8	189.07
2300.	20.786	41610.1	197.198	−411944.3	199.75
2350.	20.786	42649.4	197.645	−421815.4	210.8
2400.	20.786	43688.7	198.082	−431708.7	222.2
2450.	20.786	44728.0	198.511	−441623.5	233.9
2500.	20.786	45767.3	198.931	−451559.6	246.1

TABLE A.2. **Properties of Carbon Monoxide (CO) at 1 atm (English)**

Temp. °R	$\overline{C}_p^\circ$ Btu/lb-mol°R	$\overline{h}^\circ$ Btu/lb-mol	$\overline{s}^\circ$ Btu/lb-mol°R	$\overline{g}^\circ$ Btu/lb-mol	Pr
536.67	6.965	−47546.3	47.216	−72885.7	2.0843
600.	6.972	−47105.1	47.993	−75901.0	3.082
700.	7.003	−46406.6	49.070	−80755.5	5.298
800.	7.055	−45703.9	50.008	−85710.3	8.495
900.	7.123	−44995.1	50.843	−90753.6	12.930
1000.	7.205	−44278.8	51.597	−95876.2	18.903
1100.	7.295	−43553.9	52.288	−101071.0	26.761
1200.	7.390	−42819.7	52.927	−106332.2	36.908
1300.	7.488	−42075.8	53.522	−111655.0	49.801
1400.	7.586	−41322.1	54.081	−117035.4	65.96
1500.	7.680	−40558.7	54.608	−122470.1	85.98
1600.	7.770	−39786.2	55.106	−127956.0	110.50
1700.	7.854	−39004.9	55.580	−133490.5	140.23
1800.	7.929	−38215.7	56.031	−139071.2	175.97
1900.	7.996	−37419.4	56.461	−144696.0	218.5
2000.	8.059	−36616.5	56.873	−150362.9	268.9
2100.	8.119	−35807.6	57.268	−156070.1	327.9
2200.	8.175	−34992.9	57.647	−161815.9	396.8
2300.	8.227	−34172.8	58.011	−167599.0	476.7
2400.	8.277	−33347.5	58.363	−173417.8	568.9
2500.	8.323	−32517.5	58.701	−179271.0	674.7
2600.	8.366	−31683.1	59.029	−185157.6	795.4
2700.	8.407	−30844.4	59.345	−191076.4	932.8
2800.	8.445	−30001.7	59.652	−197026.4	1088.3
2900.	8.481	−29155.4	59.949	−203006.4	1263.7
3000.	8.514	−28305.7	60.237	−209015.8	1460.9
3100.	8.545	−27452.7	60.516	−215053.5	1681.7
3200.	8.574	−26596.8	60.788	−221118.8	1928.1
3300.	8.601	−25738.1	61.052	−227210.9	2202.3
3400.	8.626	−24876.8	61.309	−233329.0	2506.5
3500.	8.649	−24013.0	61.560	−239472.6	2843.1
3600.	8.671	−23147.0	61.804	−245640.8	3214.5
3700.	8.691	−22279.0	62.042	−251833.1	3623.2
3800.	8.709	−21409.0	62.274	−258048.9	4072.
3900.	8.727	−20537.1	62.500	−264287.7	4563.
4000.	8.743	−19663.7	62.721	−270548.8	5101.
4100.	8.758	−18788.6	62.937	−276831.8	5686.
4200.	8.772	−17912.2	63.149	−283136.1	6324.
4300.	8.785	−17034.4	63.355	−289461.3	7017.
4400.	8.797	−16155.3	63.557	−295807.0	7768.
4500.	8.808	−15275.0	63.755	−302172.6	8581.

TABLE A.2. **Properties of Carbon Monoxide (CO) at 1 atm (SI)**

Temp. K	$\overline{C}_p^\circ$ kJ/kmol K	$\overline{h}^\circ$ kJ/kmol	$\overline{s}^\circ$ kJ/kmol K	$\overline{g}^\circ$ kJ/kmol	Pr
298.15	29.143	−110518.5	197.551	−169418.4	2.0843
300.	29.143	−110464.6	197.731	−169784.0	2.130
350.	29.198	−109006.5	202.227	−179785.8	3.657
400.	29.336	−107543.5	206.134	−189996.9	5.851
450.	29.542	−106071.8	209.600	−200391.8	8.878
500.	29.804	−104588.4	212.726	−210951.3	12.930
550.	30.108	−103090.7	215.580	−221659.9	18.227
600.	30.443	−101577.0	218.214	−232505.6	25.020
650.	30.799	−100046.0	220.665	−243478.3	33.597
700.	31.166	−98496.9	222.961	−254569.5	44.282
750.	31.535	−96929.4	225.124	−265772.2	57.44
800.	31.898	−95343.5	227.171	−277080.0	73.47
850.	32.250	−93739.7	229.115	−288487.5	92.83
900.	32.583	−92118.9	230.968	−299990.0	116.00
950.	32.893	−90481.8	232.738	−311582.9	143.53
1000.	33.177	−88830.0	234.433	−323262.5	175.97
1050.	33.429	−87164.8	236.057	−335025.1	214.0
1100.	33.669	−85487.3	237.618	−346867.2	258.1
1150.	33.896	−83798.1	239.120	−358785.8	309.2
1200.	34.111	−82097.9	240.567	−370778.3	368.0
1250.	34.315	−80387.2	241.964	−382841.7	435.37
1300.	34.507	−78666.6	243.313	−394973.8	512.1
1350.	34.689	−76936.7	244.619	−407172.3	599.1
1400.	34.860	−75197.9	245.884	−419435.1	697.6
1450.	35.022	−73450.8	247.110	−431760.0	808.4
1500.	35.175	−71695.8	248.300	−444145.4	932.8
1550.	35.319	−69933.4	249.455	−456589.5	1071.9
1600.	35.454	−68164.1	250.579	−469090.4	1227.0
1650.	35.581	−66388.2	251.672	−481646.8	1399.47
1700.	35.701	−64606.1	252.736	−494257.2	1590.4
1750.	35.813	−62818.3	253.772	−506920.0	1801.5
1800.	35.918	−61025.0	254.783	−519633.9	2034.3
1850.	36.017	−59226.6	255.768	−532397.8	2290.3
1900.	36.109	−57423.4	256.730	−545210.4	2571.2
1950.	36.196	−55615.7	257.669	−558070.4	2878.6
2000.	36.277	−53803.9	258.587	−570976.9	3214.5
2050.	36.353	−51988.1	259.483	−583928.8	3580.6
2100.	36.424	−50168.6	260.360	−596924.9	3978.8
2150.	36.491	−48345.7	261.218	−609964.5	4411.
2200.	36.553	−46519.6	262.058	−623046.4	4880.
2250.	36.611	−44690.4	262.880	−636169.9	5387.
2300.	36.666	−42858.5	263.685	−649334.1	5935.
2350.	36.717	−41023.9	264.474	−662538.1	6526.
2400.	36.765	−39186.9	265.248	−675781.3	7162.
2450.	36.810	−37347.5	266.006	−689062.8	7846.
2500.	36.852	−35505.9	266.750	−702381.7	8581.

TABLE A.3. **Properties of Carbon Dioxide (CO_2) at 1 atm (English)**

Temp. °R	$\overline{C}_p^\circ$ Btu/lb-mol°R	$\overline{h}^\circ$ Btu/lb-mol	$\overline{s}^\circ$ Btu/lb-mol°R	$\overline{g}^\circ$ Btu/lb-mol	Pr
536.67	8.874	−169286.4	51.068	−196693.1	0.1448
600.	9.244	−168712.6	52.078	−199959.6	0.2408
700.	9.774	−167761.1	53.544	−205241.9	0.5034
800.	10.246	−166759.7	54.881	−210664.1	0.9864
900.	10.666	−165713.7	56.112	−216214.5	1.8330
1000.	11.040	−164628.1	57.255	−221883.6	3.259
1100.	11.376	−163507.0	58.324	−227663.1	5.579
1200.	11.678	−162354.0	59.327	−233546.2	9.242
1300.	11.951	−161172.2	60.273	−239526.6	14.875
1400.	12.199	−159964.5	61.167	−245599.0	23.336
1500.	12.423	−158733.3	62.017	−251758.6	35.782
1600.	12.628	−157480.6	62.825	−258001.0	53.75
1700.	12.813	−156208.4	63.596	−264322.4	79.23
1800.	12.981	−154918.6	64.334	−270719.1	114.82
1900.	13.115	−153613.7	65.039	−277188.0	163.75
2000.	13.241	−152295.9	65.715	−283726.0	230.1
2100.	13.360	−150965.7	66.364	−290330.2	319.0
2200.	13.472	−149624.0	66.988	−296997.9	436.6
2300.	13.577	−148271.6	67.589	−303727.0	590.9
2400.	13.675	−146908.9	68.169	−310515.1	791.2
2500.	13.767	−145536.8	68.729	−317360.2	1048.8
2600.	13.853	−144155.8	69.271	−324260.4	1377.4
2700.	13.933	−142766.4	69.795	−331213.8	1793.2
2800.	14.008	−141369.4	70.303	−338218.9	2315.7
2900.	14.078	−139965.0	70.796	−345274.0	2967.4
3000.	14.143	−138553.9	71.275	−352377.7	3775.1
3100.	14.203	−137136.6	71.739	−359528.5	4770.
3200.	14.260	−135713.4	72.191	−366725.1	5987.
3300.	14.312	−134284.8	72.631	−373966.3	7470.
3400.	14.360	−132851.1	73.059	−381250.9	9265.
3500.	14.405	−131412.8	73.476	−388577.7	11428.
3600.	14.447	−129970.1	73.882	−395945.7	14021.
3700.	14.486	−128523.5	74.278	−403353.8	17117.
3800.	14.521	−127073.1	74.665	−410801.0	20794.
3900.	14.554	−125619.3	75.043	−418286.5	25147.
4000.	14.585	−124162.3	75.412	−425809.3	30276.
4100.	14.613	−122702.4	75.772	−433368.6	36298.
4200.	14.640	−121239.7	76.125	−440963.5	43342.
4300.	14.664	−119774.5	76.469	−448593.2	51554.
4400.	14.687	−118307.0	76.807	−456257.2	61093.
4500.	14.708	−116837.3	77.137	−463954.4	72140.

TABLE A.3. **Properties of Carbon Dioxide (CO_2) at 1 atm (SI)**

Temp. K	$\bar{C}_p^\circ$ kJ/kmol K	$\bar{h}^\circ$ kJ/kmol	$\bar{s}^\circ$ kJ/kmol K	$\bar{g}^\circ$ kJ/kmol	Pr
298.15	37.128	−393495.9	213.668	−457201.0	0.1448
300.	37.212	−393427.1	213.898	−457596.5	0.1489
350.	39.370	−391511.6	219.799	−468441.3	0.3027
400.	41.309	−389493.7	225.185	−479567.8	0.5786
450.	43.053	−387384.0	230.153	−490952.9	1.0517
500.	44.624	−385191.3	234.772	−502577.4	1.8330
550.	46.044	−382924.0	239.093	−514425.1	3.082
600.	47.329	−380589.2	243.155	−526482.4	5.024
650.	48.497	−378193.1	246.990	−538736.9	7.969
700.	49.561	−375741.2	250.624	−551178.0	12.336
750.	50.534	−373238.5	254.077	−563796.3	18.688
800.	51.425	−370689.2	257.367	−576583.0	27.760
850.	52.244	−368097.2	260.510	−589530.5	40.511
900.	52.995	−365466.0	263.518	−602631.8	58.17
950.	53.683	−362798.8	266.401	−615880.3	82.28
1000.	54.310	−360098.7	269.171	−629269.9	114.82
1050.	54.818	−357370.4	271.833	−642795.5	158.15
1100.	55.298	−354617.4	274.395	−656451.6	215.2
1150.	55.753	−351840.9	276.863	−670233.4	289.6
1200.	56.183	−349042.5	279.245	−684136.6	385.7
1250.	56.589	−346223.1	281.547	−698156.6	508.7
1300.	56.972	−343384.0	283.774	−712290.0	664.9
1350.	57.334	−340526.2	285.931	−726532.8	861.8
1400.	57.674	−337650.9	288.022	−740881.9	1108.3
1450.	57.994	−334759.1	290.052	−755334.1	1414.7
1500.	58.296	−331851.8	292.023	−769886.2	1793.2
1550.	58.579	−328929.9	293.939	−784535.5	2258.0
1600.	58.845	−325994.2	295.803	−799279.3	2825.5
1650.	59.094	−323045.7	297.618	−814114.9	3514.6
1700.	59.328	−320085.0	299.385	−829040.2	4347.
1750.	59.547	−317113.1	301.108	−844052.8	5348.
1800.	59.752	−314130.6	302.789	−859150.4	6546.
1850.	59.944	−311138.1	304.429	−874330.9	7973.
1900.	60.123	−308136.4	306.030	−889592.6	9666.
1950.	60.290	−305126.0	307.593	−904933.3	11667.
2000.	60.446	−302107.6	309.122	−920351.3	14021.
2050.	60.592	−299081.6	310.616	−935844.9	16782.
2100.	60.728	−296048.5	312.078	−951412.4	20008.
2150.	60.855	−293008.9	313.508	−967052.2	23764.
2200.	60.973	−289963.2	314.909	−982762.8	28124.
2250.	61.084	−286911.8	316.280	−998542.7	33167.
2300.	61.187	−283854.9	317.624	−1014390.2	38985.
2350.	61.283	−280793.2	318.941	−1030304.4	45676.
2400.	61.373	−277726.8	320.232	−1046284.0	53350.
2450.	61.458	−274655.9	321.499	−1062327.4	62127.
2500.	61.537	−271581.1	322.741	−1078433.5	72140.

TABLE A.4. **Properties of Hydrogen atom (H) at 1 atm (English)**

Temp. °R	$\overline{C}_p^\circ$ Btu/lb-mol°R	$\overline{h}^\circ$ Btu/lb-mol	$\overline{s}^\circ$ Btu/lb-mol°R	$\overline{g}^\circ$ Btu/lb-mol	Pr
536.67	4.968	93772.2	27.391	79072.1	0.9690
600.	4.968	94086.8	27.945	77319.6	1.2806
700.	4.968	94583.6	28.711	74485.7	1.8827
800.	4.968	95080.4	29.375	71580.7	2.629
900.	4.968	95577.2	29.960	68613.4	3.529
1000.	4.968	96074.0	30.483	65590.8	4.592
1100.	4.968	96570.8	30.957	62518.4	5.828
1200.	4.968	97067.6	31.389	59400.8	7.244
1300.	4.968	97564.4	31.787	56241.8	8.849
1400.	4.968	98061.2	32.155	53044.5	10.650
1500.	4.968	98558.0	32.497	49811.7	12.655
1600.	4.968	99054.8	32.818	46545.8	14.871
1700.	4.968	99551.5	33.119	43248.7	17.305
1800.	4.968	100048.3	33.403	39922.5	19.963
1900.	4.968	100545.1	33.672	36568.6	22.852
2000.	4.968	101041.9	33.927	33188.6	25.978
2100.	4.968	101538.7	34.169	29783.7	29.349
2200.	4.968	102035.5	34.400	26355.1	32.968
2300.	4.968	102532.3	34.621	22904.0	36.843
2400.	4.968	103029.1	34.832	19431.2	40.980
2500.	4.968	103525.9	35.035	15937.8	45.383
2600.	4.968	104022.7	35.230	12424.4	50.06
2700.	4.968	104519.5	35.418	8892.0	55.01
2800.	4.968	105016.3	35.598	5341.2	60.25
2900.	4.968	105513.1	35.773	1772.6	65.77
3000.	4.968	106009.9	35.941	−1813.2	71.59
3100.	4.968	106506.7	36.104	−5415.5	77.70
3200.	4.968	107003.5	36.262	−9033.8	84.12
3300.	4.968	107500.3	36.415	−12667.6	90.85
3400.	4.968	107997.1	36.563	−16316.5	97.89
3500.	4.968	108493.9	36.707	−19980.8	105.25
3600.	4.968	108990.6	36.847	−23657.8	112.93
3700.	4.968	109487.4	36.983	−27349.3	120.93
3800.	4.968	109984.2	37.115	−31054.2	129.27
3900.	4.968	110481.0	37.244	−34772.2	137.94
4000.	4.968	110977.8	37.370	−38503.0	146.96
4100.	4.968	111474.6	37.493	−42246.2	156.31
4200.	4.968	111971.4	37.613	−46001.5	166.02
4300.	4.968	112468.2	37.729	−49768.6	176.08
4400.	4.968	112965.0	37.844	−53547.3	186.50
4500.	4.968	113461.8	37.955	−57337.3	197.27

TABLE A.4. **Properties of Hydrogen atom (H) at 1 atm (SI)**

Temp. K	$\overline{C}_p^\circ$ kJ/kmol K	$\overline{h}^\circ$ kJ/kmol	$\overline{s}^\circ$ kJ/kmol K	$\overline{g}^\circ$ kJ/kmol	Pr
298.15	20.786	217967.7	114.605	183798.3	0.9690
300.	20.786	218006.1	114.733	183586.2	0.9841
350.	20.786	219045.4	117.937	177767.3	1.4467
400.	20.786	220084.7	120.713	171799.5	2.020
450.	20.786	221124.0	123.161	165701.4	2.712
500.	20.786	222163.3	125.351	159487.7	3.529
550.	20.786	223202.6	127.332	153169.8	4.478
600.	20.786	224241.9	129.141	146757.3	5.567
650.	20.786	225281.2	130.805	140258.1	6.800
700.	20.786	226320.4	132.345	133678.9	8.184
750.	20.786	227359.8	133.779	127025.4	9.725
800.	20.786	228399.0	135.121	120302.5	11.427
850.	20.786	229438.3	136.381	113514.7	13.297
900.	20.786	230477.6	137.569	106665.6	15.340
950.	20.786	231516.9	138.693	99758.8	17.560
1000.	20.786	232556.2	139.759	92797.3	19.963
1050.	20.786	233595.5	140.773	85783.9	22.552
1100.	20.786	234634.8	141.740	78720.8	25.334
1150.	20.786	235674.1	142.664	71610.5	28.312
1200.	20.786	236713.4	143.549	64455.1	31.490
1250.	20.786	237752.7	144.397	57256.3	34.873
1300.	20.786	238792.0	145.212	50015.9	38.466
1350.	20.786	239831.3	145.997	42735.6	42.272
1400.	20.786	240870.5	146.753	35416.7	46.296
1450.	20.786	241909.8	147.482	28060.7	50.54
1500.	20.786	242949.1	148.187	20668.9	55.01
1550.	20.786	243988.4	148.868	13242.5	59.71
1600.	20.786	245027.7	149.528	5782.4	64.64
1650.	20.786	246067.0	150.168	−1710.0	69.81
1700.	20.786	247106.3	150.788	−9234.0	75.22
1750.	20.786	248145.6	151.391	−16788.6	80.88
1800.	20.786	249184.9	151.977	−24372.9	86.78
1850.	20.786	250224.2	152.546	−31986.0	92.93
1900.	20.786	251263.5	153.100	−39627.2	99.34
1950.	20.786	252302.8	153.640	−47295.8	106.00
2000.	20.786	253342.0	154.167	−54991.0	112.93
2050.	20.786	254381.3	154.680	−62712.2	120.12
2100.	20.786	255420.6	155.181	−70458.8	127.58
2150.	20.786	256459.9	155.670	−78230.1	135.31
2200.	20.786	257499.2	156.148	−86025.6	143.31
2250.	20.786	258538.5	156.615	−93844.7	151.59
2300.	20.786	259577.8	157.072	−101686.9	160.15
2350.	20.786	260617.1	157.519	−109551.7	169.00
2400.	20.786	261656.4	157.956	−117438.6	178.13
2450.	20.786	262695.7	158.385	−125347.1	187.56
2500.	20.786	263735.0	158.805	−133276.9	197.27

TABLE A.5. **Properties of Hydrogen molecule (H_2) at 1 atm (English)**

Temp. °R	$\overline{C}_p^\circ$ Btu/lb-mol°R	$\overline{h}^\circ$ Btu/lb-mol	$\overline{s}^\circ$ Btu/lb-mol°R	$\overline{g}^\circ$ Btu/lb-mol	Pr
536.67	6.891	0.0	31.208	−16748.3	6.6130
600.	6.930	437.7	31.979	−18749.5	9.747
700.	6.969	1132.8	33.050	−22002.3	16.713
800.	6.989	1830.8	33.982	−25354.9	26.714
900.	6.999	2530.3	34.806	−28795.1	40.437
1000.	7.004	3230.4	35.544	−32313.2	58.61
1100.	7.011	3931.2	36.212	−35901.5	82.03
1200.	7.023	4632.9	36.822	−39553.6	111.53
1300.	7.042	5336.1	37.385	−43264.4	148.054
1400.	7.069	6041.6	37.908	−47029.3	192.61
1500.	7.103	6750.2	38.397	−50844.8	246.3
1600.	7.141	7462.3	38.856	−54707.7	310.4
1700.	7.180	8178.4	39.290	−58615.2	386.2
1800.	7.214	8898.2	39.702	−62565.0	475.1
1900.	7.274	9622.6	40.093	−66554.9	578.5
2000.	7.333	10352.9	40.468	−70583.1	698.6
2100.	7.392	11089.2	40.827	−74648.0	837.0
2200.	7.450	11831.3	41.172	−78748.1	995.8
2300.	7.507	12579.1	41.505	−82882.0	1177.1
2400.	7.563	13332.6	41.826	−87048.6	1383.2
2500.	7.619	14091.7	42.135	−91246.8	1616.6
2600.	7.673	14856.3	42.435	−95475.4	1880.0
2700.	7.728	15626.4	42.726	−99733.5	2176.0
2800.	7.781	16401.8	43.008	−104020.3	2507.8
2900.	7.833	17182.5	43.282	−108334.8	2878.5
3000.	7.885	17968.4	43.548	−112676.4	3291.5
3100.	7.935	18759.4	43.808	−117044.2	3750.4
3200.	7.985	19555.4	44.060	−121437.7	4259.
3300.	8.034	20356.4	44.307	−125856.1	4821.
3400.	8.083	21162.3	44.547	−130298.9	5442.
3500.	8.130	21972.9	44.782	−134765.4	6125.
3600.	8.176	22788.2	45.012	−139255.2	6875.
3700.	8.222	23608.1	45.237	−143767.7	7698.
3800.	8.267	24432.6	45.457	−148302.3	8599.
3900.	8.311	25261.4	45.672	−152858.8	9583.
4000.	8.354	26094.6	45.883	−157436.6	10656.
4100.	8.396	26932.1	46.090	−162035.2	11825.
4200.	8.437	27773.7	46.292	−166654.4	13095.
4300.	8.477	28619.5	46.491	−171293.6	14475.
4400.	8.517	29469.2	46.687	−175952.2	15970.
4500.	8.556	30322.9	46.879	−180630.8	17589.

TABLE A.5. **Properties of Hydrogen molecule (H_2) at 1 atm (SI)**

Temp. K	$\bar{C}_p^\circ$ kJ/kmol K	$\bar{h}^\circ$ kJ/kmol	$\bar{s}^\circ$ kJ/kmol K	$\bar{g}^\circ$ kJ/kmol	Pr
298.15	28.833	0.0	130.573	−38930.3	6.6130
300.	28.842	53.3	130.751	−39172.1	6.756
350.	29.054	1501.1	135.215	−45823.9	11.557
400.	29.179	2957.3	139.103	−52684.0	18.449
450.	29.247	4418.1	142.544	−59726.8	27.907
500.	29.282	5881.5	145.628	−66932.5	40.437
550.	29.304	7346.1	148.420	−74284.8	56.57
600.	29.329	8811.9	150.971	−81770.4	76.89
650.	29.368	10279.3	153.320	−89378.5	101.99
700.	29.430	11749.1	155.498	−97099.6	132.54
750.	29.517	13222.7	157.531	−104925.8	169.26
800.	29.631	14701.3	159.440	−112850.6	212.9
850.	29.765	16186.1	161.240	−120868.0	264.4
900.	29.911	17677.9	162.946	−128973.1	324.6
950.	30.056	19177.1	164.567	−137161.2	394.5
1000.	30.183	20683.2	166.112	−145428.5	475.1
1050.	30.409	22198.1	167.590	−153771.2	567.5
1100.	30.633	23724.1	169.010	−162186.5	673.2
1150.	30.854	25261.3	170.376	−170671.3	793.4
1200.	31.073	26809.5	171.694	−179223.3	929.7
1250.	31.289	28368.5	172.967	−187840.0	1083.5
1300.	31.503	29938.3	174.198	−196519.3	1256.4
1350.	31.714	31518.8	175.391	−205259.1	1450.3
1400.	31.923	33109.7	176.548	−214057.8	1666.8
1450.	32.129	34711.0	177.672	−222913.4	1908.0
1500.	32.332	36322.5	178.765	−231824.5	2176.0
1550.	32.532	37944.1	179.828	−240789.4	2472.9
1600.	32.730	39575.7	180.864	−249806.8	2801.1
1650.	32.925	41217.1	181.874	−258875.4	3162.9
1700.	33.117	42868.1	182.860	−267993.8	3561.1
1750.	33.306	44528.8	183.823	−277161.0	3998.2
1800.	33.493	46198.7	184.764	−286375.7	4477.
1850.	33.676	47878.0	185.684	−295637.0	5001.
1900.	33.857	49566.3	186.584	−304943.8	5573.
1950.	34.035	51263.6	187.466	−314295.1	6197.
2000.	34.210	52969.8	188.330	−323690.1	6875.
2050.	34.381	54684.6	189.177	−333127.8	7613.
2100.	34.550	56407.9	190.007	−342607.5	8412.
2150.	34.716	58139.5	190.822	−352128.3	9279.
2200.	34.880	59879.5	191.622	−361689.4	10216.
2250.	35.040	61627.5	192.408	−371290.3	11228.
2300.	35.197	63383.4	193.180	−380930.0	12320.
2350.	35.351	65147.1	193.938	−390608.0	13497.
2400.	35.503	66918.5	194.684	−400323.7	14764.
2450.	35.652	68697.4	195.418	−410076.3	16126.
2500.	35.797	70483.6	196.140	−419865.2	17589.

TABLE A.6. **Properties of Hydroxyl (OH) at 1 atm (English)**

Temp. °R	$\bar{C}_p^\circ$ Btu/lb-mol°R	$\bar{h}^\circ$ Btu/lb-mol	$\bar{s}^\circ$ Btu/lb-mol°R	$\bar{g}^\circ$ Btu/lb-mol	Pr
536.67	7.143	16761.8	43.891	−6793.1	3.9109
600.	7.115	17213.3	44.686	−9598.4	5.835
700.	7.081	17923.0	45.780	−14123.2	10.120
800.	7.058	18629.9	46.724	−18749.5	16.274
900.	7.047	19335.1	47.555	−23464.2	24.718
1000.	7.046	20039.6	48.297	−28257.5	35.912
1100.	7.055	20744.6	48.969	−33121.3	50.36
1200.	7.072	21450.9	49.584	−38049.4	68.61
1300.	7.099	22159.3	50.151	−43036.4	91.27
1400.	7.133	22870.8	50.678	−48078.2	119.00
1500.	7.173	23586.1	51.171	−53170.9	152.54
1600.	7.220	24305.7	51.636	−58311.5	192.71
1700.	7.273	25030.4	52.075	−63497.2	240.4
1800.	7.330	25760.5	52.492	−68725.8	296.6
1900.	7.397	26496.8	52.890	−73995.0	362.3
2000.	7.462	27239.8	53.272	−79303.3	438.9
2100.	7.526	27989.3	53.637	−84648.8	527.6
2200.	7.587	28744.9	53.989	−90030.2	629.7
2300.	7.647	29506.7	54.327	−95446.1	746.7
2400.	7.705	30274.3	54.654	−100895.3	880.1
2500.	7.761	31047.7	54.970	−106376.6	1031.6
2600.	7.816	31826.5	55.275	−111888.9	1203.1
2700.	7.869	32610.8	55.571	−117431.3	1396.3
2800.	7.920	33400.2	55.858	−123002.8	1613.3
2900.	7.969	34194.6	56.137	−128602.7	1856.3
3000.	8.017	34994.0	56.408	−134230.0	2127.4
3100.	8.063	35798.0	56.672	−139884.0	2429.3
3200.	8.108	36606.6	56.928	−145564.1	2764.3
3300.	8.152	37419.6	57.179	−151269.5	3135.1
3400.	8.194	38236.9	57.422	−156999.5	3544.7
3500.	8.234	39058.3	57.661	−162753.8	3995.9
3600.	8.273	39883.7	57.893	−168531.5	4492.
3700.	8.311	40712.9	58.120	−174332.2	5036.
3800.	8.348	41545.9	58.342	−180155.4	5632.
3900.	8.383	42382.4	58.560	−186000.5	6282.
4000.	8.417	43222.4	58.772	−191867.2	6992.
4100.	8.450	44065.7	58.981	−197754.9	7764.
4200.	8.481	44912.3	59.185	−203663.1	8604.
4300.	8.512	45761.9	59.385	−209591.6	9514.
4400.	8.541	46614.6	59.581	−215539.9	10501.
4500.	8.569	47470.1	59.773	−221507.6	11568.

TABLE A.6. **Properties of Hydroxyl (OH) at 1 atm (SI)**

Temp. K	$\overline{C}_p^\circ$ kJ/kmol K	$\overline{h}^\circ$ kJ/kmol	$\overline{s}^\circ$ kJ/kmol K	$\overline{g}^\circ$ kJ/kmol	Pr
298.15	29.888	38961.8	183.639	−15790.2	3.9109
300.	29.881	39017.1	183.824	−16130.1	3.999
350.	29.722	40507.0	188.418	−25439.2	6.948
400.	29.604	41990.0	192.378	−34961.3	11.188
450.	29.526	43468.1	195.860	−44669.1	17.008
500.	29.484	44943.2	198.969	−54541.2	24.718
550.	29.478	46417.1	201.778	−64561.0	34.655
600.	29.506	47891.6	204.344	−74714.9	47.184
650.	29.565	49368.2	206.708	−84992.0	62.70
700.	29.653	50848.6	208.902	−95382.9	81.63
750.	29.768	52334.0	210.952	−105879.8	104.46
800.	29.908	53825.8	212.877	−116476.0	131.68
850.	30.070	55325.1	214.695	−127165.8	163.86
900.	30.253	56833.1	216.419	−137944.0	201.6
950.	30.453	58350.7	218.060	−148806.3	245.6
1000.	30.668	59878.6	219.627	−159748.8	296.6
1050.	30.921	61418.4	221.130	−170768.0	355.3
1100.	31.169	62970.7	222.574	−181860.8	422.7
1150.	31.409	64535.1	223.965	−193024.4	499.7
1200.	31.644	66111.5	225.307	−204256.5	587.2
1250.	31.872	67699.4	226.603	−215554.4	686.2
1300.	32.094	69298.6	227.857	−226916.0	798.0
1350.	32.310	70908.7	229.073	−238339.5	923.6
1400.	32.520	72529.4	230.252	−249822.7	1064.3
1450.	32.724	74160.5	231.396	−261364.0	1221.4
1500.	32.922	75801.7	232.509	−272961.8	1396.3
1550.	33.115	77452.7	233.592	−284614.5	1590.4
1600.	33.302	79113.1	234.646	−296320.5	1805.5
1650.	33.484	80782.8	235.674	−308078.6	2043.0
1700.	33.661	82461.4	236.676	−319887.4	2304.7
1750.	33.832	84148.8	237.654	−331745.7	2592.5
1800.	33.998	85844.5	238.609	−343652.5	2908.2
1850.	34.160	87548.5	239.543	−355606.3	3253.8
1900.	34.316	89260.4	240.456	−367606.4	3631.5
1950.	34.468	90980.1	241.350	−379651.6	4043.
2000.	34.615	92707.2	242.224	−391741.1	4492.
2050.	34.758	94441.5	243.081	−403873.8	4979.
2100.	34.896	96182.9	243.920	−416048.8	5508.
2150.	35.030	97931.1	244.743	−428265.5	6081.
2200.	35.160	99685.8	245.549	−440522.9	6701.
2250.	35.285	101447.0	246.341	−452820.2	7370.
2300.	35.407	103214.3	247.118	−465156.7	8092.
2350.	35.524	104987.5	247.881	−477531.7	8869.
2400.	35.638	106766.6	248.630	−489944.5	9706.
2450.	35.748	108551.3	249.366	−502394.5	10604.
2500.	35.854	110341.3	250.089	−514880.9	11568.

TABLE A.7. **Properties of Water (H_2O) at 1 atm (English)**

Temp. °R	$\bar{C}_p^\circ$ Btu/lb-mol°R	$\bar{h}^\circ$ Btu/lb-mol	$\bar{s}^\circ$ Btu/lb-mol°R	$\bar{g}^\circ$ Btu/lb-mol	Pr
536.67	8.025	−104032.0	45.104	−128237.8	0.7199
600.	8.072	−103522.4	46.001	−131123.1	1.1310
700.	8.167	−102710.5	47.252	−135787.3	2.123
800.	8.283	−101888.2	48.350	−140568.5	3.689
900.	8.413	−101053.5	49.333	−145453.5	6.049
1000.	8.555	−100205.2	50.227	−150432.2	9.485
1100.	8.706	−99342.2	51.049	−155496.5	14.347
1200.	8.862	−98463.8	51.814	−160640.1	21.075
1300.	9.022	−97569.6	52.529	−165857.7	30.211
1400.	9.185	−96659.3	53.204	−171144.6	42.422
1500.	9.350	−95732.5	53.843	−176497.3	58.52
1600.	9.516	−94789.3	54.452	−181912.2	79.50
1700.	9.685	−93829.2	55.034	−187386.7	106.55
1800.	9.858	−92852.1	55.592	−192918.2	141.12
1900.	10.037	−91857.3	56.130	−198504.5	184.98
2000.	10.210	−90844.9	56.649	−204143.6	240.2
2100.	10.377	−89815.6	57.152	−209833.8	309.3
2200.	10.537	−88769.8	57.638	−215573.4	395.1
2300.	10.692	−87708.3	58.110	−221360.9	500.9
2400.	10.841	−86631.6	58.568	−227194.9	630.9
2500.	10.985	−85540.2	59.014	−233074.1	789.4
2600.	11.123	−84434.8	59.447	−238997.2	981.9
2700.	11.256	−83315.8	59.869	−244963.1	1214.3
2800.	11.385	−82183.7	60.281	−250970.8	1493.9
2900.	11.508	−81039.0	60.683	−257019.0	1828.5
3000.	11.627	−79882.3	61.075	−263107.0	2227.4
3100.	11.741	−78713.9	61.458	−269233.7	2701.0
3200.	11.851	−77534.3	61.832	−275398.3	3261.2
3300.	11.956	−76343.9	62.199	−281599.9	3921.3
3400.	12.058	−75143.2	62.557	−287837.8	4696.
3500.	12.155	−73932.5	62.908	−294111.1	5604.
3600.	12.249	−72712.2	63.252	−300419.2	6662.
3700.	12.340	−71482.7	63.589	−306761.3	7892.
3800.	12.427	−70244.4	63.919	−313136.7	9319.
3900.	12.510	−68997.5	64.243	−319544.9	10969.
4000.	12.590	−67742.5	64.561	−325985.1	12871.
4100.	12.668	−66479.5	64.872	−332456.8	15058.
4200.	12.742	−65209.0	65.179	−338959.4	17566.
4300.	12.813	−63931.3	65.479	−345492.3	20435.
4400.	12.882	−62646.5	65.775	−352055.1	23710.
4500.	12.948	−61355.0	66.065	−358647.2	27439.

TABLE A.7. **Properties of Water (H_2O) at 1 atm (SI)**

Temp. K	$\bar{C}_p^\circ$ kJ/kmol K	$\bar{h}^\circ$ kJ/kmol	$\bar{s}^\circ$ kJ/kmol K	$\bar{g}^\circ$ kJ/kmol	Pr
298.15	33.577	−241816.1	188.713	−298080.8	0.7199
300.	33.586	−241753.9	188.921	−298430.1	0.7382
350.	33.884	−240067.5	194.119	−308009.2	1.3794
400.	34.263	−238364.2	198.668	−317831.2	2.384
450.	34.706	−236640.2	202.728	−327867.9	3.885
500.	35.200	−234892.7	206.410	−338097.8	6.049
550.	35.734	−233119.5	209.790	−348503.9	9.083
600.	36.297	−231318.8	212.923	−359072.6	13.240
650.	36.881	−229489.4	215.851	−369792.8	18.830
700.	37.480	−227630.5	218.606	−380654.8	26.228
750.	38.089	−225741.3	221.213	−391650.9	35.885
800.	38.705	−223821.5	223.691	−402774.0	48.345
850.	39.327	−221870.7	226.056	−414018.1	64.25
900.	39.956	−219888.6	228.321	−425378.0	84.38
950.	40.594	−217874.9	230.499	−436848.8	109.64
1000.	41.245	−215829.0	232.597	−448426.5	141.12
1050.	41.921	−213749.8	234.626	−460107.4	180.12
1100.	42.576	−211637.3	236.592	−471888.1	228.1
1150.	43.209	−209492.6	238.498	−483765.5	287.0
1200.	43.822	−207316.8	240.350	−495737.0	358.5
1250.	44.415	−205110.8	242.151	−507799.8	445.3
1300.	44.988	−202875.6	243.904	−519951.3	549.8
1350.	45.542	−200612.3	245.613	−532189.4	675.2
1400.	46.078	−198321.7	247.279	−544511.9	825.0
1450.	46.596	−196004.8	248.905	−556916.6	1003.2
1500.	47.096	−193662.4	250.493	−569401.8	1214.3
1550.	47.580	−191295.4	252.045	−581965.3	1463.6
1600.	48.047	−188904.7	253.563	−594605.6	1756.8
1650.	48.498	−186491.0	255.049	−607321.1	2100.4
1700.	48.934	−184055.1	256.503	−620109.9	2501.9
1750.	49.335	−181597.8	257.927	−632970.9	2969.5
1800.	49.762	−179119.8	259.324	−645902.3	3512.4
1850.	50.154	−176621.9	260.692	−658902.8	4141.
1900.	50.533	−174104.6	262.035	−671971.0	4867.
1950.	50.898	−171568.8	263.352	−685105.8	5702.
2000.	51.251	−169015.0	264.645	−698305.9	6662.
2050.	51.592	−166443.9	265.915	−711570.0	7761.
2100.	51.921	−163856.0	267.162	−724896.9	9017.
2150.	52.239	−161252.0	268.388	−738285.9	10449.
2200.	52.545	−158632.3	269.592	−751735.4	12078.
2250.	52.841	−155997.6	270.776	−765244.7	13927.
2300.	53.127	−153348.4	271.941	−778812.8	16020.
2350.	53.402	−150685.1	273.087	−792438.6	18387.
2400.	53.669	−148008.3	274.214	−806121.1	21056.
2450.	53.926	−145318.4	275.323	−819859.7	24062.
2500.	54.175	−142615.8	276.415	−833653.2	27439.

TABLE A.8. **Properties of Nitrogen atom (N) at 1 atm (English)**

Temp. °R	$\overline{C}_p^\circ$ Btu/lb-mol°R	$\overline{h}^\circ$ Btu/lb-mol	$\overline{s}^\circ$ Btu/lb-mol°R	$\overline{g}^\circ$ Btu/lb-mol	Pr
536.67	4.968	203436.1	36.613	183787.0	1.0039
600.	4.968	203750.7	37.167	181450.4	1.3268
700.	4.967	204247.4	37.933	177694.4	1.9506
800.	4.968	204744.2	38.596	173867.2	2.723
900.	4.968	205241.0	39.181	169977.8	3.656
1000.	4.968	205737.8	39.705	166033.0	4.758
1100.	4.969	206234.6	40.178	162038.5	6.038
1200.	4.969	206731.5	40.611	157998.7	7.505
1300.	4.968	207228.3	41.008	153917.5	9.168
1400.	4.968	207725.1	41.376	149798.0	11.035
1500.	4.968	208221.9	41.719	145643.1	13.112
1600.	4.967	208718.7	42.040	141454.9	15.407
1700.	4.968	209215.4	42.341	137235.7	17.928
1800.	4.968	209712.2	42.625	132987.3	20.682
1900.	4.969	210209.1	42.894	128711.3	23.676
2000.	4.970	210706.0	43.148	124409.0	26.917
2100.	4.971	211203.1	43.391	120082.0	30.411
2200.	4.971	211700.2	43.622	115731.2	34.164
2300.	4.971	212197.3	43.843	111357.9	38.182
2400.	4.971	212694.4	44.055	106962.9	42.471
2500.	4.971	213191.5	44.258	102547.2	47.038
2600.	4.970	213688.6	44.453	98111.6	51.89
2700.	4.970	214185.6	44.640	93656.9	57.02
2800.	4.969	214682.6	44.821	89183.8	62.45
2900.	4.969	215179.5	44.995	84692.9	68.18
3000.	4.968	215676.4	45.164	80184.9	74.21
3100.	4.968	216173.2	45.327	75660.3	80.55
3200.	4.967	216669.9	45.484	71119.7	87.20
3300.	4.967	217166.6	45.637	66563.6	94.18
3400.	4.966	217663.3	45.786	61992.4	101.47
3500.	4.966	218159.9	45.929	57406.6	109.09
3600.	4.966	218656.5	46.069	52806.7	117.05
3700.	4.966	219153.0	46.205	48192.9	125.35
3800.	4.966	219649.6	46.338	43565.7	133.99
3900.	4.967	220146.3	46.467	38925.4	142.97
4000.	4.967	220643.0	46.593	34272.4	152.31
4100.	4.968	221139.8	46.715	29607.0	162.01
4200.	4.969	221636.6	46.835	24929.5	172.07
4300.	4.971	222133.6	46.952	20240.1	182.50
4400.	4.973	222630.8	47.066	15539.2	193.31
4500.	4.975	223128.2	47.178	10826.9	204.5

TABLE A.8 **Properties of Nitrogen atom (N) at 1 atm (SI)**

Temp. K	$\overline{C}_p^\circ$ kJ/kmol K	$\overline{h}^\circ$ kJ/kmol	$\overline{s}^\circ$ kJ/kmol K	$\overline{g}^\circ$ kJ/kmol	Pr
298.15	20.786	472874.7	153.189	427201.5	1.0039
300.	20.786	472913.1	153.317	426918.0	1.0196
350.	20.784	473952.3	156.521	419170.0	1.4989
400.	20.783	474991.5	159.296	411273.0	2.093
450.	20.784	476030.7	161.744	403245.8	2.809
500.	20.786	477070.0	163.934	395102.9	3.656
550.	20.787	478109.3	165.915	386855.8	4.640
600.	20.788	479148.7	167.724	378514.2	5.767
650.	20.788	480188.1	169.388	370085.8	7.045
700.	20.788	481227.5	170.929	361577.4	8.479
750.	20.787	482266.9	172.363	352994.8	10.076
800.	20.786	483306.2	173.704	344342.7	11.840
850.	20.784	484345.4	174.964	335625.7	13.777
900.	20.784	485384.6	176.152	326847.5	15.893
950.	20.784	486423.8	177.276	318011.5	18.193
1000.	20.787	487463.1	178.342	301920.8	20.682
1050.	20.791	488502.5	179.357	300178.2	23.366
1100.	20.795	489542.2	180.324	291186.0	26.249
1150.	20.797	490582.0	181.248	282146.5	29.335
1200.	20.799	491621.9	182.133	273061.8	32.631
1250.	20.799	492661.8	182.982	263933.8	36.139
1300.	20.799	493701.8	183.798	254764.1	39.865
1350.	20.799	494741.8	184.583	245554.4	43.812
1400.	20.798	495781.7	185.340	236306.2	47.985
1450.	20.796	496821.5	186.069	227020.9	52.39
1500.	20.794	497861.3	186.774	217699.7	57.02
1550.	20.792	498901.0	187.456	208343.9	61.90
1600.	20.790	499940.5	188.116	198954.5	67.01
1650.	20.788	500979.9	188.756	189532.6	72.37
1700.	20.786	502019.3	189.377	180079.2	77.98
1750.	20.784	503058.6	189.979	170595.2	83.84
1800.	20.782	504097.7	190.565	161081.5	89.95
1850.	20.780	505136.7	191.134	151539.0	96.33
1900.	20.779	506175.7	191.688	141968.4	102.97
1950.	20.778	507214.6	192.228	132370.4	109.88
2000.	20.777	508253.5	192.754	122745.8	117.05
2050.	20.777	509292.3	193.267	113095.3	124.50
2100.	20.778	510331.2	193.768	103419.4	132.23
2150.	20.780	511370.2	194.256	93718.7	140.24
2200.	20.782	512409.2	194.734	83993.9	148.53
2250.	20.785	513448.3	195.201	74245.5	157.12
2300.	20.789	514487.7	195.658	64474.0	165.99
2350.	20.794	515527.2	196.105	54679.8	175.16
2400.	20.799	516567.1	196.543	44863.6	184.63
2450.	20.806	517607.2	196.972	35025.7	194.41
2500.	20.814	518647.7	197.392	25166.5	204.5

TABLE A.9. **Properties of Nitrogen molecule (N_2) at 1 atm (English)**

Temp. °R	$\overline{C}_p^\circ$ Btu/lb-mol°R	$\overline{h}^\circ$ Btu/lb-mol	$\overline{s}^\circ$ Btu/lb-mol°R	$\overline{g}^\circ$ Btu/lb-mol	Pr
536.67	6.961	0.0	45.769	−24563.0	1.0065
600.	6.964	440.9	46.546	−27486.6	1.4877
700.	6.983	1138.1	47.621	−32196.3	2.555
800.	7.021	1838.1	48.555	−37006.1	4.089
900.	7.073	2542.7	49.385	−41903.9	6.209
1000.	7.138	3253.1	50.134	−46880.4	9.049
1100.	7.213	3970.6	50.817	−51928.4	12.765
1200.	7.296	4696.0	51.448	−57042.1	17.537
1300.	7.384	5430.0	52.036	−62216.7	23.569
1400.	7.475	6172.9	52.586	−67448.1	31.093
1500.	7.566	6925.0	53.105	−72732.9	40.370
1600.	7.655	7686.1	53.596	−78068.2	51.69
1700.	7.738	8455.9	54.063	−83451.4	65.37
1800.	7.814	9233.5	54.508	−88880.1	81.76
1900.	7.884	10018.4	54.932	−94352.2	101.22
2000.	7.950	10810.2	55.338	−99865.8	124.17
2100.	8.013	11608.4	55.727	−105419.2	151.05
2200.	8.072	12412.7	56.102	−111010.8	182.35
2300.	8.128	13222.7	56.462	−116639.1	218.6
2400.	8.180	14038.1	56.809	−122302.7	260.3
2500.	8.229	14858.6	57.144	−128000.4	308.1
2600.	8.275	15683.8	57.467	−133731.0	362.5
2700.	8.319	16513.6	57.780	−139493.5	424.4
2800.	8.360	17347.5	58.084	−145286.8	494.4
2900.	8.398	18185.4	58.378	−151110.0	573.2
3000.	8.433	19027.0	58.663	−156962.1	661.7
3100.	8.467	19872.0	58.940	−162842.3	760.8
3200.	8.498	20720.2	59.209	−168749.8	871.2
3300.	8.527	21571.5	59.471	−174683.9	993.9
3400.	8.554	22425.6	59.726	−180643.8	1130.0
3500.	8.580	23282.3	59.975	−186628.9	1280.4
3600.	8.603	24141.5	60.217	−192638.6	1446.2
3700.	8.625	25002.9	60.453	−198672.1	1628.6
3800.	8.646	25866.5	60.683	−204728.9	1828.8
3900.	8.665	26732.1	60.908	−210808.5	2047.8
4000.	8.683	27599.5	61.127	−216910.3	2287.1
4100.	8.699	28468.6	61.342	−223033.8	2547.9
4200.	8.715	29339.3	61.552	−229178.5	2831.7
4300.	8.729	30211.6	61.757	−235344.0	3139.8
4400.	8.743	31085.2	61.958	−241529.8	3473.7
4500.	8.755	31960.1	62.155	−247735.5	3835.0

TABLE A.9. **Properties of Nitrogen molecule (N_2) at 1 atm (SI)**

Temp. K	$\overline{C}_p^\circ$ kJ/kmol K	$\overline{h}^\circ$ kJ/kmol	$\overline{s}^\circ$ kJ/kmol K	$\overline{g}^\circ$ kJ/kmol	Pr
298.15	29.125	0.0	191.498	−57095.2	1.0065
300.	29.125	53.9	191.678	−57449.7	1.0285
350.	29.152	1510.5	196.169	−67148.7	1.7652
400.	29.244	2970.2	200.067	−77056.7	2.821
450.	29.393	4435.9	203.520	−87148.0	4.273
500.	29.593	5910.3	206.627	−97403.0	6.209
550.	29.835	7395.8	209.458	−107806.1	8.728
600.	30.113	8894.4	212.066	−118345.1	11.943
650.	30.419	10407.6	214.488	−129009.6	15.983
700.	30.746	11936.7	216.754	−139791.3	20.991
750.	31.086	13482.5	218.887	−150682.8	27.129
800.	31.430	15045.3	220.904	−161678.1	34.578
850.	31.771	16625.4	222.820	−172771.6	43.538
900.	32.100	18222.2	224.645	−183958.6	54.23
950.	32.410	19835.1	226.389	−195234.8	66.88
1000.	32.692	21462.8	228.059	−206596.3	81.76
1050.	32.958	23104.1	229.661	−218039.5	99.12
1100.	33.210	24758.3	231.200	−229561.3	119.28
1150.	33.449	26424.8	232.681	−241158.5	142.55
1200.	33.676	28103.0	234.110	−252828.5	169.27
1250.	33.891	29792.2	235.489	−264568.6	199.81
1300.	34.095	31491.9	236.822	−276376.6	234.6
1350.	34.288	33201.6	238.112	−288250.1	273.9
1400.	34.471	34920.6	239.363	−300187.1	318.4
1450.	34.643	36648.5	240.575	−312185.7	368.4
1500.	34.806	38384.7	241.753	−324244.1	424.4
1550.	34.960	40128.9	242.896	−336360.4	487.0
1600.	35.105	41880.6	244.009	−348533.2	556.7
1650.	35.241	43639.3	245.091	−360760.8	634.1
1700.	35.370	45404.6	246.145	−373041.8	719.8
1750.	35.491	47176.2	247.172	−385374.9	814.5
1800.	35.605	48953.6	248.173	−397758.6	918.7
1850.	35.712	50736.6	249.150	−410191.8	1033.3
1900.	35.813	52524.7	250.104	−422673.3	1158.9
1950.	35.908	54317.7	251.036	−435201.8	1296.3
2000.	35.996	56115.4	251.946	−447776.5	1446.2
2050.	36.080	57917.3	252.836	−460396.1	1609.6
2100.	36.158	59723.2	253.706	−473059.8	1787.3
2150.	36.231	61533.0	254.558	−485766.4	1980.0
2200.	36.300	63346.3	255.392	−498515.2	2188.9
2250.	36.364	65162.9	256.208	−511305.3	2414.8
2300.	36.425	66982.6	257.008	−524135.7	2658.6
2350.	36.482	68805.3	257.792	−537005.8	2921.6
2400.	36.535	70630.8	258.561	−549914.8	3204.5
2450.	36.586	72458.8	259.314	−562861.6	3508.6
2500.	36.633	74289.3	260.054	−575845.9	3835.0

TABLE A.10. **Properties of Nitric oxide (NO) at 1 atm (English)**

Temp. °R	$\overline{C}_p^\circ$ Btu/lb-mol°R	$\overline{h}^\circ$ Btu/lb-mol	$\overline{s}^\circ$ Btu/lb-mol°R	$\overline{g}^\circ$ Btu/lb-mol	Pr
536.67	7.133	38878.4	50.344	11860.2	0.1006
600.	7.128	39329.9	51.139	8646.3	0.1501
700.	7.152	40043.6	52.240	3475.9	0.2611
800.	7.208	40761.4	53.198	−1796.9	0.4230
900.	7.286	41486.0	54.051	−7160.2	0.6498
1000.	7.380	42219.2	54.824	−12604.5	0.9585
1100.	7.483	42962.3	55.532	−18122.8	1.3689
1200.	7.589	43715.8	56.187	−23709.1	1.9040
1300.	7.694	44480.0	56.799	−29358.8	2.590
1400.	7.795	45254.5	57.373	−35067.7	3.457
1500.	7.889	46038.8	57.914	−40832.3	4.539
1600.	7.975	46832.0	58.426	−46649.5	5.873
1700.	8.052	47633.4	58.912	−52516.6	7.500
1800.	8.122	48442.2	59.374	−58431.1	9.464
1900.	8.181	49257.4	59.815	−64390.7	11.814
2000.	8.237	50078.3	60.236	−70393.4	14.603
2100.	8.289	50904.6	60.639	−76437.3	17.887
2200.	8.339	51736.1	61.026	−82520.6	21.731
2300.	8.385	52572.3	61.397	−88641.9	26.200
2400.	8.429	53413.0	61.755	−94799.7	31.369
2500.	8.469	54257.9	62.100	−100992.6	37.315
2600.	8.507	55106.8	62.433	−107219.3	44.121
2700.	8.543	55959.3	62.755	−113478.8	51.88
2800.	8.576	56815.3	63.066	−119769.9	60.67
2900.	8.607	57674.5	63.368	−126091.7	70.61
3000.	8.636	58536.7	63.660	−132443.2	81.80
3100.	8.663	59401.7	63.944	−138823.4	94.35
3200.	8.688	60269.2	64.219	−145231.6	108.38
3300.	8.711	61139.2	64.487	−151667.0	124.01
3400.	8.733	62011.4	64.747	−158128.7	141.37
3500.	8.753	62885.7	65.001	−164616.1	160.60
3600.	8.772	63762.0	65.247	−171128.6	181.84
3700.	8.789	64640.0	65.488	−177665.4	205.2
3800.	8.805	65519.7	65.723	−184226.0	231.0
3900.	8.820	66400.9	65.951	−190809.7	259.2
4000.	8.833	67283.6	66.175	−197416.1	290.0
4100.	8.846	68167.6	66.393	−204044.6	323.7
4200.	8.858	69052.8	66.607	−210694.6	360.4
4300.	8.869	69939.2	66.815	−217365.7	400.2
4400.	8.879	70826.6	67.019	−224057.5	443.5
4500.	8.889	71715.1	67.219	−230769.4	490.4

TABLE A.10. **Properties of Nitric oxide (NO) at 1 atm (SI)**

Temp. K	$\overline{C}_p^\circ$ kJ/kmol K	$\overline{h}^\circ$ kJ/kmol	$\overline{s}^\circ$ kJ/kmol K	$\overline{g}^\circ$ kJ/kmol	Pr
298.15	29.844	90370.5	210.640	27568.3	0.1006
300.	29.841	90425.7	210.824	27178.5	0.1029
350.	29.838	91917.1	215.422	16519.4	0.1788
400.	29.962	93411.6	219.413	5646.3	0.2890
450.	30.187	94915.0	222.954	−5414.5	0.4425
500.	30.486	96431.5	226.150	−16643.4	0.6498
550.	30.837	97964.4	229.072	−28024.9	0.9234
600.	31.220	99515.8	231.771	−39546.8	1.2777
650.	31.618	101086.7	234.286	−51198.9	1.7289
700.	32.017	102677.6	236.643	−62972.8	2.296
750.	32.406	104288.2	238.866	−74861.0	2.999
800.	32.775	105917.9	240.969	−86857.4	3.863
850.	33.119	107565.3	242.966	−98956.2	4.911
900.	33.434	109229.2	244.868	−111152.4	6.174
950.	33.721	110908.2	246.684	−123441.6	7.680
1000.	33.981	112600.9	248.420	−135819.5	9.464
1050.	34.205	114305.6	250.084	−148282.4	11.560
1100.	34.481	116021.2	251.680	−160826.8	14.007
1150.	34.619	117747.2	253.214	−173449.3	16.846
1200.	34.809	119482.9	254.692	−186147.2	20.122
1250.	34.988	121227.8	256.116	−198917.7	23.883
1300.	35.158	122981.5	257.492	−211758.1	28.180
1350.	35.318	124743.4	258.822	−224666.1	33.068
1400.	35.468	126513.1	260.109	−237639.5	38.604
1450.	35.610	128290.1	261.356	−250676.3	44.852
1500.	35.744	130074.0	262.566	−263774.5	51.88
1550.	35.869	131864.4	263.740	−276932.3	59.74
1600.	35.987	133660.8	264.880	−290147.9	68.53
1650.	36.098	135463.0	265.990	−303419.8	78.31
1700.	36.202	137270.5	267.069	−316746.4	89.16
1750.	36.299	139083.0	268.120	−330126.2	101.17
1800.	36.390	140900.3	269.143	−343557.9	114.43
1850.	36.476	142722.0	270.142	−357040.2	129.03
1900.	36.556	144547.8	271.115	−370571.7	145.06
1950.	36.630	146377.5	272.066	−384151.3	162.63
2000.	36.700	148210.8	272.994	−397777.9	181.84
2050.	36.766	150047.4	273.901	−411450.4	202.8
2100.	36.826	151887.3	274.788	−425167.7	225.6
2150.	36.883	153730.0	275.655	−438928.9	250.4
2200.	36.937	155575.5	276.504	−452732.9	277.3
2250.	36.986	157423.6	277.334	−466579.0	306.5
2300.	37.033	159274.1	278.148	−480466.1	338.0
2350.	37.077	161126.9	278.945	−494393.5	372.0
2400.	37.117	162981.8	279.726	−508360.3	408.6
2450.	37.156	164838.6	280.492	−522365.8	448.0
2500.	37.192	166697.3	281.243	−536409.3	490.4

TABLE A.11. **Properties of Oxygen atom (O) at 1 atm (English)**

Temp. °R	$\overline{C}_p^\circ$ Btu/lb-mol°R	$\overline{h}^\circ$ Btu/lb-mol	$\overline{s}^\circ$ Btu/lb-mol°R	$\overline{g}^\circ$ Btu/lb-mol	Pr
536.67	5.237	107197.7	38.467	86553.5	2.5521
600.	5.196	107528.0	39.049	84098.6	3.420
700.	5.145	108044.9	39.846	80152.8	5.108
800.	5.107	108557.4	40.530	76133.2	7.208
900.	5.079	109066.6	41.130	72049.6	9.747
1000.	5.059	109573.5	41.664	67909.4	12.753
1100.	5.045	110078.6	42.146	63718.5	16.249
1200.	5.034	110582.5	42.584	59481.7	20.261
1300.	5.026	111085.5	42.987	55202.9	24.811
1400.	5.019	111587.7	43.359	50885.4	29.922
1500.	5.012	112089.2	43.705	46532.0	35.613
1600.	5.007	112590.2	44.028	42145.2	41.905
1700.	5.002	113090.6	44.332	37727.0	48.817
1800.	4.999	113590.6	44.617	33279.4	56.37
1900.	4.996	114090.3	44.887	28804.1	64.58
2000.	4.994	114589.9	45.144	24302.4	73.46
2100.	4.992	115089.2	45.387	19775.8	83.05
2200.	4.990	115588.3	45.620	15225.3	93.34
2300.	4.988	116087.3	45.841	10652.2	104.36
2400.	4.987	116586.0	46.054	6057.4	116.13
2500.	4.985	117084.6	46.257	1441.8	128.65
2600.	4.984	117583.0	46.453	−3193.8	141.95
2700.	4.982	118081.3	46.641	−7848.5	156.04
2800.	4.981	118579.4	46.822	−12521.7	170.93
2900.	4.980	119077.5	46.997	−17212.7	186.65
3000.	4.979	119575.4	47.165	−21920.8	203.2
3100.	4.978	120073.2	47.329	−26645.6	220.6
3200.	4.977	120571.0	47.487	−31386.4	238.8
3300.	4.977	121068.7	47.640	−36142.7	258.0
3400.	4.976	121566.3	47.788	−40914.2	278.0
3500.	4.976	122063.9	47.933	−45700.2	298.9
3600.	4.976	122561.5	48.073	−50500.6	320.8
3700.	4.976	123059.1	48.209	−55314.7	343.6
3800.	4.976	123556.7	48.342	−60142.3	367.3
3900.	4.977	124054.3	48.471	−64982.9	392.0
4000.	4.977	124552.0	48.597	−69836.4	417.6
4100.	4.978	125049.8	48.720	−74702.3	444.3
4200.	4.979	125547.7	48.840	−79580.3	471.9
4300.	4.980	126045.6	48.957	−84470.2	500.6
4400.	4.982	126543.7	49.072	−89371.6	530.3
4500.	4.983	127042.0	49.184	−94284.4	561.0

TABLE A.11. **Properties of Oxygen atom (O) at 1 atm (SI)**

Temp. K	$\overline{C}_p^\circ$ kJ/kmol K	$\overline{h}^\circ$ kJ/kmol	$\overline{s}^\circ$ kJ/kmol K	$\overline{g}^\circ$ kJ/kmol	Pr
298.15	21.911	249174.4	160.946	201188.4	2.5521
300.	21.901	249214.9	161.081	200890.5	2.594
350.	21.669	250303.9	164.439	192750.2	3.885
400.	21.490	251382.6	167.320	184454.5	5.494
450.	21.353	252453.6	169.843	176024.1	7.442
500.	21.251	253518.6	172.088	167474.8	9.747
550.	21.175	254579.1	174.109	158819.0	12.431
600.	21.118	255636.4	175.949	150066.9	15.510
650.	21.075	256691.2	177.638	141226.6	19.002
700.	21.040	257744.0	179.198	132305.2	22.925
750.	21.012	258795.3	180.649	123308.6	27.295
800.	20.987	259845.2	182.004	114241.9	32.128
850.	20.964	260894.0	183.276	105109.6	37.437
900.	20.943	261941.7	184.473	95915.6	43.237
950.	20.926	262988.3	185.605	86663.3	49.543
1000.	20.914	264034.3	186.678	77356.0	56.37
1050.	20.906	265079.8	187.698	67996.4	63.73
1100.	20.898	266124.9	188.671	58587.0	71.63
1150.	20.890	267169.6	189.600	49130.0	80.10
1200.	20.883	268213.9	190.489	39627.7	89.14
1250.	20.875	269257.9	191.341	30081.8	98.76
1300.	20.869	270301.4	192.159	20494.2	108.98
1350.	20.862	271344.7	192.947	10866.4	119.80
1400.	20.856	272387.7	193.706	1200.0	131.25
1450.	20.850	273430.3	194.437	−8503.7	143.32
1500.	20.845	274472.7	195.144	−18243.3	156.04
1550.	20.840	275514.8	195.827	−28017.7	169.41
1600.	20.836	276556.8	196.489	−37825.7	183.44
1650.	20.832	277598.5	197.130	−47666.3	198.14
1700.	20.829	278640.0	197.752	−57538.4	213.5
1750.	20.826	279681.3	198.356	−67441.2	229.6
1800.	20.823	280722.5	198.942	−77373.7	246.4
1850.	20.821	281763.7	199.513	−87335.2	263.9
1900.	20.820	282804.7	200.068	−97324.8	282.1
1950.	20.819	283845.7	200.609	−107341.7	301.1
2000.	20.819	284886.7	201.136	−117385.4	320.8
2050.	20.819	285927.6	201.650	−127455.1	341.2
2100.	20.820	286968.6	202.152	−137550.2	362.5
2150.	20.822	288009.6	202.642	−147670.1	384.5
2200.	20.824	289050.8	203.120	−157814.2	407.3
2250.	20.827	290092.1	203.588	−167981.9	430.8
2300.	20.830	291133.4	204.046	−178172.9	452.2
2350.	20.834	292175.0	204.494	−188386.5	480.4
2400.	20.838	293216.9	204.933	−198622.2	506.5
2450.	20.843	294258.9	205.363	−208879.5	533.3
2500.	20.849	295301.2	205.784	−219158.3	561.0

TABLE A.12. **Properties of Oxygen molecule (O_2) at 1 atm (English)**

Temp. °R	$\bar{C}_p^\circ$ Btu/lb-mol°R	$\bar{h}^\circ$ Btu/lb-mol	$\bar{s}^\circ$ Btu/lb-mol°R	$\bar{g}^\circ$ Btu/lb-mol	Pr
536.67	7.021	0.0	49.005	−26299.5	5.1281
600.	7.074	446.3	49.791	−29428.3	7.616
700.	7.178	1158.7	50.889	−34463.6	13.234
800.	7.299	1882.4	51.855	−39601.8	21.521
900.	7.429	2618.8	52.722	−44831.4	33.295
1000.	7.561	3368.3	53.512	−50143.6	49.537
1100.	7.691	4130.9	54.239	−55531.7	71.41
1200.	7.813	4906.2	54.913	−60989.7	100.27
1300.	7.927	5693.3	55.543	−66512.8	137.67
1400.	8.028	6491.1	56.134	−72097.0	185.38
1500.	8.119	7298.6	56.691	−77738.5	245.4
1600.	8.198	8114.5	57.218	−83434.3	319.8
1700.	8.269	8937.9	57.717	−89181.2	411.1
1800.	8.335	9768.2	58.192	−94976.9	522.0
1900.	8.384	10604.1	58.644	−100818.8	655.3
2000.	8.431	11444.8	59.075	−106704.9	814.1
2100.	8.477	12290.3	59.487	−112633.1	1002.0
2200.	8.522	13140.2	59.883	−118601.8	1222.5
2300.	8.565	13994.6	60.263	−124609.2	1480.0
2400.	8.608	14853.3	60.628	−130653.8	1778.8
2500.	8.649	15716.1	60.980	−136734.3	2123.8
2600.	8.690	16583.1	61.320	−142849.4	2520.1
2700.	8.729	17454.0	61.649	−148998.0	2973.4
2800.	8.767	18328.8	61.967	−155178.9	3489.6
2900.	8.805	19207.4	62.275	−161391.1	4075.
3000.	8.841	20089.7	62.574	−167633.6	4737.
3100.	8.877	20975.6	62.865	−173905.7	5483.
3200.	8.912	21865.0	63.147	−180206.4	6320.
3300.	8.946	22757.9	63.422	−186534.9	7257.
3400.	8.980	23654.2	63.690	−192890.5	8303.
3500.	9.012	24533.8	63.950	−199272.6	9468.
3600.	9.045	25456.7	64.205	−205680.4	10760.
3700.	9.076	26362.7	64.453	−212113.3	12192.
3800.	9.107	27271.9	64.695	−218570.8	13774.
3900.	9.138	28184.1	64.932	−225052.2	15519.
4000.	9.168	29099.4	65.164	−231557.1	17438.
4100.	9.197	30017.6	65.391	−238084.9	19546.
4200.	9.226	30938.8	65.613	−244635.1	21856.
4300.	9.255	31862.8	65.830	−251207.3	24383.
4400.	9.283	32789.7	66.043	−257801.0	27142.
4500.	9.311	33719.3	66.252	−264415.8	30151.

TABLE A.12. **Properties of Oxygen molecule (O_2) at 1 atm (SI)**

Temp. K	$\overline{C}_p^\circ$ kJ/kmol K	$\overline{h}^\circ$ kJ/kmol	$\overline{s}^\circ$ kJ/kmol K	$\overline{g}^\circ$ kJ/kmol	Pr
298.15	29.375	0.0	205.036	−61131.6	5.1281
300.	29.386	54.4	205.218	−61511.1	5.241
350.	29.719	1531.6	209.772	−71888.6	9.064
400.	30.130	3027.5	213.766	−82479.0	14.654
450.	30.592	4545.4	217.341	−93258.3	22.527
500.	31.082	6087.2	220.590	−104207.7	33.295
550.	31.580	7653.8	223.576	−115312.9	47.680
600.	32.071	9245.1	226.345	−126561.7	66.52
650.	32.541	10860.5	228.930	−137944.3	90.79
700.	32.981	12498.7	231.358	−149452.1	121.58
750.	33.384	14157.9	233.648	−161077.8	160.12
800.	33.748	15836.4	235.814	−172814.8	207.8
850.	34.073	17532.1	237.870	−184657.3	266.1
900.	34.363	19243.1	239.826	−196600.1	336.6
950.	34.627	20967.9	241.691	−208638.5	421.3
1000.	34.873	22705.5	243.473	−220767.9	522.0
1050.	35.057	24453.7	245.179	−232984.5	640.9
1100.	35.236	26211.1	246.814	−245284.6	780.2
1150.	35.411	27977.3	248.384	−257664.8	942.3
1200.	35.581	29752.1	249.895	−270122.1	1130.1
1250.	35.747	31535.3	251.351	−282653.5	1346.4
1300.	35.909	33326.7	252.756	−295256.3	1594.3
1350.	36.068	35126.2	254.114	−307928.3	1877.2
1400.	36.222	36933.4	255.429	−320667.0	2198.7
1450.	36.373	38748.4	256.703	−333470.5	2562.8
1500.	36.521	40570.7	257.938	−346336.7	2973.4
1550.	36.666	42400.4	259.138	−359263.8	3435.0
1600.	36.807	44237.3	260.305	−372249.9	3952.3
1650.	36.946	46081.1	261.439	−385293.6	4530.
1700.	37.081	47931.8	262.544	−398393.3	5174.
1750.	37.214	49789.2	263.621	−411547.6	5890.
1800.	37.345	51653.2	264.671	−424755.0	6683.
1850.	37.472	53523.6	265.696	−438014.3	7559.
1900.	37.598	55400.4	266.697	−451324.2	8526.
1950.	37.721	57283.3	267.675	−464683.6	9591.
2000.	37.842	59172.4	268.632	−478091.4	10760.
2050.	37.961	61067.5	269.568	−491546.5	12042.
2100.	38.078	62968.5	270.484	−505047.8	13445.
2150.	38.194	64875.3	271.381	−518594.6	14978.
2200.	38.307	66787.9	272.261	−532185.7	16648.
2250.	38.419	68706.0	273.123	−545820.4	18467.
2300.	38.529	70629.7	273.968	−559497.6	20444.
2350.	38.638	72558.9	274.798	−573216.9	22590.
2400.	38.745	74493.5	275.613	−586977.3	24915.
2450.	38.851	76433.4	276.413	−600778.0	27432.
2500.	38.955	78378.5	277.199	−614618.4	30151.

TABLE B.1. **Properties of Dry Air at 1 atm (SI)**

Temp K	C_p° kJ/kg K	h° kJ/kg	s° kJ/kg K	Pr
210.	1.006	−92.7	6.3468	0.3987
215.	1.006	−87.6	6.3705	0.4330
220.	1.006	−82.6	6.3936	0.4693
225.	1.005	−77.6	6.4162	0.5077
230.	1.005	−72.5	6.4383	0.5483
235.	1.005	−67.5	6.4599	0.5912
240.	1.005	−62.5	6.4811	0.6364
245.	1.005	−57.5	6.5018	0.6841
250.	1.005	−52.4	6.5221	0.7342
255.	1.004	−47.4	6.5420	0.7868
260.	1.004	−42.4	6.5615	0.8421
265.	1.004	−37.4	6.5806	0.9002
270.	1.004	−32.4	6.5994	0.9610
275.	1.004	−27.3	6.6178	1.0247
280.	1.004	−22.3	6.6359	1.0914
285.	1.005	−17.3	6.6537	1.1611
290.	1.005	−12.3	6.6712	1.2340
295.	1.005	−7.2	6.6883	1.3100
300.	1.005	−2.2	6.7052	1.3894
305.	1.005	2.8	6.7219	1.4722
310.	1.005	7.8	6.7382	1.5585
315.	1.006	12.9	6.7543	1.6483
320.	1.006	17.9	6.7701	1.7418
325.	1.006	22.9	6.7857	1.8390
330.	1.007	28.0	6.8011	1.9401
335.	1.007	33.0	6.8162	2.045
340.	1.007	38.0	6.8312	2.154
345.	1.008	43.1	6.8459	2.268
350.	1.008	48.1	6.8604	2.385
355.	1.009	53.1	6.8747	2.507
360.	1.009	58.2	6.8888	2.633
365.	1.010	63.2	6.9027	2.764
370.	1.010	68.3	6.9164	2.899
375.	1.011	73.3	6.9300	3.040

TABLE B.1. **Properties of Dry Air at 1 atm (SI)**

Temp K	C_p° kJ/kg K	h° kJ/kg	s° kJ/kg K	Pr
380.	1.011	78.4	6.9434	3.185
385.	1.012	83.4	6.9566	3.335
390.	1.012	88.5	6.9697	3.490
395.	1.013	93.6	6.9826	3.651
400.	1.014	98.6	6.9953	3.816
405.	1.014	103.7	7.0079	3.987
410.	1.015	108.8	7.0204	4.164
415.	1.016	113.9	7.0327	4.347
420.	1.016	118.9	7.0448	4.535
425.	1.017	124.0	7.0569	4.729
430.	1.018	129.1	7.0688	4.929
435.	1.019	134.2	7.0806	5.135
440.	1.019	139.3	7.0922	5.348
445.	1.020	144.4	7.1037	5.567
450.	1.021	149.5	7.1151	5.792
455.	1.022	154.6	7.1264	6.025
460.	1.023	159.7	7.1376	6.264
465.	1.024	164.8	7.1486	6.510
470.	1.024	170.0	7.1596	6.763
475.	1.025	175.1	7.1704	7.023
480.	1.026	180.2	7.1812	7.291
485.	1.027	185.3	7.1918	7.566
490.	1.028	190.5	7.2024	7.849
495.	1.029	195.6	7.2128	8.140
500.	1.030	200.8	7.2232	8.439
505.	1.031	205.9	7.2334	8.745
510.	1.032	211.1	7.2436	9.060
515.	1.033	216.2	7.2536	9.384
520.	1.034	221.4	7.2636	9.716
525.	1.035	226.6	7.2735	10.057
530.	1.036	231.8	7.2833	10.407
535.	1.037	236.9	7.2931	10.765
540.	1.038	242.1	7.3027	11.133
545.	1.039	247.3	7.3123	11.511
550.	1.040	252.5	7.3218	11.898

TABLE B.1. **Properties of Dry Air at 1 atm (SI)**

Temp K	C_p° kJ/kg K	h° kJ/kg	s° kJ/kg K	Pr
555.	1.041	257.7	7.3312	12.295
560.	1.042	262.9	7.3405	12.701
565.	1.043	268.1	7.3498	13.118
570.	1.044	273.4	7.3590	13.545
575.	1.046	278.6	7.3681	13.983
580.	1.047	283.8	7.3772	14.431
585.	1.048	289.1	7.3862	14.890
590.	1.049	294.3	7.3951	15.360
595.	1.050	299.5	7.4040	15.841
600.	1.051	304.8	7.4128	16.334
605.	1.052	310.1	7.4215	16.838
610.	1.053	315.3	7.4302	17.354
615.	1.055	320.6	7.4388	17.882
620.	1.056	325.9	7.4473	18.422
625.	1.057	331.1	7.4558	18.974
630.	1.058	336.4	7.4642	19.540
635.	1.059	341.7	7.4726	20.118
640.	1.060	347.0	7.4809	20.709
645.	1.062	352.3	7.4892	21.313
650.	1.063	357.6	7.4974	21.931
655.	1.064	363.0	7.5055	22.562
660.	1.065	368.3	7.5136	23.207
665.	1.066	373.6	7.5216	23.867
670.	1.068	378.9	7.5296	24.540
675.	1.069	384.3	7.5376	25.229
680.	1.070	389.6	7.5455	25.932
685.	1.071	395.0	7.5533	26.650
690.	1.072	400.4	7.5611	27.384
695.	1.074	405.7	7.5689	28.133
700.	1.075	411.1	7.5766	28.897
705.	1.076	416.5	7.5842	29.678
710.	1.077	421.8	7.5918	30.475
715.	1.078	427.2	7.5994	31.289
720.	1.080	432.6	7.6069	32.119
725.	1.081	438.0	7.6144	32.967

TABLE B.1. **Properties of Dry Air at 1 atm (SI)**

Temp K	C_p° kJ/kg K	h° kJ/kg	s° kJ/kg K	Pr
730.	1.082	443.4	7.6218	33.831
735.	1.083	448.9	7.6292	34.713
740.	1.085	454.3	7.6366	35.613
745.	1.086	459.7	7.6439	36.531
750.	1.087	465.1	7.6511	37.468
755.	1.088	470.6	7.6584	38.423
760.	1.089	476.0	7.6655	39.397
765.	1.091	481.5	7.6727	40.390
770.	1.092	486.9	7.6798	41.402
775.	1.093	492.4	7.6869	42.434
780.	1.094	497.8	7.6939	43.487
785.	1.095	503.3	7.7009	44.559
790.	1.096	508.8	7.7079	45.652
795.	1.098	514.3	7.7148	46.766
800.	1.099	519.8	7.7217	47.902
805.	1.100	525.3	7.7285	49.058
810.	1.101	530.8	7.7353	50.24
815.	1.102	536.3	7.7421	51.44
820.	1.104	541.8	7.7489	52.66
825.	1.105	547.3	7.7556	53.91
830.	1.106	552.8	7.7622	55.17
835.	1.107	558.4	7.7689	56.47
840.	1.108	563.9	7.7755	57.78
845.	1.109	569.5	7.7821	59.12
850.	1.110	575.0	7.7886	60.49
855.	1.112	580.6	7.7951	61.87
860.	1.113	586.1	7.8016	63.29
865.	1.114	591.7	7.8081	64.73
870.	1.115	597.3	7.8145	66.19
875.	1.116	602.8	7.8209	67.68
880.	1.117	608.4	7.8273	69.20
885.	1.118	614.0	7.8336	70.74
890.	1.119	619.6	7.8399	72.31
895.	1.120	525.2	7.8462	73.91
900.	1.121	630.8	7.8524	75.53

TABLE B.1. **Properties of Dry Air at 1 atm (SI)**

Temp K	C_p° kJ/kg K	h° kJ/kg	s° kJ/kg K	Pr
905.	1.122	636.4	7.8586	77.19
910.	1.124	642.0	7.8648	78.87
915.	1.125	647.7	7.8710	80.58
920.	1.126	653.3	7.8771	82.32
925.	1.127	658.9	7.8832	84.09
930.	1.128	664.5	7.8893	85.88
935.	1.129	670.2	7.8953	87.71
940.	1.130	675.8	7.9014	89.57
945.	1.131	681.5	7.9074	91.46
950.	1.132	687.1	7.9133	93.38
955.	1.133	692.8	7.9193	95.34
960.	1.134	698.5	7.9252	97.32
965.	1.135	704.1	7.9311	99.34
970.	1.136	709.8	7.9369	101.39
975.	1.136	715.5	7.9428	103.47
980.	1.137	721.2	7.9486	105.59
985.	1.138	726.9	7.9544	107.74
990.	1.139	732.6	7.9602	109.93
995.	1.140	738.3	7.9659	112.15
1000.	1.141	744.0	7.9716	114.41
1005.	1.142	749.7	7.9773	116.70
1010.	1.143	755.4	7.9830	119.02
1015.	1.144	761.1	7.9886	121.39
1020.	1.145	766.8	7.9942	123.79
1025.	1.145	772.5	7.9998	126.23
1030.	1.146	778.3	8.0054	128.70
1035.	1.147	784.0	8.0110	131.21
1040.	1.148	789.7	8.0165	133.77
1045.	1.149	795.5	8.0220	136.36
1050.	1.150	801.2	8.0275	138.99
1055.	1.150	807.0	8.0330	141.66
1060.	1.151	812.7	8.0384	144.37
1065.	1.152	818.5	8.0438	147.12
1070.	1.153	824.3	8.0492	149.91
1075.	1.154	830.0	8.0546	152.74

TABLE B.1. **Properties of Dry Air at 1 atm (SI)**

Temp K	C_p° kJ/kg K	h° kJ/kg	s° kJ/kg K	Pr
1080.	1.154	835.8	8.0599	155.62
1085.	1.155	841.6	8.0653	158.54
1090.	1.156	847.3	8.0706	161.50
1095.	1.157	853.1	8.0759	164.50
1100.	1.158	858.9	8.0812	167.55
1105.	1.158	864.7	8.0864	170.65
1110.	1.159	870.5	8.0916	173.79
1115.	1.160	876.3	8.0969	176.97
1120.	1.161	882.1	8.1020	180.20
1125.	1.162	887.9	8.1072	183.47
1130.	1.162	893.7	8.1124	186.80
1135.	1.163	899.5	8.1175	190.17
1140.	1.164	905.3	8.1226	193.59
1145.	1.165	911.2	8.1277	197.05
1150.	1.165	917.0	8.1328	200.6
1155.	1.166	922.8	8.1379	204.1
1160.	1.167	928.7	8.1429	207.7
1165.	1.168	934.5	8.1479	211.4
1170.	1.168	940.3	8.1529	215.1
1175.	1.169	946.2	8.1579	218.9
1180.	1.170	952.0	8.1629	222.7
1185.	1.171	957.9	8.1678	226.6
1190.	1.171	963.7	8.1727	230.5
1195.	1.172	969.6	8.1777	234.5
1200.	1.173	975.4	8.1825	238.5
1205.	1.173	981.3	8.1874	242.6
1210.	1.174	987.2	8.1923	246.7
1215.	1.175	993.1	8.1971	250.9
1220.	1.176	998.9	8.2020	255.2
1225.	1.176	1004.8	8.2068	259.5
1230.	1.177	1010.7	8.2116	263.9
1235.	1.178	1016.6	8.2163	268.3
1240.	1.178	1022.5	8.2211	272.8
1245.	1.179	1028.4	8.2258	277.3
1250.	1.180	1034.3	8.2306	281.9

TABLE B.1. **Properties of Dry Air at 1 atm (SI)**

Temp K	C_p° kJ/kg K	h° kJ/kg	s° kJ/kg K	Pr
1255.	1.180	1040.2	8.2353	286.6
1260.	1.181	1046.1	8.2400	291.3
1265.	1.182	1052.0	8.2447	296.1
1270.	1.182	1057.9	8.2493	301.0
1275.	1.183	1063.8	8.2540	305.9
1280.	1.184	1069.7	8.2586	310.9
1285.	1.184	1075.6	8.2632	315.9
1290.	1.185	1081.6	8.2678	321.0
1295.	1.186	1087.5	8.2724	326.2
1300.	1.186	1093.4	8.2770	331.4
1305.	1.187	1099.4	8.2815	336.7
1310.	1.188	1105.3	8.2861	342.1
1315.	1.188	1111.2	8.2906	347.5
1320.	1.189	1117.2	8.2951	353.0
1325.	1.190	1123.1	8.2996	358.6
1330.	1.190	1129.1	8.3041	364.2
1335.	1.191	1135.0	8.3085	369.9
1340.	1.192	1141.0	8.3130	375.7
1345.	1.192	1146.9	8.3174	381.6
1350.	1.193	1152.9	8.3219	387.5
1355.	1.193	1158.9	8.3263	393.5
1360.	1.194	1164.8	8.3307	399.6
1365.	1.195	1170.8	8.3351	405.7
1370.	1.195	1176.8	8.3394	411.9
1375.	1.196	1182.8	8.3438	418.2
1380.	1.196	1188.7	8.3481	424.6
1385.	1.197	1194.7	8.3524	431.0
1390.	1.198	1200.7	8.3568	437.6
1395.	1.198	1206.7	8.3611	444.2
1400.	1.199	1212.7	8.3654	450.9
1405.	1.199	1218.7	8.3696	457.6
1410.	1.200	1224.7	8.3739	464.5
1415.	1.201	1230.7	8.3781	471.4
1420.	1.201	1236.7	8.3824	478.4
1425.	1.202	1242.7	8.3866	485.5

TABLE B.1. **Properties of Dry Air at 1 atm (SI)**

Temp K	C_p° kJ/kg K	h° kJ/kg	s° kJ/kg K	Pr
1430.	1.202	1248.7	8.3908	492.7
1435.	1.203	1254.7	8.3950	499.9
1440.	1.203	1260.7	8.3992	507.3
1445.	1.204	1266.8	8.4034	514.7
1450.	1.205	1272.8	8.4075	522.2
1455.	1.205	1278.8	8.4117	529.8
1460.	1.206	1284.8	8.4158	537.5
1465.	1.206	1290.9	8.4199	545.3
1470.	1.207	1296.9	8.4240	553.1
1475.	1.207	1302.9	8.4281	561.1
1480.	1.208	1309.0	8.4322	569.1
1485.	1.208	1315.0	8.4363	577.2
1490.	1.209	1321.0	8.4404	585.5
1495.	1.210	1327.1	8.4444	593.8
1500.	1.210	1333.1	8.4485	602.2
1505.	1.211	1339.2	8.4525	610.7
1510.	1.211	1345.3	8.4565	619.3
1515.	1.212	1351.3	8.4605	628.0
1520.	1.212	1357.4	8.4645	636.8
1525.	1.213	1363.4	8.4685	645.7
1530.	1.213	1369.5	8.4724	654.7
1535.	1.214	1375.6	8.4764	663.8
1540.	1.214	1381.6	8.4804	673.0
1545.	1.215	1387.7	8.4843	682.3
1550.	1.215	1393.8	8.4882	691.7
1555.	1.216	1399.9	8.4921	701.2
1560.	1.216	1405.9	8.4960	710.8
1565.	1.217	1412.0	8.4999	720.5
1570.	1.217	1418.1	8.5038	730.3
1575.	1.218	1424.2	8.5077	740.2
1580.	1.218	1430.3	8.5115	750.2
1585.	1.219	1436.4	8.5154	760.3
1590.	1.219	1442.5	8.5192	770.6
1595.	1.220	1448.6	8.5231	780.9
1600.	1.220	1454.7	8.5269	791.4

TABLE B.1. **Properties of Dry Air at 1 atm (SI)**

Temp K	C_p° kJ/kg K	h° kJ/kg	s° kJ/kg K	Pr
1605.	1.221	1460.8	8.5307	801.9
1610.	1.221	1466.9	8.5345	812.6
1615.	1.222	1473.0	8.5383	823.4
1620.	1.222	1479.1	8.5420	834.3
1625.	1.223	1485.2	8.5458	845.3
1630.	1.223	1491.3	8.5496	856.5
1635.	1.223	1497.4	8.5533	867.7
1640.	1.224	1503.5	8.5571	879.1
1645.	1.224	1509.7	8.5608	890.6
1650.	1.225	1515.8	8.5645	902.2
1655.	1.225	1521.9	8.5682	913.9
1660.	1.226	1528.0	8.5719	925.7
1665.	1.226	1534.2	8.5756	937.7
1670.	1.227	1540.3	8.5793	949.8
1675.	1.227	1546.4	8.5829	962.0
1680.	1.228	1552.6	8.5866	974.3
1685.	1.228	1558.7	8.5902	986.8
1690.	1.228	1564.9	8.5939	999.4
1695.	1.229	1571.0	8.5975	1012.1
1700.	1.229	1577.1	8.6011	1024.9
1705.	1.230	1583.3	8.6047	1037.9
1710.	1.230	1589.4	8.6083	1051.0
1715.	1.231	1595.6	8.6119	1064.2
1720.	1.231	1601.8	8.6155	1077.6
1725.	1.231	1607.9	8.6191	1091.1
1730.	1.232	1614.1	8.6227	1104.7
1735.	1.232	1620.2	8.6262	1118.5
1740.	1.233	1626.4	8.6298	1132.4
1745.	1.233	1632.6	8.6333	1146.4
1750.	1.234	1638.7	8.6368	1160.6
1755.	1.234	1644.9	8.6403	1174.9
1760.	1.234	1651.1	8.6439	1189.4
1765.	1.235	1657.2	8.6474	1204.0
1770.	1.235	1663.4	8.6509	1218.7
1775.	1.236	1669.6	8.6543	1233.6

TABLE B.1. **Properties of Dry Air at 1 atm (SI)**

Temp K	C_p° kJ/kg K	h° kJ/kg	s° kJ/kg K	Pr
1780.	1.236	1675.8	8.6578	1248.6
1785.	1.236	1681.9	8.6613	1263.8
1790.	1.237	1688.1	8.6647	1279.1
1795.	1.237	1694.3	8.6682	1294.6
1800.	1.238	1700.5	8.6716	1310.2
1805.	1.238	1706.7	8.6751	1326.0
1810.	1.238	1712.9	8.6785	1341.9
1815.	1.239	1719.1	8.6819	1358.0
1820.	1.239	1725.3	8.6853	1374.2
1825.	1.239	1731.5	8.6887	1390.5
1830.	1.240	1737.7	8.6921	1407.1
1835.	1.240	1743.9	8.6955	1423.7
1840.	1.241	1750.1	8.6989	1440.6
1845.	1.241	1756.3	8.7022	1457.6
1850.	1.241	1762.5	8.7056	1474.7
1855.	1.242	1768.7	8.7089	1492.1
1860.	1.242	1774.9	8.7123	1509.5
1865.	1.242	1781.1	8.7156	1527.2
1870.	1.243	1787.3	8.7189	1545.0
1875.	1.243	1793.5	8.7223	1562.9
1880.	1.244	1799.7	8.7256	1581.1
1885.	1.244	1806.0	8.7289	1599.4
1890.	1.244	1812.2	8.7322	1617.8
1895.	1.245	1818.4	8.7355	1636.5
1900.	1.245	1824.6	8.7387	1655.3
1905.	1.245	1830.9	8.7420	1674.2
1910.	1.246	1837.1	8.7453	1693.4
1915.	1.246	1843.3	8.7485	1712.7
1920.	1.246	1849.5	8.7518	1732.2
1925.	1.247	1855.8	8.7550	1751.9
1930.	1.247	1862.0	8.7583	1771.7
1935.	1.247	1868.3	8.7615	1791.7
1940.	1.248	1874.5	8.7647	1811.9
1945.	1.248	1880.7	8.7679	1832.3
1950.	1.248	1887.0	8.7711	1852.9

TABLE B.1. **Properties of Dry Air at 1 atm (SI)**

Temp K	C_p° kJ/kg K	h° kJ/kg	s° kJ/kg K	Pr
1955.	1.249	1893.2	8.7743	1873.7
1960.	1.249	1899.5	8.7775	1894.6
1965.	1.249	1905.7	8.7807	1915.7
1970.	1.250	1912.0	8.7839	1937.0
1975.	1.250	1918.2	8.7870	1958.5
1980.	1.250	1924.5	8.7902	1980.2
1985.	1.251	1930.7	8.7934	2002.1
1990.	1.251	1937.0	8.7965	2024.1
1995.	1.251	1943.2	8.7996	2046.4
2000.	1.252	1949.5	8.8028	2068.8
2005.	1.252	1955.7	8.8059	2091.5
2010.	1.252	1962.0	8.8090	2114.3
2015.	1.253	1968.3	8.8121	2137.4
2020.	1.253	1974.5	8.8152	2160.6
2025.	1.253	1980.8	8.8183	2184.1
2030.	1.254	1987.1	8.8214	2207.7
2035.	1.254	1993.3	8.8245	2231.5
2040.	1.254	1999.6	8.8276	2255.6
2045.	1.255	2005.9	8.8307	2279.8
2050.	1.255	2012.1	8.8337	2304.3
2055.	1.255	2018.4	8.8368	2329.0
2060.	1.255	2024.7	8.8398	2353.9
2065.	1.256	2031.0	8.8429	2378.9
2070.	1.256	2037.2	8.8459	2404.2
2075.	1.256	2043.5	8.8489	2429.8
2080.	1.257	2049.8	8.8520	2455.5
2085.	1.257	2056.1	8.8550	2481.4
2090.	1.257	2062.4	8.8580	2507.6
2095.	1.257	2068.7	8.8610	2534.0
2100.	1.258	2075.0	8.8640	2560.6
2105.	1.258	2081.2	8.8670	2587.4
2110.	1.258	2087.5	8.8700	2614.4
2115.	1.259	2093.8	8.8730	2641.7
2120.	1.259	2100.1	8.8759	2669.2
2125.	1.259	2106.4	8.8789	2696.6

TABLE B.1. **Properties of Dry Air at 1 atm (SI)**

Temp K	C_p° kJ/kg K	h° kJ/kg	s° kJ/kg K	Pr
2130.	1.259	2112.7	8.8819	2724.8
2135.	1.260	2119.0	8.8848	2753.0
2140.	1.260	2125.3	8.8878	2781.4
2145.	1.260	2131.6	8.8907	2810.1
2150.	1.261	2137.9	8.8936	2838.9
2155.	1.261	2144.2	8.8966	2868.0
2160.	1.261	2150.5	8.8995	2897.4
2165.	1.261	2156.8	8.9024	2927.0
2170.	1.262	2163.1	8.9053	2956.8
2175.	1.262	2169.4	8.9082	2986.8
2180.	1.262	2175.8	8.9111	3017.1
2185.	1.262	2182.1	8.9140	3047.7
2190.	1.263	2188.4	8.9169	3078.5
2195.	1.263	2194.7	8.9198	3109.5
2200.	1.263	2201.0	8.9226	3140.8
2205.	1.264	2207.3	8.9255	3172.3
2210.	1.264	2213.6	8.9284	3204.1
2215.	1.264	2220.0	8.9312	3236.1
2220.	1.264	2226.3	8.9341	3268.4
2225.	1.265	2232.6	8.9369	3301.0
2230.	1.265	2238.9	8.9398	3333.8
2235.	1.265	2245.3	8.9426	3366.8
2240.	1.265	2251.6	8.9454	3400.2
2245.	1.266	2257.9	8.9482	3433.7
2250.	1.266	2264.2	8.9511	3467.6
2255.	1.266	2270.6	8.9539	3501.7
2260.	1.266	2276.9	8.9567	3536.1
2265.	1.267	2283.2	8.9595	3570.7
2270.	1.267	2289.6	8.9623	3605.6
2275.	1.267	2295.9	8.9651	3640.8
2280.	1.267	2302.2	8.9678	3676.2
2285.	1.268	2308.6	8.9706	3712.0
2290.	1.268	2314.9	8.9734	3748.0
2295.	1.268	2321.3	8.9761	3784.2
2300.	1.268	2327.6	8.9789	3820.8

TABLE B.1. **Properties of Dry Air at 1 atm (SI)**

Temp K	C_p° kJ/kg K	h° kJ/kg	s° kJ/kg K	Pr
2305.	1.268	2333.9	8.9817	3857.6
2310.	1.269	2340.3	8.9844	3894.7
2315.	1.269	2346.6	8.9872	3932.1
2320.	1.269	2353.0	8.9899	3969.8
2325.	1.269	2359.3	8.9926	4008.
2330.	1.270	2365.7	8.9954	4046.
2335.	1.270	2372.0	8.9981	4085.
2340.	1.270	2378.4	9.0008	4123.
2345.	1.270	2384.7	9.0035	4163.
2350.	1.271	2391.1	9.0062	4202.
2355.	1.271	2397.4	9.0089	4242.
2360.	1.271	2403.8	9.0116	4282.
2365.	1.271	2410.1	9.0143	4322.
2370.	1.271	2416.5	9.0170	4363.
2375.	1.272	2422.8	9.0197	4403.
2380.	1.272	2429.2	9.0223	4445.
2385.	1.272	2435.6	9.0250	4486.
2390.	1.272	2441.9	9.0277	4528.
2395.	1.273	2448.3	9.0303	4570.
2400.	1.273	2454.6	9.0330	4613.
2405.	1.273	2461.0	9.0356	4655.
2410.	1.273	2467.4	9.0383	4698.
2415.	1.273	2473.7	9.0409	4742.
2420.	1.274	2480.1	9.0435	4786.
2425.	1.274	2486.5	9.0462	4830.
2430.	1.274	2492.9	9.0488	4874.
2435.	1.274	2499.2	9.0514	4919.
2440.	1.274	2505.6	9.0540	4964.
2445.	1.275	2512.0	9.0566	5009.
2450.	1.275	2518.3	9.0592	5055.
2455.	1.275	2524.7	9.0618	5101.
2460.	1.275	2531.1	9.0644	5147.
2465.	1.276	2537.5	9.0670	5193.
2470.	1.276	2543.8	9.0696	5240.
2475.	1.276	2550.2	9.0722	5288.

TABLE B.1. **Properties of Dry Air at 1 atm (SI)**

Temp K	C_p° kJ/kg K	h° kJ/kg	s° kJ/kg K	Pr
2480.	1.276	2556.6	9.0748	5335.
2485.	1.276	2563.0	9.0773	5383.
2490.	1.277	2569.4	9.0799	5432.
2495.	1.277	2575.8	9.0825	5480.
2500.	1.277	2582.1	9.0850	5529.
2505.	1.277	2588.5	9.0876	5579.
2510.	1.277	2594.9	9.0901	5628.
2515.	1.278	2601.3	9.0927	5678.
2520.	1.278	2607.7	9.0952	5729.
2525.	1.278	2614.1	9.0977	5780.
2530.	1.278	2620.5	9.1003	5831.
2535.	1.278	2626.9	9.1028	5882.
2540.	1.279	2633.2	9.1053	5934.
2545.	1.279	2639.6	9.1078	5986.
2550.	1.279	2646.0	9.1103	6039.
2555.	1.279	2652.4	9.1128	6092.
2560.	1.279	2658.8	9.1153	6145.
2565.	1.279	2665.2	9.1178	6199.
2570.	1.280	2671.6	9.1203	6253.
2575.	1.280	2678.0	9.1228	6307.
2580.	1.280	2684.4	9.1253	6362.
2585.	1.280	2690.8	9.1278	6417.
2590.	1.280	2697.2	9.1302	6473.
2595.	1.281	2703.6	9.1327	6529.
2600.	1.281	2710.0	9.1352	6585.
2605.	1.281	2716.4	9.1376	6642.
2610.	1.281	2722.8	9.1401	6699.
2615.	1.281	2729.2	9.1426	6756.
2620.	1.281	2735.6	9.1450	6814.
2625.	1.282	2742.1	9.1474	6872.
2630.	1.282	2748.5	9.1499	6931.
2635.	1.282	2754.9	9.1523	6990.
2640.	1.282	2761.3	9.1547	7049.
2645.	1.282	2767.7	9.1572	7109.
2650.	1.283	2774.1	9.1596	7169.

TABLE B.2. **Properties of Dry Air at 1 atm (English)**

Temp °R	C_p° Btu/lb°R	h° Btu/lb	s° Btu/lb°R	Pr
390.	0.240	−37.0	1.5244	0.4449
400.	0.240	−34.6	1.5305	0.4861
410.	0.240	−32.2	1.5365	0.5300
420.	0.240	−29.8	1.5423	0.5767
430.	0.240	−27.4	1.5479	0.6262
440.	0.240	−25.0	1.5534	0.6786
450.	0.240	−22.6	1.5588	0.7342
460.	0.240	−20.2	1.5641	0.7929
470.	0.240	−17.8	1.5693	0.8548
480.	0.240	−15.4	1.5743	0.9201
490.	0.240	−13.0	1.5793	0.9890
500.	0.240	−10.6	1.5841	1.0614
510.	0.240	−8.2	1.5889	1.1375
520.	0.240	−5.8	1.5935	1.2175
530.	0.240	−3.4	1.5981	1.3014
540.	0.240	−1.0	1.6026	1.3894
550.	0.240	1.4	1.6070	1.4816
560.	0.240	3.9	1.6113	1.5781
570.	0.240	6.3	1.6156	1.6790
580.	0.240	8.7	1.6198	1.7845
590.	0.241	11.1	1.6239	1.8947
600.	0.241	13.5	1.6279	2.010
610.	0.241	15.9	1.6319	2.130
620.	0.241	18.3	1.6358	2.255
630.	0.241	20.7	1.6397	2.385
640.	0.241	23.1	1.6435	2.521
650.	0.241	25.5	1.6472	2.662
660.	0.241	27.9	1.6509	2.809
670.	0.241	30.3	1.6545	2.961
680.	0.242	32.8	1.6581	3.120
690.	0.242	35.2	1.6616	3.284
700.	0.242	37.6	1.6651	3.455

TABLE B.2. **Properties of Dry Air at 1 atm (English)**

Temp °R	C_p° Btu/lb°R	h° Btu/lb	s° Btu/lb°R	Pr
710.	0.242	40.0	1.6685	3.632
720.	0.242	42.4	1.6719	3.816
730.	0.242	44.9	1.6753	4.007
740.	0.243	47.3	1.6786	4.204
750.	0.243	49.7	1.6818	4.409
760.	0.243	52.1	1.6850	4.620
770.	0.243	54.6	1.6882	4.839
780.	0.243	57.0	1.6914	5.066
790.	0.244	59.4	1.6945	5.300
800.	0.244	61.9	1.6975	5.542
810.	0.244	64.3	1.7006	5.792
820.	0.244	66.8	1.7036	6.051
830.	0.244	69.2	1.7065	6.318
840.	0.245	71.6	1.7094	6.593
850.	0.245	74.1	1.7123	6.878
860.	0.245	76.5	1.7152	7.171
870.	0.245	79.0	1.7180	7.474
880.	0.246	81.5	1.7209	7.786
890.	0.246	83.9	1.7236	8.107
900.	0.246	86.4	1.7264	8.439
910.	0.246	88.8	1.7291	8.780
920.	0.247	91.3	1.7318	9.132
930.	0.247	93.8	1.7345	9.494
940.	0.247	96.2	1.7371	9.866
950.	0.247	98.7	1.7397	10.250
960.	0.248	101.2	1.7423	10.645
970.	0.248	103.7	1.7449	11.051
980.	0.248	106.2	1.7474	11.468
990.	0.249	108.6	1.7500	11.898
1000.	0.249	111.1	1.7525	12.339
1010.	0.249	113.6	1.7549	12.793
1020.	0.249	116.1	1.7574	13.259
1030.	0.250	118.6	1.7598	13.738
1040.	0.250	121.1	1.7622	14.230
1050.	0.250	123.6	1.7646	14.736

TABLE B.2. **Properties of Dry Air at 1 atm (English)**

Temp °R	C_p° Btu/lb°R	h° Btu/lb	s° Btu/lb°R	Pr
1060.	0.251	126.1	1.7670	15.254
1070.	0.251	128.6	1.7694	15.787
1080.	0.251	131.1	1.7717	16.334
1090.	0.252	133.6	1.7740	16.894
1100.	0.252	136.2	1.7763	17.470
1110.	0.252	138.7	1.7786	18.060
1120.	0.252	141.2	1.7809	18.666
1130.	0.253	143.7	1.7831	19.287
1140.	0.253	146.3	1.7853	19.923
1150.	0.253	148.8	1.7875	20.576
1160.	0.254	151.3	1.7897	21.245
1170.	0.254	153.9	1.7919	21.930
1180.	0.254	156.4	1.7941	22.633
1190.	0.255	158.9	1.7962	23.352
1200.	0.255	161.5	1.7984	24.090
1210.	0.255	164.0	1.8005	24.844
1220.	0.256	166.6	1.8026	25.617
1230.	0.256	169.2	1.8047	26.409
1240.	0.256	171.7	1.8067	27.219
1250.	0.257	174.3	1.8088	28.049
1260.	0.257	176.9	1.8108	28.897
1270.	0.257	179.4	1.8129	29.766
1280.	0.258	182.0	1.8149	30.655
1290.	0.258	184.6	1.8169	31.564
1300.	0.258	187.2	1.8189	32.494
1310.	0.259	189.7	1.8209	33.445
1320.	0.259	192.3	1.8228	34.417
1330.	0.259	194.9	1.8248	35.412
1340.	0.259	197.5	1.8267	36.429
1350.	0.260	200.1	1.8287	37.468
1360.	0.260	202.7	1.8306	38.530
1370.	0.260	205.3	1.8325	39.616
1380.	0.261	207.9	1.8344	40.725
1390.	0.261	210.5	1.8363	41.859
1400.	0.261	213.1	1.8381	43.016

TABLE B.2. **Properties of Dry Air at 1 atm (English)**

Temp °R	C_p° Btu/lb°R	h° Btu/lb	s° Btu/lb°R	Pr
1410.	0.262	215.7	1.8400	44.199
1420.	0.262	218.4	1.8419	45.408
1430.	0.262	221.0	1.8437	46.642
1440.	0.263	223.6	1.8455	47.902
1450.	0.263	226.2	1.8473	49.188
1460.	0.263	228.9	1.8492	50.50
1470.	0.264	231.5	1.8510	51.84
1480.	0.264	234.1	1.8527	53.21
1490.	0.264	236.8	1.8545	54.61
1500.	0.264	239.4	1.8563	56.03
1510.	0.265	242.1	1.8580	57.49
1520.	0.265	244.7	1.8598	58.97
1530.	0.265	247.4	1.8615	60.49
1540.	0.266	250.0	1.8633	62.03
1550.	0.266	252.7	1.8650	63.60
1560.	0.266	255.4	1.8667	65.21
1570.	0.267	258.0	1.8684	66.85
1580.	0.267	260.7	1.8701	68.52
1590.	0.267	263.4	1.8718	70.22
1600.	0.267	266.0	1.8735	71.96
1610.	0.268	268.7	1.8751	73.73
1620.	0.268	271.4	1.8768	75.53
1630.	0.268	274.1	1.8784	77.37
1640.	0.269	276.7	1.8801	79.24
1650.	0.269	279.4	1.8817	81.15
1660.	0.269	282.1	1.8833	83.10
1670.	0.269	284.8	1.8849	85.08
1680.	0.270	287.5	1.8866	87.10
1690.	0.270	290.2	1.8882	89.16
1700.	0.270	292.9	1.8897	91.25
1710.	0.270	295.6	1.8913	93.38
1720.	0.271	298.3	1.8929	95.56
1730.	0.271	301.0	1.8945	97.77
1740.	0.271	303.7	1.8960	100.02
1750.	0.272	306.5	1.8976	102.31

TABLE B.2. **Properties of Dry Air at 1 atm (English)**

Temp °R	C_p° Btu/lb°R	h° Btu/lb	s° Btu/lb°R	Pr
1760.	0.272	309.2	1.8991	104.65
1770.	0.272	311.9	1.9007	107.02
1780.	0.272	314.6	1.9022	109.44
1790.	0.272	317.3	1.9037	111.90
1800.	0.273	320.1	1.9053	114.41
1810.	0.273	322.8	1.9068	116.95
1820.	0.273	325.5	1.9083	119.55
1830.	0.273	328.3	1.9098	122.18
1840.	0.274	331.0	1.9113	124.87
1850.	0.274	333.7	1.9128	127.60
1860.	0.274	336.5	1.9142	130.37
1870.	0.274	339.2	1.9157	133.20
1880.	0.275	342.0	1.9172	136.07
1890.	0.275	344.7	1.9186	138.99
1900.	0.275	347.4	1.9201	141.96
1910.	0.275	350.2	1.9215	144.98
1920.	0.275	353.0	1.9230	148.04
1930.	0.276	355.7	1.9244	151.17
1940.	0.276	358.5	1.9258	154.34
1950.	0.276	361.2	1.9272	157.56
1960.	0.276	364.0	1.9286	160.84
1970.	0.276	366.7	1.9300	164.17
1980.	0.277	369.5	1.9314	167.55
1990.	0.277	372.3	1.9328	170.99
2000.	0.277	375.1	1.9342	174.49
2010.	0.277	377.8	1.9356	178.04
2020.	0.278	380.6	1.9370	181.65
2030.	0.278	383.4	1.9384	185.31
2040.	0.278	386.2	1.9397	189.04
2050.	0.278	388.9	1.9411	192.82
2060.	0.278	391.7	1.9424	196.66
2070.	0.279	394.5	1.9438	200.6
2080.	0.279	397.3	1.9451	204.5
2090.	0.279	400.1	1.9465	208.6
2100.	0.279	402.9	1.9478	212.6

TABLE B.2. **Properties of Dry Air at 1 atm (English)**

Temp °R	C_p° Btu/lb°R	h° Btu/lb	s° Btu/lb°R	Pr
2110.	0.279	405.7	1.9491	216.8
2120.	0.280	408.5	1.9504	221.0
2130.	0.280	411.2	1.9518	225.3
2140.	0.280	414.0	1.9531	229.6
2150.	0.280	416.8	1.9544	234.0
2160.	0.280	419.6	1.9557	238.5
2170.	0.280	422.5	1.9570	243.1
2180.	0.281	425.3	1.9583	247.7
2190.	0.281	428.1	1.9596	252.4
2200.	0.281	430.9	1.9608	257.1
2210.	0.281	433.7	1.9621	261.9
2220.	0.281	436.5	1.9634	266.8
2230.	0.282	439.3	1.9646	271.8
2240.	0.282	442.1	1.9659	276.8
2250.	0.282	445.0	1.9672	281.9
2260.	0.282	447.8	1.9684	287.1
2270.	0.282	450.6	1.9697	292.4
2280.	0.283	453.4	1.9709	297.7
2290.	0.283	456.2	1.9721	303.1
2300.	0.283	459.1	1.9734	308.6
2310.	0.283	461.9	1.9746	314.2
2320.	0.283	464.7	1.9758	319.9
2330.	0.283	467.6	1.9770	325.6
2340.	0.284	470.4	1.9782	331.4
2350.	0.284	473.2	1.9795	337.3
2360.	0.284	476.1	1.9807	343.3
2370.	0.284	478.9	1.9819	349.3
2380.	0.284	481.8	1.9831	355.5
2390.	0.284	484.6	1.9843	361.7
2400.	0.285	487.4	1.9854	368.0
2410.	0.285	490.3	1.9866	374.4
2420.	0.285	493.1	1.9878	380.9
2430.	0.285	496.0	1.9890	387.5
2440.	0.285	498.8	1.9901	394.2
2450.	0.285	501.7	1.9913	400.9

TABLE B.2. **Properties of Dry Air at 1 atm (English)**

Temp °R	C_p° Btu/lb°R	h° Btu/lb	s° Btu/lb°R	Pr
2460.	0.286	504.6	1.9925	407.8
2470.	0.286	507.4	1.9936	414.7
2480.	0.286	510.3	1.9948	421.8
2490.	0.286	513.1	1.9959	428.9
2500.	0.286	516.0	1.9971	436.1
2510.	0.286	518.8	1.9982	443.4
2520.	0.287	521.7	1.9994	450.9
2530.	0.287	524.6	2.0005	458.4
2540.	0.287	527.4	2.0016	466.0
2550.	0.287	530.3	2.0028	473.7
2560.	0.287	533.2	2.0039	481.5
2570.	0.287	536.1	2.0050	489.5
2580.	0.287	538.9	2.0061	497.5
2590.	0.288	541.8	2.0071	505.6
2600.	0.288	544.7	2.0083	513.9
2610.	0.288	547.6	2.0095	522.2
2620.	0.288	550.4	2.0106	530.6
2630.	0.288	553.3	2.0117	539.2
2640.	0.288	556.2	2.0127	547.9
2650.	0.288	559.1	2.0138	556.6
2660.	0.289	562.0	2.0149	565.5
2670.	0.289	564.9	2.0160	574.5
2680.	0.289	567.8	2.0171	583.6
2690.	0.289	570.6	2.0182	592.9
2700.	0.289	573.5	2.0192	602.2
2710.	0.289	576.4	2.0203	611.7
2720.	0.289	579.3	2.0214	621.2
2730.	0.290	582.2	2.0224	630.9
2740.	0.290	585.1	2.0235	640.8
2750.	0.290	588.0	2.0245	650.7
2760.	0.290	590.9	2.0256	660.7
2770.	0.290	593.8	2.0266	670.9
2780.	0.290	596.7	2.0277	681.2
2790.	0.290	599.6	2.0287	691.7
2800.	0.291	602.5	2.0298	702.2

TABLE B.2. **Properties of Dry Air at 1 atm (English)**

Temp °R	C_p° Btu/lb°R	h° Btu/lb	s° Btu/lb°R	Pr
2810.	0.291	605.4	2.0308	712.9
2820.	0.291	608.3	2.0318	723.7
2830.	0.291	611.2	2.0329	734.7
2840.	0.291	614.2	2.0339	745.7
2850.	0.291	617.1	2.0349	756.9
2860.	0.291	620.0	2.0359	768.3
2870.	0.292	622.9	2.0370	779.8
2880.	0.292	625.8	2.0380	791.4
2890.	0.292	628.7	2.0390	803.1
2900.	0.292	631.6	2.0400	815.0
2910.	0.292	634.6	2.0410	827.0
2920.	0.292	637.5	2.0420	839.2
2930.	0.292	640.4	2.0430	851.5
2940.	0.292	643.3	2.0440	863.9
2950.	0.293	646.3	2.0450	876.5
2960.	0.293	649.2	2.0460	889.3
2970.	0.293	652.1	2.0470	902.2
2980.	0.293	655.0	2.0480	915.2
2990.	0.293	658.0	2.0489	928.4
3000.	0.293	660.9	2.0499	941.7
3010.	0.293	663.8	2.0509	955.2
3020.	0.293	666.8	2.0519	968.8
3030.	0.293	669.7	2.0528	982.6
3040.	0.294	672.6	2.0538	996.6
3050.	0.294	675.6	2.0548	1010.7
3060.	0.294	678.5	2.0557	1024.9
3070.	0.294	681.4	2.0567	1039.4
3080.	0.294	684.4	2.0576	1053.9
3090.	0.294	687.3	2.0586	1068.7
3100.	0.294	690.3	2.0595	1083.6
3110.	0.294	693.2	2.0605	1098.7
3120.	0.294	696.2	2.0614	1113.9
3130.	0.295	699.1	2.0624	1129.3
3140.	0.295	702.0	2.0633	1144.9
3150.	0.295	705.0	2.0643	1160.6

TABLE B.2. **Properties of Dry Air at 1 atm (English)**

Temp °R	C_p° Btu/lb°R	h° Btu/lb	s° Btu/lb°R	Pr
3160.	0.295	707.9	2.0652	1176.5
3170.	0.295	710.9	2.0661	1192.6
3180.	0.295	713.8	2.0671	1208.9
3190.	0.295	716.8	2.0680	1225.3
3200.	0.295	719.8	2.0689	1242.0
3210.	0.295	722.7	2.0698	1258.7
3220.	0.296	725.7	2.0707	1275.7
3230.	0.296	728.6	2.0717	1292.9
3240.	0.296	731.6	2.0726	1310.2
3250.	0.296	734.5	2.0735	1327.7
3260.	0.296	737.5	2.0744	1345.5
3270.	0.296	740.5	2.0753	1363.3
3280.	0.296	743.4	2.0762	1381.4
3290.	0.296	746.4	2.0771	1399.7
3300.	0.296	749.3	2.0780	1418.2
3310.	0.296	752.3	2.0789	1436.8
3320.	0.297	755.3	2.0798	1455.7
3330.	0.297	758.2	2.0807	1474.7
3340.	0.297	761.2	2.0816	1494.0
3350.	0.297	764.2	2.0825	1513.4
3360.	0.297	767.1	2.0834	1533.1
3370.	0.297	770.1	2.0842	1552.9
3380.	0.297	773.1	2.0851	1573.0
3390.	0.297	776.1	2.0860	1593.2
3400.	0.297	779.0	2.0869	1613.7
3410.	0.297	782.0	2.0877	1634.4
3420.	0.298	785.0	2.0886	1655.3
3430.	0.298	788.0	2.0895	1676.4
3440.	0.298	790.9	2.0904	1697.7
3450.	0.298	793.9	2.0912	1719.2
3460.	0.298	796.9	2.0921	1740.9
3470.	0.298	799.9	2.0929	1762.9
3480.	0.298	802.8	2.0938	1785.0
3490.	0.298	805.8	2.0947	1807.4
3500.	0.298	808.8	2.0955	1830.1

TABLE B.2. **Properties of Dry Air at 1 atm (English)**

Temp °R	C_p° Btu/lb°R	h° Btu/lb	s° Btu/lb°R	Pr
3510.	0.298	811.8	2.0964	1852.9
3520.	0.298	814.8	2.0972	1876.0
3530.	0.299	817.8	2.0981	1899.3
3540.	0.299	820.8	2.0989	1922.8
3550.	0.299	823.7	2.0997	1946.5
3560.	0.299	826.7	2.1006	1970.5
3570.	0.299	829.7	2.1014	1994.8
3580.	0.299	832.7	2.1023	2019.2
3590.	0.299	835.7	2.1031	2043.9
3600.	0.299	838.7	2.1039	2068.8
3610.	0.299	841.7	2.1048	2094.0
3620.	0.299	844.7	2.1056	2119.4
3630.	0.299	847.7	2.1064	2145.1
3640.	0.300	850.7	2.1072	2171.0
3650.	0.300	853.7	2.1080	2197.2
3660.	0.300	856.7	2.1089	2223.6
3670.	0.300	859.6	2.1097	2250.2
3680.	0.300	862.6	2.1105	2277.1
3690.	0.300	865.6	2.1113	2304.3
3700.	0.300	868.6	2.1121	2331.7
3710.	0.300	871.6	2.1129	2359.4
3720.	0.300	874.6	2.1137	2387.3
3730.	0.300	877.6	2.1146	2415.6
3740.	0.300	880.7	2.1154	2444.0
3750.	0.300	883.7	2.1162	2472.7
3760.	0.300	886.7	2.1170	2501.8
3770.	0.301	889.7	2.1178	2531.0
3780.	0.301	892.7	2.1186	2560.6
3790.	0.301	895.7	2.1193	2590.4
3800.	0.301	898.7	2.1201	2620.5
3810.	0.301	901.7	2.1209	2650.8
3820.	0.301	904.7	2.1217	2681.5
3830.	0.301	907.7	2.1225	2712.4
3840.	0.301	910.7	2.1233	2743.6
3850.	0.301	913.7	2.1241	2775.1

TABLE B.2. **Properties of Dry Air at 1 atm (English)**

Temp °R	C_p° Btu/lb°R	h° Btu/lb	s° Btu/lb°R	Pr
3860.	0.301	916.7	2.1249	2806.9
3870.	0.301	919.8	2.1256	2838.9
3880.	0.301	922.8	2.1264	2871.3
3890.	0.301	925.8	2.1272	2903.9
3900.	0.302	928.8	2.1280	2936.9
3910.	0.302	931.8	2.1287	2970.1
3920.	0.302	934.8	2.1295	3003.6
3930.	0.302	937.8	2.1303	3037.5
3940.	0.302	940.9	2.1310	3071.6
3950.	0.302	943.9	2.1318	3106.0
3960.	0.302	946.9	2.1326	3140.8
3970.	0.302	949.9	2.1333	3175.9
3980.	0.302	952.9	2.1341	3211.2
3990.	0.302	956.0	2.1348	3246.9
4000.	0.302	959.0	2.1356	3282.9
4010.	0.302	962.0	2.1364	3319.2
4020.	0.302	965.0	2.1371	3355.8
4030.	0.302	968.1	2.1379	3392.7
4040.	0.302	971.1	2.1386	3430.9
4050.	0.303	974.1	2.1394	3467.6
4060.	0.303	977.1	2.1401	3505.5
4070.	0.303	980.2	2.1409	3543.7
4080.	0.303	983.2	2.1416	3582.3
4090.	0.303	986.2	2.1423	3621.2
4100.	0.303	989.2	2.1431	3660.4
4110.	0.303	992.3	2.1438	3700.0
4120.	0.303	995.3	2.1445	3739.9
4130.	0.303	998.3	2.1453	3780.2
4140.	0.303	1001.4	2.1460	3820.8
4150.	0.303	1004.4	2.1467	3861.7
4160.	0.303	1007.4	2.1475	3903.0
4170.	0.303	1010.5	2.1482	3944.6
4180.	0.303	1013.5	2.1489	3986.6
4190.	0.303	1016.5	2.1497	4029.
4200.	0.303	1019.6	2.1504	4072.

TABLE B.2. **Properties of Dry Air at 1 atm (English)**

Temp °R	C_p° Btu/lb°R	h° Btu/lb	s° Btu/lb°R	Pr
4210.	0.304	1022.6	2.1511	4115.
4220.	0.304	1025.6	2.1518	4158.
4230.	0.304	1028.7	2.1525	4202.
4240.	0.304	1031.7	2.1533	4246.
4250.	0.304	1034.7	2.1540	4291.
4260.	0.304	1037.8	2.1547	4335.
4270.	0.304	1040.8	2.1554	4381.
4280.	0.304	1043.9	2.1561	4426.
4290.	0.304	1046.9	2.1568	4472.
4300.	0.304	1049.9	2.1575	4519.
4310.	0.304	1053.0	2.1582	4565.
4320.	0.304	1056.0	2.1589	4613.
4330.	0.304	1059.1	2.1596	4660.
4340.	0.304	1062.1	2.1603	4708.
4350.	0.304	1065.1	2.1610	4756.
4360.	0.304	1068.2	2.1617	4805.
4370.	0.304	1071.2	2.1624	4854.
4380.	0.305	1074.3	2.1631	4904.
4390.	0.305	1077.3	2.1638	4954.
4400.	0.305	1080.4	2.1645	5004.
4410.	0.305	1083.4	2.1652	5055.
4420.	0.305	1086.5	2.1659	5106.
4430.	0.305	1089.5	2.1666	5157.
4440.	0.305	1092.6	2.1673	5209.
4450.	0.305	1095.6	2.1680	5261.
4460.	0.305	1098.7	2.1687	5314.
4470.	0.305	1101.7	2.1693	5367.
4480.	0.305	1104.8	2.1700	5421.
4490.	0.305	1107.8	2.1707	5475.
4500.	0.305	1110.9	2.1714	5529.

TABLE B.3. **Dry Air at 1 atm**

Temp. °R	k	Temp. K	k
400	1.400		
500	1.400	250	1.400
600	1.399	300	1.400
700	1.396	350	1.398
800	1.392	400	1.395
900	1.386	450	1.391
1000	1.381	500	1.386
1100	1.374	550	1.381
1200	1.368	600	1.376
1300	1.362	650	1.370
1400	1.356	700	1.364
1500	1.350	750	1.359
1600	1.345	800	1.354
1700	1.340	850	1.349
1800	1.336	900	1.344
1900	1.332	950	1.340
2000	1.329	1000	1.336
2100	1.326	1050	1.333
2200	1.323	1100	1.330
2300	1.320	1150	1.327
2400	1.318	1200	1.324
2500	1.315	1250	1.322
2600	1.313	1300	1.319
2700	1.311	1350	1.317
2800	1.309	1400	1.315
2900	1.307	1450	1.313
3000	1.306	1500	1.311
3100	1.304	1550	1.309
3200	1.303	1600	1.308
3300	1.301		
3400	1.300		
3500	1.299		

TABLE C.1. **U.S. Standard Atmosphere, 1962**
(Geopotential Altitude)
(English Units)

Altitude feet	Temperature °R	Pressure psia	Altitude feet	Temperature °R	Pressure psia
0	518.670	14.696	25000	429.516	5.454
1000	515.104	14.173	26000	425.950	5.220
2000	511.538	13.664	27000	422.384	4.994
3000	507.972	13.171	28000	418.818	4.776
4000	504.405	12.692	29000	415.251	4.567
5000	500.839	12.228	30000	411.685	4.364
6000	497.273	11.770	31000	408.119	4.169
7000	493.707	11.340	32000	404.553	3.981
8000	490.141	10.916	33000	400.987	3.800
9000	486.575	10.505	34000	397.421	3.626
10000	483.008	10.106	35000	393.854	3.458
11000	479.442	9.720	*36089	389.970	3.283
12000	475.876	9.346	37000	389.970	3.142
13000	472.310	8.984	38000	389.970	2.994
14000	468.744	8.633	39000	389.970	2.854
15000	465.178	8.294	40000	389.970	2.720
16000	461.611	7.965	41000	389.970	2.592
17000	458.045	7.647	42000	389.970	2.471
18000	454.479	7.339	43000	389.970	2.355
19000	450.913	7.041	44000	389.970	2.244
20000	447.347	6.753	45000	389.970	2.139
21000	443.781	6.475	46000	389.970	2.039
22000	440.214	6.206	47000	389.970	1.943
23000	436.648	5.947	48000	389.970	1.852
24000	433.082	5.696	49000	389.970	1.765

*Boundary between troposphere and stratosphere.

TABLE C.2. **U.S. Standard Atmosphere, 1962**
(Geopotential Altitude)
(SI Units)

Altitude meters	Temperature K	Pressure kPa
0	288.150	101.325
500	284.900	95.461
1000	281.650	89.875
1500	278.400	84.556
2000	275.150	79.495
2500	271.900	74.683
3000	268.650	70.109
3500	265.400	65.764
4000	262.150	61.640
4500	258.900	57.728
5000	255.650	54.020
5500	252.400	50.507
6000	249.150	47.181
6500	245.900	44.035
7000	242.650	41.061
7500	239.400	38.251
8000	236.150	35.600
8500	232.900	33.099
9000	229.650	30.742
9500	226.400	28.524
10000	223.150	26.436
10500	219.900	24.474
11000	216.650	22.632
11500	216.650	20.916
12000	216.650	19.330
12500	216.650	17.865
13000	216.650	16.510
13500	216.650	15.259
14000	216.650	14.102
14500	216.650	13.033
15000	216.650	12.045
15500	216.650	11.131
16000	216.650	10.287
16500	216.650	9.507
17000	216.650	8.787
17500	216.650	8.120
18000	216.650	7.505
18500	216.650	6.936
19000	216.650	6.410
19500	216.650	5.924

INDEX

Lightning Source UK Ltd.
Milton Keynes UK
UKOW07f1853280415

250503UK00001B/2/P

9 780471 311225